Springer

Berlin
Heidelberg
New York
Hong Kong
London
Milan
Paris
Tokyo

New Results in Numerical and Experimental Fluid Mechanics IV

Contributions to the 13th STAB/DGLR Symposium
Munich, Germany 2002

Christian Breitsamter, Boris Laschka,
Hans-Joachim Heinemann
and Reinhard Hilbig (Editors)

Springer

Dr. Christian Breitsamter
Prof. Dr. Boris Laschka
Technische Universität München
Lehrstuhl für Fluidmechanik
Bolzmannstr. 15
85748 Garching
Germany

Dipl.-Ing. Reinhard Hilbig
Deutsche Airbus GmbH
Hünefeldstr. 1-5
28199 Bremen
Germany

Dr. Hans-Joachim Heinemann
Deutsches Zentrum für
Luft- und Raumfahrt e. V. (DLR)
Bunsenstr. 10
37073 Göttingen
Germany

ISBN 3-540-20258-7 Springer-Verlag Berlin Heidelberg New York

Library of Congress Cataloging-in-Publication-Data

STAB/DGLR Symposium (13th : 2002 : Munich, Germany)
New results in numerical and experimental fluid mechanics IV :
contributions to the 13th STAB/DGLR Symposium, Munich, Germany 2002 /
Christian Breitsamter ... [et al.] (editors).
p. cm. -- (Notes on numerical fluid mechanics and multidisciplinary design, ISSN 0179-9614 ; 87)
Includes bibliographical references. ISBN 3-540-20258-7 (alk. paper)
1. Aerodynamics--Congresses. 2. Airplanes--Design and construction--Congresses.
3. Fluid dynamics--Congresses. 4. Turbulence--Congresses. I. Breitsamter, Christian, 1962-II. Title.
III. Notes on numerical fluid mechanics and multidisciplinary design ; v. 87.
TL570.S65 2004
629.132'3--dc22

Springer-Verlag is a part of Springer Science+Business Media
springeronline.com

Typesetting: camera-ready by editors
Cover design: deblik Berlin
Printed on acid free paper 62/3020/M - 5 4 3 2 1 0

NNFM Editor Addresses

Prof. Dr. Ernst Heinrich Hirschel
(General editor)
Herzog-Heinrich-Weg 6
D-85604 Zorneding
Germany
E-mail: e.h.hirschel@t-online.de

Prof. Dr. Kozo Fujii
Space Transportation Research
Division
The Institute of Space
and Astronautical Science
3-1-1, Yoshinodai, Sagamihara,
Kanagawa, 229-8510
Japan
E-mail: fujii@flab.eng.isas.ac.jp

Dr. Werner Haase
Höhenkirchener Str. 19d
D-85662 Hohenbrunn
Germany
E-mail: wug.haase@t-online.de

Prof. Dr. Bram van Leer
Department of Aerospace Engineering
The University of Michigan
Ann Arbor, MI 48109-2140
USA
E-mail: bram@engin.umich.edu

Prof. Dr. Michael A. Leschziner
Imperial College of Science,
Technology and Medicine
Aeronautics Department
Prince Consort Road
London SW7 2BY
U. K.
E-mail: mike.leschziner@ic.ac.uk

Prof. Dr. Maurizio Pandolfi
Politecnico di Torino
Dipartimento di Ingegneria
Aeronautica e Spaziale
Corso Duca degli Abruzzi, 24
I - 10129 Torino
Italy
E-mail: pandolfi@polito.it

Prof. Dr. Jacques Periaux
Dassault Aviation
78, Quai Marcel Dassault
F-92552 St. Cloud Cedex
France
E-mail: periaux@rascasse.inria.fr

Prof. Dr. Arthur Rizzi
Department of Aeronautics
KTH Royal Institute of Technology
Teknikringen 8
S-10044 Stockholm
Sweden
E-mail: rizzi@aero.kth.se

Dr. Bernard Roux
L3M – IMT La Jetée
Technopole de Chateau-Gombert
F-13451 Marseille Cedex 20
France
E-mail: broux@l3m.univ-mrs.fr

FOREWORD

This volume contains the paper presented at the 13th DGLR/STAB-Symposium held at the Technische Universität München, November 12 to 14, 2002. STAB is the German Aerospace Aerodynamics Association, founded towards the end of the 70's, whereas DGLR is the German Society for Aeronautics and Astronautics (Deutsche Gesellschaft für Luft- und Raumfahrt - Lilienthal Oberth e.V.).

The mission of STAB is to foster development and acceptance of the discipline "Aerodynamics" in Germany. One of its general guidelines is to concentrate resources and know-how in the involved institutions and to avoid duplication in research work as much as possible. Nowadays, this is more necessary than ever. The experience made in the past makes it easier now, to obtain new knowledge for solving today's and tomorrow's problems.

STAB unites German scientists and engineers from universities, research-establishments and industry doing research and project work in numerical and experimental fluid mechanics and aerodynamics for aerospace and other applications. This has always been the basis of numerous common research activities sponsored by different funding agencies.

Since 1986 the symposium has taken place at different locations in Germany every two years. In between STAB workshops regularly take place at the DLR in Göttingen. The changing meeting places were established as focal points in Germany's Aerospace Fluid Mechanics Community for a continuous exchange of scientific results and their discussion. Moreover, they are a forum where new research activities can be presented, often resulting in new commonly organised research- and technology projects.

It is the forth time now that the contributions to the Symposium are published after being subjected to a peer review. The material highlights the key item of integrated research and development based on fruitful collaboration of industry, research establishments and universities. Some of the contributions still present results from the "Luftfahrtforschungsprogramm der Bundesregierung (German Aeronautical Research Programme) 1995-1998". Some of the papers report on work sponsored by the Deutsche Forschungsgemeinschaft (German Research Council), DFG or by the European Community within 5th Framework "Critical Technology Projects and Targeted Platforms". In more detail one part of the present Symposium has focused on "Laminar/Turbulent Boundary-Layer Transition" and "Hypersonic Flows", two programmes sponsored by the DFG in a Priority

Program (Verbundschwerpunkt-Programm) and in Collaborative Research Centres (Sonderforschungsbereiche 253, 255, 259). Other main areas include numerical simulation and mathematics, aeroelasticity, small and large aspect ratio wings in the context of leading-edge vortices, wake vortices, high lift systems and propulsion integration, and new developments in wind tunnel facilities and measurement techniques. Therefore, this volume gives an almost complete review of the ongoing aerodynamics research work in Germany.

The order of the papers in this book corresponds closely to that of the sessions of the Symposium.

The Review-Board, partly identical with the Program-Committee, consisted of J. Ballmann (Aachen), K. Becker (Bremen), Chr. Breitsamter (München), K. Broichhauen (München), G. Eitelberg (Emmeloord), R. Friedrich (München), F.-R. Grosche (Göttingen), R. Grundmann (Dresden), P. Hakenesch (München), St. Hein (Göttingen), D. K. Hennecke (Darmstadt), P. Hennig (Unterschleißheim), H. Hönlinger (Göttingen), D. Jacob (Aachen), W. Koch (Göttingen), W. Kordulla (Noordwijk), H. Körner (Braunschweig), H.-P. Kreplin (Göttingen), D. Kröner (Freiburg), Th. Lutz (Stuttgart), E. Meyer (München), W. Nitsche (Berlin), H. Olivier (Aachen), K. Pahlke (Braunschweig), R. Radespiel (Braunschweig), M. Raffel (Göttingen), G. Redeker (Braunschweig), U. Rist (Stuttgart), C.-C. Rossow (Braunschweig), G. Schewe (Göttingen), D. Schmitt (Toulouse), U. Schöttle (Stuttgart), G. Schrauf (Bremen), H.-A. Schreiber (Köln), W. Schröder (Aachen), P. Thiede (Bremen), C. Tropea (Darmstadt), C. Wagner (Göttingen), C. Weiland (München), C. Weishäupl (München) and H.-B. Weyer (Köln). Nevertheless, the authors sign responsible for the contents of their contributions.

The editors are grateful to Prof. Dr. E. H. Hirschel as the General Editor of the "Notes on Numerical Fluid Mechanics and Multidisciplinary Design" and to the Springer-Verlag for the opportunity to publish the results of the Symposium.

Chr. Breitsamter, München
B. Laschka, München
H.-J. Heinemann, Göttingen
R. Hilbig, Bremen

August 2003

CONTENTS

Wake vortex and Configurations:

High-Lift:

Engine-Integration:

Numerical Simulation:

Low Aspect-Ratio Wings:

Bluff Bodies:

Laminar Flow Control and Transition:

Hypersonic Flows:

Design and Configurations:

Transition and Fluidmechanical Modelling:

Fluidmechanical Modelling; Numerical Modelling: LES and DNS:

Numerical Simulation:

Mathematical Fundamentals:

Grid Generation and Flow Solvers:

Measuring Techniques:

Propulsion:

LARGE TRANSPORT AIRCRAFT WAKE VORTEX AFFECTED by VORTEX DEVICES

C. Bellastrada, C. Breitsamter

Lehrstuhl für Fluidmechanik - Technische Universität München
Boltzmannstrasse 15 - 85748 Garching - Germany
claubell@flm.mw.tum.de

Summary

The results of a wind tunnel investigation on the wake vortex evolution behind a half–model of a four–engined large transport aircraft in high–lift configuration, with and without vortex devices, are presented. The wake is measured by means of advanced hot–wire anemometry up to 5.6 span distances downstream of the model. Results obtained include velocity, vorticity and turbulence intensity fields. The measured data are evaluated in order to establish and test possible wake vortex strength alleviation strategies.

1 Introduction

The trailing wake of a lifting wing, after rolling up into a pair of strong counter–rotating longitudinal vortices, persists still several span dimensions downstream and poses a serious danger factor for a following aircraft, especially during landing and take–off operations. The necessity to obey standard separation distances to avoid wake vortex encounter has resulted in a progressive saturation of airport capabilities. Undoubtedly, a reduction of these separation standards, without compromising the current safety level, would be helpful in order to increase airports efficiency. This goal is attainable only if the strength of the wake is weakened to a tolerable level within the desired spacing, [9]. The advent of very large transport aircraft (VLTA) implies even stronger demands on developing effective means to minimise or to eliminate the trailing vortex hazard, [2] and [4]. Manifold research activities have found out several design concepts to get this objective and to make it a practical tool to improve air traffic capacity, [3], [5] and [10].

2 Wind Tunnel Testing

2.1 Model and Facility

The tests have been performed using a 1:22.5 scale half–model of a typical Large Transport Aircraft (Fig. 1). The half–span is 1.301 m, the wing mean aerodynamic chord is 0.323 m and the fuselage length is 2.605 m. The wing has sets of fully deployable slats, ailerons and flaps to perform the investigation of approach, landing and taking–off configurations. The wing equipment includes two through–flow–nacelles (TFN). Two strakes are mounted on the inboard nacelle and one strake on the inner side of the outboard nacelle. The model is also equipped with the horizontal stabiliser. Landing gear, vertical tailplane and exhaust simulation devices are not installed. The measurements were carried out in the FLM wind tunnel C. It has a closed test section with height of 1.8 m, a width of 2.7 m and a length of 21 m, which enables to investigate the wake up to 5.6 spans downstream of the model. The maximum attainable velocity is 30 m/s with a turbulence level less than 0.5 %. The model is positioned on the tunnel floor with the wing reference point (WRP) 2.6 m downstream of the nozzle exit and the wing tip pointing to the roof of the wind tunnel. The test section is equipped with a three-degree-of-freedom probe traversing system giving minimum steps of ± 0.2 mm in axial, lateral and vertical directions.

2.2 Measurement Technique

Triple–sensor hot–wire probes have been used to measure the instantaneous flowfield velocities. The sensors consist of 5–μm–diam. platinum–plated wires having a length/diam. ratio of 250. The measuring volume is approximately 1 mm in diameter. The hot–wire probe was operated by a multi–channel constant-temperature anemometer. The output signals of the anemometer bridges have been low–pass filtered at 820 Hz before digitization (16–bit) and amplified for optimal signal level. The sampling rate for each channel was set to 3000 Hz (Nyquist frequency: 1500 Hz) to ensure that all significant flowfield phenomena are detected. The sampling time was 6.4 s resulting in a sample block of 19200 values for each survey point. Anemometer output signals are converted into time–dependent velocity components u, v, and w using a look–up table previously obtained from the velocity and angle dependent calibration of the hot–wire probe.

2.3 Measurement Conditions

The aircraft half–model has been tested in a typical high–lift configuration. The high lift devices are set as follows: (i) inboard slat at 19.6°, (ii) mid-board slat at 23.0°, (iii) outboard slat at 26.5°, (iv) both inboard and outboard flaps at 26°. The horizontal tailplane is fixed at −9.5° and the ailerons are set at 10°. The angle of attack is 9.5° giving a lift coefficient of 1.76 (reference configuration). The experiments have been done at a freestream reference velocity of $U_\infty = 25$ m/s corresponding to a Reynolds number of $Re = 0.471 \cdot 10^6$ (based on the wing mean aerodynamic chord). Transition strips are put on the wing nose section to avoid any laminar separation. Inspecting the surface flow with tufts it has been proven that attached flow is present on wing and control surfaces. The wake vortex has been measured in planes perpendicular to the freestream direction at four distances from the WRP ranging from $x/b = 0.02 \div 5.56$, where b is the full wingspan and x the axial coordinate with the origin at the WRP (Fig. 1). The survey points are closely spaced in the wake area with a relative grid resolution of 0.006 in spanwise (y^*axis) and 0.004 (z^*axis) in vertical direction based on the wing span. Outside these main areas a relative grid resolution of 0.035 in spanwise and 0.023 in vertical direction was used. The high–lift configuration has been tested with and without a flap–plate, [8], in order to validate the wake alleviation capability of such a vortex device. The flap–plate geometry is related to the outboard flap dimensions as shown in Figure 1. Its planform and size is derived from the test results on a generic configuration reported in [8].

3 Results and Discussion

3.1 Mean Axial Vorticity Fields

The flowfield characteristics of the wake vortex are described by the contour–plots of the normalized axial vorticity $\xi = \overline{\omega}_x \cdot (b/2)/U_\infty$, which is the quantity used to identify significant vortical structures and to define their location on each cross flow plane. The coordinates y and z are normalized using the wing half span $b/2$. Figures 2 and 3 show the development of the wake in the near field at $x/b = 0.02$ and $x/b = 0.37$, respectively. Abbreviations of single vortices are given in Table 1. High levels of peak vorticity show six dominant vortices at $x/b = 0.37$ (Fig. 3a): the wing tip vortex (WTV), the outboard flap vortex (OFV), the outboard and the inboard nacelle vortex (ONV and INV), the wing–fuselage junction vortex (WFV) and the horizontal tail tip vortex (HTV). Except the HTV, all these vortices are embedded in the wing vortex layer. The comparison of all peak vorticity values reveals that the strongest vortical structure of the wake is the OFV ($\xi = 39$), which is located approximately in–line with the outboard engine nacelle and the outer edge of the outboard flap. The ONV is attributed to the rolled up shear layer of the outboard TFN and nacelle strake. The investigation of the wake at $x/b = 0.02$ reveals that this vortex, labelled as ONV at $x/b = 0.37$, is not actually formed as an individual vortex, but from the combination of two vortices: the *original* ONV ($\xi = 22.0$) and the NSV ($\xi = 30.7$), which is a strong and concentrated vortex attributed to the strake of the outboard nacelle (Fig. 2a). Figure 2a also shows several other distinct vortices attributed to the different geometrical features of the wing as sketched in Table 1. The effect of the flap–plate vortex device is mainly evident on the NSV peak vorticity, producing for all other vortices only subtle differences. The flap–plate device, a triangular surface mounted at the outer edge of the outboard flap, is intended to introduce a strong vortex of opposite rotation with respect to the OFV, whose vorticity is shown to be positive (Fig. 1). The position where the NSV leaves the trailing edge of the wing, after passing over the wing, is seen to be very close to the position of the flap–plate tip. Hence, the 50% reduction in the peak vorticity of the NSV (co-rotating to the OFV) is attributed to the interaction of the NSV with the counter–rotating flap–plate vortex (not directly detectable in the plot). The topology of other vortices is largely unchanged, with only a slight shift in vorticity peak levels (Figs. 2a, 2b). The results at $x/b = 0.37$ correspond well with those obtained for $x/b = 0.02$: the strength of the ONV for the flap–plate configuration is reduced as expected, due to the change on the NSV introduced by the flap–plate (Figs. 3a, 3b). At $x/b = 0.37$ the OFV and the ONV start to merge leading to the dominant vortical structure, which acts as the rolling center of the whole wake. This merging process is already completed within 3 wingspan lengths downstream of the model ($x/b = 3.0$) and the resulting OFV–ONV system will be referred to as the OFV (Fig. 4a). The OFV strength of the flap–plate configuration at this stage ($x/b = 3.0$) is still lower than that of the reference configuration, as might be expected due to the reduction of the strength of the ONV. Further on, the more distorted shape of the flap–plate–ONV (Fig. 3b), as it approaches the merging at $x/b = 0.37$ and the lower spacing d between the OFV and the ONV (flap–plate $d = 90$ mm, reference $d = 110$ mm) indicates a more advanced merging in the flap–plate case, as well. Consequently, the most important merging process between the OFV and the WTV, which is occurring at $x/b = 3.0$, should also be in advance for the flap–plate configuration. Moving from $x/b = 0.37$ to $x/b = 3.0$ the external radius r_{ext} (defined as the radius of the region containing nearly the total amount of vorticity of the vortex) of the OFV is augmented by a factor of 3 for both configurations. On the other hand the decrease of OFV–WTV spacing $(d_{0.37} - d_{3.0})/d_{0.37}$ for the flap–plate configuration (12.2%) is greater than that observed for the reference configuration (2.6%). Therefore, the ratio $2r_{ext}/d$, which is an important parameter to describe the merging onset, [6], is much higher for the flap–plate configuration than for the reference one. This shows that the merging of the OFV with the WTV is more advanced for the flap–plate case. The results for the last plane ($x/b = 5.56$) show that the merging process between the OFV and the WTV is almost completed. Here the WTV can still be identified but its vorticity level is very low (Fig. 5). The OFV peak vorticity has risen from $x/b = 3.0$ due to the vorticity feeding mechanism ascribed to the merging: the peak of the flap–plate–OFV is higher than that of the reference configuration. The calculation of the induced rolling moment coefficient $c_R = (c_{L_\alpha}/U_\infty \cdot b_f^2) \cdot \int_{-b_f/2}^{b_f/2} w(y)\, y\, dy$ on a following aircraft with rectangular wing planform (wingspan $b_f = 0.2b$, lift gradient $c_{L_\alpha} = 2\pi$, [4] and [7]) also gives, for a horizontal encounter at $x/b = 5.56$, a higher value for the flap–plate configuration (Fig. 6). These results at this plane were unexpected. On the basis of the previous observations about the merging onset between $x/b = 0.37$ and $x/b = 3.0$, this result can be attributed to the fact that the merging process is much more completed and more vorticity has been already fed into the OFV for the flap–plate configuration than for the reference one. The changes produced by the flap–plate on the spanwise load factor $s = \Gamma/\Gamma_0$ ($s = \pi/4$ is the typical value for an elliptical loaded wing and Γ and Γ_0 are the circulation and the root circulation, respectively) and on the lift coefficient c_L (a slight increase of about 0.5%) may cause an increase of the non–dimensional time t^* giving an "older" wake than in the case of the reference configuration, [1] and [4].

3.2 Conclusions and Outlook

The wake near field shows six significant vortical structures, namely the wing tip vortex (WTV), the outboard flap vortex (OFV), the outboard nacelle vortex (ONV), the inboard nacelle vortex (INV), the horizontal tail tip vortex (HTV) and the wing–fuselage junction vortex (WFV). Except the HTV all other vortices are connected by the vorticity sheet shed from the wing trailing edge. The highest level of axial peak vorticity is registered inside the OFV, which is the strongest and most persistent vortex of the wake. The vortex labelled as the ONV results from the merging of the *original* ONV and the NSV (outboard Nacelle strake vortex). At $x/b = 0.37$ the OFV and the ONV merge to form the vortical structure which works as the roll up center of the wake. The merging between the WTV and the OFV starts further downstream, at $x/b = 3.0$. These results apply to both configurations tested. The additional counter–rotating (to the OFV) vortex introduced by the flap–plate is not detectable in the axial vorticity plots. The effect of this vortex is evident on the strength of the NSV which is reduced. Consequently, the ONV (combination of the *original* ONV and the NSV) is also reduced in strength. The vorticity distribution and the strength of the OFV is not affected by the flap–plate vortex at all. On the other hand, the reduction of the ONV is still perceivable at $x/b = 3.0$, where the OFV of the flap–plate configuration is still weaker than the reference OFV, due to the reduced vorticity feeding from the ONV. The alteration of wing lift parameter by the introduction of the flap–plate and the observation of the development of the external OFV radius and the spacing between the WTV and OFV shows that the merging process is in a more advanced stage in the case of the flap–plate configuration. This may explain that the peak vorticity of the flap–plate OFV is slightly higher than that of the reference OFV at $x/b = 5.56$. The evaluation of the induced rolling moment coefficient has shown that no appreciable benefit, at least within $x/b = 5.56$, have been brought by the use of the flap–plate.

References

[1] Breitsamter C., Bellastrada C., Laschka B.: Investigations on the Turbulent Wake Vortex Flow of Large Transport Aircraft. Proc. ICAS 2002 Congress, Toronto, Canada, September, 2002, pp. 382.1–382.13.

[2] Coustols E., de Saint-Victor X.: Wake Vortex Dynamics: from Characterisation to Control. Proc. 3rd ONERA-DLR Aerospace Symposium, Paris, June, 2001, S1-2 pp. 1–14.
[3] Crouch J.D., Miller G., Spalart P.R.: Airplane Trailing Vortices. Boeing Aero–Magazine, No. **14**, April, 2001, pp. 1–8.
[4] Gerz T., Holzaepfel F., Darracq D.: Aircraft Wake Vortices - A Position Paper. WakeNet - The European Thematic Network on Wake Vortex, April, 2001.
[5] Hinton D.A.: An Aircraft Vortex Spacing System (AVOSS) for Dynamical Wake Vortex Spacing Criteria. AGARD-CP-584, Trondheim, Norway, May, 1996, pp. 23-1–23-12.
[6] Leweke T., Meunier P., Laporte F., Darracq D.: Controlled Interaction of Co-rotating Vortices. Proc. 3rd ONERA-DLR Aerospace Symposium, Paris, June, 2001, S1-2 pp. 1–14.
[7] Özger E., Schell I., Jacob D.: On the Structure and Attenuation of an Aircraft Wake. J. Aircraft, Vol. **38**, No. **5**, 2001, pp. 878–887.
[8] Ortega J.M., Bristol R.L., Savas Ö.: Wake Alleviation Properties of Triangular–Flapped Wings, AIAA Journal, Vol. **40**, No. **4**, 2002, pp. 709–721.
[9] Rossow V.J.: Reduction of Uncertainties in Prediction of Wake-Vortex Locations. J. Aircraft, Vol. **39**, No. **4**, 2002, pp. 587–596.
[10] Spalart P.R.: Airplane Trailing Vortices. Ann. Rev. Fluid Mech., Vol. **30**, 1998, pp. 107–138.

Tables

Table 1 Vortex designation and vortex origin

Abbreviation	Designation	Vortex origin
HTV	Horizontal tail tip vortex	Horizontal Stabilizer
IAV	Inboard aileron vortex	Aileron inboard edge
INV	Inboard nacelle vortex	Inboard nacelle and strakes
NSV	Nacelle strake vortex	Inboard strake on the outboard nacelle
OFV	Outboard flap vortex	Outboard edge of outboard flap
ONV	Outboard nacelle vortex	Outboard nacelle (and strake)
OSV	Over wing slat vortex	Mid-board slat edge area above the wing leading edge
PFV	Pylon fairing vortex	Outboard pylon fairings
USV	Under wing slat vortex	Mid–board slat edge area below the wing leading edge
WFV	Wing–fuselage junction vortex	Inboard flap edge and wing–fuselage junction
WTV	Wing tip vortex	Winglet trailing edge and winglet–wing junction

Figures

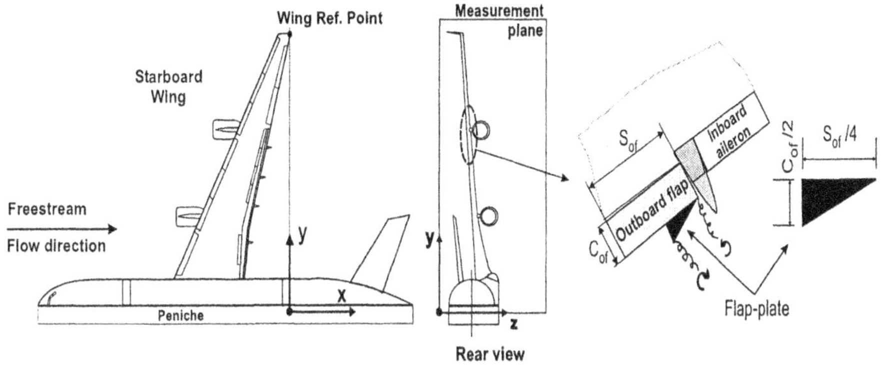

Figure 1 The half-model geometry and the flap-plate set-up.

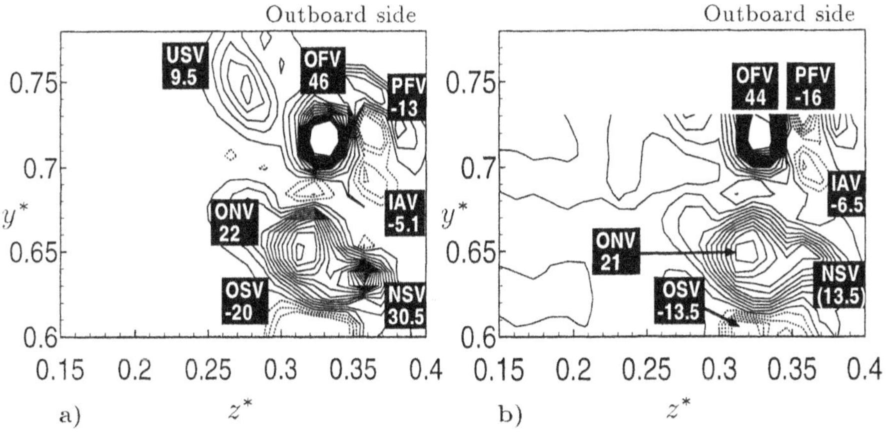

Figure 2 The non-dimensional axial vorticity distribution ξ at $x/b = 0.02$: a) reference and b) flap-plate configuration. Dotted contour lines show negative vorticity regions. Numbers in the black boxes indicate maximum values of vorticity.

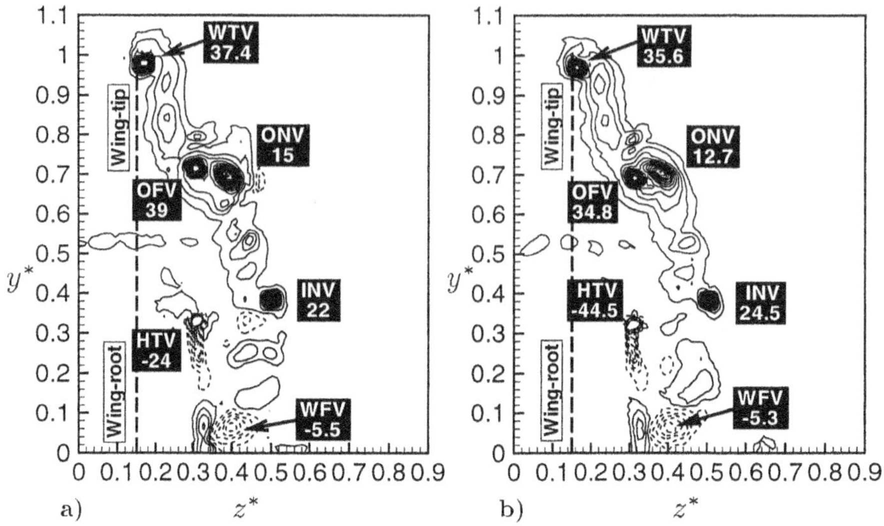

Figure 3 The non-dimensional axial vorticity distribution ξ at $x/b = 0.37$: a) reference and b) flap–plate configuration. Dotted contour lines show negative vorticity regions. The wing position is indicated by the dashed line. Numbers in the black boxes indicate maximum values of vorticity.

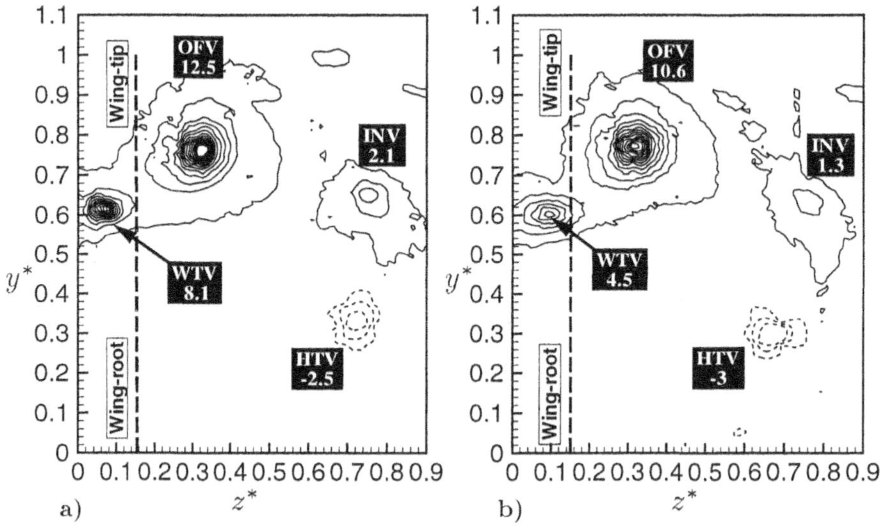

Figure 4 The non-dimensional axial vorticity distribution ξ at $x/b = 3.0$: a) reference and b) flap–plate configuration. Dotted contour lines show negative vorticity regions. The wing position is indicated by the dashed line. Numbers in the black boxes indicate maximum values of vorticity.

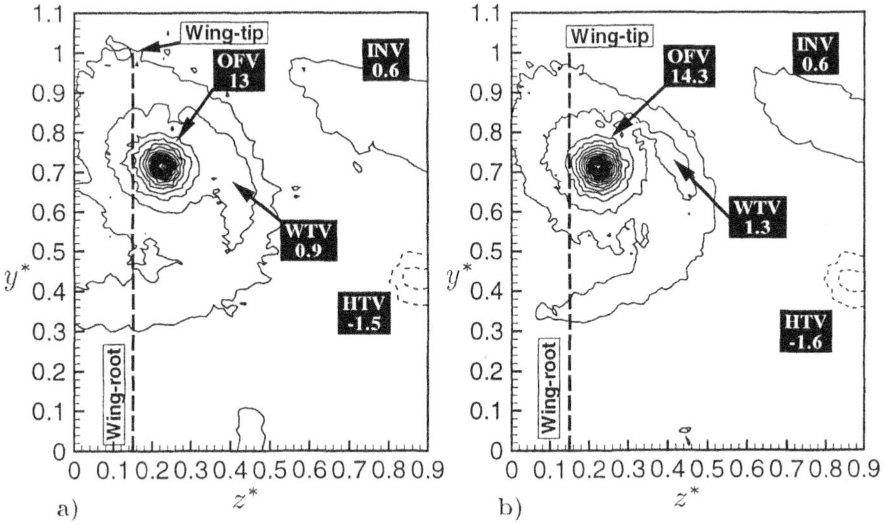

Figure 5 The non-dimensional axial vorticity distribution ξ at $x/b = 5.56$: a) reference and b) flap–plate configuration. Dotted contour lines show negative vorticity regions. The wing position is indicated by the dashed line. Numbers in the black boxes indicate maximum values of vorticity.

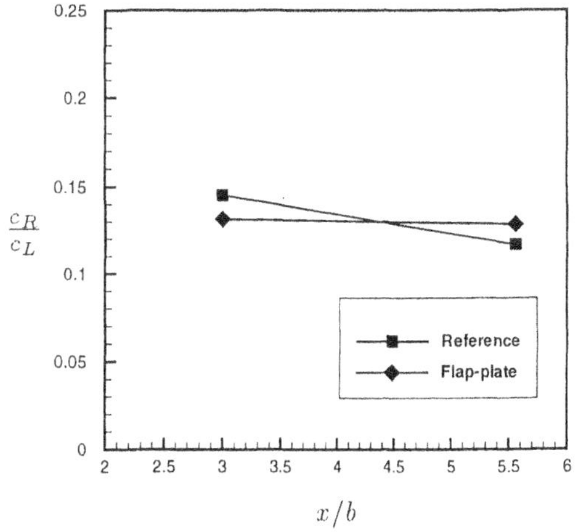

Figure 6 The OFV induced rolling moment coefficient c_R related to the lift coefficient c_L of the generating aircraft.

Numerical Study of Four-Vortex Aircraft Wakes

Eike Stumpf

German Aerospace Center DLR, Institute of Aerodynamics and Flow Technology, Lilienthalplatz 7, 38108 Braunschweig, Germany, `eike.stumpf@dlr.de`

Summary

Instabilities evolving on simplified vortex wakes, built up by two pairs of counter-rotating vortices, are investigated numerically. Two configurations are considered: the first in circulation and spacing ratio corresponding to a wide-body airliner in landing configuration and a modified set-up with a strong inner vortex pair. The investigation shows that a strong inner pair is necessary for the instability to cause a large increase of vortex radii of the outer vortex pair. This increase in turn leads to a rapid decay of rolling moment.

1 Introduction

World airports face a long-term capacity problem due to the separation standards presently in use. These air traffic control (ATC) separation regulations determine the minimum longitudinal distance between two aircraft on the same flight path dependent on the maximum take-off weights of both aircraft in order to prevent the following aircraft from encountering potentially hazardous wake turbulence. For an alleviation of wake vortices of a subsonic transport aircraft it seems most promising to make use of inherent vortex system instabilities: Either by triggering actively the most unstable wavelength or by designing a highly unstable vortex system for which natural perturbations from the aircraft boundary layer or weak atmospheric turbulence are sufficient for an onset of instability. For the investigation of the latter approach the complex topology of a high-lift configuration vortex wake is generally reduced to a four-vortex system, consisting of two pairs of counter-rotating vortices, see Fig. 1. These represent on either side of the symmetry plane respectively the merged vortices of the wing tip and outer flap side-edge and the merged vortices of the inner flap side-edge and horizontal tail. Following numerical simulations by Crouch [3] and Fabre & Jacquin [5] the amplification rate of inherent instabilities on such vortex systems is up to one order of magnitude higher and the wave length nearly one order of magnitude smaller compared to the Crow instability on a single counter-rotating vortex pair. The underlying instability mechanism is a resonance between a normal mode of oscillations of a vortex and the external straining field imposed by a closely spaced second vortex. The instability is invariably three-dimensional and has linear growth rates that scale with the strain rate.
The specific interest on the temporal development of 4-vortex systems arose since the knowledge might contribute to the understanding of wind tunnel and free-flight tests by Coton [2], Stuff [13] and Rossow [7] which were successful regarding the

alleviation of the wake vortex hazard. The experiments by Coton [2] and Stuff [13] were conducted in the ONERA catapult facility with an Airbus A300 model with outer flaps deflected by $\delta_{flap} = 15°$ and retracted inner flaps. Rossow [7] used a Boeing 747 in the 80 × 120ft NASA Ames wind tunnel and in flight tests with retracted outer flaps and by $\delta_{flap} = 30°$ deflected inner flaps. In either case a local 4-vortex system with two counter-rotating vortex pairs develops which experiences a topology change due to instability within 40 spans downstream of the aircraft wing tip resulting in highly reduced rolling moments. It is claimed in [13] and [7] that a reduced hazard is obtained only if the circulation ratio of the counter-rotating vortex pairs $\Gamma_{inner}/\Gamma_{outer}$ exceeds 20%.

2 Flow Solver

The flow solver used is NCAR[1]-EuLag [4]. EuLag was developed at NCAR for meteorological and geophysical applications and is based on the nonoscillatory advection scheme MPDATA (multidimensional positive definite advection transport algorithm) by Smolarkiewicz [8]. EuLag is a pure advection scheme, no diffusive fluxes are explicitly considered. The advection is calculated in a first pass by a first order accurate upstream differencing predictor step. A second pass, as a corrector step, increases the accuracy of the calculation to second order by estimating and compensating the truncation error of the first pass. In the present paper EuLag is used for monotonically integrated large eddy simulations (MILES), as originally proposed by Boris et al. [1]. In MILES the incompressible Navier-Stokes equations are discretized on a coarse LES mesh without introducing neither a filtering operation nor a subgrid scale stress tensor. The use of EuLag for MILES has been validated for decaying isotropic turbulence [4] and various geophysical flows [9]. For the results presented in the following sections the Adams-Bashforth time integration scheme [11] is used, the boundary conditions specified in all calculations are periodic in axial direction and free slip walls in vertical and horizontal direction.

3 Validation

For the validation two cases are calculated: first a Crow instability on a pair of counter-rotating vortices with a prescribed sinusoidal displacement of $1/10$ of the vortex spacing on both vortices, secondly the same vortex pair without displacement but superimposed white noise. The calculations are initialized in analogy to water tank experiments performed by Leweke & Williamson [6] with a vortex spacing of $b = 0.025\,m$, a circulation of $\Gamma = 0.0026\,m^2/s$ and vortex radii of $r_c = 0.2\,b$. The distribution of circumferential velocity is calculated using the Lamb-Oseen vortex model [10], no axial velocity is prescribed. Following an analysis by Steijl [12] the Crow wave length corresponding to $r_c/b = 0.2$ is $\lambda = 7.428\,b$. The simulation domain comprises one Crow wave length in axial direction, resolved by 55 points. While in vertical direction 151 points are equidistantly distributed, in lateral direction there are 201 points, slightly clustered in the center region. Vertical and lateral

[1] National Center for Atmospheric Research, Boulder, Colorado

edge lengths of the domain are 8 spans. Fig. 2(a) to 2(d) show iso-surfaces of vorticity (displayed vorticity level $|\omega| = 0.25 \times max(|\omega_{ini}|))$ as a function of normalized time $t^* = (\Gamma/(2\pi b^2))\, t$ for case 1. The Crow instability is amplified and the vortex pair breaks up into two half rings. Fig. 3 shows the very good agreement between theory and simulation result. The theoretical amplification rate is taken from [12] with $\alpha = 0.807$ assuming a time-dependence of the perturbation as $e^{\alpha t^*}$.
The second validation test case is a vortex pair with the same vortex radii, circulation and vortex spacing as above. However, no displacement is prescribed but white noise is superimposed. The RMS-value of the perturbation is one percent of the maximum initial circumferential velocity $v_{RMS} = 0.01 \times max(v_\Theta)$. It is known from the water tank experiments [6] that besides the long wave Crow instability a short wave instability has to be expected. Thus, for this simulation the axial resolution is doubled to 109 points, the domain size and the vertical and lateral resolution remain unchanged. The theoretical Crow instability amplification rate reads as above $\alpha_{lw} = 0.807$ and the simulation yields again a very good agreement between theoretical prediction and calculation result, shown in Fig. 4. Following [6] the theoretical amplification rate of the short wave length instability in case of inviscid flow is given by $\alpha_{sw} = 18/16 = 1.125$. The simulation shows varying amplification rates of the short wave instability along the vortices. In Fig. 4 only the maximum amplification rate of the short wave instability found along the vortices is displayed. The predicted rate is with $\alpha_{sw-sim} = 1.14$ slightly larger than the theoretical average value of 1.125.
From the two validation test cases it can be concluded that EuLag is capable to simulate the temporal development of cooperative elliptical instabilities in good agreement with theoretical predictions and experiments both with prescribed sinusoidal perturbation and triggered by white noise.

4 Results

Two 4-vortex configurations are chosen to be investigated in the present paper. As shown in Fig. 1 a 4-vortex system is defined by four parameters: the vortex spacing ratio $B = b_2/b_1$, with b_1 resp. b_2 being the spacing of the counter-rotating vortices of the two pairs, $G = \Gamma_2/\Gamma_1$ the ratio of their circulations and the normalized vortex core radii r_1/b_1 and r_2/b_1. Identical in both calculations are the normalized vortex core radii $r_1/b_1 = 5\%$ and $r_2/b_1 = 2.5\%$, the main vortex spacing $b_1 = 2.8m$, the spacing ratio $B = 0.35$ and the total circulation $\Gamma = \Gamma_1 + \Gamma_2 = 30.0\, m^2/s$. These values correspond roughly to a scaled down Airbus A340 in landing configuration ($b = 3.56\, m$, $u_\infty = 143kts$, $Ma = 0.21$, $c_L \approx 1.8$, ambient pressure and temperature). The loading of the horizontal tail plane (HTP) of a subsonic transport aircraft varies typically in the range $-0.5 < c_{L-HTP} < 0$. This translates into a shed HTP-vortex with a maximum circulation of $\Gamma_{HTP} \approx -0.175\, \Gamma_{wing}$. Accordingly, the parameters for configuration 1 are given by $B = 0.35$; $G = -0.175$. In order to study the potential benefit of a strong inner vortex pair in the wake evolution for configuration 2 the circulation ratio has been doubled ($B = 0.35$; $G = -0.35$).
Taking again the analysis by Steijl [12] the Crow wave length corresponding to the

radius of the outer vortex pair $r_1/b_1 = 0.05$ is $\lambda = 9.4285\,b$. The simulation domain comprises one Crow wave length in axial direction resolved by 151 points. In vertical direction 201 points (domain height 2.5b) and in lateral direction 401 points (domain width 5b) are distributed, slightly clustered in the center region. White noise with an RMS-value of $v_{RMS} = 0.01 \times max(v_\Theta)$ is superimposed.
Fig. 5 & 6 show vorticity iso-surfaces (displayed vorticity level is $|\omega| = 0.05 \times max(|\omega_{ini}|)$) of the two generic wake vortex models evolving in time. The rotation rate of the inner around the outer vortices is identical in both configurations, see Fig. 5(b) & Fig. 6(b). The upcoming instability develops at the same time and shows in the linear phase a comparable topology. On the inner vortices Ω-loops develop which wrap around the outer vortices. The main difference is the vortex interaction. While in configuration 1 the Ω-loops are small and confined to each side of the symmetry plane (Fig. 5(d)) in configuration 2 the vorticity stretching produces large loops that collide in the symmetry plane, Fig. 6(d). After the dissipation of the scattered vorticity (Fig. 6(e)) it becomes visible that, just as in configuration 1, the main vortices survive the chaotic process.
An analysis of the temporal rolling moment evolution is given for configuration 1 in Fig. 7 up to $t^* = 3.17$ ($\widehat{=}$ 5.19 NM for a real size Airbus A340 at 143 kts) and respectively for configuration 2 in Fig. 8 up to $t^* = 1.76$ ($\widehat{=}$ 2.89 NM). For vortex encounter here the considered scenario is a medium-class airplane (with half the span of the wake generator) flying into the vortex center of one of the two main vortices (which represent a heavy-class airplane wake). The ATC separation in this case is 5 NM. In order to account for numerical dissipation effects the calculated rolling moment is not given in absolute values, but analyzed relative to the simulation result for a 2-vortex reference system which was obtained on a mesh with the same resolution as used for the 4-vortex configurations. Total circulation and vortex spacing b_1 of the reference 2-vortex system and the 4-vortex configurations are identical. The rolling moment c_l of configuration 1 and 2 are on the one hand normalized with the value of the rolling moment $|[c_l]_2|$ of the reference 2-vortex system at a downstream distance equal to the separation distance (i.e. 5 NM) and on the other hand with the ratio $c_L/[c_L]_2$ of the lift coefficient c_L of configuration 1 or 2 and the lift coefficient $[c_L]_2$ of the reference configuration. The rolling moment is evaluated in 10 planes equidistantly distributed along the axial direction of the numerical domain. The magnitude of normalized maximum rolling moment coefficient changes periodically, i.e. it is maximal whenever the orbiting inner vortices are in the horizontal plane with the outer vortices. In Fig. 7 & 8 the upper curves correspond to the maximal rolling moment found in the left main vortex, the lower curves respectively to the right main vortex. The lower abscissa shows the normalized time t^*, the upper one gives the traveled distance for a real size Airbus A340. In the simulation result for configuration 1 no axial gradient is found up to $t^* \approx 0.8$, Fig. 7 . This changes with the onset of 3D instabilities where a slightly accelerated decrease, different in magnitude in each analyzed plane, is observable. Until $t^* = 3.17$ the maximum rolling moment of configuration 1 has decreased to 76% of the maximum rolling moment of the reference 2-vortex system. As known from the vorticity iso-surfaces

the instability develops more rigorous in configuration 2. Here, the sudden decrease of rolling moment occurs at $t^* = 0.6$ and leads to a drop down to 44%-52% of the maximum rolling moment of the reference 2-vortex system until $t^* = 1.76$, see Fig. 8. An analysis of the vorticity in the axial planes reveals that the reason for the strong rolling moment decay of configuration 2 is the laminar-turbulent transition of the vortex.

5 Conclusion

The present investigation has shown that low intensity white noise is sufficient to trigger instabilities on 4-vortex systems. Doubling the circulation of the inner vortices increases strongly the vortex interaction in the course of the instability. A configuration corresponding to a standard wide-body-type airliner in landing configuration ($\Gamma_2/\Gamma_1 = -0.175$) showed a moderate reduction of maximum rolling moment of 24% compared to a reference 2-vortex system whereas a modified configuration with doubled circulation ratio ($\Gamma_2/\Gamma_1 = -0.35$) leads to a maximum reduction of 56%. The underlying mechanism for the strong rolling moment decay of the modified configuration is the transition to a turbulent vortex. The transition fails to appear if the circulation of the inner vortices is too low. This supports the prediction of Rossow [7] and Stuff [13] that the circulation ratio of a 4-vortex system has to exceed 20% to be effective.

References

[1] J.P. Boris, F.F. Grinstein, E.S. Oran, R.L. Kolbe: "*New insights into large eddy simulation*", Fluid Dyn. Research, Vol. 10, pp. 199-228, 1992

[2] P. Coton: "*Charactérisation et Modélisation du Sillage d'un Avion à partir d'Essais en Vol de Marquettes en Laboratoire*", AGARD CP-584, pp. 29-1 to 29-12, 1996

[3] J.D. Crouch: "*Instability and transient growth of two trailing-vortex pairs*", J. Fluid Mech., Vol. 350, pp. 311-330, 1997

[4] J.A. Domaradzki, Z. Xiao, P.K. Smolarkiewicz: "*Effective viscosities in implicit modeling of turbulent flows*", submitted to Phys. Fluids, 2003

[5] D. Fabre, L. Jacquin: "*Stability of a four-vortex aircraft wake model*", Phys.Fluids, Vol. 12, No. 10, pp. 2438-2443, 2000

[6] T. Leweke, C.H.K. Williamson: "*Cooperative elliptic instability of a vortex pair*", J. Fluid Mech., Vol. 360, pp. 85-119, 1998

[7] V.J. Rossow: "*Lift-Generated Vortex Wakes of Subsonic Transport Aircraft*", Progr. Aerospace Sci., Vol. 35, No. 6, pp. 507-660, 1999

[8] P.K. Smolarkiewicz, L.G. Margolin: "*MPDATA: A finite-difference solver for geophysical flows*", J.Computational Physics, No. 140, pp. 459 - 479, 1998

[9] P.K. Smolarkiewicz, J.M. Prusa: "*Forward-in-time differencing for fluids: Simulations of geophysical turbulence*", in "Turbulent Flow Computations", Eds. D. Drikakis, B.J. Guertz, Kluwer Academic Publ., 2002

[10] P.G. Saffman: "*Vortex Dynamics*", Cambride Univ. Press, 1992

[11] L.F. Sampine, M.K. Gordon: "*Computer Solution of Ordinary Differential Equations: the Initial Value Problem*", Freeman, 1975

[12] R. Steijl: "*Computational Study of Vortex Pair Dynamics*", Ph.D. thesis, University of Twente, Netherlands, 2001

[13] R. Stuff: "*Relationship between Near- and Farfield of Vortex Wakes from Aircraft with High-Aspect Ratio Wings*", Notes Num.Fluid Mech., Vol. 77, pp. 58-65, 2002

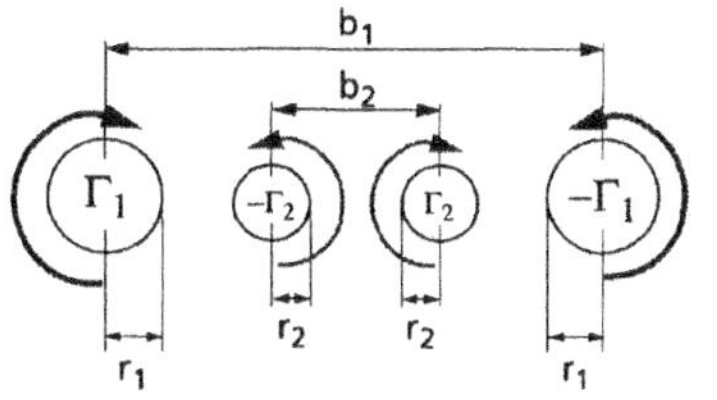

Figure 1 Sketch of 4-vortex set-up

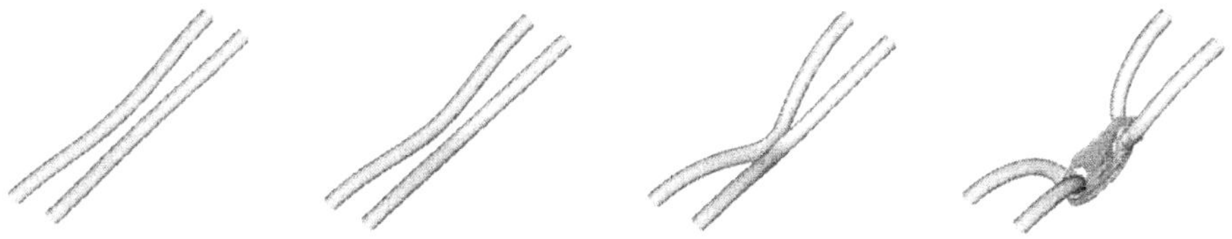

(a) Initial (b) t*=0.63 (c) t*=1.27 (d) t*=1.9

Figure 2 Vorticity iso-surfaces of temporal development of Crow instability

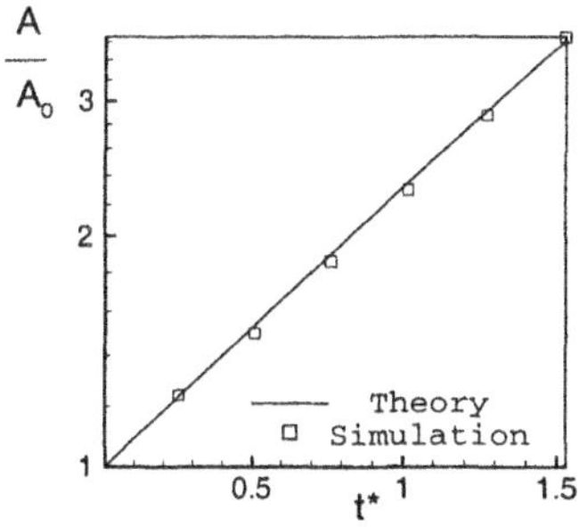

Figure 3 Amplification of Crow instability with prescribed vortex displacement

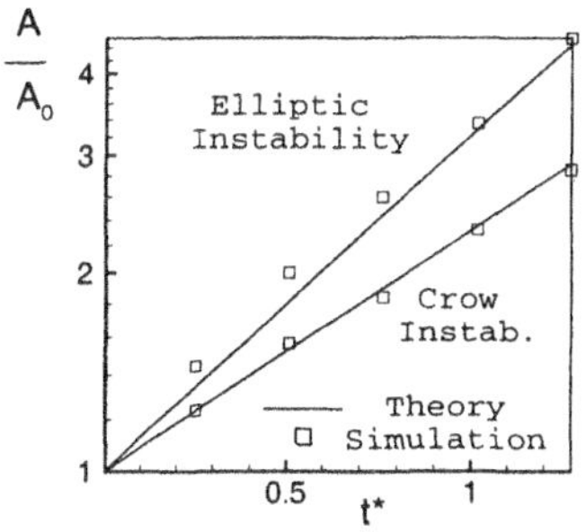

Figure 4 Amplification of instabilities triggered by white noise

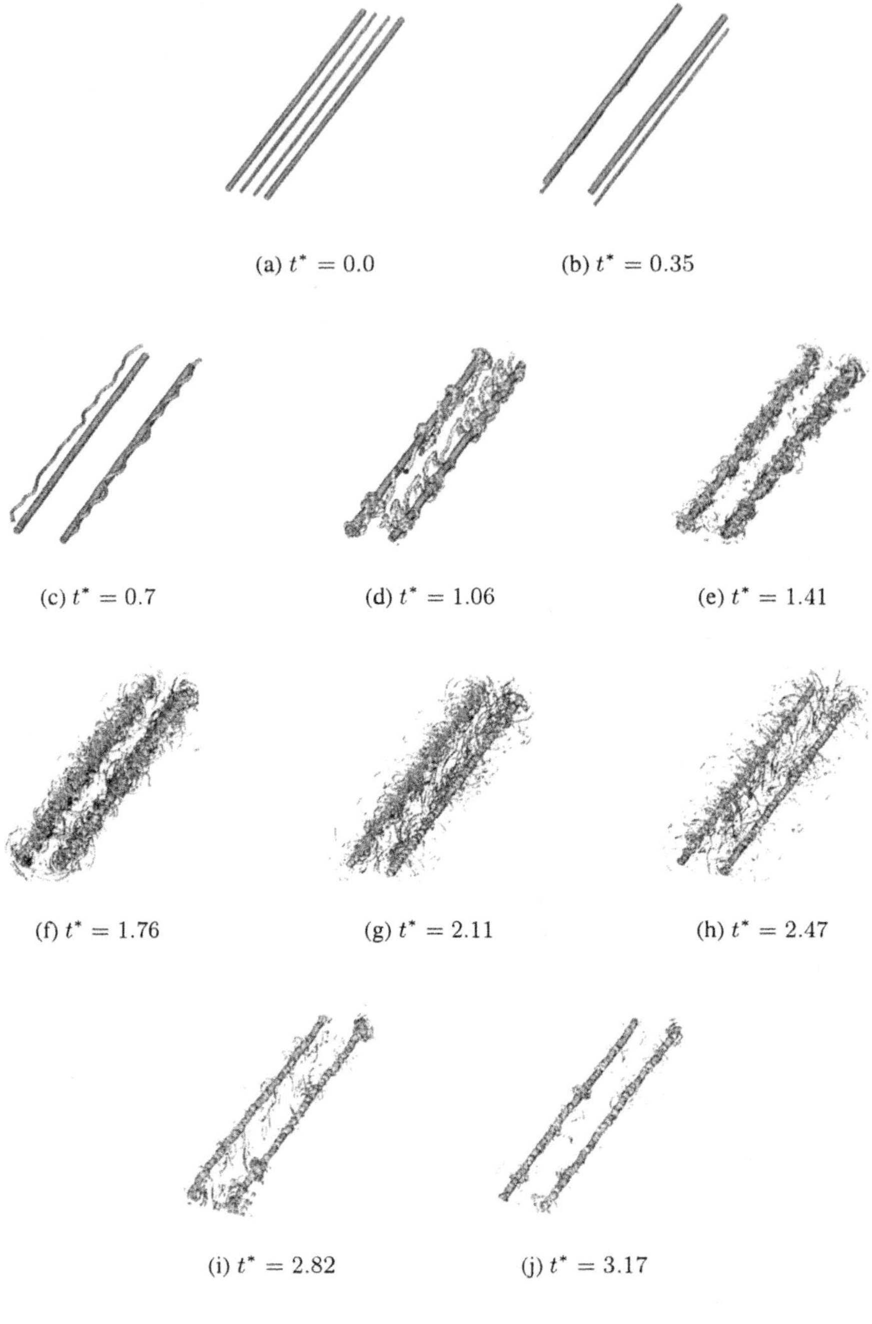

Figure 5 Temporal development of iso-surfaces of vorticity of configuration 1

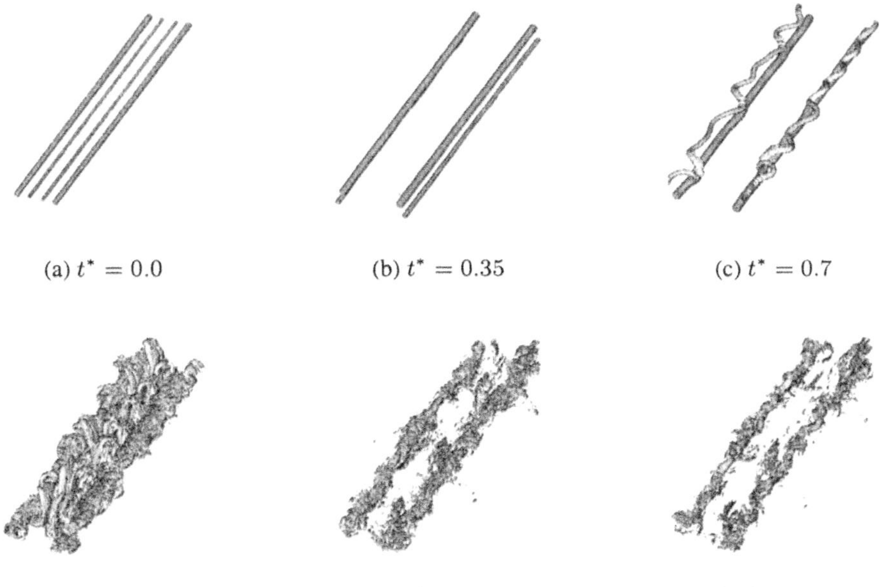

(a) $t^* = 0.0$ (b) $t^* = 0.35$ (c) $t^* = 0.7$

(d) $t^* = 1.06$ (e) $t^* = 1.41$ (f) $t^* = 1.76$

Figure 6 Temporal development of iso-surfaces of vorticity of configuration 2

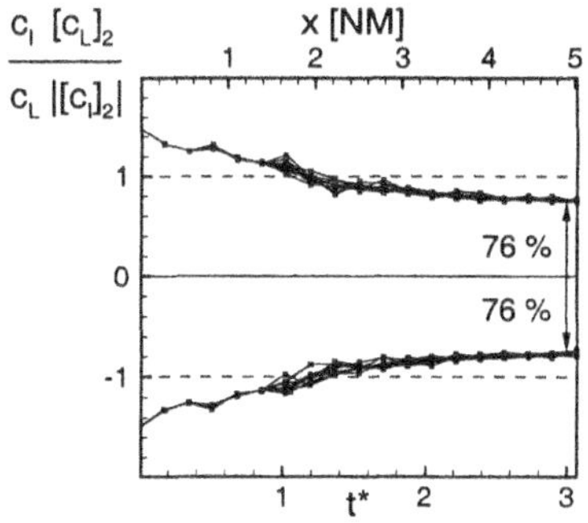

Figure 7 Rolling moment config. 1

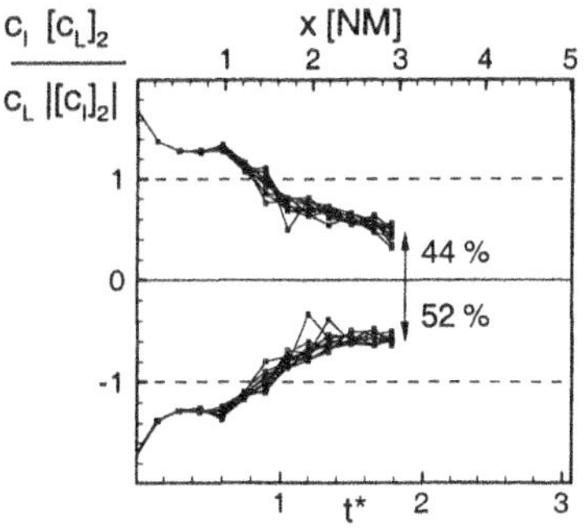

Figure 8 Rolling moment config. 2

Analysis of the Oblique Flying Wing Concept's Aerodynamic Potential by Means of Parameterization and Optimization

B. LIEBRECHT, G. NEUWERTH, D. JACOB
Institut für Luft- und Raumfahrt, RWTH-Aachen
Wüllnerstraße 7, 52062 Aachen, Germany

Summary

A parameterized description of a widely arbitrary Oblique Flying Wing geometry has been developed. In extension to previous studies with more or less symmetric geometries, asymmetric planform (chord distribution and sweep), dihedral bending and profile distribution were allowed. The aerodynamic properties of these geometries have been calculated with a higher order panel method. A parallel genetic algorithm has been developed and applied to determine geometries with optimized lift to drag ratios under realistic constraints concerning the total aircraft size and the available cabin space.

1 Introduction

Studying the linearized potential theory R.T. Jones showed in the 1950's, that an oblique wing minimizes the sum of induced drag and wave drag in supersonic flight at moderate Mach Numbers [8][9]. The oblique wing is basically a straight wing, flying with a sweep angle 'oblique' to the flight direction. Variation of the sweep angle allows to minimize induced and wave drag for different Mach Numbers and lift coefficients. Aircraft configurations taking advantage of this effect were Oblique Wing Body Configurations, beginning with R. Vogt's 1942 fighter design [12]. These configurations use a conventional fuselage and a straight wing attached with a single pivot to the fuselage to allow sweep angle variations.

An Oblique Flying Wing (OFW) without fuselage, first proposed by H. Lee of Handley Page in 1961, benefits vastly from the Oblique Wing Concepts' aerodynamic advantage. Additionally, the OFW enables a variable geometry without pivots and a further reduction of structural weights due to span loading effects. The OFW concept was subsequently studied more detailed [1][7][13]. The investigations showed, that an OFW was promising for very large aircraft flying at Mach numbers between 1.4 and 1.6. Aerodynamic studies of such configurations [10][11][14] showed, that a thorough design process is indispensable to prevent cross-flow shocks and to exploit the concept's high aerodynamic potential. Due to the complexity of an OFW and the strong coupling between the different disciplines, a multidisciplinary approach is necessary to design a competitive OFW.

All the OFW configurations studied until now suffer from two major disadvantages. First, these aircraft are far too large to be compatible with current airport infrastructure and market's needs. Second, these configurations are, though flying with an asymmetric sweep angle, more or less symmetric. This restriction prevents further improvements of aerodynamic efficiency as an investigation [10] with more asymmetric configurations has shown.

2 Realistic OFW

To overcome the above mentioned disadvantages of previous OFW designs, the present paper focuses on realistic OFW configurations. Realistic in this context means that the resulting OFW is designed to operate under current airport conditions with a passenger cabin for approximately 350 passengers. Nevertheless, the aerodynamic efficiency should be maximized to demonstrate the OFW concept's aerodynamic potential in comparison to actual subsonic transports aircraft.

2.1 Geometry Generation

As shown in Figure 1, we chose a 'span' (maximum extent between forward and aft tip) of 120 m, that although slightly larger than a theoretic 80 m × 80 m box, should allow handling on most conventional airports [15]. The profiles (perpendicular to the span direction) are parameterized as described in [10]. Profile parameters and twist vary in span direction as polynomials of fourth order. The extension of the asymmetric approach in [10] demanded a wing planform definition that covers a broad range of possible asymmetric shapes. The chord distribution (aircraft dimension perpendicular to the span direction) is elliptic plus additional parameters that enable a symmetric/asymmetric distortion. The inherent sweep of the wing is reached through a polynomial variation of third order of the quarter chord line. Dihedral is a polynomial of second order. With the sweep angle as additional parameter, the total number of parameters is 53.
Additionally, the wing tips are blended to thin, symmetric profiles and the planform is transformed to tips oriented towards the freestream direction. This definition is necessary to prevent numeric instabilities in the panel code during the optimization process. Special subroutines check the entire geometry for consistency to the limitations [2] of the aerodynamic code.
The number of surface points is 29 in profile direction and 37 in span direction, which is a good compromise of accuracy and execution speed. Calculations with a test geometry (not fulfilling the cabin constraint of section 2.2) have shown (Table 1), that higher numbers of grid points change the inviscid lift to drag ratio by 2.2%.

2.2 Cabin

The traditional approach in the design of an OFW was to create a geometry with a given volume, which was then scaled to attain the required cabin space. This aerodynamic approach usually leads to configurations with a Sears-Haack area

distribution. Combined with prismatic cabins with a constant section, the scaling process led to slender, but very large geometries.
In the current approach the cabin is integrated in the optimization process as a constraint. A cabin cross section comparable to a A340 section (Figure 2) is integrated to the wing on a length of at least 54m which results to a cabin comparable to an A340-500 cabin.

2.3 Aerodynamic Model

The aerodynamic coefficients of the OFW are calculated with the higher order panel method PanAir [6], commonly used in preliminary design. The panel code has been chosen because of the good prediction of inviscid lift and drag [5] and the execution speed. The large number of variables causes a high number of different configurations to be calculated during the optimization process. The application of linearized theory is valid as the geometry can be considered as slender and all shocks are assumed to be weak. A computation with the geometry from [10] reproduces the inviscid results of the Euler Calculations presented there with good accuracy.
At the current stage viscous effects are not yet taken into account. It can be expected that the inclusion of a simple, flat plate based, friction model would not lead to significantly different geometries.
For the application with an optimization algorithm, the panel code had to be modified. The input and output routines have been coupled to the geometry generator and major parts of the code have been adapted to allow continuous execution of the panel code. In addition to PanAir's internal error detection specific routines analyze the resulting pressure and Mach number distributions to verify compliance with linearized potential flow theory. Memory intensive subroutines have been rewritten to allow a significant reduction in execution time.

2.4 Genetic Algorithm

Because the target function (lift to drag ratio) was expected to be rough, a genetic algorithm has been chosen for the optimization process. Starting from the Pikaia algorithm [4] with its advantages of simplicity and modularity at the cost of poor performance, the different operators have subsequently been replaced and adapted to the special problem. The most important modifications may be shortly described:

- The Algorithm has been transformed to real coding, thus an individual (i.e. a configuration) is represented by a vector of 53 real numbers.
- One of three following crossover operators is randomly chosen: Multipoint Crossover, Intermediate Crossover, and exchange of parameters.
- Mutation is effected with variances and covariances interpreted as rotation angles as described in [1].
- Replacement is done under steady state reproduction plan. The individual to be replaced is chosen by the Euclidean distance between the parameter vectors of the new and the existing individual. This scheme ensures population diversity since only those individuals are replaced, whose parameters are

similar to the new individual's parameters and whose fitness is inferior to those of the new individual.

- The genetic algorithm has been parallelized with MPI. The use of multiple processors enables the parallel calculation of multiple individuals and thus large improvements in execution speed.

The computation of the genetic Algorithm is usually performed on RWTH-Aachen's SunFire SMP-Cluster. Starting from random parameters the optimizer needs about 1000 function evaluations to find valid geometries; after some 125000 function evaluations convergence is reached (Figure 2). The average execution time on 24 processors is slightly less than 8 minutes for 1000 function evaluations.

3 Optimization Results

The axis definition chosen for the OFW is shown in Figure 1, where v is the incoming flow, β the sweep angle and x, y and z the reference axes. The best configuration obtained by optimization is shown in Figure 4. It has an inviscid lift to drag ratio of 24.9 at a lift coefficient of 0.083 at cruise conditions with Ma=1.4 and β=20.5°. The reference surface is 1092 m^2. The cabin area is 2 m closer to the front tip than to the rear tip. The difference of the inviscid lift to drag ratio to those published in literature is due to the realistic constraints presented above.
The geometry and pressure distribution in different streamwise cuts are shown in Figure 5. The cuts show that the assumption of a slender body for the choice of the aerodynamic model is valid. The pressure distributions are smooth with increasing suction maximum from the front to the aft tip. The pressure gradients are smooth, thus the assumption of weak shocks will probably be fullfilled. Nevertheless, further calculations with more sophisticated aerodynamic models for selected geometries at different operating conditions are necessary to validate the results.
The aerodynamic behavior at off design points is shown in Figure 6. The curves show the lift to drag ratio at different lift coefficient for different sweep angles for the optimized configuration. Here the aerodynamic advantage of an OFW becomes clearly visible. With an adjustment of the sweep angle it is possible to maximize the lift to drag ratio at a given lift coefficient. Thus an OFW not only has an excellent lift to drag ratio at the design point, but also very good lift to drag ratios over a broad band of lift coefficients.

Conclusion

A realistic OFW configuration has been optimized with linearized potential theory. The results show, that the aerodynamic potential of the oblique wing concept may be utilized for a comparatively small OFW. Further work will focus on the implementation of a friction model and the effects of constraint variations on optimization results. The optimized geometry will also be the basis for further studies on flight mechanics, structures, refined aerodynamics and propulsion. Additional constraints resulting from other disciplines can easily be integrated into the optimization algorithm.

References

[1] Bäck, T.; Schwefel, H.-P.: Evolution Strategies and their computational implementation, Genetic Algorithms in Engineering and Computer Science, John Wiley & Sons, 1995

[2] Boeing Commercial Airplane Co.: A502I User's Manual-PAN AIR Technology Program for Solving Problems of Potential Flow about Arbitrary Configurations, Boeing D6-54703-TN, 1992

[3] Boeing Commercial Airplane Co.: Oblique Wing transonic transport configuration development, final report, NASA CR 151928

[4] Charbonneau, P; Knapp, B.: A User's guide to PIKAIA 1.0, NCAR Technical Note 418+IA, 1995

[5] Covert, E. E. (Ed.): Thrust and Drag: Its Prediction and Verification, Progress In Astronautics and Aeronautics Vol. 98, AIAA 1985

[6] Ehlers, F.E., et al.: A Higher Order Panel Method for Linearized Supersonic flow, NASA°CR-3062, 1979

[7] Galloway, T.L., Phillips, J.A., Kennelly, R.A.jr., Waters, M.H.: Large Capacity Oblique All-Wing transport Aircraft, NASA Doc. ID 19960023626

[8] Jones, R.T.: The Minimum Drag of Thin Wings in Frictionless Flow, Journal of The Aeronautical Sciences, Feb. 1951, pp. 75-81

[9] Jones, R.T.: Theoretical Determination of the Minimum Drag of Airfoils at Supersonic Speeds, Journal of The Aeronautical Sciences, Dec. 1952, pp. 813-822

[10] Li, P.; Seebass, R., Sobieczky, H.: Manual Aerodynamic Optimization of an Oblique Wing Supersonic Transport, Journal of Aircraft Vol. 36, No.6. Nov.-Dec. 1999

[11] Li, P.; Seebass, R., Sobieczky, H.: Oblique flying wing aerodynamics, AIAA 96-2120

[12] Pohlmann, H.: Blohm & Voss, Chronik eines Flugzeugwerkes 1932-1945, Motorbuch Verlag Stuttgart, 1979

[13] Waters, M.H.; Ardema, M.D.; Roberts, C.; Kroo, I.: Structural and Aerodynamic Considerations for an Oblique All-Wing Aircraft, AIAA 92-4420, Aug. 1992

[14] Van der Velden, A.: Aerodynamic Design and Synthesis of the Oblique Flying Wing Supersonic Transport, Stanford University Report SUDDAR 621, June 1992

[15] Van der Velden, A.: Supersonic Aircraft Shape Optimization, in Sobieczky, H. (ed.): New Design Concepts For High Speed Air Transport, Springer Verlag 1997

Table 1: surface grid test

Points in profile direction	Points in span direction	Inviscid lift to drag ratio	Execution time [s]
29	37	25.2263	18.76
57	37	25.1855	88.73
85	37	24.9892	233.14
57	73	24.7321	451.86
57	109	24.7651	1315.84
85	73	24.6868	1331.38

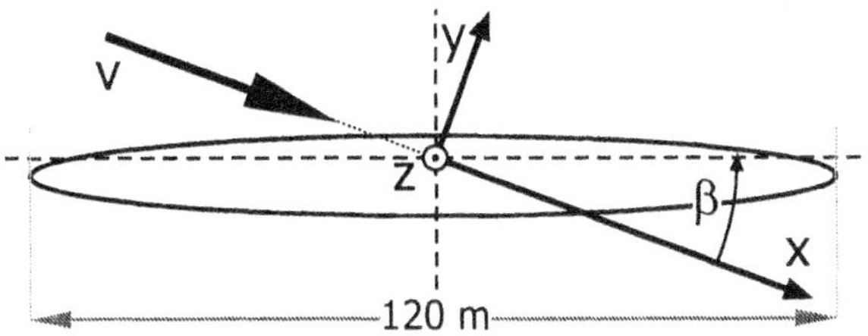

Figure 1: Definition of axes

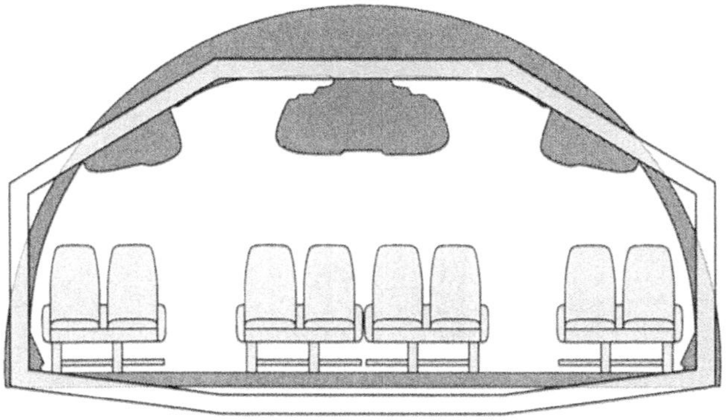

Figure 2: Cabin Cross Section: A340 - OFW

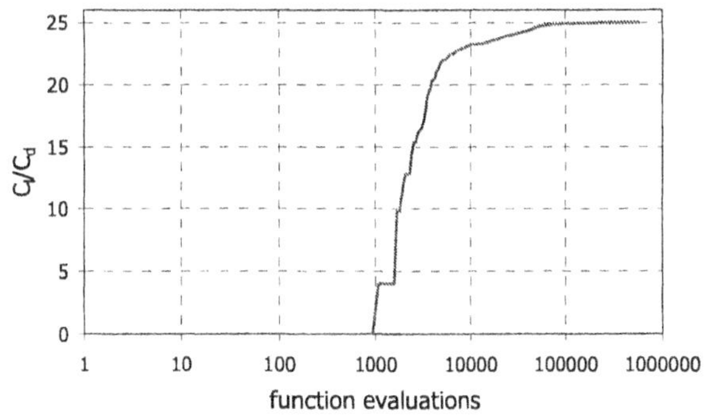

Figure 3: Convergence History of Inviscid Lift to Drag Ratio

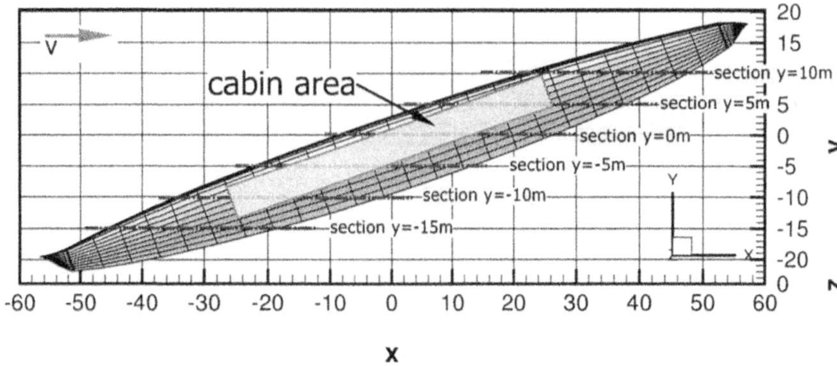

Figure 4: **OFW Geometry**

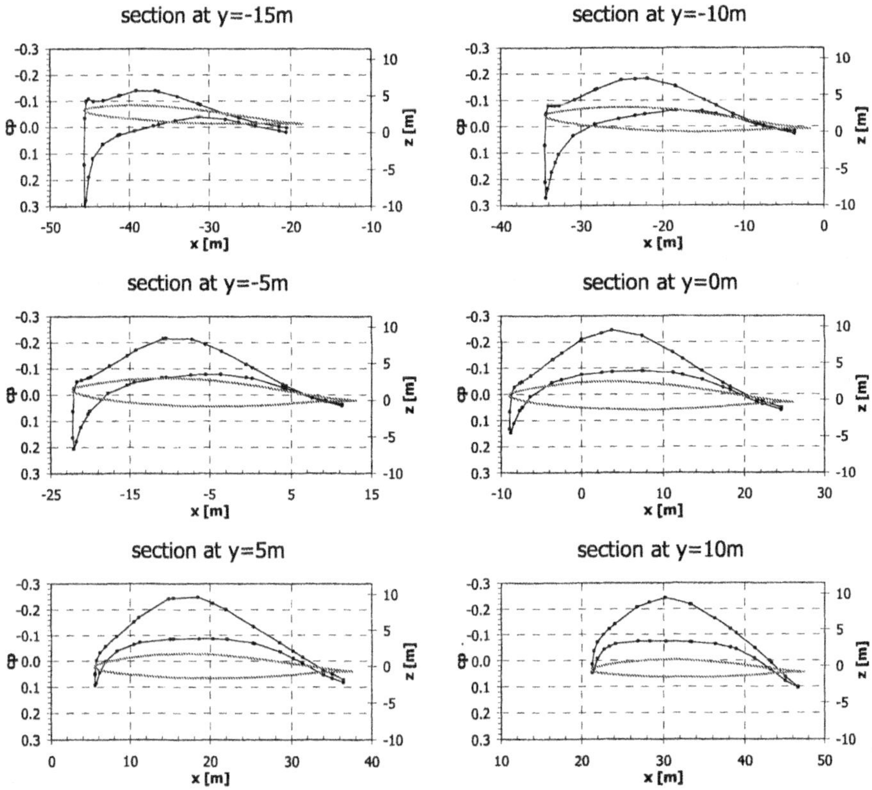

Figure 5: **Cut and Pressure Distribution at different position of y**

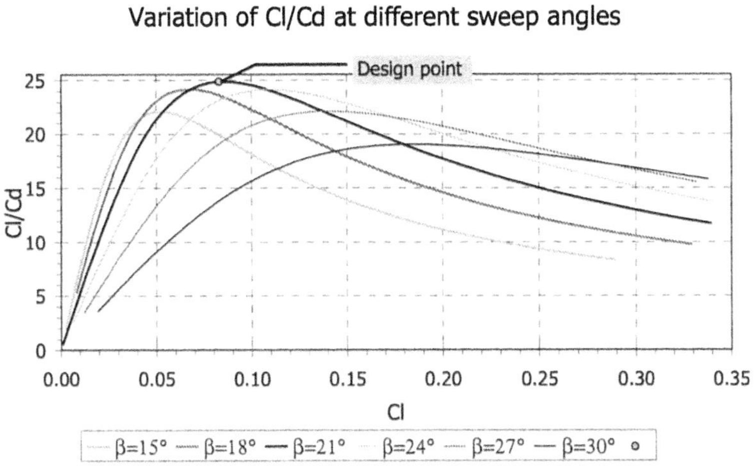

Figure 6: Inviscid Lift to Drag Ratio at different Lift Coefficients

Environmental Friendly Transport Aircraft

MARTIN HEPPERLE

German Aerospace Center, Institute of Aerodynamics and Flow Technology, Lilienthalplatz 7, D-38108 Braunschweig, Germany

Summary

The paper discusses some key elements of aircraft configurations which are intended to meet future requirements of environmental friendly transportation. Increased transport capacity is achieved by enhancing takeoff and landing performance in order to reduce the separation between subsequent flights. Additionally, overall productivity is improved by adapting the configuration to future noise restrictions allowing for extended daily operating times.

1 Introduction

Almost all scenarios for the development of air traffic show that a further growth of transport capacity can be anticipated in the future. Solutions to handle the increased traffic are constrained by economical and environmental issues which may lead to quite different or even completely new aircraft configurations. In the past, it was tried to minimize the environmental impact after the configuration had already been developed mainly based on cost and performance requirements. In the future, environmental issues will have a stronger influence at the component level, but also on the arrangement of the aircraft components and thus on the configurations. Increased transport capacity can be provided by:

1. increasing the number of aircraft,
2. reducing the turn around time at the airport and separation between flights,
3. enlarging the aircraft capacity, and
4. increasing the cruise speed of aircraft.

The first path has been followed in the past by simply using a larger number of available aircraft. It is severely limited by the number of airport slots and by environmental issues. The second possibility seems to offer a small potential for further improvements only. Over the years, the last two solutions have been the subject of conceptual and preliminary design studies which have lead to proposals for new configurations.

Enlarging the capacity of configurations by scaling has been a traditional and quite successful strategy during the history of flight. The latest sample of this method is the Airbus A380, which is scheduled to enter Service in 2008. While refinements in all disciplines have made such a design possible, it seems to be close to the limits of the classical airplane configuration. Therefore new aircraft configurations have been proposed, which promise higher transport capacity at reduced cost.

Figure 1 presents the passenger capacity of a few current and future configurations over their cruise Mach number regime. It can be seen that the direction of most developments is directed into one single direction: either an increase of passenger capacity or an increase in speed. Only the OFW concept tries to extend the operating range in speed as well as capacity.

One of the configurations shown in the graph seems to be rather unspectacular in terms of cruise speed and passenger capacity, but may offer more challenges than expected. This configuration, dubbed "The Green 24 Hour Aircraft" (24HAC) will be discussed in more detail below.

2 The Green 24 Hour Aircraft

Currently, many airports have already reached the limits of their capacity. Increasing the passenger capacity of airplanes helps to limit the number of takeoffs and landings for long range flights, but the huge number of short and medium range flights cannot be reduced easily. Each week, nearly 70% of all flights take less than 5 hours but carry 50% of the travelers (see figure 2, derived from [6]).

One design target of a new configuration design must be to minimize the time between subsequent takeoffs and landings. This can be done by introducing the capability to climb and approach at steeper angles and to select different angles for the flight path depending on the current traffic situation (Figure 4). Thus it is possible to avoid the trailing wakes of preceding aircraft and to reduce the distance between flights. Additional flexibility can be achieved by developing flight control systems suitable for dynamic takeoff and approach patterns. This aim is augmented by a considerable reduction of the perceived noise level so that the aircraft can be operated 24 hours each day. These requirements favor certain design features and can lead to configurations with significant changes compared to current aircraft [9]. The following paragraphs will discuss some of these features and how they affect the configuration.

2.1 Improving the Takeoff and Landing Performance

The performance during these two mission segments can be characterized by two parameters: the climb and sink angle and the flight speed. A steeper angle can be used to adapt the flight pattern to the current traffic density at the airport. By varying flight path angle and speed from flight to flight it is possible to reduce the distance between aircraft without increasing the danger of wake vortex interaction.

As noise scales approximately with velocity to the power of 5, it is desirable to fly as slow as possible. On the other hand, the time spent close to the ground and the area covered by the noise footprint during takeoff should be kept at a minimum. Therefore an assessment of the perceived noise must take place already during the conceptual design phase of new aircraft.

The equilibrium of the forces during climb and glide can be written in the form

$$\sin\theta = \frac{T}{W} - \underbrace{\frac{\rho_\infty}{2} \cdot v_\infty^2 \cdot \frac{S}{W} C_{D0}}_{\frac{\text{zero lift drag}}{\text{weight}}} - \underbrace{\frac{W}{S} \cdot \frac{2 \cdot k \cdot \cos^2\theta}{\rho_\infty \cdot v_\infty^2}}_{\frac{\text{induced drag}}{\text{weight}}} \quad . \tag{1}$$

A differentiation of (1) leads to the maximum climb angle θ_{max} and the following expression for the associated flight speed [2]:

$$v_{\infty,\theta_{max}} = \sqrt{\frac{2}{\rho_\infty} \sqrt{\frac{k}{C_{D0}} \frac{W}{S}} \cos\theta_{max}} \;, \quad \text{where} \quad C_L = \sqrt{\frac{C_{D0}}{k}} \quad . \tag{2}$$

Figure 3 illustrates how the operating point during climb can be controlled by a variation of the design variables. The climb angle can be increased by:

- increasing the ratio of T/W (favoring twin jets),
- increasing the aspect ratio Λ, or
- increasing the ratio L/D, reduction of C_{D0}.

The desired reduction of flight speed can be achieved by the following measures:

- reduction of the wing loading W/S (not affecting the climb angle), or
- increasing the usable lift coefficient C_L.

A simple numerical study shows that the T/W ratio is the most powerful driver to increase the climb angle. The most important driver for a reduction of the flight speed is the wing loading W/S.

A high thrust to weight ratio favors twin jet configurations, due to the FAR takeoff requirements, which result in 50% reserve power compared to 33% for an aircraft equipped with four engines.

The usual methods to adapt the cruise design for takeoff and landing are a reduction of the wing loading by enlarging the lifting area and an increase in lift coefficient by cambering of the airfoil sections. This can be achieved with a slotted high lift system, which typically consists of three elements at least (slat, main wing, flap). Drawbacks of such a slotted multi-element flap system are mechanical complexity, weight, costs and a rather high noise level.

The actually usable lift coefficient can also be raised by altering the wing planform (e.g. reduced sweep angle). Additionally, the chord length could be increased in order to maintain the absolute wing thickness, mass and cruise performance. This would also have the beneficial effect of increasing the wing area, thus lowering W/S and helping to recover any lift loss which occurs when the slotted high lift system is replaced by a single element system. Segmented full span ailerons provide means to adapt the lift and drag distribution to the current flight condition.

2.2 Reducing the Noise during Takeoff and Landing

In order to extend daily operating hours, the noise during takeoff and landing must be reduced considerably (-10 ... -20 dB). During these two flight segments different sources are responsible for the noise generation:

takeoff:

- engine noise (fan noise and jet noise, depending on BPR),
- airframe noise (landing gear, high lift systems),

landing:

- airframe noise (landing gear, high lift systems),
- engine noise (mainly fan noise).

Engine noise is dominant during takeoff and is composed mainly of fan and jet noise. The high frequency fan noise can be reduced by improving the absorption inside the nacelle and by using longer inlets to achieve a shielding effect – for a new configuration an additional shielding effect could be achieved by positioning the engine inlet above the wing or by embedding the engines in the rear fuselage [8]. In the latter case the long and curved inlet ducts must be carefully optimized to keep pressure losses low and to limit flow distortion at the fan entry. Also maintenance costs of such buried engines will be higher than on conventional arrangements.

Jet noise levels can be lowered by increasing the bypass ratio (thus reducing the jet velocity) and by modifying the structure of the jet shear layer (e.g. by chevrons or mixers). While ultra high bypass ratio engines offer great benefits in noise (Figure 5) as well as propulsive efficiency, their development requires the addition of a gearbox. Such a geared engine could also be used to raise the turbine RPM towards the higher frequency range, where the human ear is less sensitive.

The takeoff and landing configuration of a conventional aircraft is characterized by slotted high lift systems. Minimizing the number of high lift elements and their track systems can lead to a considerable reduction of noise and weight. The weight savings (≈2-3% of the wing weight) can be spent in a larger wing area to achieve the same lift capability with a plain flap droop nose system. Any drag penalty during cruise due to the larger wing area can be compensated by adapting the flight altitude [5].

One possible candidate for a low noise high lift system could consist of a clean leading edge and a full span single slotted flap. If required, a simple droop nose could be added. A comparison of achievable maximum lift coefficients indicates, that such a plain wing would require about 40% more chord length when compared to a slotted configuration (Figure 6).

Additional noise is generated by extended landing gears and open cavities like wheel wells. Modern aircraft already cover their wheel wells while the gear is extended, but further improvements can be made. Struts can be streamlined and covers can be added to the wheels and any cavities. The length of the main landing gear can be reduced by attaching it to the fuselage and by moving the engines away from their typical location under the wing. Moving the engines to the tail can results in a shorter rear fuselage so that the landing gear can be shorter without reducing the required rotation angle.

2.3 Selecting Elements for a new Configuration

Based on the requirements and effects discussed above, several design options for a new configuration can be identified. By combining these elements, various configurations can then be developed and analyzed (Figure 7).

Lifting surfaces:

- increased wing span b and aspect ratio Λ to improve L/D,
- increased wing area S to reduce the wing loading W/S, and
- simple high lift system without slats,
- laminar flow technology to reduce cruise drag.

Landing gear:

- short and few legs, add fairings, covers and close all open cavities.

Engines:

- twin jets (high T/W ratio due to "one engine out" takeoff requirement),
- shielded fan face (inlet shape, placement above wings or fuselages, buried engines),
- low jet velocity (increased BPR, geared fan),
- improve absorption and reduce mechanical noises.

3 Conclusions

Currently, several European research programs explore single aspects of environmental friendly aircraft (e.g. noise, airport operations, engine technology, aerodynamics, etc.). Their findings can partially be applied to the current fleet of conventional aircraft in terms of changed operating procedures and retrofit solutions in certain areas. Only a combination of these results together with some specific elements of aircraft configurations may lead to new aircraft configurations with reduced environmental impact. These will also have the additional benefit of reduced operating costs. In the long term, the huge transport volume of short- and medium range flights can only be handled by introducing new technologies together with a revised arrangement of aircraft components. For these specific missions, this does not necessarily mean a completely new configuration like a flying wing aircraft, but a clever arrangement and sizing of the key components.

Future work will focus on the modeling of these key elements at preliminary aircraft design level, the development of the most promising configurations and the assessment and comparison of the complete aircraft configurations. This will also include new models for the modeling of airframe and engine noise and suitable additions to cost models. Additional work should address the cost for manufacturing and operating such aircraft.

4 Symbols and Abbreviations

BWB	Blended Wing Body
BPR	bypass ratio (UHBR = ultra high bypass ratio)
OFW	Oblique Flying Wing
IWFC	Integrated Wing Fuselage Configuration
SST	Supersonic Transport
SC	Sonic Cruiser

Symbol	*Description*	*Unit*
D	Drag Force	N
L	Lift Force	N
M	Mach number	-
S	wing area	m^2
T	engine thrust	N
W	aircraft weight	N
b	wing span	m
k	k-Factor	-

Symbol	*Description*	*Unit*
v_∞	aircraft speed	m/s
C_D	drag coefficient	-
C_{D0}	C_D at C_L=0	-
C_L	lift coefficient	-
θ	climb angle	°
ρ_∞	density of air	kg/m^3
Λ, AR	aspect ratio (b^2/S)	-

5 References

[1] Ira H. Abbott, Albert E. Doenhoff. "Theory of Wing Sections", Dover Publications, New York, 1959.

[2] John D. Anderson, Jr. "Aircraft Performance and Design". McGraw-Hill, ISBN 0-07-001971-1, 1999.

[3] J. E. Green. "Greener by Design – the technology challenge", The Aeronautical Journal, Vol. 106 No. 1056, February 2002.

[4] Jeffrey J. Berton. "Noise Reduction Potential of Large, Over-the-Wing Mounted, Advanced Turbofan Engines", 14th International Symposium on Airbreathing Engines, Florence, Italy, 1999.

[5] Martin J. Hepperle. "The Sonic Cruiser – A Concept Analysis", International Symposium "Aviation Technologies of the XXI Century: New Aircraft Concepts and Flight Simulation", ILA-2002, Berlin, 7-8 May, 2002.

[6] Lufthansa Flight Plan, 2002-2003.

[7] Mark A. Potsdam, Mark A. Page, Robert H. Liebeck. "Blended Wing Body Analysis and Design", AIAA-97-2317, 1997.

[8] Steven F. Yaros et al. "Synergistic Airframe-Propulsion Interactions and Integrations", NASA TM-1998-207644, 1998.

[9] Nicolas E. Antoine, Ilan M. Kroo. "Aircraft Optimization for minimal Environmental Impact", 9th AIAA/ISSMO Symposium on Multidisciplinary Analysis and Optimization, September 2002, Atlanta.

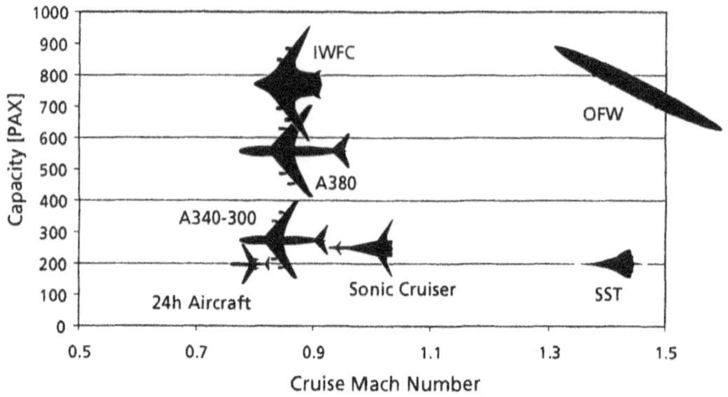

Figure 1 Comparison of various aircraft configurations in terms of their passenger capacity over cruise speed.

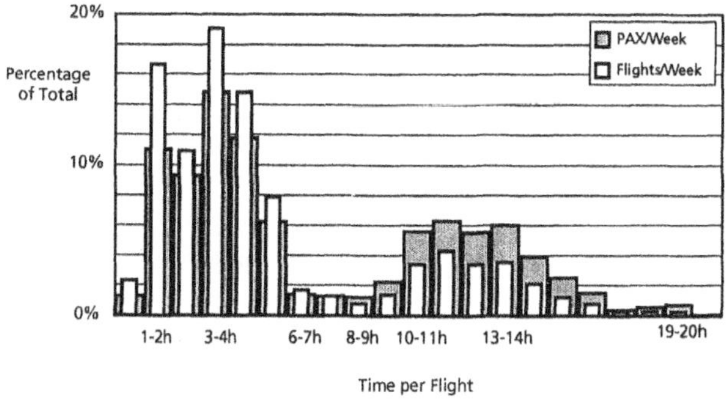

Figure 2 Distribution of number of flights and passengers versus overall flight time.

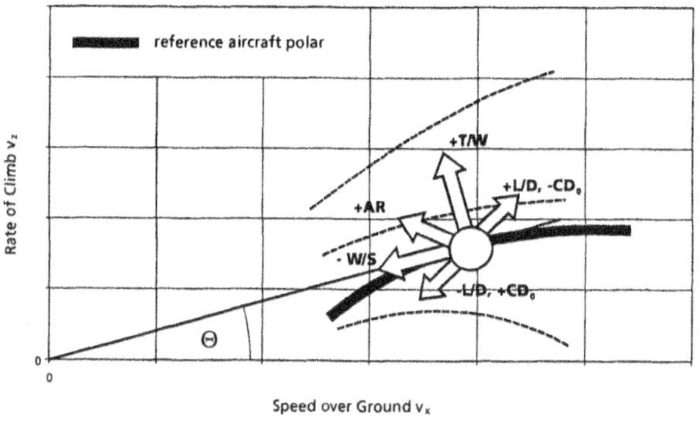

Figure 3 Effect of a variation of parameters on climb performance.

Figure 4 Improved takeoff and landing performance allows for adaptation of the flight path according to the current traffic density.

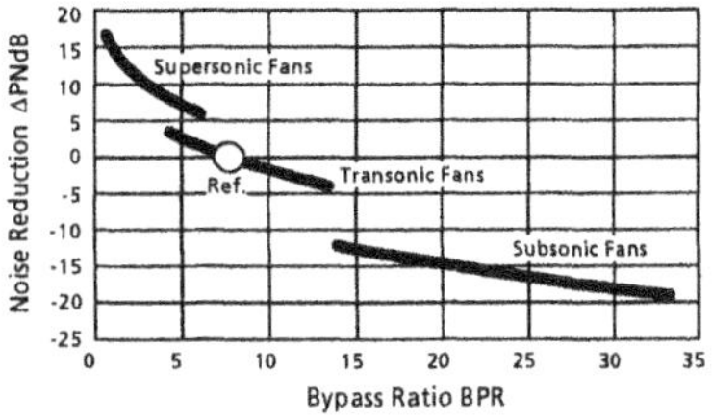

Figure 5 Potential of noise reduction by increasing the bypass ratio.

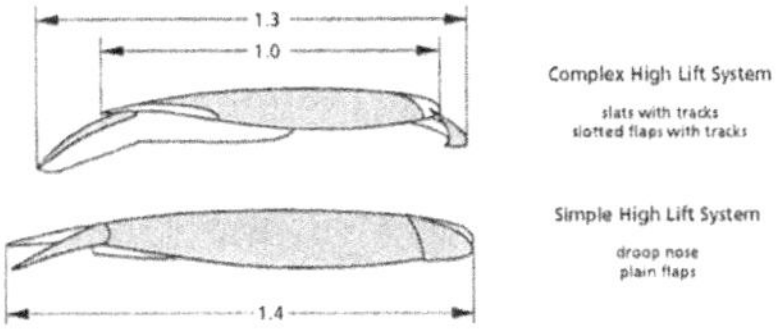

Figure 6 Two high lift systems having similar maximum lift capabilities.

a) UHBR engines, upper surface blowing.

b) fan shielded by wings, high wing.

c) fan shielded by wings, low wing.

d) buried engines, low wing.

Figure 7 Several possible configurations for a *Green 24h Aircraft*.

Application of fast and advanced CFD for high-lift configuration design tasks

D. RECKZEH[1], K. SCHINDLER[2]

Airbus Deutschland GmbH
Aerodynamic Design & Data Department EGAG
Hünefeldstr. 1-5
28183 Bremen
Germany

Summary

Today, CFD is a standard tool for the aerodynamic design of high-lift systems at Airbus. Fast and reliable CFD-methods allowing a quick evaluation of numerous design modifications as well as advanced methods capable to predict the complex flow around a complete aircraft in high-lift configuration are part of the design process. Examples for the application of CFD-tools suitable for the respective design task are presented, i.e. 2D-contour design with a Navier-Stokes method, trailing edge layout studies with a quasi-3D method and prediction of Reynolds number and engine installation effects with 3D-viscous-Panel, 3D-Euler- and 3D-Navier-Stokes methods.

High-lift configuration design with the means of CFD-tools

The design of Airbus high-lift configurations is characterised by several needs and constraints [1],[5]. The aerodynamic design is integrated in interdisciplinary relations with other disciplines as systems or structures design, which imply requirements and limitations to the opportunities in aero design. The solution to be found has to be the optimum for the overall aircraft, fulfilling the projects requirements and it has to be delivered in time for the projects schedules.

In the past the required time for the verification of the aerodynamic design was characterised by the turn-around-time for the needed wind-tunnel model design, manufacture and testing. Besides the wind-tunnel test being also today the backbone for high-lift design verification, the developments made in CFD-methods capable to cope with complex flows [4] allows their integration into the high-lift design process.

The essential advantage of the use of CFD is the significant reduction in the time required for design assessment. Based on CFD-predictions the interaction in the iterative

---

[1] Head of High-Lift Design Group, contact: Daniel.Reckzeh@airbus.com
[2] High-Lift Wing Design Engineer, contact: Klaus.Schindler@airbus.com

- Prediction of wake vortex effects and downstream flow quality
- *Current turn-around-time for a typical verification of a design task: < 2 weeks (Panel & Euler) or < 4 weeks (Navier-Stokes)*

The description of following tasks shall give short examples for the design work being conducted with the use of CFD in recent Airbus projects [5], however no detailed discussion and validation of the results can be given in this overview.

Examples for design tasks with a 2D-Navier-Stokes method: 2D-design, comparison and maximum performance prediction of trailing edge systems

The contour design of 2D-sections is today dominated by CFD due to the availability of fast and reliable methods, which allow the assessment of a large number of different designs in a short time. Panel methods with viscous coupling and lately also Navier-Stokes methods are used for this task. With increasing computer power and therefore significantly reduced calculation time Navier-Stokes calculations have become a standard method for 2D-section design of multi-element airfoils. The capability to capture viscous effects like flow separation and confluent boundary layers, which are maximum lift limiting effects, allows a more detailed analysis and prediction of the profiles characteristics. The Navier-Stokes method *TAU* [7] is being used for 2D-analysis of multi-element profiles. Validation work has shown good accuracy in the prediction of setting-effects and flow separation onset on the elements.

One application of *TAU* for the design of a double slotted flap for a transport aircraft is illustrated in Fig. 2. Streamlines are shown for two different flap/tab deflection angles at a constant angle of attack of 2°. The very limited flow separation on the tab element for the lower deflection (flap 34°, tab 19.4°) occurring at low to medium angle of attack spreads over large parts of the tab with increased deflection of the elements (flap 35°, tab 20°). The resulting lift loss is reflected in the corresponding lift vs. incidence curves. With increasing angle of attack the separation is progressively reduced for the 35°/20° configuration, which leads to the characteristic non-linear dependency between lift and incidence and finally to a higher lift in the maximum lift area of the polar relative to 34°/19.4°. Consequently the next step could be to analyse the influence of the flap and tab setting parameters (gap, overlap) on the flow behaviour in order to reduce the separation on the tab for the higher deflection and improve the linear lift performance. The predicted effect is in line with observations made in wind-tunnel test of this configuration.

Comparisons of different high-lift layouts and trade studies are key elements of the design process. Due to the large number of required variations necessary to find a well-balanced design, CFD predictions are a valuable tool in addition to wind-tunnel testing, where extensive studies are often limited by time and cost.

One example for the prediction of the maximum flap performance, i.e. maximum lift in this case, for different flap layouts is given in Fig. 3. The maximum lift coefficient versus the flap deflection angle calculated by *TAU* is shown for double and single slotted flap layouts. As expected, the double slotted flaps lead to higher maximum lift in comparison to the single slotted flap for a given flap deflection. The layout variations (flap chord and shroud-line) of the double slotted flaps indicate a significantly stronger influence of the flap chord on the maximum lift compared to variation of the shroud-line position. Furthermore these results give qualitative information on the maximum effective flap deflection angle, i.e. the deflection angle giving the highest maximum lift. However, these results obtained for modifications on 2D-sections can only be seen as an indication for the behaviour of the 3D-wing. It is therefore necessary to make use of 3D-methods at this stage of the design process.

Example for design tasks with a quasi-3D method:
3D-assessment of wings with different trailing edge layouts

As mentioned before the effect of design modifications calculated for 2D-sections cannot simply be transferred to a 3D-wing. With the quasi-3D tool *Q3D* [3], which couples calculated 2D sectional results with a lifting surface method; an initial evaluation of the high-lift wing layout considering also the wing and flap planform can be performed. The short turn-around-time for a typical design evaluation with *Q3D* is a decisive factor in a design phase, where multi-disciplinary constraints still leave enough degree of freedom for a large number of layout variations. The different impact of design modifications predicted by the 2D-NS and the quasi-3D method is presented in Fig. 4. Results of a flap layout study using both methods show the effect of flap chord and shroud-line position variations on the lift curves. While the results show good agreement in a qualitative way the absolute lift increments predicted by the quasi-3D computation are significantly smaller compared to the 2D-NS computation. Wind-tunnel tests confirm well the trends obtained with *Q3D*, however the absolute increments may differ.

Example for design tasks with a 3D-viscous-Panel method:
3D-analysis of Reynolds number effects

Once the number of trailing edge layouts is narrowed down to a few selected candidates, which are feasible from the aerodynamic and also from the system and structural point of view, more detailed studies can be performed with a 3D-panel model of the high-lift wing. Using available CAD geometry as a basis such a model can be built up in sufficiently short time to include the 3D-panel method in the design loop. With a coupled integral boundary layer calculation it is possible to capture viscous effects and obtain exclusive information on e.g. transition locations. Fig. 5 presents the results of a 3D-panel computation with coupled integral boundary-layer method with the method *VSAERO* [9] for a high-lift wing with a clean leading edge. The Reynolds number effect on the transition location is predicted for wind-tunnel and full-scale conditions. The transition is indicated by a strong local rise of the skin friction coefficient calculated along on-body streamlines. Fig. 5 shows strongly developed laminar flow on the wing upper surface for Re=1.25 Mio. and low angle of attack. With increased Reynolds number (Re=5 Mio.) the transition moves significantly upstream and finally occurs directly behind the leading edge at full-scale Reynolds number (Re=22.8 Mio.). Even with the method simulating only the transition effects with a rather simple approach (details see [9]) these results can serve as good qualitative indications how Reynolds-number may affect the flow behaviour on the wing and are especially useful for the application of transition fixing for wind-tunnel testing at low Reynolds-numbers.

Example for design tasks with 3D Euler & Navier-Stokes methods: 3D-analysis of engine integration effects

The maximum performance of a high-lift design on the aircraft level is often strongly influenced by other components attached to the wing, e.g. engine pylons. For the evaluation of the performance of the complete configuration it is therefore necessary to consider the effect of these components, which requires complex and expensive CFD models. 3D-Euler and 3D-Navier-Stokes methods are capable to capture these effects and allow deeper insight into the flow features. Despite the progress made in grid generators and solver codes these methods are still very time consuming and are therefore mainly used in the design process for detailed studies of already refined configurations.

One example for the application of the Euler method (*TAU* in 3D-Euler mode) is presented in Fig. 6. Different local wing leading edge modifications designed to minimise the engine installation effect on the maximum lift are evaluated. As viscous effects are not captured with this calculation, an empirical method is used to predict the effect on the maximum lift [8]. A criterion based on the local pressure distributions is utilised for the assessment of the potential maximum lift improvement. This procedure is suitable if only incremental results are required and takes advantage of the reduced CPU time of the Euler method.

Fig. 7 demonstrates the capability of the Navier-Stokes method (*TAU* in 3D-NS-mode) to predict the complex viscous flow around nacelle-wing geometry. Very strong minimum pressure peaks on the leading edge at the nacelle-wing junction lead to a strong bending of the surface streamlines in that area. A characteristic streamline pattern is displayed which corresponds well with oil flow visualisations from wind-tunnel tests. The deeper understanding of the flow behaviour in this critical area is the key for the development of effective design modifications and the improvement of the maximum lift performance of such a configuration. Moreover, the fact, that these flow features are captured by the method shows a lot of promise for the evaluation of adapted designs aimed at influencing these features for the better.

Conclusions & Outlook

The ongoing development in grid-generators' and CFD-codes' capabilities to cope with complex configurations in combination with available computing power is currently leading to a significantly increased use of CFD as a tool for high-lift design tasks. A lot of complex tasks even with 3D Navier-Stokes methods can already be carried out from the design department of Airbus [2]. The impressive development in CFD enabled a significant step ahead compared to the Airbus design environment described in [6] in 1999.

Major importance in future demands for the next steps in CFD tools development will be laid on the user-friendliness of the methods but also especially on the careful benchmarking and validation of the application cases in order to know best about the uncertainties and limitations in prediction to consider them when selecting the design based on CFD-results. A focus has to be made on these issues to feed them carefully back out of the use in production to the tools developers.

With the increasing capability to give satisfying answers to daily design tasks a stronger demand will come up also for the prediction of effects "beyond" the results of standard wind-tunnel-testing, such as the detailed prediction and analysis of boundary layer transition effects or instationary flow features. For these tasks the CFD development will be stronger than today challenged by the lack of suited validation data and by the fact that the quality of the prediction may well reach the uncertainty level of the experimental result. A main focus should be given to the development in turbulence models and transition modelling used in the Navier-Stokes methods, as they deal with key effects of high-lift configurations.

References

[1] Flaig A., Hilbig R., *High-Lift Design for Large Civil Aircraft*, AGARD-CP-515, 1992

[2] Hansing J., Sutcliffe M., Knobloch O., *Numerical Simulation of the Propeller Flow around a Four-engined Aircraft in High-Lift Configuration,* 13th AG-STAB -DGLR Symposium, 2002

[3] Jakob H, *Zur rechnerischen Vorhersage des Maximalauftriebs an Flügel-Rumpf-Konfigurationen bei Unterschallströmung,* Deutsche Airbus Bericht EF-1993, 1994

[4] Kroll N. et al., Megaflow – a numerical flow simulation tool for transport aircraft design, ICAS 2002-1.10.5

[5] Reckzeh D., *Design Work for the A3XX High-Lift Wing*, (12th AG-STAB -DGLR Symposium 2000), New Results on Numerical and Experimental Fluid Mechnics Volume 77, pp 3-10, Vieweg Verlag, 2002

[6] Reckzeh D., *CFD-Methods for the Design Process of High-Lift Configurations* (11th AG-STAB -DGLR Symposium 1998), New Results on Numerical and Experimental Fluid Mechnics Volume 72, pp 347-354, Vieweg Verlag, 1999

[7] TAU-code user guide, DLR, 2001

[8] Valarezo W.O., Chin V.D.: Method for the Prediction of Wing Maximum Lift, Journal of Aircraft, Vol. 31, No. 1, Jan.-Feb. 1994

[9] *VSAERO Users' Manual Version 6.2*, Analytical Methods Inc. Redmont, Washington, USA, 2001

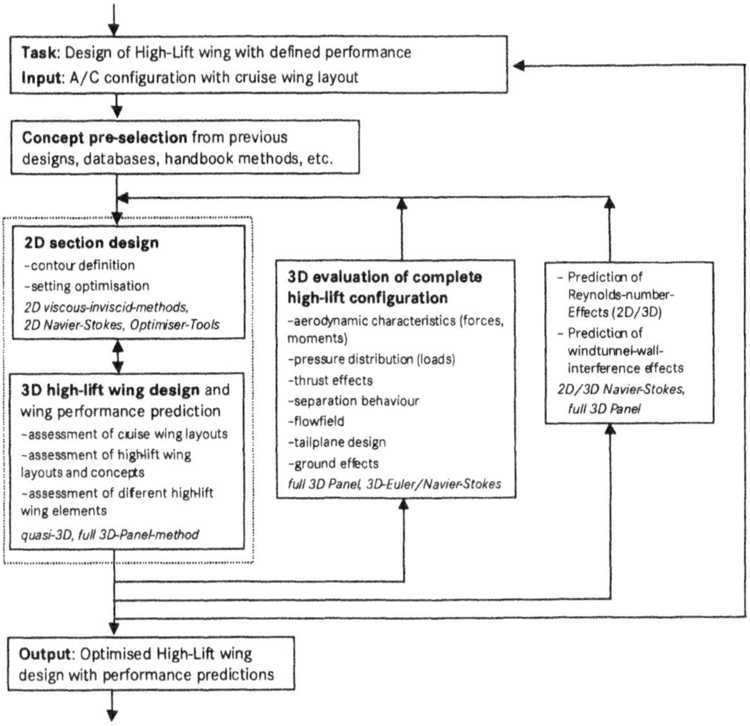

Fig. 1: The CFD-based high-lift wing design process

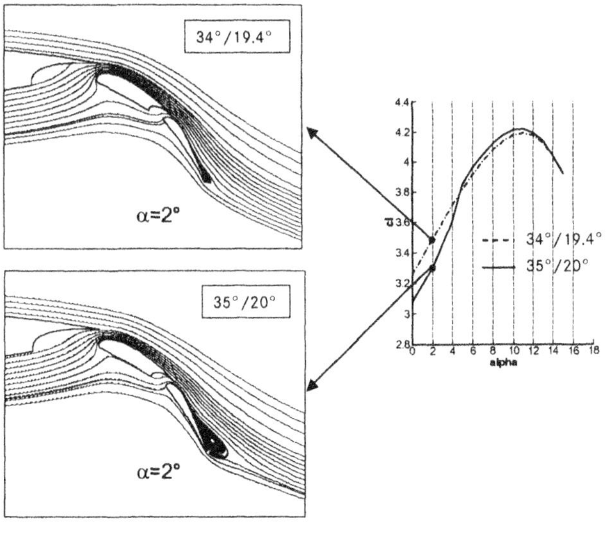

Fig. 2: Prediction of separation effects limiting maximum lift of a double slotted flap with the 2D Navier-Stokes method *TAU*

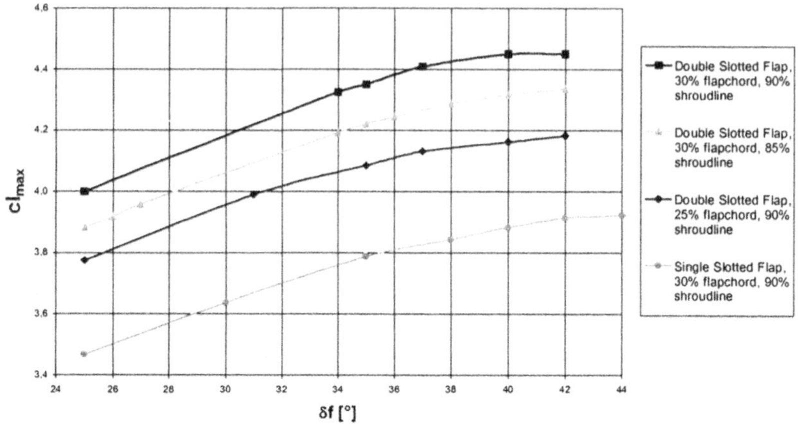

Fig. 3: Prediction of maximum flap performance of different flap layouts with the 2D Navier-Stokes method *TAU*

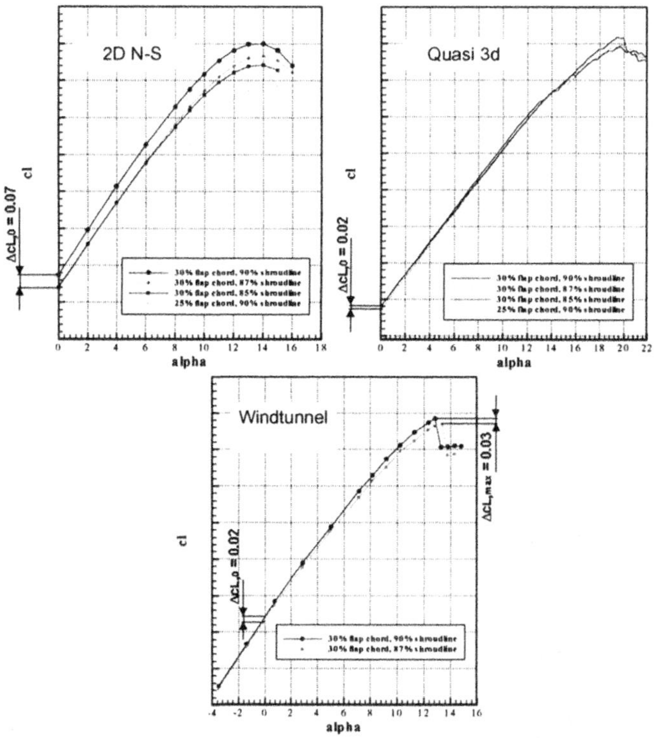

Fig. 4: Evaluation of high-lift layout modifications with the 2D Navier-Stokes method *TAU* and the quasi-3D method *Q3D* and comparison with W/T-results

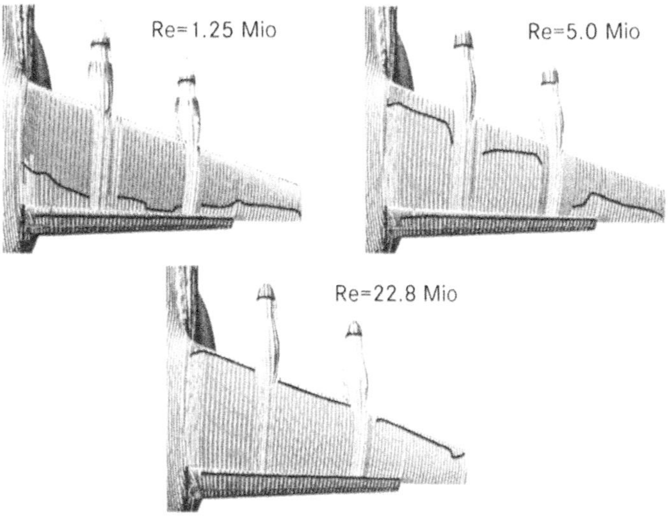

Fig. 5: Prediction of the Reynolds number effect on transition locations with the 3D-viscous-Panel method VSAERO

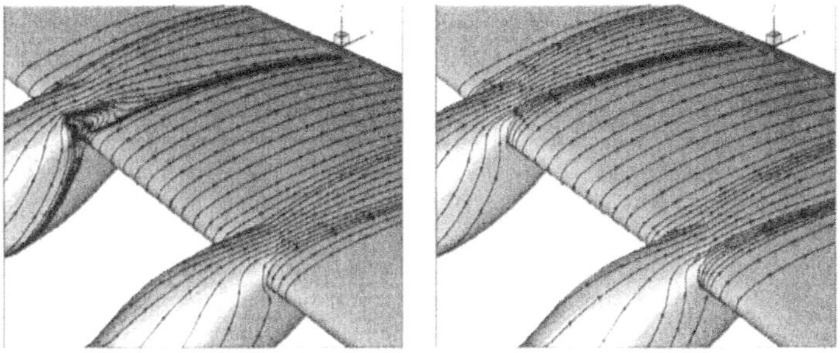

Fig. 6: Evaluation of local wing leading edge modifications with the 3D-Euler method *TAU*

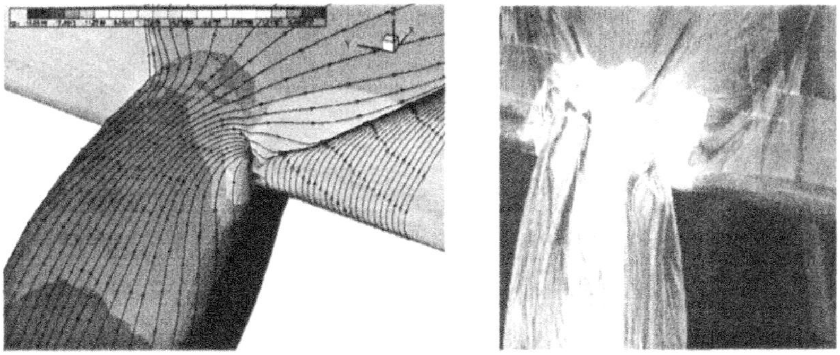

Fig. 7: Analysis of the viscous flow in the area of the nacelle-wing junction with the 3D-NS method *TAU* and comparison with experimental oilflow visualization

DLR Contributions to Code Validation Activities within the European High Lift Project EUROLIFT

R. Rudnik, R. Heinrich, B. Eisfeld, Th. Schwarz
DLR, Institute of Aerodynamics and Flow Technology,
Lilienthalplatz 7, D-38108 Braunschweig, Germany

Summary

The contribution of DLR to the numerical flow simulation activities 'assessment and improvement of CFD tools' of the European high lift project EUROLIFT is outlined. For the code assessment the DLR capabilities with respect to block-structured and Chimera grid generation for 2D and 3D high lift configurations are demonstrated, as well as to generate flow adapted hybrid unstructured grids for a 3D configuration. The corresponding results of 3D Navier-Stokes computations of aircraft configurations of different complexity show a good agreement of the DLR block-structured FLOWer code and the unstructured TAU code in comparison to experimental wind tunnel data up to maximum lift. Improvements of accuracy and efficiency in Navier-Stokes solvers are demonstrated by implementation of an advanced upwind discretisation scheme and low speed preconditioning in the TAU code. The Chimera technique is used to evaluate the influence of the outer grid extension on the solution accuracy, and to improve the grid resolution in critical areas of complex 3D aircraft high lift configurations.

1 Introduction

A major objective of the European 5th framework project EUROLIFT [6] is to enhance the understanding of dominant flow phenomena and scale effects of high lift aerodynamics, and to integrate modern CFD based analysis tools into the design process for transport aircraft. A prerequisite is the validation and verification of the numerical methods. The required test data for low Reynolds-Number (LSWT, Airbus-D) and high Re-No. conditions (ETW) are provided by experiments defined and carried out in the EUROLIFT project using an Airbus-type wing/fuselage high lift model, denoted as KH3Y or DLR-F11 model. In the numerical work package, which is coordinated by DLR [4], a consortium of 10 European partners of industry and research establishments carries out common computations to assess the high lift capabilities of state-of-the art CFD methods. The majority of methods features either block-structured or unstructured Navier-Stokes algorithms. The computations on the 3D configurations are accompanied by 2D computations of multielement airfoils. The present article describes the technical contribution of DLR to the numerical tasks assessment and improvement of CFD methods for high lift flow computations.

2 CFD code assessment

The main focus of the task 'Assessment of CFD tools for high lift application' lies on the generation of computational grids around the KH3Y model for different high lift settings and flow conditions, as well as on the corresponding viscous flow computations. In addition, special aspects of high lift computations are investigated by incorporating two existing 2D test cases, an A310 airfoil in landing configuration, denoted as test case 11 (TC11) and the more modern H8Y airfoil [5] (TC12), which stems from the national German high lift project.
In order to address the item of grid resolution a successive grid refinement study has been carried out using the block-structured FLOWer code and the k – ω model of Wilcox in DLR modification [3] for the A310 airfoil, TC11. The 3 level successive grids and a mandatory grid have been generated using the MEGACADS software [1]. *Figure 1* depicts the slat area of the coarsest and finest grid. The dependency of the resulting force coefficients on the total grid point number is highlighted in *Figure 2* for two angles of attack in the linear lift range, and close to maximum lift. For both conditions a linear increase of the lift coefficient with grid point number is detected, proving the second order character of the central spatial discretisation scheme even for the higher angle of attack, where supersonic flow is present at the slat nose. A force extrapolation to zero mesh size is carried out revealing, that grid resolution influence of the solution on the mandatory grid amounts to 20 lift cts. in the linear range, and 25 lift cts. at maximum lift. The drag evaluation underlines the dominance of the pressure drag, which is strongly reduced on refined grids due to the higher resolution of the suction peaks at the element leading edges. The friction drag is nearly unaffected by the resolution. The same test case serves as the basis for the assessment of different transport equation eddy viscosity turbulence models. In *Figure 3* results of four different turbulence models of the FLOWer code [7] on the mandatory grid are compared to the experimental lift polar. The uncertainty due to turbulence modelling amounts to 10 lift cts. in the linear lift range and 25 lift cts. at maximum lift, if the result of the k-ω LLR model is disregarded. Among the considered models, the modified Wilcox k–ω model [3] yields the best agreement to the experimental data on the present grid. Yet, as indicated in *Figure 2*, it has to be taken into account, that the results are still considerably depending on the grid resolution. Different from the one equation model, the k-ω models tend to produce unsteady solutions for the highest angle of attack of 25°. This is demonstrated by the time accurate force history in *Figure 4* using the dual time-stepping approach. This type of unsteadiness is inherently caused by the flow field and not imposed from outside, so the dominating frequency and the required temporal resolution is not known a priori. A variation of the physical

time step of two orders of magnitude causes basically no changes in the unsteadiness, proving that the temporal resolution is appropriate.

A major concern of DLR has been to compare the block-structured and the unstructured approach under as identical conditions as possible. Therefore polars of the KH3Y model are computed on a block-structured grid of DLR and a hybrid unstructured grid of the Swedish research establishment FOI, using basically the same numerical and physical modelling in both approaches. Both grids obey an upper grid point number limit of 3 million due to project related reasons. The surface grids are compared in *Figure 5*, and a cut through the slat areas is depicted in *Figure 6*. The normal boundary layer resolution is about 20 cells for the block-structured grid, and about 15 cells for the hybrid unstructured grid, due to better anisotropy features of the block-structured grid. The resulting computed lift polars, *Figure 7*, compare well with each other and to the experimental data of the low speed wind tunnel LSWT of Airbus-D including the maximum lift simulation. The post maximum lift range cannot be assessed at present due to uncertainties in the experimental set-up. The over estimation of the total drag has to be attributed to the fact, that a fully turbulent flow is simulated. Pressure distributions at mid span are shown in *Figure 8* for incidences in the linear lift range and at maximum lift. In both cases a good agreement to the experimental evidence is achieved with small deviation between both approaches. As a variation of the configuration complexity, a part span flap configuration is investigated using the TAU code. Again the hybrid unstructured grid with 4.5 million grid points is generated by FOI. The lift and drag polars are plotted vs. the LSWT data in *Figure 9*. The nearly constant lift decrease of about 15 lift cts. due to the reduced flap area is captured quite well by the computations. Yet, in contrast to the experiments, the TAU computations do not reproduce the decreased maximum lift of the part span flap configuration. Despite a different total drag level, the fact, that drag is not affected by the flap reduction is captured quite well.

Another configuration variation is the change from the full span flap take-off to the landing configuration. *Figure 10* shows the corresponding lift polar and a pressure distribution at mid span. The hybrid unstructured grid with 8 million grid points has been generated by DLR. The grid is adapted to obtain a y+ value below 1. 24 grid cells are distributed normal to the surface in a contant height prismatic layer. As the lift polar reveals, there is a considerable influence of the slat pressure tube bundles, which are attached outside at the slat tracks, on the overall data. Therefore an assessment of the maximum lift predictive capability is at present not conclusive. As the pressure distribution proves, a good agreement to the experimental data is obtained for low angles of attack.

3 CFD code improvement

DLR has contributed to the code improvement activities using both numerical approaches. For the structured method the application of the Chimera technique to a full 3-dimensional high lift configuration has been investigated, with the objective to simplify the grid generation. Cutting planes of the chimera grid and a lift polar for the KH3Y take-off configuration are shown in *Figure 11*. The grid consists of 3.9 million grid points in 333 blocks. The Chimera processes and the generation of the cartesian backgound grids are widely automated. For the flow computations with the FLOWer code the k-ω model has been selected, due to its fully local formulation. Despite the different turbulence model and the finer grid, the total lift agrees quite well with the baseline FLOWer computations for the three considered incidences. Finally the influence of an advanced hybrid upwind spatial discretization scheme MAPS+ [2] is investigated for the 3-element airfoil, TC11. The scheme has been implemented in the TAU code and is compared to the standard central differencing scheme with and without preconditioning. The convergence history proves, that for both disctretization schemes the lift convergence is strongly enhanced using preconditioning. While for the standard scheme the application of preconditioning reveals high frequency oscillations of the main flow residual, the MAPS+ scheme with preconditioning shows the best overall convergence properties for this case. As the evaluation of the slat pressure peaks proves, the application of the MAPS+ scheme results also in a superior agreement to the experimental pressure distributions.

4 Conclusion

The DLR contribution to the European high lift project EUROLIFT has served basically to assess and verify the flow simulation capabilities of the DLR MEGAFLOW code system with respect to the computation of 3D realistic high lift configurations. Both, the structured and the unstructured code versions are able to simulate the flow field around such type of configurations including the maximum lift range in good agreement to the experimental evidence. The EUROLIFT assessment process and the evaluation of the experimental results have underlined the necessity to consider comparatively small scaled details of the configuration, e.g. the device tracks, as they may have a considerable influence on the overall aerodynamic properties. This is an indication, that the hybrid unstructured approach will probably play the dominant role for future transport aircraft high lift applications, due to its highly automated grid generation and adaptation capabilities. Algorithmic features like robust preconditioning or advanced accurate and efficient spatial discretization schemes are successfully implemented. They represent a prerequisite to satisfy the industrial constraints for routine applications of such type of configurations.

References

[1] Kroll, N., Rossow, C.-C., Becker, K., Thiele, F, "MEGAFLOW-A Numerical Flow Simulation System", *ICAS-98-2.7.4*, 1998

[2] Rossow, C.-.C. "A Flux Splitting Scheme for Compressible and Incompressible Flows", Journal of Computational Physics, Vol. 164, pp. 104-122, 2000
[3] Rudnik, R., "Untersuchung der Leistungsfähigkeit von Zweigleichungs-Turbulenzmodellen bei Profilumströmungen", DLR-FB 49-97, 1997
[4] Rudnik, R.. "Towards CFD Validation For 3D High Lift Flows–EUROLIFT". In ECCOMAS 2001, Conference Proceedings, IMA, ISBN 0 905 091, ECCOMAS 124 (CD-Rom), 2001, pp. 1-20
[5] Rudnik R., "Common European CFD-Code Validation for Complex High Lift Configurations – the challenge of coming from observations to conclusions", DGLR JT2002-015, DGLR Jahrbuch 2002, Bd. 3, to be published (2002)
[6] Thiede, P., "EUROLIFT - Advanced High Lift Aerodynamics for Transport Aircraft". In AIR & SPACE EUROPE, VOL. 3, No 3 / 4, 2001
[7] Franke, M., Rung, Th., Schatz, M., Thiele, F., "Numerical Simulation of High-Lift Flows Employing Improved Turbulence Modelling", ECCOMAS 2000, Conference Proceedings, ISBN 84-89925-70-4, (CD-Rom), 2000, pp. 1-20

Figures

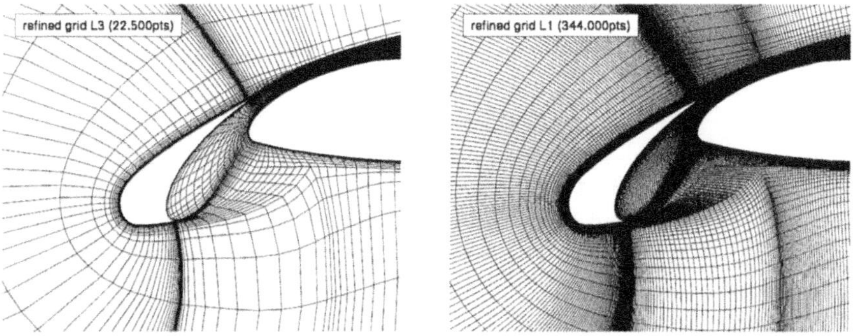

Figure 1. Finest and coarsest grid resolution levels study, TC11

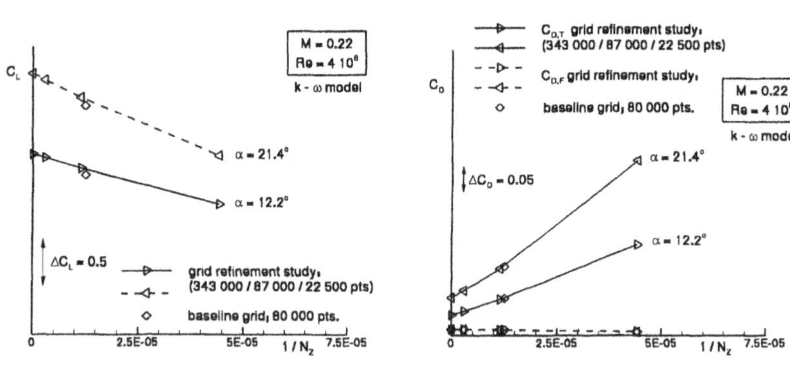

Figure 2. Force coefficients for successive grid refinement, TC11

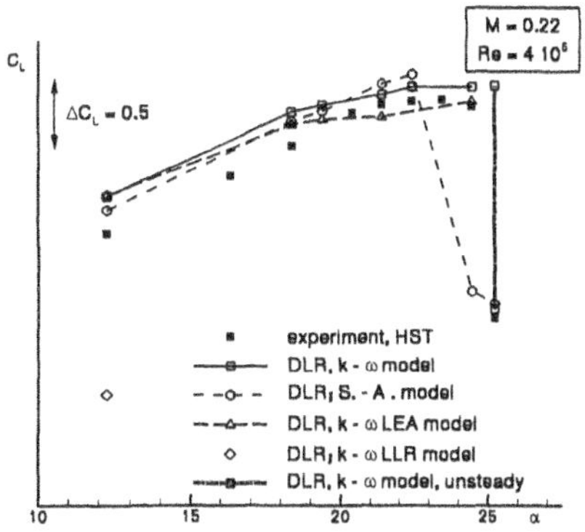

Figure 3. Lift polar, TC11

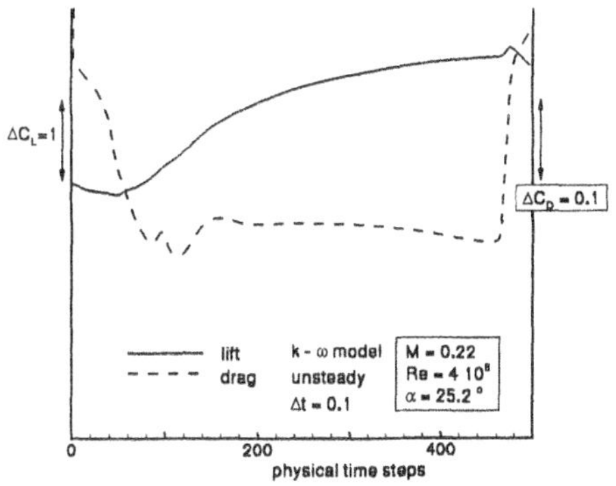

Figure 4. Unsteady convergence, TC11

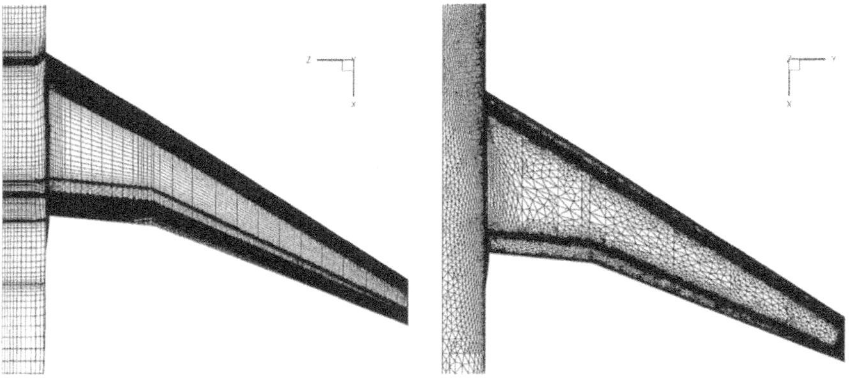

Figure 5. Mandatory grids for the KH3Y take-off configuration, TC211

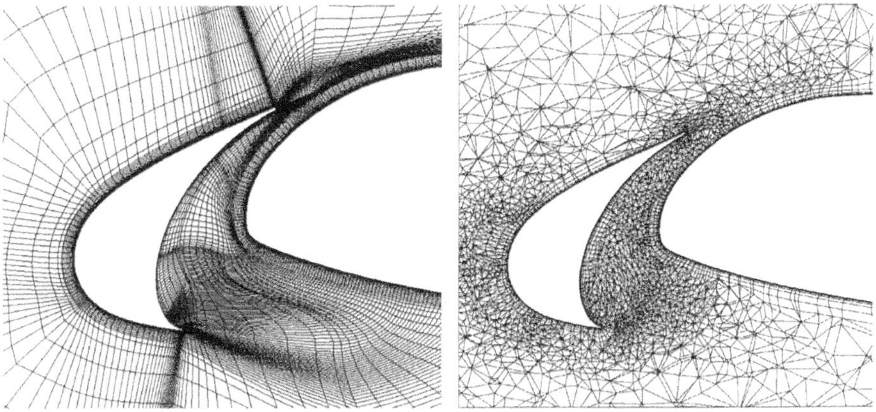

Figure 6. Cut through grids around slat at 14.7% half span , TC211

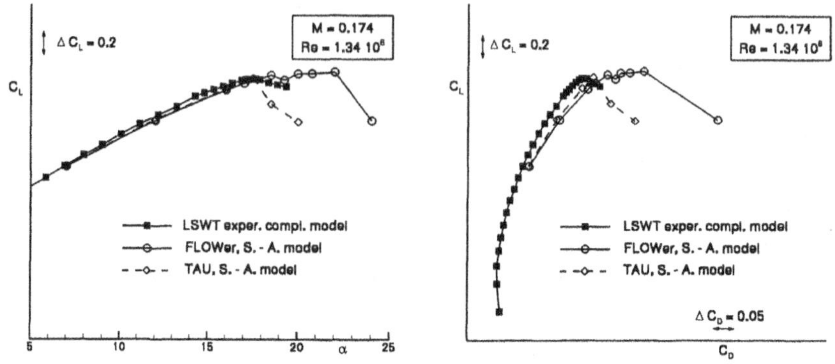

Figure 7. Lift and drag polar for KH3Y, take-off configuration, TC211

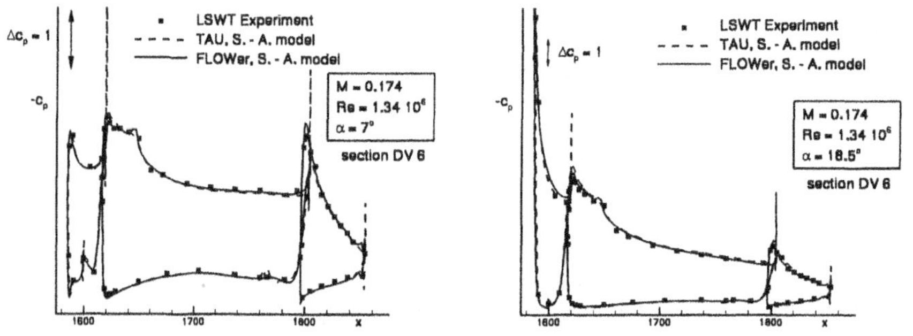

Figure 8. c_p distributions at 68% half span for linear and maximum lift condition, TC211

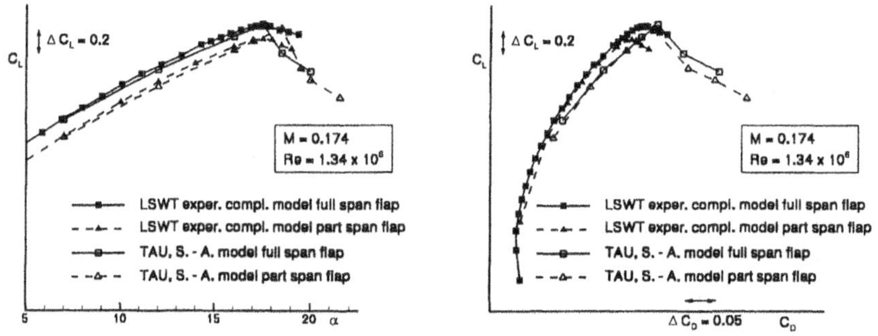

Figure 9. Lift and drag polar for KH3Y, take-off part span flap configuration, TC212

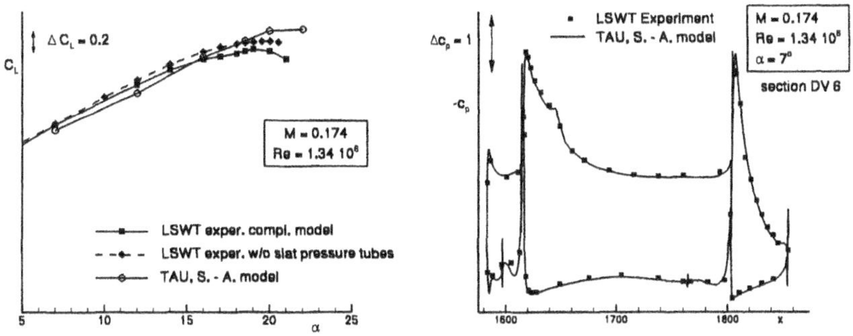

Figure 10. Lift polar and c_p distributions at 68% half span for KH3Y, landing configuration with full span flap, TC214

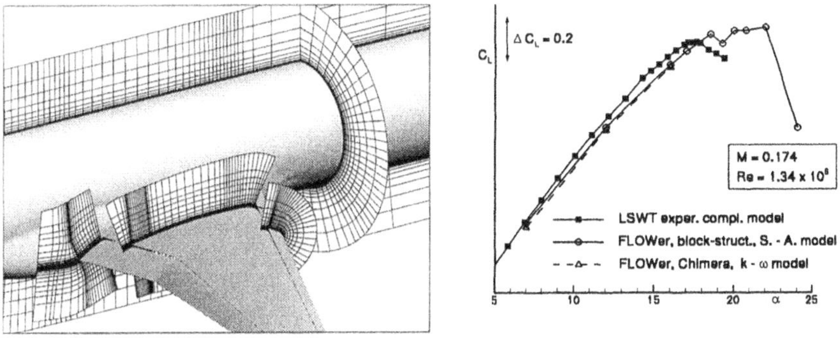

Figure 11. Chimera grid and lift polar for KH3Y, take-off configuration, TC211

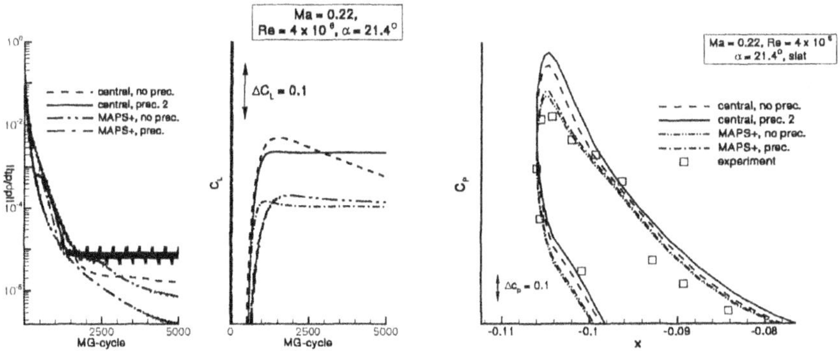

Figure 12. Advanced spatial discretisation in combination with preconditioning , TC11

Calculation of attached and separated flow for a tailplane airfoil with rudder

A. GROTE, M. KRUSE and R. RADESPIEL

Institute of Fluid Mechanics, Technical University Braunschweig
Bienroder Weg 3, 38106 Braunschweig, Germany
Internet: www.tu-braunschweig.de/ism, E-mail: arne.grote@tu-braunschweig.de

Summary

The accuracy of an unstructured Navier-Stokes solver in predicting the airflow around the common tailplane airfoil NACA 64A010 at low Reynolds numbers is investigated. Since the performance of this airfoil is strongly affected by the transition of the boundary layer, the calculations using the Spalart-Allmaras, k-ω and k-ω SST models are carried out with specified transition positions on the airfoil's surface. It turns out, that the Navier-Stokes solver can predict leading edge stall via the appearance of laminar separation bubbles. However, the code is less accurate in predicting the flow around airfoils with deflected rudders, where the onset of boundary layer separation as a result of two consecutive regions of adverse pressure gradients is predicted at too high angles of attack.

1 Introduction

Tailplanes are designed to provide stability and control over the whole flight trajectory. While the sizing of the horizontal tailplane for existing transport aircrafts is mainly driven by the requirement of a certain amount of longitudinal stability at cruise conditions, there will be more emphasis in the future in aerodynamic designs for safe low-speed manoevring. This makes detailed numerical prediction capabilities for tail stall at low speed necessary in order to allow fast aerodynamic design cycles and to keep the development risk sufficiently low. An obvious initial step is to demonstrate airfoil prediction capabilities. Unfortunately, a review of public literature reveals that the knowledge base of experimental data for 2D tailplane airfoils is quite outdated. Similarily, systematic validation results of Reynolds-averaged Navier-Stokes solvers for typical tailplane airfoils with and without rudder do not exist. Here it is worth noting that the 2D flows over tailplane airfoils can be quite complex. Depending on the Reynolds number and airfoil shape there may exist leading edge, trailing edge or combined stall [1]. Airfoils with deflected rudders exhibit suction peaks at both the leading edge and the hinge line and it seems quite difficult to predict separation onset under these conditions.

The present work presents recent results in the flow prediction for these airfoils. The flow computations involve the well-known airfoil code XFOIL [3], release 6.92, and the unstructured RANS solver TAU, release 2002.2.0, developed by the German Aerospace Center (DLR) [5]. A large number of computations are performed for the NACA 64A010 airfoil with rudder for which some experimental data exists ([4], [7] and [9]). The computations address requirements for numerical accuracy, variations of turbulence modelling, and they cover broad ranges of Re, angle of attack, and rudder deflection. Special attention is paid on simulating the partially laminar boundary layers, since this has a strong effect on stall characteristics.

2 Numerical methods

2.1 XFOIL

The XFOIL code [3] provides an efficient tool for viscous and inviscid analysis of subcritical airfoils. It follows the interacted viscous/inviscid zonal approach and therefore couples an inviscid linear-vorticity panel method with a two-equation integral method to represent the viscous layer. The integration of an e^n-type amplification formulation enables the calculation of transition points. By using a Newton method to solve the boundary layer and the transition equation simultanously with the inviscid flowfield, the XFOIL code is suitable to analyse low Reynolds number airfoil flows with transitional separation bubbles.

2.2 Navier Stokes method

The unstructured TAU code [5] requires unstructured hybrid grids in order to resolve viscous flows over airfoils. These are generated using the *Centaur™* software package provided by CentaurSoft [2]. A standard grid in this study as shown in Fig.2 contains about 320 points allong the airfoils contour and 48 prism layers perpendicular to it for resolving the attached boundary layer. The initial spacing perpendicular to the wall is $7 \cdot 10^{-6}$, which is based on the requirement to obtain a y^+ of approximately one. A prismatic stretching factor of 1.09 is used. The farfield is positioned as a circle of radius 100 times the reference chord length $c = 1$ around the trailing edge, yielding a grid with a total of about 24000 nodes ($\approx$ 32000 cells). Grid refinement studies during the investigation were carried out to investigate that the influence of the grid on the numerical results of c_d does not vary more than 0.0005.

Starting from the initial hybrid grid a pre-processing module of the TAU code computes dual grids independent of the element types in the hybrid grid with an edge-based data structure. Coarse grids for the multi-grid algorithm are constructed recursively by agglomerating the control volumes at the finer grid level. The flow solver is based on a finite-volume scheme integrating the RANS equations. In this study the inviscid fluxes are calculated by employing a central method with scalar dissipation and standard parameter settings. The viscous fluxes are discretized using central differences. The compressible flow solver employs an explicit multi-step Runge-Kutta scheme for the discretization of the temporal gradients. In order to accelerate the convergence to steady state, residual smoothing and a multi-grid technique (4w cycle in case of the SA model and 3v cycle in case of the two-equation models) are employed.

3 Turbulence Modelling

The one-equation turbulence model by Spalart and Allmaras [8] adds an additional transport equation for the eddy viscosity written in terms of the kinematic eddy viscosity. The turbulent kinetic energy is not explicitly modelled. The model has been calibrated for airfoil flows and is numerically robust. It is therefore attractive for routine applications.

In the k-ω model, proposed by Wilcox [10] in 1998, transport equations are formulated for the turbulent kinetic energy k and the specific dissipation rate ω. In comparison to its older version an additional production term in the ω equation was added, which improves the predictive accuracy for free shear flows.

The k-ω SST (shear-stress-transport) model by Menter [6] combines different elements of existing models. To eliminate the dependency of the Wilcox k-ω model from the initial value of ω, the SST model switches to the standard k-ε model in the outer region of the boundary layer. In the inner region the original Wilcox k-ω model is used. Although a smooth transition between the two models is obtained by a control-function, the results still keep their zonal behaviour. In contrast to Menter's baseline model the definition of the eddy-viscosity is modified in order to model the effect of transport of the principal turbulent shear stress. These modifications lead to improvements in the prediction of adverse pressure gradient flows.

The implementation of all turbulence models in the TAU code include a numerical transition function, which yields a smooth transition from laminar to turbulent flow state at user-specified points on the surface.

4 Results

The calculations for the NACA 64A010 airfoil in Fig.1 were carried out first without and afterwards with a deflection of the rudder.

4.1 Airfoil without rudder deflection

A first series of computations was carried out for Reynolds numbers of $0.9 \cdot 10^6$ and $4.1 \cdot 10^6$ with the assumption of a fully turbulent boundary layer (BL) and using the Spalart-Allmaras turbulence model (SA). As shown in Fig.3 the results are in good aggreement with the measurements by Peterson [7] below the maximum lift coefficient, but they turn out badly in the prediction of stall. Also calculations with the XFOIL code with a fully turbulent boundary layer and with free transition (using the e^n-type amplification formulation to compute the point of transition for both sides of the airfoil) show good aggreement below maximum lift. Thereby it appears that with increasing angle of attack the point of transition on the airfoil's upper surface suddenly moves forward to the leading edge where the transition occurs via a laminar separation bubble. As shown in Tab.1 these transition points can not directly be used for the TAU calculations, if a laminar separation bubble appears. For every turbulence model turbulence growth from laminar to turbulent needs a certain length, so that the transition-point has to be adjusted to produce a steady laminar separation bubble with a complete numerical transition. The results with specified transition points in Fig.4 show for the SA and the k-w model, that the lift coefficient is reduced by the appearance of the laminar separation bubble. These results are obtained with the standard mesh. Numerical accuracy was checked by using a fine mesh with four times the nodes which gave very similar force coefficients (less then 4% difference). The sensitivity of the boundary layer to adverse pressure gradients with laminar separation bubbles which finally burst and lead into stall are computed in good aggreement with the measurements. For lift coefficients higher than maximum the turbulence model reactions to the position of transition at constant angle of attack are very sensitiv. A shift of the transition point of only 0.05% chord length downstream changes the state of flow from attached to separated. In the case of the attached flow the calculated pressure distribution is very similar to the result of a simulation with fully turbulent BL without laminar separation bubbles. Good aggreement with measurements (by Stivers [9]) can also be achieved for the drag coefficient by considering the transition of the boundary layer as shown in Fig.5. The influence of the transition setting can be seen in the distribution of the pressure coefficient over the airfoil plotted in Fig.6 for α=8.0°. The front part of the airfoil is drawn out in the detail showing the typical distribution of a laminar separation bubble. For this case the boundary layer velocity distribution in Fig.7 shows some differences between the results experimentally obtained by a boundary layer rake and the calculated distribution. All computations show some influence of the interface between the structured and the unstructured part of the grid on the velocity at $\eta/c \approx 0.005$.

4.2 Airfoil with rudder deflection

Deflections of the rudder lead to a higher suction peak at the leading edge and an additional suction peak at the hinge line. Measurements by Dods [4] for these cases show, that the flow over the suction side of the rudder separates at small angles of attack, leading to a certain loss of lift. As seen in Fig.8 for a deflection angle of 15.0° all turbulence models yield false predictions of the separation onset on the rudder. The XFOIL code calculates the correct loss of lift initiated by a beginning flow separation at the trailing edge of the rudder. This point of separation moves forward with increasing angle of attack right as seen in the pressure distribution of Fig.9. The lack of flow separation on the rudder leads to lower drag coefficients of about $\Delta c_d \approx 0.003$ for all TAU results in comparison to the values given by XFOIL, which is not shown here. The different turbulence models produce an attached turbulent velocity profile in the boundary layer allowing too strong adverse pressure gradients. In Fig.10 the velocity profiles calculated with the three turbulence models are plotted at the location, where XFOIL calculates the point of flow separation. Here the SST model predicts a distribution which is slightly closer to separation compared to SA and k-ω. This is confirmed by the distribution of the eddy-viscosity in Fig.11, where the SST model calculates the lowest values. Similar discrepancies between the results of TAU in comparison with the measurements are found for a deflection angle of 25.0° for the SA and the k-ω model in Fig.12. Only the SST model succeeds in predicting the onset of rudder stall. Downstream of the second pressure rise the flow inside the turbulent boundary layer is strongly decelerated, so that it separates from the rudder's contour. Depending on the angle of attack this point of separation moves forward but stays downstream of the hinge line. Even XFOIL shows in this case some discrepancies with the measurements.

5 Conclusions

The present computational results indicate that, assuming an accurate transition prediction model is available, it will be possible to simulate the flow phenomena around the leading edge at high angles of attack including the laminar separation bubble. In terms of rudder stall the prediction of the location and the angle of attack of incipient flow separation on the rudder is not satisfactory for all tested turbulence models. So for airfoils with rudder there is a need for further investigations into the prediction capabilities of flow separation onset using advanced turbulence models. This study also indicates that there is a need for detailed flow measurements at modern tailplane airfoils with rudder to provide a suitable database for flow validation of numerical prediction methods.

Acknowledgement

The authors gratefully acknowledge the DFG (Deutsche Forschungsgemeinschaft) for the financial support.

References

[1] van den Berg, B.: "Reynolds number and Mach number effects on the maximum lift and the stalling characteristics of wings at low speeds". NLR TR 69025 U, Amsterdam, Netherlands, 1969.

[2] Centaur^{TM} by Centaursoft. www.centaursoft.com

[3] Drela, M.: XFOIL: An analysis and design system for low Reynolds number airfoils. In Mueller, T.J.(Ed.): "Low Reynolds number aerodynamics". Lecture notes in engineering. Springer, Germany, 1989, pp. 1–12.

[4] Dods Jr., J.B.: "Wind-tunnel investigations of horizontal tails. IV — unswept plan form of aspect ratio 2 and a two-dimensional model". NACA RM A8J21, Washington, USA, 1948.

[5] Galle, M.; Gerhold, T.; Evans, J.: Parallel computation of turbulent flows around complex geometries on hybrid grids with the DLR-TAU code. In
Keyes, D. et al. (Ed.): "Parallel computational fluid dynamics". Proceedings of the parallel CFD '99 conference. Amsterdam, Netherlands, 2000.

[6] Menter, F.R.: "Two-equation eddy-viscosity turbulence models for engineering applications". AIAA Journal, Vol. 32 (1994), No. 8, pp. 1598–1605.

[7] Peterson, R.F.: "The boundary-layer and stalling characteristics of the NACA 64A010 airfoil section". NACA TN 2235, Washington, USA, 1950.

[8] Spalart, P.R.; Allmaras, S.R.: "A one-equation turbulence model for aerodynamic flows". AIAA Paper 92-0439, 1992.

[9] Stivers Jr., L.S.: "Effects of subsonic Mach number on the forces and pressure distributions on four NACA 64A-series airfoil sections at angles of attack as high as 28°". NACA TN 3162, Washington, USA, 1954.

[10] Wilcox, D.C.: "Turbulence modeling for CFD". 2nd Ed., DCW Industries Inc., 2000.

Table 1 Point of transition on the upper surface of the NACA 64A010 airfoil ($Re_\infty = 4.1 \cdot 10^6$, $Ma = 0.167$)

α	$(x/c)_{\text{trans.,upper surface}}$ XFOIL	SA	Wilcox
0.0°	0.6312	0.6312	0.6312
4.0°	0.0200	0.0200	0.0200
8.0°	0.0072	0.0058	0.0058
9.0°	0.0062	0.0050	0.0046
9.5°	0.0058	0.0041	0.0041

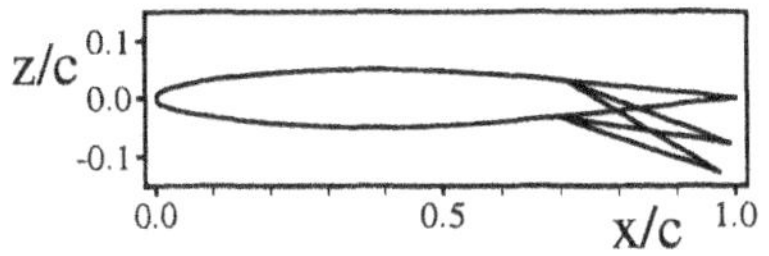

Figure 1 NACA 64A010 airfoil, rudder deflection $\delta = 15.0°$ and $25.0°$

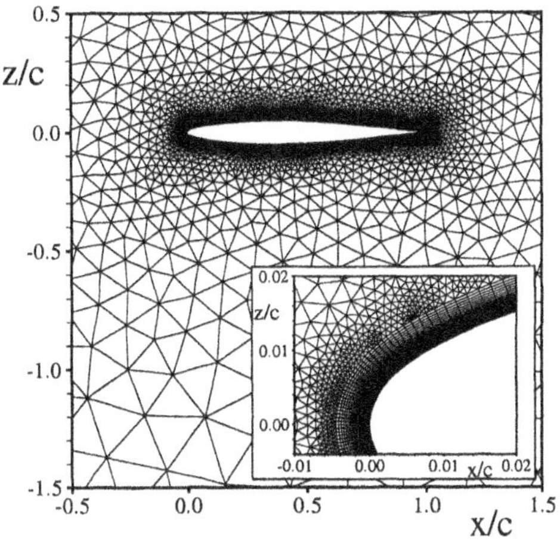

Figure 2 Unstructured hybrid grid around the NACA 64A010 airfoil without rudder deflection

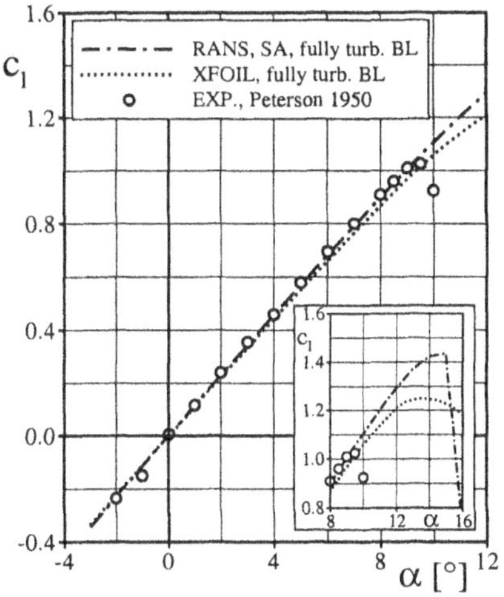

Figure 3 NACA 64A010 without rudder deflection, fully turbulent boundary layer ($Re_\infty = 4.1 \cdot 10^6$, $Ma = 0.167$)

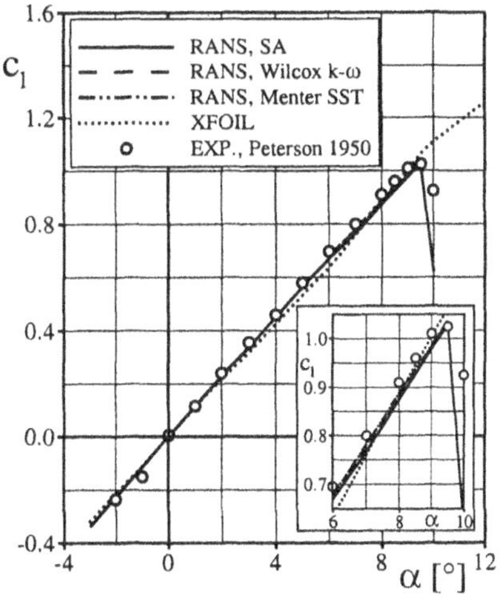

Figure 4 NACA 64A010 without rudder deflection, calculation with transition ($Re_\infty = 4.1 \cdot 10^6$, $Ma = 0.167$)

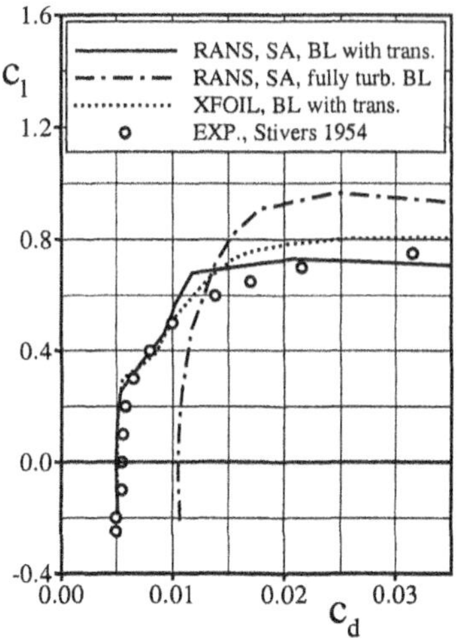

Figure 5 NACA 64A010 without rudder deflection
($Re_\infty = 0.9 \cdot 10^6$, $Ma = 0.31$)

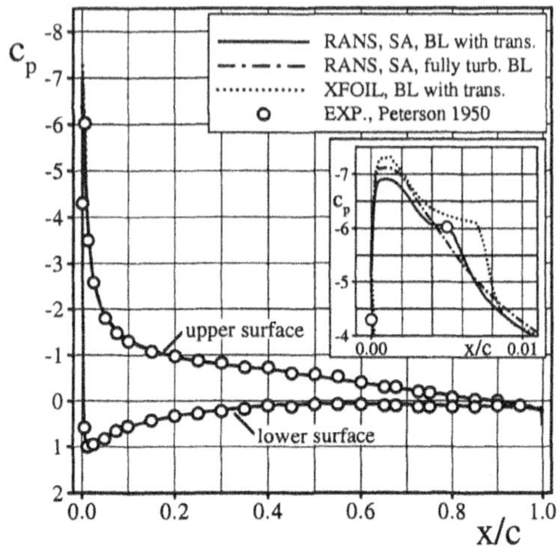

Figure 6 Pressure distribution at $\alpha = 8.0°$ without rudder deflection
($Re_\infty = 4.1 \cdot 10^6$, $Ma = 0.167$)

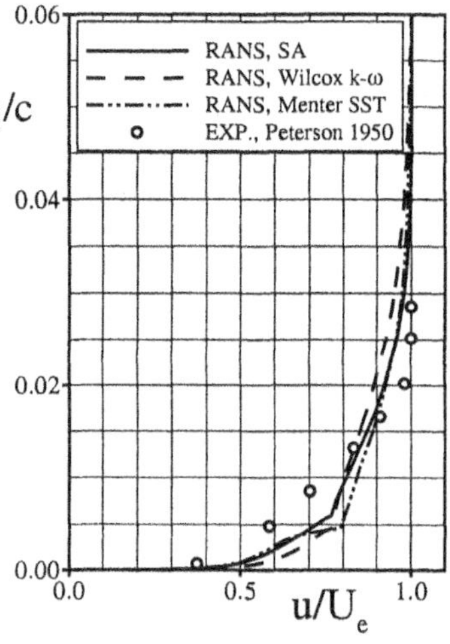

Figure 7 Boundary layer velocity distribution on the upper surface at $x/c = 0.8$ and $\alpha = 8.0°$, calculation with transition
($Re_\infty = 4.1 \cdot 10^6$, $Ma = 0.167$)

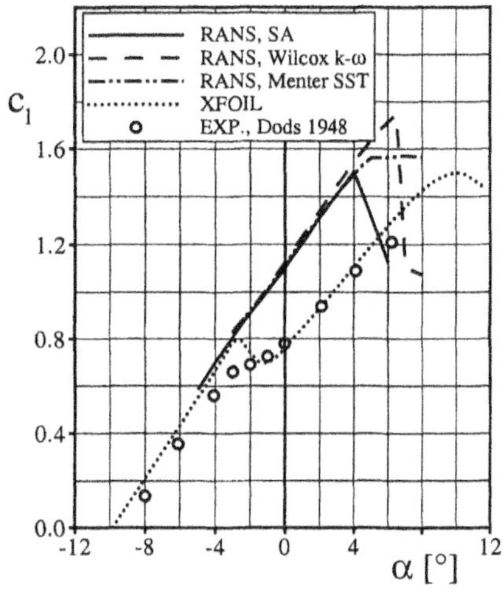

Figure 8 NACA 64A010 with a rudder deflection of $\delta = 15.0°$, calculation with transition
($Re_\infty = 3.0 \cdot 10^6$, $Ma = 0.2$)

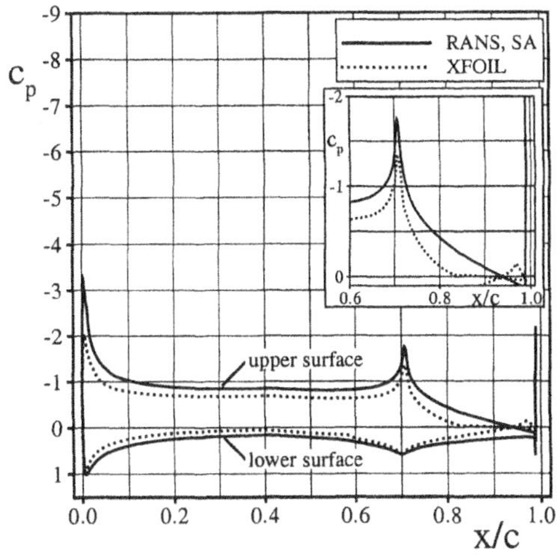

Figure 9 Pressure distribution at $\alpha = 0.0°$ with a rudder deflection of $\delta = 15.0°$, calculation with transition ($Re_\infty = 3.0 \cdot 10^6$, $Ma = 0.2$)

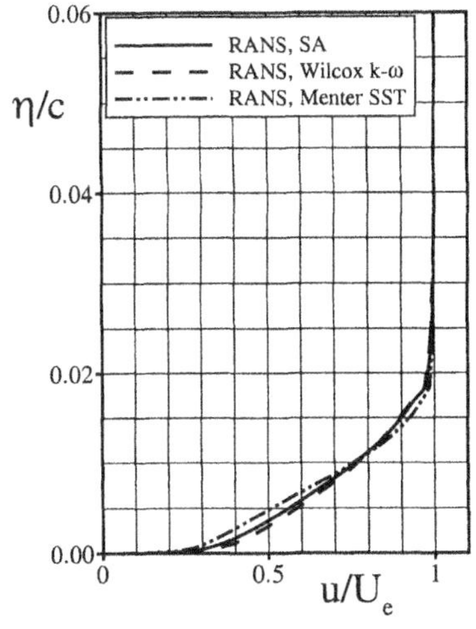

Figure 10 Boundary layer velocity distribution on the upper surface at $x/c = 0.863$, $\delta = 15.0°$ and $\alpha = 0.0°$, calculation with transition ($Re_\infty = 3.0 \cdot 10^6$, $Ma = 0.2$)

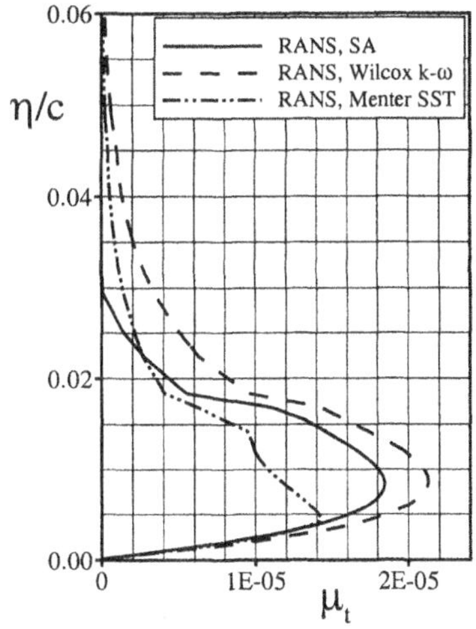

Figure 11 Distribution of the eddy viscosity on the upper surface at $x/c = 0.863$, $\delta = 15.0°$ and $\alpha = 0.0°$, calculation with transition ($Re_\infty = 3.0 \cdot 10^6$, $Ma = 0.2$)

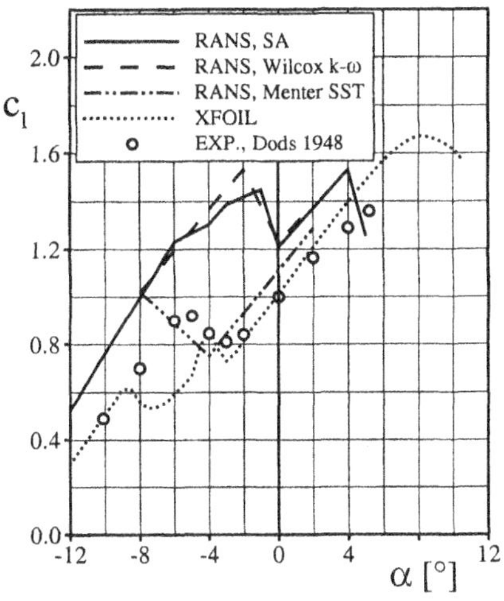

Figure 12 NACA 64A010 with a rudder deflection of $\delta = 25.0°$, calculation with transition ($Re_\infty = 3.0 \cdot 10^6$, $Ma = 0.2$)

Numerical Investigation of the Flow Around Different Slat Tracks on a Swept High Lift Wing Segment

R. Rudnik, A. Ronzheimer
DLR, Institute of Aerodynamics and Flow Technology,
Lilienthalplatz 7, D-38108 Braunschweig, Germany

Summary

A numerical investigation of a high lift wing segment using an unstructured RANS method shows, that a slat track can have a significant influence on the overall aerodynamic properties. Two track types are considered, one which is representative of a wind tunnel model, and a second one derived from an aircraft track. The major differences are the larger cross-section area of the model track, and the intersection with the fixed wing. The high lift devices are positioned in a landing setting. The onflow conditions refer to wind tunnel conditions. Periodic lateral boundary conditions are applied. Compared to a reference high lift wing segment without track, a 10% increase in lift and a 20% increase in drag is being computed for the model track. For the aircraft track only a minor lift change is found, but also a 20% drag increase. Based on the outcome of the present investigations, the major impact on the lift is caused by the downstream shift of the stagnation line in the vicinity of the junction of the model track and the fixed wing.

1 Introduction

The objective of the present study is to investigate the aerodynamic impact of slat tracks on the overall flow field of a high lift configuration. In contrast to [6], where the influence of flap tracks has been simulated on a nominally 2-dimensional model set-up, the motivation for the present activity is basically twofold. On the one hand it is found in the course of aero acoustic investigations on airframe noise reduction [1], that the flow around the slat tracks is a dominant source of noise for certain configurations. Attempts to shield the track resulted in a considerable loss of maximum lift. On the other hand experimental investigations within the framework of the European high lift project EUROLIFT show, that the blockage in the slot between gap and fixed wing due to slat tracks and attached pressure tube bundles leads to considerable changes in the aerodynamic performance close to maximum lift [4]. The influence has been verified by repeated measurements with removed pressure tube bundles. The bundles account for roughly two/third of the total cross-flow blockage of slat and tubes. Both studies raised the demand for a deeper understanding of the effects and sensitivities of the flow around slat tracks on a swept high lift wing. As a first step in this direction the present investigaton is concerned with Navier-Stokes computations of two different slat track geometries on a swept wing segment with deployed high lift devices using the unstructured TAU code of DLR.

2 Geometry and Computational Grids

Although it is meanwhile feasible to compute complete high lift aircraft configurations with the present numerical method [4], a different approach is chosen for the track study. In order to be able to realise an appropriate grid resolution, only a wing segment is regarded. It is assumed, that the baseline effects are already captured by this set-up. The baseline wing is an Airbus-type three element wing in landing configuration. A sketch of the numerical set-up is given in *Figure 1*. In order to simplify the grid generation, the span wise cuts have been placed normal to the leading edge and not in line of flight. The distance between two tracks on the full-scale aircraft wing defines the width of the segment. The arrangement is representative for the midboard portion of the wing. The location of the evaluated pressure sections (PS) and cross flow planes is also depicted in *Figure 1*. Periodic boundary conditions are applied at each of the lateral cuts. The geometries of the slat tracks are depicted in *Figure 2*. The size and shape of the so-called 'model track' is derived directly from a wind tunnel model. The aircraft track on the left side is simplified in several aspects. First the pinion drive inside the track is neglected and the H-type cross-section is transformed to a square section. Moreover, the fastening devices at the slat/track junction are only considered by their outer extensions. The width of the model track is 10% larger than that of the scaled aircraft track .

The hybrid unstructured grids have been generated with the commercial grid generation software system of CENTAURSOFT™. Cuts through the grid at the track centreline are shown in *Figure 3*. The first wall spacing results in y^+ values below 2 on all aerodynamic surfaces. The boundary layers are resolved with roughly 20 grid points. The total grid point number amounts to 2.8 million. In addition to the track-configuration, a clean configuration without tracks is investigated to provide the reference data.

3 Numerical Approach and Results

The investigations have been carried out with the unstructured finite volume flow solver TAU of DLR and the Spalart-Allmaras turbulence model with Edwards modification [2]. Baseline features of the code are described in [3]. The standard solver adjustment is used, featuring a node based central spatial discretisation for the mean flow equations, and a first order upwind AUSMDV discretisation for the turbulence equation. A 3-stage Runge-Kutta time stepping with three level full multigrid and preconditioning is applied for the time integration. The flow conditions have been selected to Mach number M=0.22, Re-No. = 2.44×10^6 and an angle of attack of $\alpha = 7^{\circ}$, corresponding to wind tunnel conditions on a large-scale model in the linear lift range. Computed surface isobars and streamlines for both tracks are evaluated under different views in *Figure 4*, *Figure 5*, and *Figure 6*. It is quite obvious, that the larger size of the model track leads to considerably higher disturbances of the surface flow pattern. For the model track the displacement of the wall streamlines on the fixed wing lower side in *Figure 4* extends about 5% reference chord length further downstream than that of the aircraft track. As the aircraft track ends close to the stagnation line, the lower side of the fixed wing is widely unaffected. A stagnation lines is also visible on the windward surface of the tracks. Independent of the track type, its streamwise position close to the fixed wing lower surface appears to coincide with the fixed wing stagnation line.

A similar impact of the different tracks is found for the flow in the slat cove, *Figure 5*. For the aircraft track the footprint of the separation in the slat cove is only weakly disturbed. In contrast to this, for the model track a complex separation pattern exhibits outboard of the slat track junction. Especially for the model track the isobars reveal a large low-pressure area outboard of the track close to the slat trailing edge. This pressure reduction is also reflected at the fixed wing leading edge, see *Figure 6*. On the fixed wing upper surface the presence of the tracks leads to an inboard shift of the wall streamlines due to the track wakes. Again, the lateral extension of the disturbance is larger for the model track.

Figure 7 and *8* show the streamlines in two cross flow planes, the first located at the centre of the track and the second on its lee side. Due to the 3-dimensional character of the flow, the streamlines are discontinuous in the considered planes. In the track centreline cuts, *Figure 7*, the disturbance of the main slat cove separation bubble is quite obvious, when compared to the configuration without track in *Figure 9*. Due to the special intersection of the model track lower side and the slat, two co-rotating vortices exhibit above and below the model track in the slat cove. Because of the larger cross-section, the vortex above the model track is smaller, than that found above the aircraft track. Further outboard in cross flow plane 2, the 3-vortex system of the model track is still present in the slat cove, while for the aircraft track only two vortices are visible, see *Figure 8*. The most important difference appears to be the fourth vortex, which is present only for the model track configuration at the leeward side of the track / fixed wing intersection. This vortex clearly leads to a downstream shift of the fixed wing stagnation line, see also *Figure 6*. As a consequence, a stronger flow acceleration develops for this configuration in the slot between slat and fixed wing. This is confirmed by the slat pressure distributions in pressure section 2, *Figure 10*. While the distribution of the reference configuration and of the aircraft track are quite similar, the pressure level along the cove is considerably lower for the model track. This influence extends until the slat trailing edge and thus influences the complete slat upper side. The pressure distributions at the inboard section 1 and at the outboard section 3 in *Figure 11* show, that these effects influence the complete slat upper side. Moreover, the lower pressure level on the slat of the model track configuration affects also the fixed wing and the flap. For the aircraft track and the configuration

without track the flap pressure distribution indicates a separation of about 45% local chord. This separation has nearly vanished for the model track configuration. On the other side, the pressure level at the stagnation line is higher for the model track, indicating, that the presence of the large model decelerates the stagnation line flow.
The impact of the discussed flow field effects on the overall aerodynamic properties is shown in *Figure 12*. The presence of the model track leads to a lift increase of about 10%. For the aircraft track configuration the lift doesn't differ significantly. For both track types, the track presence causes a total drag increase of about 20%. In *Figure 13* a convergence history for the model track is plotted to domenstrate, that the aerodynamic coefficients have reached a converged status.

4 Conclusion

Numerical investigations have been carried out using the unstructured Navier-Stokes flow solver TAU of DLR on two different slat track types, representing a wind tunnel model, and a simplified aircraft track. For the computations a swept wing segment in high lift configuration with periodic lateral boundary conditions is considered. The investigations reveal, that although the cross-section and stream wise extension of the model track is considerable larger than that of the aircraft track, a 10% increase in total lift is produced by this configuration. The lift increase is primarily attributed to a local downstream shift of the stagnation line of the fixed wing that leads to an increased circulation on all elements of the configuration. In contrast to this the aircraft track does not cause significant changes in lift compared to the reference configuration without track. The track installation leads to an increase in total drag of about 20% compared to the reference configuration. For a final assessment of the rather strong absolute changes in lift on the considered wing segment, the influence of the periodic boundary condition and the lateral extension of the segment has to be evaluated.
The current computations indicate, that the vortices at the lee side of the track are confined to the vicinity of the tracks. Yet, a conclusive statement concerning the vortex development requires a grid adaptation, which is subject of current investigations. In addition to the validation of the periodic boundary condition approach and the adaptation issue, future activities wil be devoted to the computation of the Re-No. influence, the variation of the track width, corresponding to the effects, that have been detected in EUROLIFT, and the variation of the high lift device setting from landing to a take-off configuration.

References

[1] Dobrzynski, W., Pott-Polenske, M., „Slat Noise Source Studies for Farfield Noise Predictions", AIAA paper 2001-2158, 2001

[2] Edwards, J. R., Chandra, S., "Comparison of Eddy Viscosity-Transport turbulence Models for three-Dimensional, Shock Separated Flowfields", AIAA Journal, No. 4, April 1996
[3] Kroll, N., Rossow, C.-C., Becker, K., Thiele, F, "MEGAFLOW-A Numerical Flow Simulation System", *ICAS-98-2.7.4*, 1998
[4] Rudnik, R., Melber, S., Ronzheimer, A., Brodersen, O., „Three-Dimensional Navier-Stokes Simulations for Transport Aircraft High Lift Configurations", Journal of Aircraft, Vol. 38, No. 5, Sept.-Oct.. 2001, pp .895-903 (2001)
[5] Rudnik, R., "Common European CFD-Code Validation for Complex High Lift Configurations – the challenge of coming from observations to conclusions", DGLR JT2002_015, DGLR Jahrbuch 2002, Bd. 3, to be published, 2002
[6] Rumsey, C. L., et.al., "Three-Dimensional Effects on Multi-Element High Lift Computations", AIAA paper 2002-0845, 2002

Figures

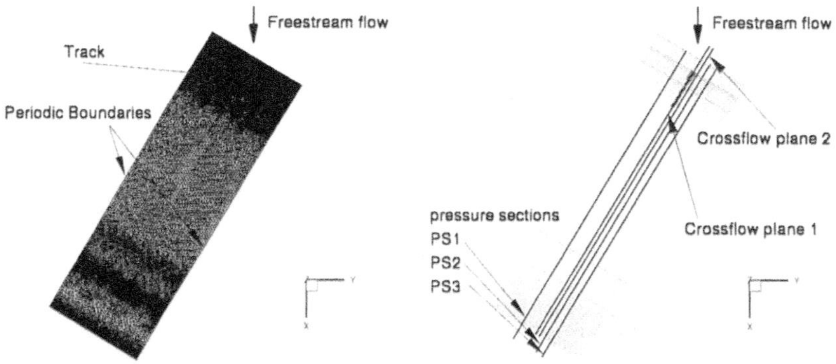

Figure 1. Numerical set-up and evaluation planes

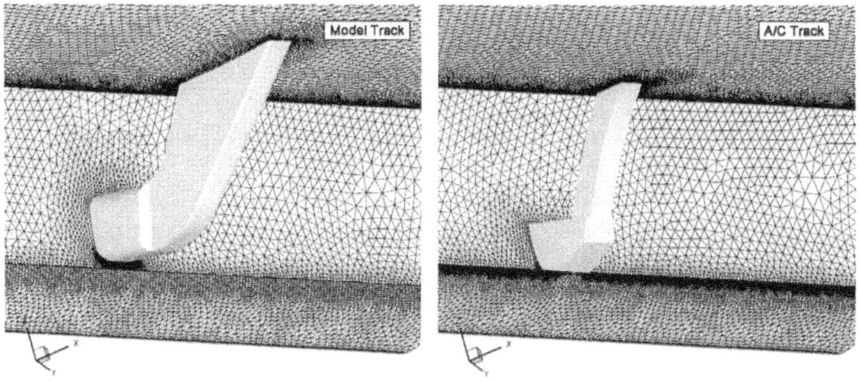

Figure 2. Geometry of model and aircraft slat track, view from below in the slat cove

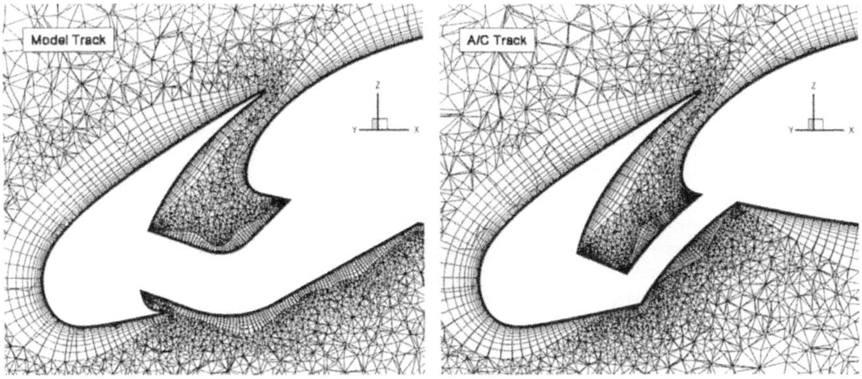

Figure 3. Grid cut normal to leading edge at slat area

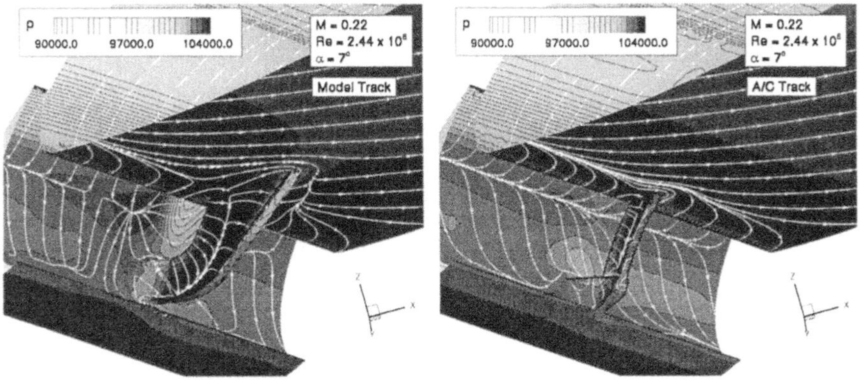

Figure 4. Isobars and streamlines, view from i/b, below

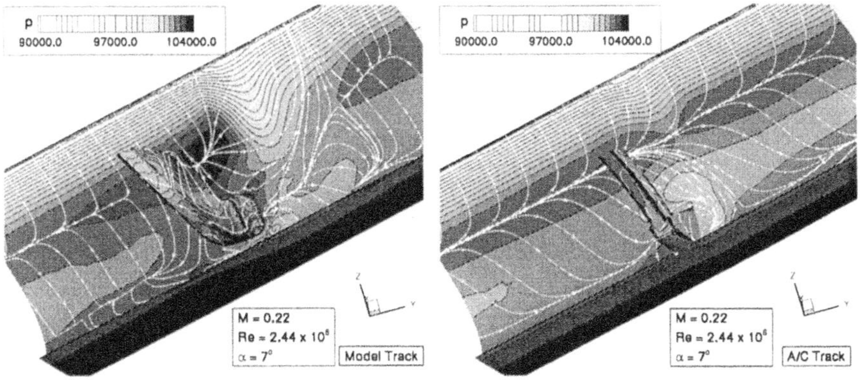

Figure 5. Isobars and streamlines, view in slat cove from o/b, below

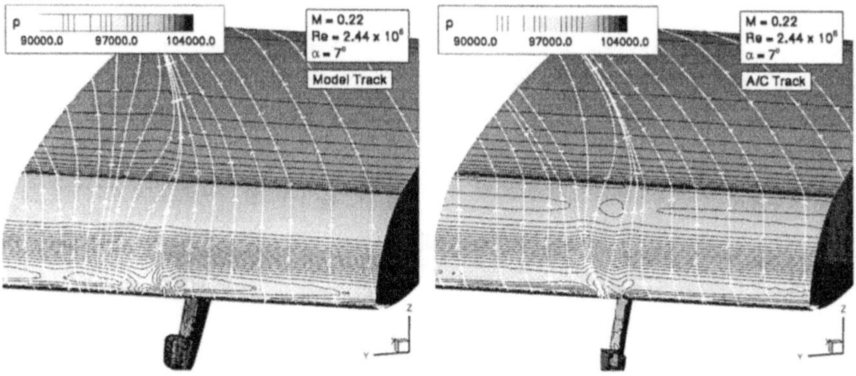

Figure 6. Isobars and streamlines, view on fixed leading edge, slat blinded out

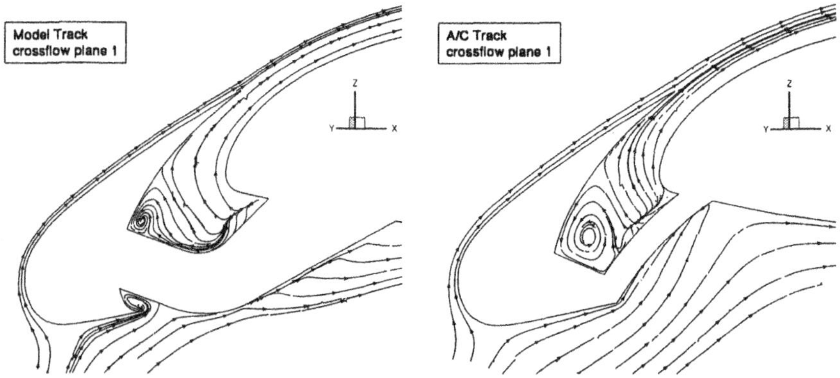

Figure 7. Streamlines at cross flow plane 1

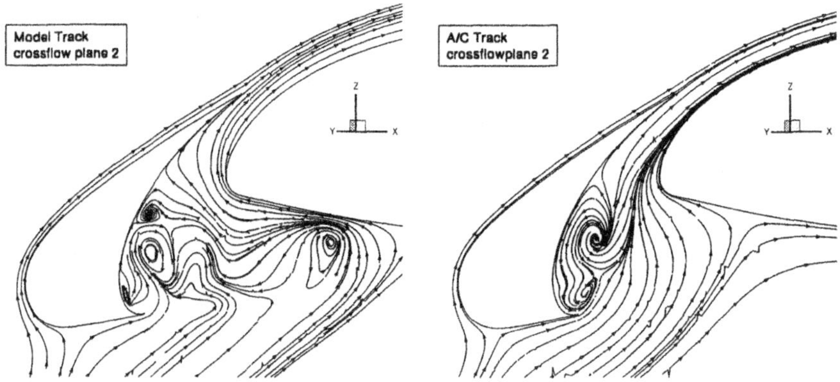

Figure 8. Streamlines at cross flow plane 2

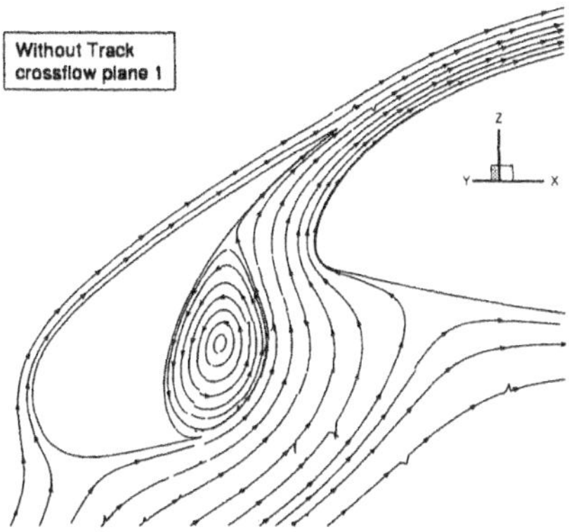

Figure 9. Streamlines at cross flow plane 1

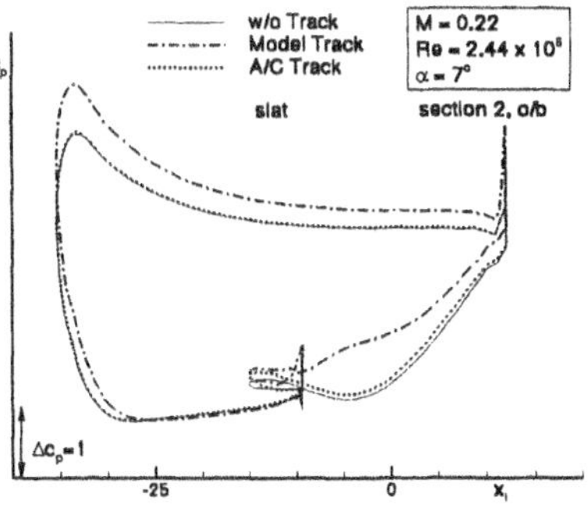

Figure 10. Pressure distributions at sections o/b of the track

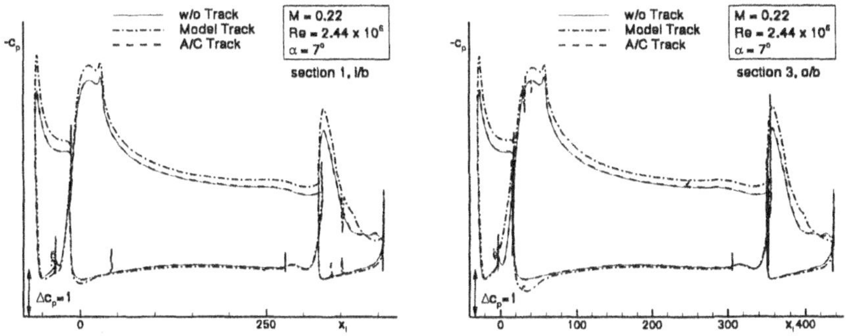

Figure 11. Pressure distributions at sections i/b and o/b of the track

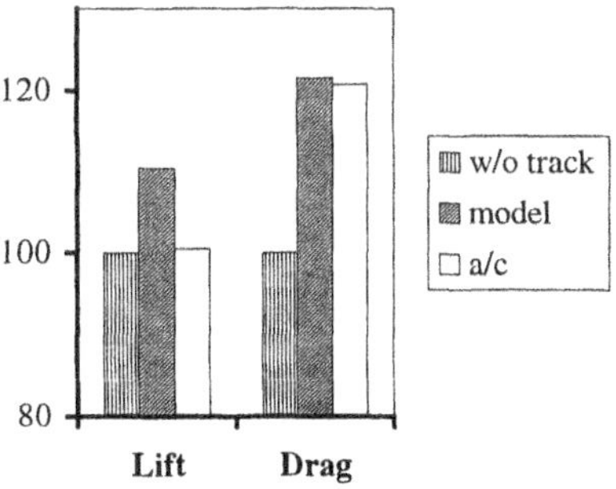

Figure 12. Lift and drag evaluation

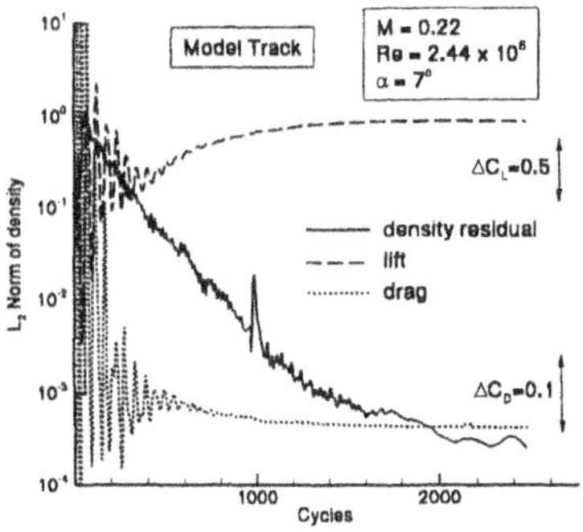

Figure 13. Convergence history, model track

Integrated Design of Installed Engine Nacelles Using an Inverse Design Method

R. Wilhelm

German Aerospace Center DLR, Institute of Aerodynamics and Flow Technology, Lilienthalplatz 7, 38108 Braunschweig, Germany, `roland.wilhelm@dlr.de`

Summary

An iterative Inverse Design System is presented for the aerodynamic design of isolated and wing-mounted engine nacelles. The flow solver used is the DLR Tau code, a finite-volume flow solver for the solution of the Reynolds-Averaged Navier-Stokes equations on arbitrary grids. The inverse design method is based on an iterative residual-correction type approach and attempts to generate a geometry which satisfies a user-prescribed target pressure distribution. Results are presented for the inverse design of isolated and wing-mounted engine nacelles under consideration of the pylon.

1 Introduction

The aerodynamic design of engine nacelles is often carried out for the isolated nacelle only. The interaction between wing, nacelle and pylon flow fields and the resulting flow phenomena are not taken into account within the design process. With the increasing size of engine bypass ratio and therefore engine nacelle size and the closer coupling between wing and nacelle, this design methodology contains a high risk for an unsufficient design solution. Therefore, an iterative Inverse Design System is applied for designing engine nacelles which are mounted at the aircrafts wing. Using this system, an initial nacelle design can be carried out for the isolated nacelle. Then, after integrating this nacelle with the aircraft, a detailed design can be done which takes into account the aerodynamic installation effects.

2 Modular Inverse Design System

The Inverse Design System used in this paper [6] links an inverse design method formerly combined with a block-structured flow solver to the DLR Tau code[4], a flow solver using unstructured grids. Doing so, a higher flexibility concerning configuration changes can be achieved because the time consuming initial grid generation process known for block-structured grids around complex configurations can be reduced significantly. Figure 1 gives an overview of the design system. As can be seen the design system includes four main modules being the *flow solver*, the *solution interpolation*, the *inverse design*, and the *grid deformation* module. In the following subsections, the three modules *flow solver*, *inverse design* and *grid deformation* will be explained in detail. The *interpolation* module, which transfers the flow field solution from the triangular (e.g. unstructured) surface mesh to the structured design surface grid, will not be considered for further explanations.

Flow Solver. The flow solver used within the Inverse Design System is the DLR Tau Code [4]. The code solves the three-dimensional Euler/Navier-Stokes equations using control volumes consisting of triangular or quadrilateral surfaces. The discretization in space is done using a central differencing scheme. Therefore, additional second and fourth order dissipative terms are added to the flux balance. The time integration is performed using a 3-stage Runge-Kutta scheme. Acceleration techniques like multigrid and local time stepping are applied, too. The flow field calculations performed within the design loop are done in an inviscid flow regime. Therefore, all primary grids consist of tetrahedral control volumes only. The reason for this is the time saving gained by applying the Euler instead of the Navier-Stokes equations. Since for nacelle flows the main flow phenomena at cruise are sufficiently resolved in an inviscid flow regime, this simplification is justified. Nevertheless, for a final design check viscous flow field calculations should be applied to estimate the configuration improvements. Note that since the Inverse Design System handles the flow solver as a black box, this decision depends solely on the design strategy of the aerodynamicist and the time available.

Inverse Method. The inverse design method presented in this paper is based on an iterative residual-correction type approach. The residual Δc_p being the pressure difference between the actual and the target pressure distribution is used to calculate a geometry difference Δz. The correction step is done by solving an inverse formulation of the transonic small perturbation equation (TSP equation) as derived by Takanashi [5].

The three-dimensional potential equation can be written in terms of a perturbation velocity potential Φ as

$$(1 - M_\infty^2)\Phi_{xx} + \Phi_{yy} + \Phi_{zz} = K \cdot \Phi_x \Phi_{xx} \qquad \text{with} \qquad K = (\kappa + 1)M_\infty^2 \tag{1}$$

where Φ is defined as

$$\phi = U_\infty(x + \Phi) \ . \tag{2}$$

Two simplified boundary conditions can be applied to (1). The first one uses the tangency condition on the surface while the second one is a simplified pressure relation. In order to eliminate the dependency on the free stream Mach number in $\beta^2 = 1 - M_\infty^2$ a Prandtl-Glauert transformation is performed and new coordinates $\overline{x}$, $\overline{y}$, $\overline{z}$ are introduced. Assuming that for an initial geometry $\overline{z}(\overline{x}, \overline{y}, \pm 0)$ a flow field solution $\overline{\Phi}(\overline{x}, \overline{y}, \overline{z})$ exists, a differential perturbation potential $\Delta\overline{\Phi}(\overline{x}, \overline{y}, \overline{z})$ can be introduced into (1)

$$\Delta\overline{\Phi}_{\overline{x}\,\overline{x}} + \Delta\overline{\Phi}_{\overline{y}\,\overline{y}} + \Delta\overline{\Phi}_{\overline{z}\,\overline{z}} = \frac{\partial}{\partial x}\left[\frac{1}{2}\left(\overline{\Phi}_{\overline{x}} + \Delta\overline{\Phi}_{\overline{x}}\right)^2 - \frac{1}{2}\left(\overline{\Phi}_{\overline{x}}\right)^2\right] \ . \tag{3}$$

The two boundary conditions become

$$\Delta\overline{\Phi}_{\overline{z}}(\overline{x},\overline{y},\pm 0) = \frac{K}{\beta^3}\frac{\partial \Delta\overline{z}(\overline{x},\overline{y},\pm 0)}{\partial \overline{x}} \tag{4}$$

$$\Delta\overline{\Phi}_{\overline{x}}(\overline{x},\overline{y},\pm 0) = -\frac{K}{2\beta^2}\Delta c_p(\overline{x},\overline{y},\pm 0) \ . \tag{5}$$

The $\pm$ sign denotes the upper or lower side of the surface. The pressure difference $\Delta c_p(\overline{x},\overline{y},\pm 0)$ in (5) can be computed using the calculated flow field solution and the prescribed target pressure distribution

$$\Delta c_p = c_p^{target} - c_p^{calculated} \ . \tag{6}$$

The right hand side of (3) is known and can be solved for the unknown geometry difference $\overline{z}(\overline{x},\overline{y},\pm 0)$ using Green's theorem as described by Bartelheimer [1]. In a final step the computed geometry correction has to be transfered back into the original x, y, z-coordinate system.

The formulas above are given in a general coordinate system applicable for wing design, i.e. the x-coordinate running in streamwise, the y-coordinate running in spanwise direction. In order to use one single inverse design formulation for both wing and nacelle design this formulation should be kept unchanged. Therefore, in case of a nacelle design, the coordinate system is changed with y now being the circumferential coordinate direction and z running in the radial direction (both taken with respect to the engine axis). The x-coordinate remains the streamwise direction. In case of nacelle design, the calculated geometry difference Δz is treated as a radial geometry difference Δr.

Radial displacements of the design surface are allowed to take place in the design method. This procedure ensures that the nacelle planform remains constant throughout the whole design process. The size of the nacelle in terms of nacelle length and intake droop angle have to be specified by the initial geometry. Two additional fix points of the nacelle which are kept constant during the inverse design step are the fan radius and the fan nozzle radius.

Grid Deformation. The grid manipulation is necessary in order to generate a computational grid for the modified nacelle geometry. Besides the possibility to generate the finite volume grid including the modified nacelle in each design iteration from scratch, the more efficient way is to apply a grid manipulation method which fits an existing finite volume grid to a modified configuration. Also, in case only one finite volume grid consisting of a constant number of grid points is used for all flow field calculations in the design loop, a restart ability of the flow solver contributes to the systems overall efficiency, too. Finally, since the grid manipulation process runs in an iterative design loop, all geometry and grid modification steps have to be done automatically, in a closed process chain without any intermediate user input.

In order to fulfill the above mentioned demands, the grid generation package *Centaur* [2] is used to generate the initial finite volume grid for the start configuration. *Centaur* is capable of designing hybrid grids consisting of prismatic, hexahedral, pyramidal or tetrahedral control volumes. In case of the presented Inverse

Design System, an initial grid has to be generated before the main inverse design run can start. Then, after each new nacelle design, a grid deformation algorithm is used to fit the initial grid to the modified configuration. Since the initial grid will be the baseline finite volume grid for all flow field calculations performed in the design loop, its quality should be analyzed carefully.

3 Results

In order to test the presented inverse design method, a first step is to do a redesign, i.e. a design of an a priori known geometry. Doing so, it can be avoided to try to solve a so-called ill-posed problem, where the prescribed target pressure distribution can not be reached by a physical flow solution. The aircraft used is the DLR ALVAST configuration, a twin-engine, subsonic transport aircraft model of Airbus A320 type. The aircraft is equipped with a very high bypass ratio (VHBR) engine nacelle, see figure 2.

The installed nacelle redesign is performed with a free stream Mach number of $M_\infty = 0.75$ and an angle of attack of $\alpha = 0.5°$. The engine intake mass flow is defined by a stream tube area ratio of $\epsilon_{Fan} = 0.92$. The initial aircraft is the ALVAST configuration equipped with a modified engine nacelle which consists of the scaled side profile ($\Theta = 90°$) of the original VHBR engine nacelle in each circumferential section only. The nacelle target pressure distribution is taken from the original ALVAST VHBR configuration. Figure 3 shows surface pressure distributions and nacelle cross sections in three circumferential sections ($\Theta = 45°$, $180°$ and $315°$) for the initial and the designed nacelle. For comparison purposes, data of the target nacelle is shown, too. As can be seen, the designed nacelle matches the target data in all sections.

In the following case a pure nacelle design shall be performed where it is unknown if a solution exists. The aim is to design a typical hybrid-laminar flow (HLF) nacelle at a higher free stream Mach number, i.e. at $M_\infty = 0.82$, than in the before mentioned redesign case. In general, hybrid-laminar flow is established by both boundary layer suction (up to $x/c \approx 0.20$) and contouring of the nacelle. A typical resulting pressure distribution shows two acceleration peaks in the nose and mid-chord nacelle region. It should be mentioned that due to the inviscid flow regime within this design run it is not possible to evaluate if the boundary layer is really laminar. The aim of this calculation is to demonstrate the principle capability of the Inverse Design System to design engine nacelles with various types of pressure distributions. In a first step an isolated HLF nacelle is designed where the influence of wing and pylon is not taken into account. Then, in the second step, the design will be carried on as an installed nacelle design where the aircrafts influence on the nacelle is taken into account.

The initial geometry for the isolated nacelle design is the original VHBR engine nacelle designed for a free stream Mach number of $M_\infty = 0.75$. The angle of attack is $\alpha = 2.0°$. The result of this isolated nacelle design is shown in figure 4. As can be seen the target pressure distribution (showing the typical HLF outline) is met in all three sections. The maximum nacelle diameter has slightly increased in the section $\Theta = 90°$ whereas in section $\Theta = 180°$ its axial location is shifted downstream. The highlight diameter is unchanged in order to preserve the intake flow field.

The previously generated pressure distribution for the isolated HLF nacelle is used as start distribution for an installed nacelle design run. Nevertheless, due to engine installation effects (e.g. toe-in and nose-up angle as well as pylon and wing interference effects) the target pressure distribution has to be adapted. The free stream Mach number for this installed nacelle design run is $M_\infty = 0.82$ and the angle of attack is $\alpha = 0.5°$. The initial nacelle geometry is the VHBR engine nacelle. The resulting surface pressure distributions and nacelle cross sections are shown in figure 5. In both sections $\Theta = 45°$ and $315°$ the prescribed target pressure distribution is met by increasing the maximum nacelle diameter. For section $\Theta = 180°$ again the maximum diameter location is shifted downstream. Also shown is the surface pressure distribution of the isolated HLF nacelle design. Especially in section $\Theta = 315°$ it can be seen that the target pressure distribution had to be adapted to avoid an extensive increase in maximum nacelle diameter by shifting the second acceleration peak to lower c_p values.

A second application should demonstrate the ability of the Inverse Design System to more focus on the engine installation effects taking place on the wing. For this purpose the DLR F6 configuration is used, a twin engine wide body aircraft of Airbus type. The configuration is equipped with a long duct throughflow nacelle of CFM56 type which is closely coupled to the wing. Due to the close coupling, the wing lower side inboard of the pylon is subjected to a narrow channel flow resulting in a strong shock on the surrounding wing, pylon, and nacelle surfaces. Figure 6 shows the F6 configuration with iso-c_p lines in the wing, pylon, nacelle region. Also shown is the hybrid grid in the symmetry plane used for the analysis calculation. Figure 7 shows the wing pressure distribution inboard of the pylon at $\eta = 0.331$ for the initial F6 CFM56 configuration. The surface pressure data is gained by a viscous flow field calculation using the Tau Code and the Spalart-Allmaras turbulence model with Edwards modification [3]. The initial hybrid grid used consists of 2.3 million grid points and is adapted once with respect to y^+ and the initial flow field solution. For the design cruise Mach number of $M_\infty = 0.75$ the lift coefficient is set to $C_L = 0.50$. The Reynolds number is $Re = 3 \cdot 10^6$. The resulting drag coefficient is $C_D = 0.0316$. As comparison, experimental data of a test campaign performed in the ONERA S2MA wind tunnel in 1993 is shown, too.

The result of the inverse design calculation can be seen in figure 8 where surface pressure distributions and nacelle cross section profiles are shown for three circumferential locations. Clearly visible is the reduction of the acceleration peak on the designed nacelle contour in the sections $\Theta = 315°$ and $330°$. In order to reduce the shock in the wing, nacelle, pylon region the local nacelle diameter is decreased by the design method. Due to a user specified constraint in the design algorithm, the nacelle surface is allowed to change only at $x/c_\Theta > 0.4$. This limits the geometric surface modifications and, as can be seen in figure 8, keeps the nacelle intake region unchanged. In order to increase the design system efficiency, inviscid flow field calculations are performed within this design run only.

In order to quantify the aerodynamic gains of the modified configuration, again a viscous flow field analysis is done using the same parameter for grid and flow solution generation as for the initial configuration. For the spanwise wing section $\eta = 0.331$ figure 7 also shows the surface pressure distribution for the modified configuration. As can be seen the suction peak on the lower wing surface is considerably reduced. For the targeted lift coefficient of $C_L = 0.50$ the drag coefficient for the modified configuration is $C_D = 0.0311$. Compared to the initial configuration this is a reduction of 5 drag counts ($dc = 10^{-4}$).

4 Conclusion

An algorithm for the iterative inverse design of isolated and wing-mounted engine nacelles has been presented. The method uses a residual-correction type approach to design a nacelle surface which fulfills a prescribed target pressure distribution. The presented Inverse Design System uses the DLR flow solver Tau for the flow field analysis on arbitrary grids. Due to the modular setup of the Inverse Design System, it is possible to include the commercial grid generation and modification package *Centaur* as grid deformation tool into the design loop.

The results presented show that the method is capable of designing engine nacelles. A redesign test case demonstrates the complete design functionality for wing-mounted engine nacelles. A design of a HLF nacelle shows an application without an a-priori known solution. A second application demonstrates the ability to reduce engine installation effects via nacelle surface modifications.

References

[1] Bartelheimer, W.: "*An Improved Integral Equation Method for the Design of Transonic Airfoils and Wings*". AIAA Paper 95–1688 (1995)

[2] CentaurSoft: "*The Grid Generation Package Centaur*". http://www.centaursoft.com/. (2002)

[3] Edwards, J. R., Chandra, S.: "*Comparison of Eddy Viscosity-Transport Turbulence Models for Three-Dimensional, Shock Separated Flowfields*". AIAA Journal, Vol. 34, No. 4 (1996), pp. 756–763

[4] Gerhold, T., Friedrich, O., Evans, J., Galle, M.: "*Calculation of Complex Three-Dimensional Configurations Employing the DLR Tau-Code*". AIAA Paper 97–0167 (1997)

[5] Takanashi, S.: "*Iterative Three-Dimensional Transonic Wing Design Using Integral Equations*". Journal of Aircraft, Vol. 22, No. 8 (1985), pp. 655–660

[6] Wilhelm, R.: "*An Inverse Design Method for Designing Isolated and Wing-Mounted Engine Nacelles*". AIAA Paper 2002–0104 (2002)

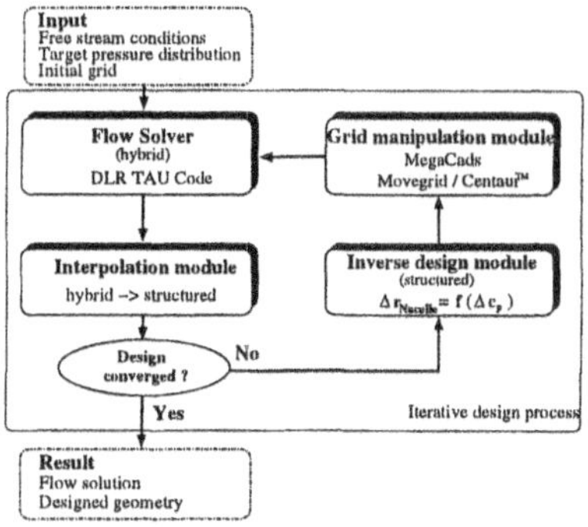

Figure 1 Flow chart of the Inverse Design System

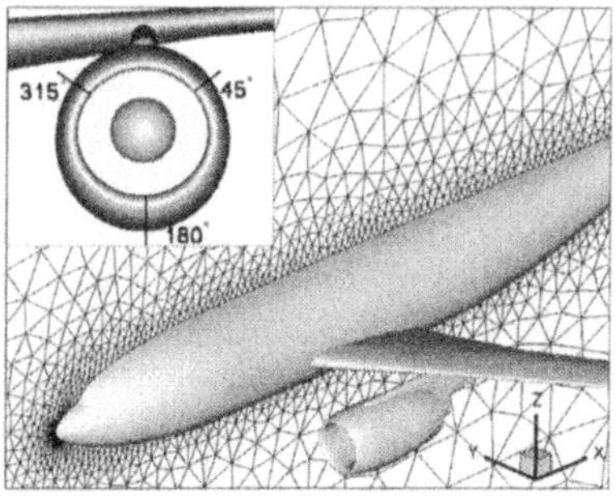

Figure 2 ALVAST VHBR configuration

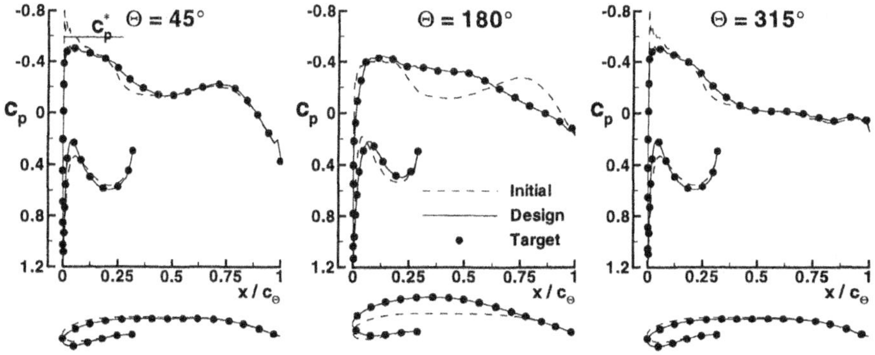

Figure 3 Redesign of an installed engine nacelle: surface pressure distributions and nacelle profiles in three circumferential sections

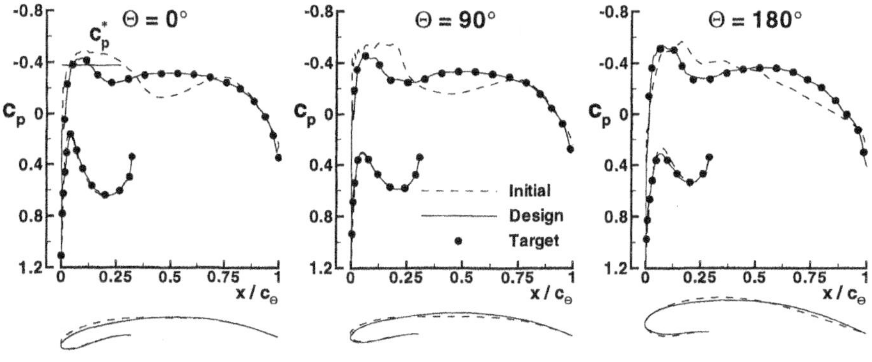

Figure 4 Design of an isolated HLF engine nacelle: surface pressure distributions and nacelle profiles in three circumferential sections

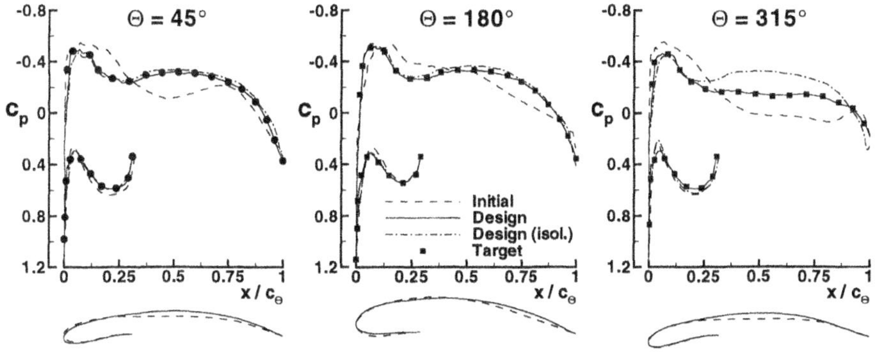

Figure 5 Design of an installed HLF engine nacelle: surface pressure distributions and nacelle profiles in three circumferential sections

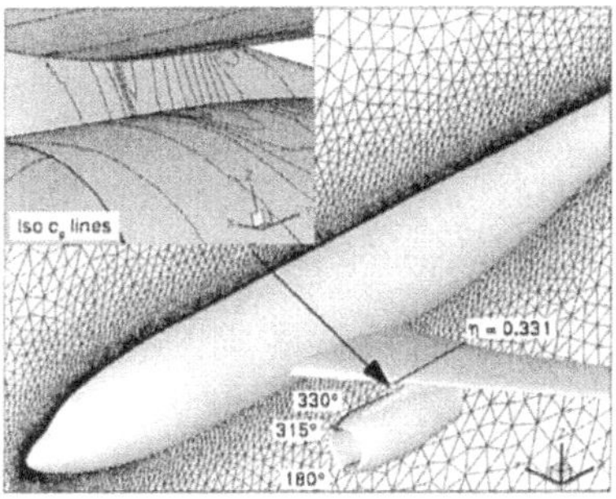

Figure 6 F6 configuration with iso-c_p lines in wing, pylon, nacelle region

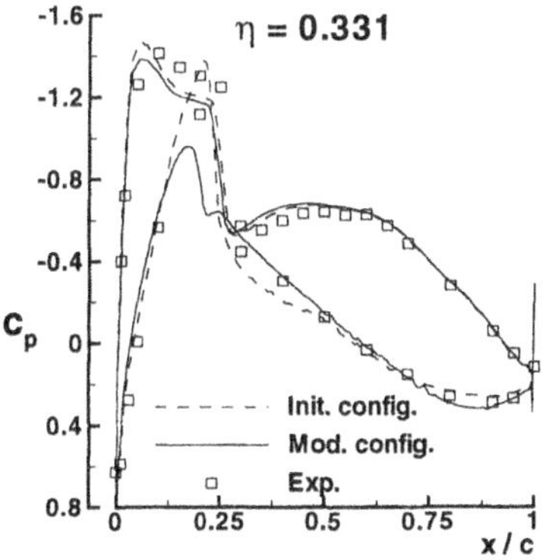

Figure 7 F6 wing pressure distribution for initial and modified configuration

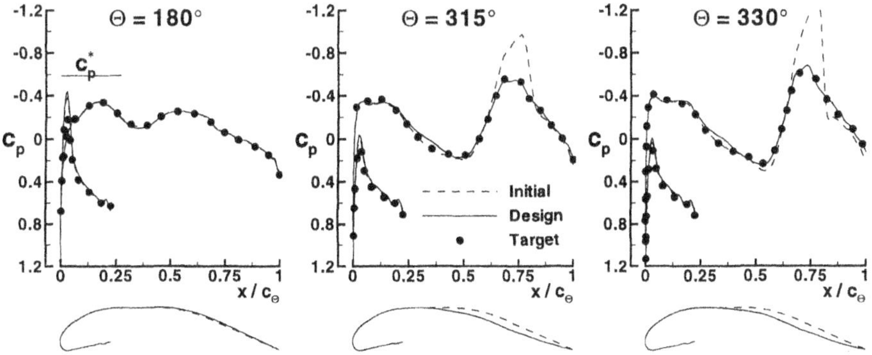

Figure 8 Design of an installed throughflow nacelle for reduction of installation effects

Numerical simulation of the propeller flow around a four-engined aircraft in high-lift configuration.

Jörg Hansing[1], Mark Sutcliffe[2,3], and Olaf Knobloch[4]

[1] University of Applied Sciences, Berliner Tor 5, 20099 Hamburg, Germany
[2] DLR Göttingen, Institute of Aerodynamics and Flow Technology, Bunsenstr. 10, 37073 Göttingen, Germany
[3] Airbus Deutschland, Methods and Tools, Hünefelstr. 1-5, 28199 Bremen, Germany
[4] Airbus Deutschland, High-Lift Design, Hünefelstr. 1-5, 28199 Bremen, Germany

Summary

Operating propellers, especially wing mounted propellers have significant and sometimes subtle effects on the aerodynamics of an airplane. In order to shed more light on these effects an actuator disc model to simulate propeller flow has been developed, implemented into the DLR Navier-Stokes code TAU and tested on an actual four-engined transport aircraft in high-lift configuration. Based on a wind tunnel experiment with an isolated propeller, the flow defining parameters of the actuator discs could be evaluated using a numerical model of the isolated propeller. The experiences made with the isolated propeller computations were then utilized to adjust the actuator discs to the aircraft. Low-speed computations of the configuration were then performed for $Re = 1.36 \times 10^6$, $Ma = 0.176$ and specific thrust conditions at low and moderate incidences, showing good agreement between the numerical and experimental results in the area of the expected propeller flow. Hence the capability of the new propeller module could be proved.

1 Introduction

Operating propellers, especially wing mounted propellers have significant effects on the aerodynamics of an airplane. There are two sources of propeller effects, those due to the direct forces and moments created by the propeller and those due to the propeller slipstream. The direct forces and moments are more or less predictable, therefore this work concentrates on the slipstream effects wich are difficult to predict and their effects are sometimes subtle. For the analysis of the propeller effects in airframe integration studies, it is desirable to have a computational method available that can predict these effects on the flow field around complex aircraft configurations. Therefore, a method to simulate the propeller flow aerodynamics has been developed and implemented into the DLR Navier-Stokes code TAU [1, 2]. The method is based on an actuator disc model [3–5] and tested on an actual four-engined transport aircraft in high-lift configuration at low speed.

The flow in the propeller area is defined using flow parameters on inflow and outflow surfaces on the actuator disc. The inflow surface is the upstream facing side of the propeller (i.e. a fluid particle flows into the propeller), whereas the outflow

surface is facing downstream. In order to simulate the effects of a propeller, the method allows a total pressure ratio (i.e. the total pressure of the outflow surface in relation to the total pressure in the farfield) on the propeller outflow surface to be defined using a polynomial as a function of the propeller radius. Several options exist for controlling the direction of the flow both into and out of an actuator disc. By default, it is assumed that the flow enters and leaves the actuator disc normal to the inflow and outflow surfaces. If necessary, the direction of the flow leaving the outflow surface may be set via a vector, which may be desirable for example, when trying to simulate the effect of an angle of attack. In addition to this a swirl angle as a function of the propeller radius can be defined using a polynomial function on the outflow surface. Using a combination of swirl and outflow direction (i.e. the two are added together) a large range of outflow conditions can be simulated. It is additionaly possible to extrapolate the flow direction for each point on the inflow surface from an upstream point. This option was used on all the aircraft computations.
These inflow and outflow parameters could be partly validated using experimental data of a so called 'isolated propeller' wind tunnel test and a simple 'isolated propeller' CFD model of the same geometry. The final hybrid grid of this model generated using the CENTAUR^{TM} software package provided by CentaurSoft can be seen in Fig.1(left). It includes the inflow disc, the outflow disc and a bounding surface of Euler type connecting both discs. The discs are located on a body of rotation, which simulates a nacelle. The test rig which usually keeps the model connected with the wind tunnel facility has been neglected.
Computations were intended to simulate a slipstream of a thrust coefficient of $C_T = 2T/(\rho u_u^2 A) = 0.1$, with T as the thrust, ρ the density, u_u the azimuthal velocity, A the reference area and an advance ratio of $J = 30u_c/(\pi * r * n_p) = 0.8$, with u_c as the relative onset velocity, r the radius, n_p the revolutions per minute. In order to find the parameters on the inflow and outflow disc surfaces of the isolated propeller model to approximate the propeller flow behaviour of $C_T = 0.1$ and $J = 0.8$, the flow properties (direction and magnitude) in a plane behind the propeller were evaluated and compared with experimental results. During the investigation parameters for total pressure, swirl and the desired in and outflow direction on the disc surfaces as well as grid properties, such as cell sizes and topology, were altered until the numerical results were in good agreement with the wind tunnel data. The resulting Polynomials for swirl and pressure ratio were then implemented into the aircraft configuration without further modification.
The computed four-engined high-lift configuration is a right half model with two nacelles and with clockwise co-rotating propellers with flaps deployed by $\delta_F = 25^\circ$ and tabs deployed by $\delta_T = 14.3^\circ$, without vertical and horizontal stabilizers. For simulating the left half model, the grid is mirrored while the propeller rotating direction is maintained. Effects due to the non-symmetrical flow on the symmetry plane are assumed to be small. The parts of the final mesh, including the surface grid, the field grid on the wing and in the symmetry plane can be seen in Fig.2. The grid consists of 400k points on the surface, 30 prisms layers close to the surface to capture the boundary-layer effects and has a total size of 13×10^6 points. The

farfield boundary is 25 semispans upstream and 25 semispans downstream the body, respectively.
Computations were performed for two incidences ($\alpha = 0^\circ$ and 7.6°) with the parameters for total pressure, swirl and in and outflow direction defined on the disc surfaces to simulate $C_T = 0.1$ and $J = 0.8$.

2 Numerical Method

The hybrid unstructured DLR flow solver TAU is applied to solve the Reynolds-averaged Navier-Stokes equations (RANS). The flow calculation algorithm is based on a dual grid approach. The three-dimensional Navier-Stokes equations are solved in time towards a steady state by Runge-Kutta time stepping method. A classic central Jameson scheme [6] is used for the spatial discretization of the inviscid and viscous terms on all grid levels. A multigrid strategy and local time stepping are employed to improve the convergence rate. The investigated four-engined high-lift configuration is calculated using the Spalart-Allmaras [7] turbulence model with Edwards [8] modification. Calculations were performed in 'full turbulent' mode on a HP SD32000 machine with 16 processors.
Since the flow characteristics on the outflow discs are prescribed by the polynomials for swirl, total pressure and the general value for the outflow direction, the flow on the inflow discs have to be adapted to obtain equal mass flow rates compared to the outflow discs. Thus, during the iteration process the static pressure and hence the inflow velocity on an inflow disc is changed until the mass flow of the inflow disc and of the outflow disc are balanced. This adaptation procedure was performed only on the finest grid level for reasons of stability.

3 Discussion and Results

Different wind tunnel test results [9–12] were available for the validation of the simulated propeller flow. The isolated propeller experiment included a thrust variation and wake measurements in a plane at a distance of 212mm downstream of the propeller, while the wind-tunnel tests with the aircraft right half-model included the investigation of force behaviour as well as measurements of wing pressure distributions at different incidences and different thrust coefficients. In the experiment, the left half-model is simulated by changing the rotational direction of the propellers. The wind tunnel tests were performed at the LSWT Bremen for a Reynolds number of $Re = 1.36 \times 10^6$ and a Mach number of $Ma = 0.176$.
The effect of the isolated propeller model on the flow is qualitatively shown in Fig.1(right), with stream ribbons highlighting the rotating flow. The preliminary test with the isolated propeller has revealed some difficulties simulating the propeller aerodynamics using actuator disc surfaces. Although the flow properties within most parts of the propeller wake could be recreated with this method, close to the surface of the nacelle, in the propeller tip area and on the bounding surface connecting the inflow and outflow disc, the calculated flow properties exhibit significant differences to the experiment, shown in Fig.3. This is expected since inflow and outflow discs are physically independent of each other, the boundary Euler type surface does not

allow the flow to penetrate (to be seen in Fig.4) and the near-wall flow is wrongly influenced by the outflow parameters.

With the observation made with the isolated propeller, the adjustments were then implemented into the aircraft configuration with its two propellers on each side. Preliminary results are shown in the following figures. In Fig.5 the flow released from the outboard engine outflow disc is shown utilizing velocity coloured stream ribbons while the surface contours represent the pressure distribution.

The sectional surface pressure distributions represent the main validation data. Fig.6 illustrates the locations of the investigated pressure distribution sections. A comparison of pressure distributions between experiment and calculation for two wing sections PD500 and PD600 is shown in Figs.7 and 8. PD500 lies within the propeller wake with an up-moving rotation and PD600 in the down-moving rotation of the outboard propeller. Generally the computed curves of both sections are in good agreement with the measured pressure distribution. Particularly, the rear main wing, upper and lower flap and the upper tab pressures are sufficiently predicted. Minor differences exist close to the main wing leading edge and the pressure side of the tab. One reason for the discrepancy in the pressure distributions around the wing leading-edge may be the not properly simulated effect of the wake in the up and down-moving side of the propeller. It has been observed that the outflow direction parameter defined on the outflow disc has a strong effect on the pressure distribution; variations of this parameter lead to different results for the pressure in the vicinity of the leading edge. Since an average value for the local flow angle has been estimated from the power-off case, where the local flow vectors were evaluated in the area of the propeller discs, more investigations are needed here to find suitable values for the outflow direction.

Fig.9 shows the lift behaviour of computed and measured model. Although the numerical simulation slightly exaggerates the lift for all the cases by 2%-4%, the gradients are comparable with the experiment. It is interesting to note that the predicted lift of the left half model is slightly greater than that of the right half model, a result which can also be observed in the experimental results. This is due to the improved flow characteristics in the nacelle-wing junction of the left half model, which are caused by the opposite rotational directions of the propellers. The strong effect of the relatively high thrust coefficient C_T can also be clearly seen when comparing the differences between the power on/off results, producing a strong increase in lift slope and reaching a higher maximum angle and maximum lift, respectively.

Fig.10 discloses the source of the improved performance of the powered configuration. Here the sectional lift over the dimensionless semispan of the calculated right half model with the two clockwise-rotating propellers can be seen. Affected by the accelerated flow and the changed onset condition due to the propeller, most of the wing sections produce a significant higher amount of lift. In particular the lift behaviour within the outboard engine stream remarkably shows the differences between the effects on the wing of the up-moving and down-moving rotation of the propeller. The local losses of lift are caused by the presence of nacelles and flap track fairings (FTF's).

4 Conclusions and further Developments

The current methodology used to simulate propeller aerodynamics using 'actuator discs' has proved its capability to predict global flow features also seen in experiments. The parameters for controlling the actuator discs could be evaluated by using the isolated propeller model and then applied to the actuator discs on the aircraft model. The predicted flow properties, the lift coefficients as well as the local pressure distributions are in good agreement with the experimental data.
It is planned to investigate more thrust parameters than the considered one and compare the simulated axial force with the measured thrust from the experiment. Further work intends to extend the propeller values for swirl and total pressure in that way that they are not only radius dependent but also angle dependent. This may improve the computation of the effects of the up and down-moving rotation.
A second wind tunnel test using the isolated propeller is planned to give more information about the propeller stream, such as flow properties closer to the actuator discs, which will be useful for further CFD validation exercises.
It is also planned to examine results from unsteady computations of propeller blades.

References

[1] unknown. *Technical Report - Documentation of the DLR TAU code.* DLR, 2001.

[2] T. Gerhold et al. *TAU code User guide.* DLR, 2001.

[3] N.J. Yu; H.C. Chen. Flow Simulations for Nacelle-Propeller Configurations using the Euler Equations. *AIAA*, 84-2143, 1984.

[4] N.J. Yu; S.S. Samant; P.E. Rubbert. Flow Prediction for Propfan Configurations Using the Euler Equations. *AIAA*, 84-2143, 1984.

[5] R. Wilhelm. Development and Testing of an Aktuator Disk Boundary Condition implemented into the DLR FLOWer Code. Technical report, DLR Braunschweig, 1999. IB 129-99/22.

[6] A. Jameson. Transonic flow calculation. MAE Report 1651, Princeton University, Princeton, New Jersey, 1983.

[7] P.R. Spalart; S.R. Allmaras. A one-equation turbulence model for aerodynamic flows. *AIAA*, 92-0439, 1992.

[8] J.R. Edwards; S. Chandra. Comparison of eddy viscosity-transport turbulence models for three-dimensional, shock-separated flowfields. *AIAA Journal*, 34(4):756–763, 1996.

[9] M. Huhnd. Test Report of Isolated FLA 5.2 Propeller Performance Measurement in LSWT-Bremen. Technical report, Airbus Deutschland GmbH, Bremen, 2002. Report EF-029/01.

[10] S. Schaber. Test Programme A400M FLA 5.2-03, Final for Testing. Technical report, Airbus Deutschland GmbH, Bremen, 2002. Report EGXG-080/02.

[11] D. Lejeau. High-Power Effects on Lift and Drag and Comparison of alternative Flap Systems. Technical report, Airbus Deutschland GmbH, Bremen, 2002. Report EGAG 593-02.

[12] S. Schmid-Göller M. Huhnd. Aspects of low-speed Wind Tunnel Testing on a A400M model with Propeller Simulation. *Notes on numerical Fluid Mechanics*, 77:82–91, 2002. Springer-Verlag.

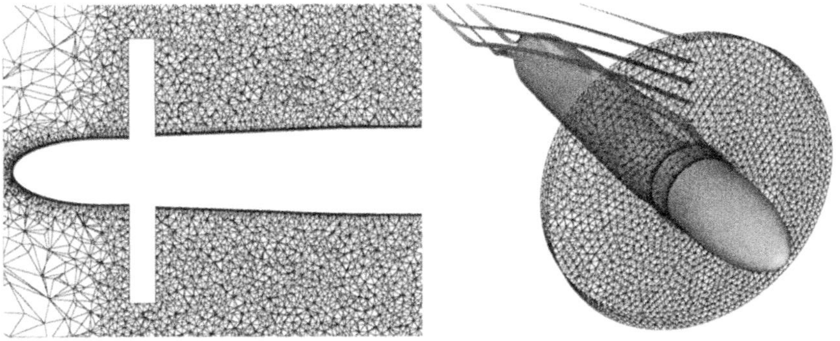

Figure 1: (left) A slice through the field mesh and (right) the surface grid with stream ribbons representing the velocity behaviour downstream of the outflow disc.

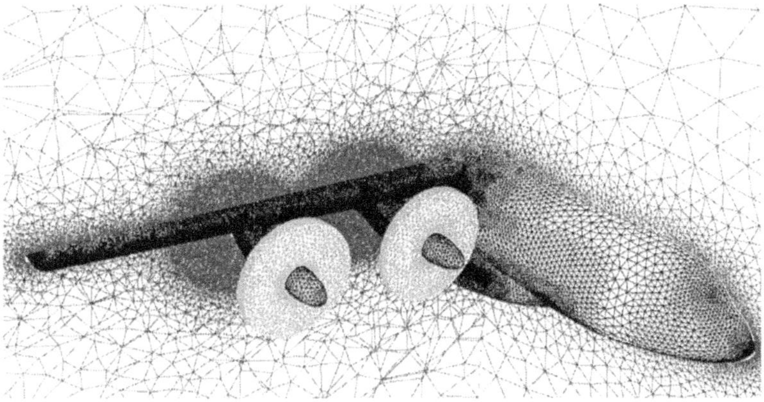

Figure 2: Surface grid of front part of aircraft, field grid on the wing and in the symmetry plane.

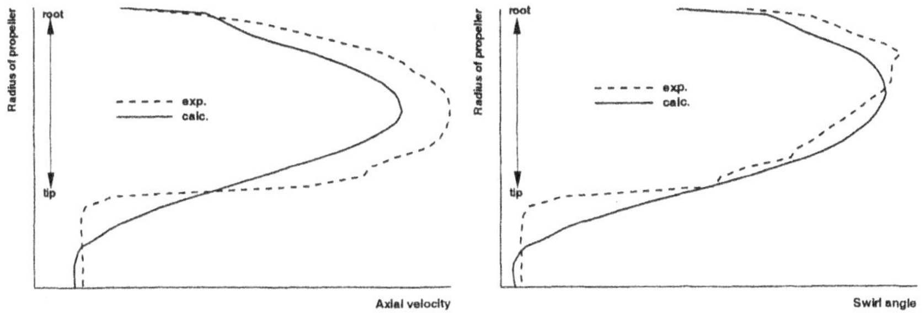

Figure 3: Comparison between the real flow properties of the isolated propeller and the numerical results. Propeller radius vs. axial velocity component (left) and vs. swirl angle component (right).

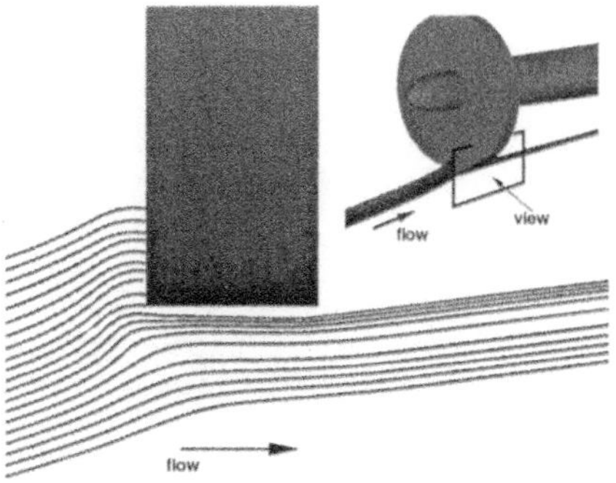

Figure 4: Stream lines in the vicinity of the boundary panel of Euler type, which prevents penetration in this area between inflow and outflow discs. The inflow direction of the inlet is set to the surface normal.

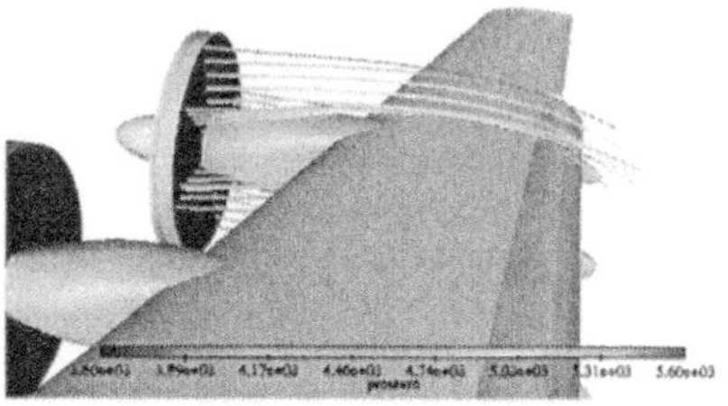

Figure 5: Computed result of the configuration at $Ma = 0.176$, $\alpha = 7.58^\circ$ and $Re = 1.36 \times 10^6$, with $C_T = 0.1$ and $J = 0.8$. Stream ribbons are released from the outboard engine outfl ow disc representing the local velocity vectors, contours on the surface illustrating the pressure distribution.

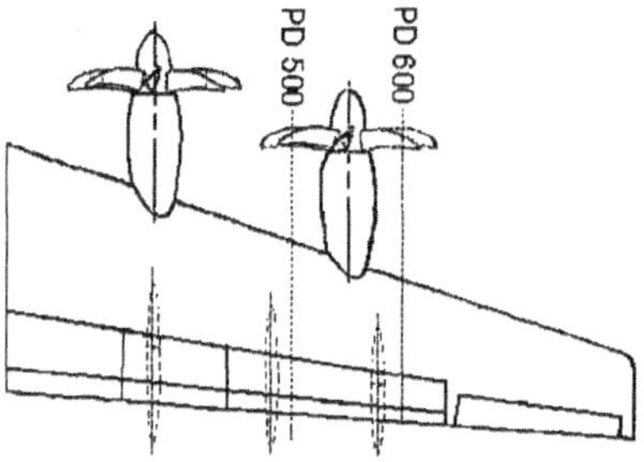

Figure 6: Locations of the investigated pressure distribution sections.

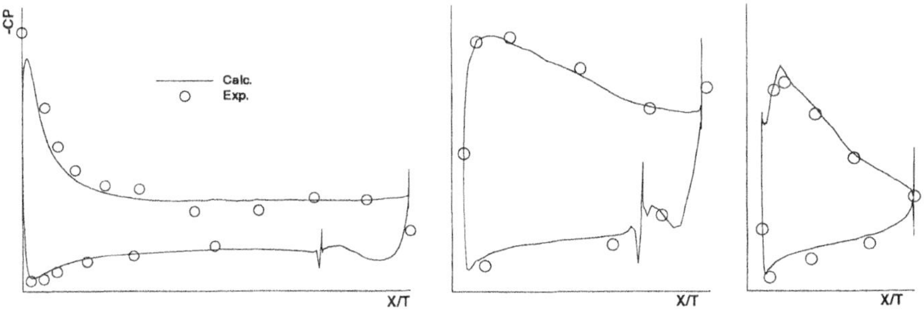

(a) main-wing (b) flap (c) tab

Figure 7: Computed result of the configuration at $Ma = 0.176$, $\alpha = 7.58^\circ$ and $Re = 1.36 \times 10^6$, with $C_T = 0.1$ and $J = 0.8$. Pressure distribution of plane PD500.

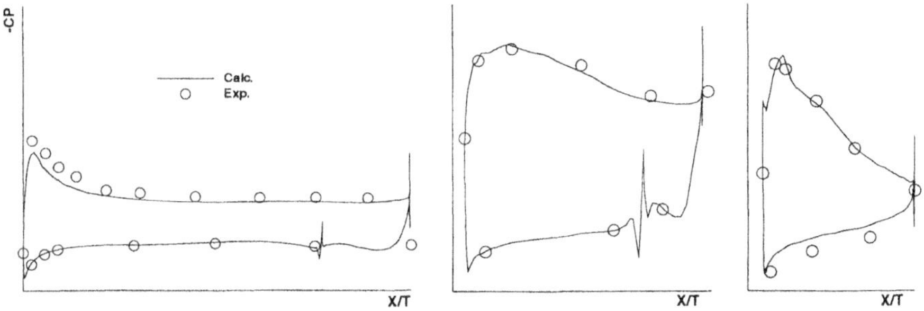

(a) main-wing (b) flap (c) tab

Figure 8: Computed result of the configuration at $Ma = 0.176$, $\alpha = 7.58^\circ$ and $Re = 1.36 \times 10^6$, with $C_T = 0.1$ and $J = 0.8$. Pressure distribution of plane PD600.

Figure 9: Lift behaviour vs. incidence of the computed and measured model for $Re = 1.36 \times 10^6$ and $Ma = 0.176$.

Figure 10: Lift behaviour vs. dimensionless semispan of the computed right-half model with clockwise-rotating propellers for $Re = 1.36 \times 10^6$ and $Ma = 0.176$. The fuselage is on the left, the wing tip on the right-hand side.

Numerical Investigation of Wing-Nacelle Interference Effects at Transonic Flow Conditions for a Generic Transport Aircraft Configuration

Ante Soda and Tiana Tefy

DLR Institute of Aeroelasticity, Bunsenstrasse 10, D-37073 Goettingen, Germany

Summary

A detailed numerical and experimental investigation of the steady/unsteady wing-pylon-nacelle interference effects on flutter characteristics has been initiated between DLR and ONERA. This paper presents preliminary numerical simulation results of steady flow phenomena occurring in the engine region of a generic aircraft configuration. Main issues of this project phase are parameter investigation and definition of a wind-tunnel model geometry which is used for the forthcoming experimental measurements. Several geometry and flow parameters are varied and their influence on interference phenomena is investigated in order to design the final model geometry which satisfies given wind-tunnel and production constraints. Results show that, for the development and strength of interference effects, the nacelle yaw angle and the wing-nacelle overlapping parameters play the crucial role while the effect of wing sweep angle parameter is similar to that of nacelle yaw angle. Thus interference effects of a swept wing - nacelle configuration can be simulated by a rectangular wing with nacelle and pylon yaw degree of freedom.

1 Introduction

Modern transport aircrafts fly at transonic cruise conditions where local supersonic flow regions and shock waves occur. This may lead to non-linear aeroelastic problems (non-linear flutter, LCO) which can cause structural fatigue and it is important to get an accurate understanding of the complex aerodynamic phenomena for the transonic flow conditions [6]. For flutter related phenomena the integration of large engine nacelles plays an increasingly important role [1, 3]. Therefore, a project called HighPerFlex, joint effort of DLR and ONERA, has been initiated in order to numerically and experimentally investigate the steady/unsteady flow physics and complex interference effects on a generic transport aircraft wing-pylon-nacelle configuration. In a wind-tunnel experiment a 100% accurate scaled version of the realistic geometry can seldom be used. One typical geometry constraint is the wing sweep angle limitation for unsteady experiments. i.e., a pitch rotation axis usually has to be orthogonal to the wind-tunnel walls and this does not permit large sweep angles in the wind-tunnel. Therefore, a similar generic geometry, which can reproduce the most important realistic flow effects (shock waves, flow separation), has to be designed. In the first stage of the HighPerFlex project main goals are definition and optimization of the wind-tunnel model geometry and improvement of the physical understanding of steady interference phenomena. This paper therefore presents results and conclusions of extensive and detailed steady flow field numerical simulations performed with the various combinations of geometrical and flow parameters (Figure 1). Additional objective is to investigate if a fast and relatively simple CFD code (DLR-AE EUVIMB) is capable of predicting details of interfering flows with significant accuracy when compared to a highly sophisticated and established CFD code (DLR TAU).

2 Geometry Definition and Grid Generation

Configuration used in the current investigation (so-called HighPerFlex configuration) is a generic assembly of three different geometries. For the geometry of HighPerFlex wing entity a HPF7 aerofoil (derived from the Airbus A340 aerofoil) is used. The geometry of HighPerFlex nacelle-pylon cluster is taken from the generic DLR-F6 configuration [1] with modifications regarding the D/L ratio of the standard DLR-F6 through-flow nacelle. These changes are necessary in order to accommodate the shape of the HPF7 aerofoil and to preserve the physical resemblance between the generic HighPerFlex configuration and the standard configurations for which there is either experimental or numerical flow field data available. Figure 2 gives the main geometrical data for the generic HighPerFlex geometry and shows how its characteristic dimensions compare to dimensions of two standard configurations. The aspect ratio of the HighPerFlex wing is set to 2.8 (1.0 m span and 0.36 m constant chord) and the pylon is positioned at 50% span. Two different computational geometries of the HighPerFlex configuration are used for the steady state calculations, the main difference being the shape of the generic pylon. A 2-D pylon approximation is employed for the structured approach while a full 3-D pylon shape is used with the unstructured approach. In all grids the front and rear boundaries are modelled as inlet/outlet boundary conditions whereas top/bottom and left/right box boundaries have solid wall boundary conditions to simulate the wind-tunnel walls (no farfield or symmetry boundary conditions). Upper/lower wall boundaries are distanced at 20 chord lengths away from the configuration surface in order to account for the test-section pressure wall adaptation capabilties which will be used in the actual experimental investigation. An in-house developed 3-D multi-block grid generator Aengus_MB [4] is used to generate structured body-fitted grids about the HighPerFlex geometry. Topologically simple 5-block grids are generated for the Euler flow simulations with the structured CFD code. All the grid blocks are of 'H' topological type (in both streamwise and spanwise directions) and the complete grid contains just over 520,000 nodes in total (Figure 3). A commercial grid generation software Centaur [2] is used to generate unstructured grids about the HighPerFlex configuration. It enables creation of high quality meshes in an automatic and efficient manner and it provides the user with almost complete control over point placement. Euler calculations are performed with pure unstructured grids consisting of tetrahedra elements (Figure 4). Typically, an Euler grid consists of 360,000 nodes and about 2 million computational cells.

3 Flow Simulations

Two CFD codes based on different numerical approaches are used for numerical flow simulations. Both solvers are based on finite volume method, one utilizing the structured approach while the other is based on the unstructured approach. This paper presents and compares numerical results obtained with both of them. The simulation of inviscid flow has proved to be capable of capturing all the important physical flow phenomena described in Section 1 [3]. Thus, the preliminary analysis of steady interference effects is performed by numerically solving the Euler equations. The first solver used is a viscous-inviscid multi-block structured code, EUVIMB [5], developed at the DLR Institute of Aeroelasticity. This solver uses a centered scheme corrected with the second and fourth order artificial dissipation terms and an explicit multi-stage Runge-Kutta scheme for the time integration. This is an unsteady CFD code adopting the boundary layer coupling and it has been developed primarily for use in aeroelastic simulations. The well known DLR TAU code [7], an unstructured/hybrid CFD code, is the second code used to obtain Euler steady solutions on unstructured grids for the HighPerFlex configuration.

4 Discussion of Numerical Results

First, the Euler simulations with a HPF7 clean wing are performed. After the incorporation of pylon and nacelle the steady interference effects are studied. The influence of the longitudal position of pylon-nacelle cluster with respect to the wing (wing-nacelle overlapping, DX, Figure 2), the nacelle yaw angle (toe-in/-out angle, Figure 1) and the wing sweep angle are investigated. The calculations are completed with the analysis of the influence of flow parameters such as Mach number and angle of attack.

Effect of Pylon-Nacelle Cluster

Results obtained with the complete generic HighPerFlex configuration (Figure 6) showed, when compared to the clean wing results (Figure 5), strong acceleration of the flow on the lower wing side. The effect gets stronger with the increase of wing-nacelle overlap. Due to channel effects in the wing-pylon-nacelle region local supersonic region occurs closed by a relatively strong shock wave. Flow on the upper wing side, in contrast, is only slightly affected.

Influence of Geometrical Parameters

The initial HighPerFlex geometry is defined with a 10 deg sweep angle wing. The overlapping length between the wing and the nacelle (parameter DX, Figure 2) is the first parameter that is varied and investigated. Three values are chosen: no overlapping (DX=0), 15% of chord length overlapping (DX=0.15*chord) and a quarter-chord overlapping (DX=0.25*chord). For these three cases the flow conditions are the same (M=0.82, $\alpha = 0.6$ deg). Both codes are successful in capturing important flow features, especially in the critical region between nacelle, pylon and wing lower surface (Figure 6). On the upper wing surface, the variation of the position of the pylon-nacelle with respect to the wing shows no significant influence. Therefore, the discussion is focused on the wing lower surface where results show that the geometry with 0.25*chord overlapping generates the strongest shock wave in the region of pylon-nacelle cluster (Figure 6e and 6f). The differences in the suction peak on the lower surface between both codes are due to the different pylon shape (thin vs. thick pylon). This geometry is chosen as a starting point for the variation of next parameter, the nacelle-pylon yaw angle. Three yaw angle values are investigated, i.e., the parameter value is varied between +3 deg (toe-in), 0 deg and, -3 deg (toe-out). Results show that the toe-out nacelle-pylon angle (reduction of the angle between wing and nacelle axes, Figure 7a) reduces the strength of shock wave and it causes the shock wave to move to the other side of the pylon. The toe-in angle (increase of the angle, Figure 7b), however, increases the interference effect leading to a stronger shock wave on the original side of the pylon. Last parameter to be investigated is the wing sweep angle. Swept wing in the original HighPerFlex configuration is replaced with the rectangular wing of the same profile. This new (modified) HighPerFlex geometry, with the constant 0.25*chord wing-nacelle overlapping parameter, is used to repeat the calculations with the varying nacelle-pylon yaw angle. Three different yaw angle values are, of course, not necessary because the toe-in and toe-out cases are symmetrical when a rectangular wing is employed. Results (Figure 8) show that the general effect of wing sweep angle parameter on the wing-nacelle-pylon interference can successfully be reproduced with the rectangular wing and the -/+3 deg nacelle-pylon yaw angle although the flow is by no means physically equal. The yaw angle completely compensates for the absence of the sweep angle (compare Figure 8d to Figure 6f). Combining the experimental constraints (see Section 1), the complexity of wind tunnel model and the calculations results, the geometry with zero sweep angle, 0.25*chord wing-nacelle overlapping and 3 deg yaw angle represents the optimal choice and is therefore retained as the final HighPerFlex geometry for the experimental model.

Influence of Flow Parameters

Finally, the investigation of influence of flow parameters is performed with the DLR-AE EUVIMB code. Calculations with two different values of Mach number and angle of attack are performed with the modified HighPerFlex geometry (sweep=0 deg, DX=0.25*chord, nacelle yaw=0 deg). Figure 9a shows the c_p section distributions for two different Mach number values, 0.82 and 0.86. As expected, the wing lower surface supersonic region for a higher Mach number is larger and is terminated with the stronger shock wave. The influence of the variation of incidence angle is illustrated in Figure 9b. The shock wave in the wing-nacelle-pylon region is stronger when the value of angle of attack is smaller. On the upper wing surface the reverse phenomenon occurs, i.e. weaker shock corresponds to the smaller angle of attack.

5 Conclusions

Within a DLR-ONERA cooperation an investigation of the wing-pylon-nacelle interference effects on the non-linear aeroelastic phenomena is under way. In the introductory investigation two CFD codes were employed to simulate the interference effects on the steady flow field. During intensive numerical simulation campaign values of several important geometrical parameters were varied and their influence was investigated. As a result of these calculations the geometrical parameters of the wind-tunnel model and the typical flow values for the forthcoming experimental research were defined. Both codes, multi-block structured and unstructured, were run in the Euler mode and the analysis of results showed good agreement (taking into considerations simplified pylon shape in the structured approach) in predicting surface pressure distributions. The profile shape of the pylon is of course important but does not dominate the desired interference effects. It has been shown that the strength of local supersonic region on the wing lower side is strongly influenced by the wing/nacelle overlapping and the nacelle yaw angle parameters. The latter is especially important because it can yield similar interference effects like the wing sweep angle (which must be kept small in the planned wind tunnel test for technical reasons). Based on these results the configuration with zero wing sweep angle, DX=0.25*chord, and the nacelle with the variable yaw angle is chosen as the HighPerFlex wind tunnel model for the forthcoming unsteady experimental investigation.

Acknowledgments

Authors would like to thank Dr. T. Gerhold and Dr. K. Weinman, DLR Institute of Aerodynamics, Goettingen, and Dr. A. Ronzheimer, DLR Institute of Aerodynamics, Braunschweig, for their help regarding the TAU calculations and the generation of Centaur grids.

References

[1] Brodersen, O., and Rossow, C.C., "Calculation of Interference Phenomena For a Transport Aircraft Configuration Considering Viscous Effects", 1993 European Forum: Recent Development and Applications in Aeronautical CFD, Bristol, UK, 1993

[2] McMorris, H., Centaur Online User's Manual - Centaur v5.0, USA, 2003 http://www.centaursoft.com

[3] Rossow, C.C., Godard, J.L., Hoheisel, H., and Schmitt, V., "Investigation of Propulsion Integration Interference Effects on a Transport Aircraft Configuration", AIAA/SAE/ASME/ASEE 28th Joint Propulsion Conference and Exhibit, USA, 1992

[4] Soda, A., "The Multi-Block Grid Generator Aengus_MB - Development and Apllications For Complete Transport Aircraft Configurations", DLR-IB 232-2001 C 05, Germany, 2001

[5] Tefy, T., "EUVIMB - The Euler Viscous-Inviscid Multi-Block CFD Code", DLR-IB to be published in 2003

[6] Tijdeman, H., "Investigation of the Transonic Flow Aroud Oscillating Airfoils", PhD Thesis, NLR TR 77090 U, Netherlands, 1977

[7] Weinman, K., et al, "Tau-Code User Guide", Release Feb 18, 2003, DLR, Germany http://www.taurus.dlr.de/taurus_main.htm

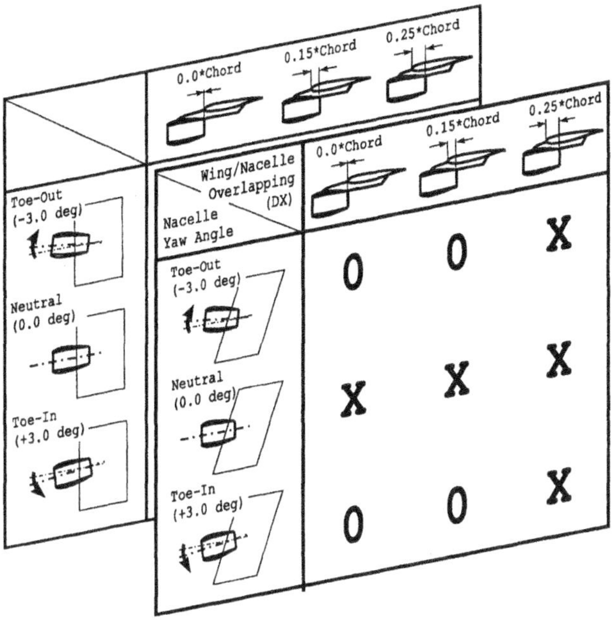

Figure 1 3-D matrix of computational test cases with variation of geometric parameters. X-marks show combinations of parameters (different geometries) used in calculations.

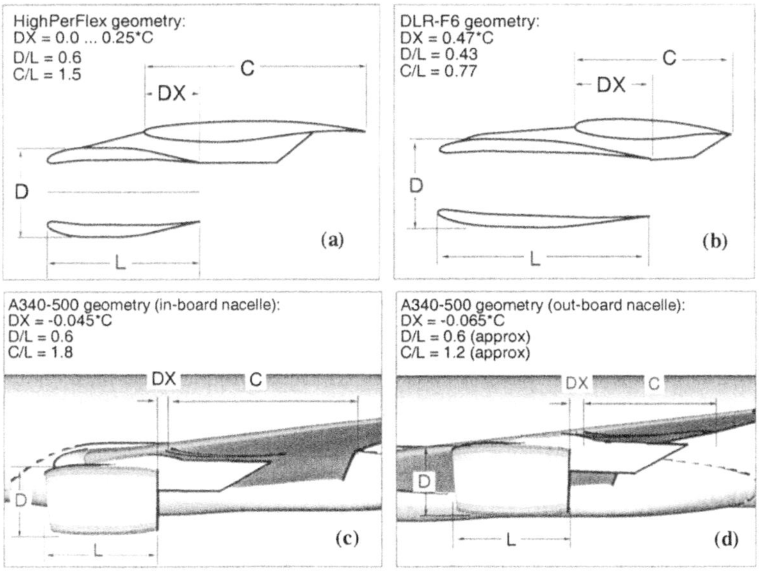

Figure 2 Comparison of geometrical parameters for HighPerFlex geometry (a) and two standard geometries (b, c, d).

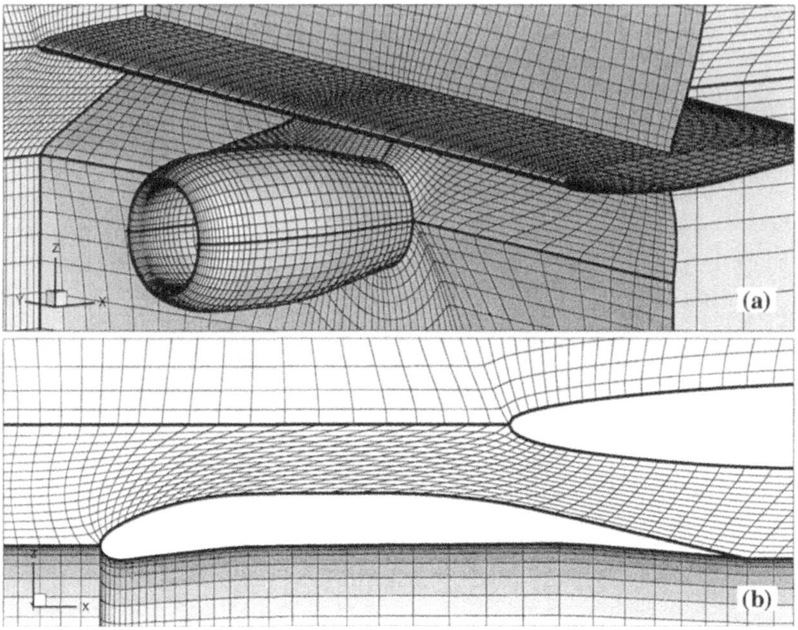

Figure 3 Structured grid about the HighPerFlex configuration (500,000 grid nodes). Cut through the field mesh (a) and side view of the wing/nacelle leading edge (b) are shown.

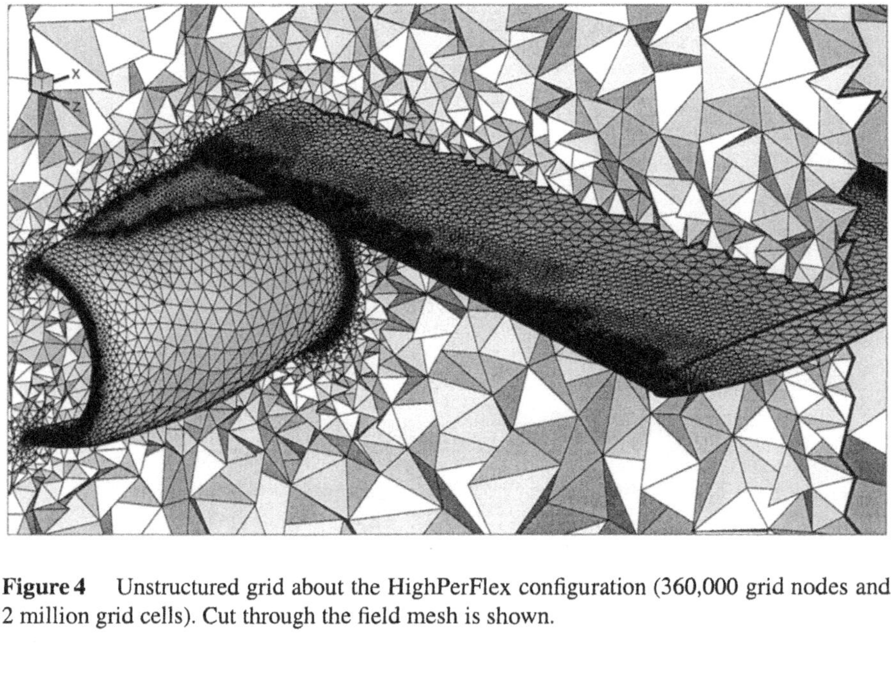

Figure 4 Unstructured grid about the HighPerFlex configuration (360,000 grid nodes and 2 million grid cells). Cut through the field mesh is shown.

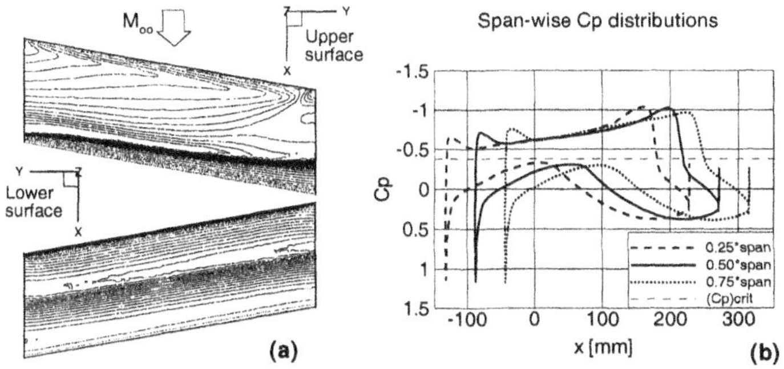

Figure 5 Clean wing (HPF7 aerofoil) configuration. Shown are Cp distributions on both wing surfaces (a) and three span-wise sections (b). Results obtained with the TAU code. Case: sweep=10 deg, Ma=0.82, incidence=0.6 deg.

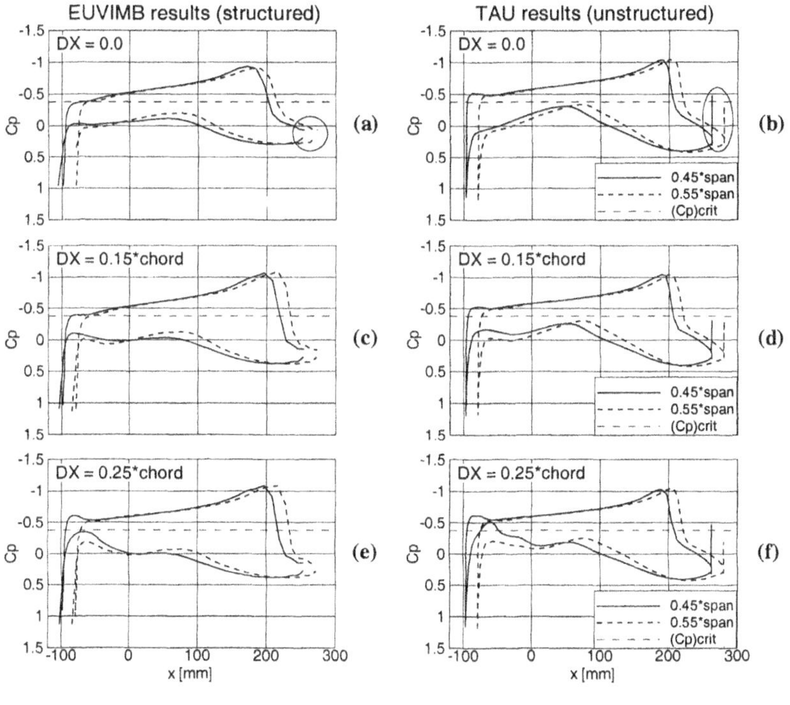

Figure 6 Results for initial HighPerFlex configuration. Cp distributions on two span-wise cut-planes for three different values of the wing-nacelle overlapping (DX) geometrical parameter. Simulation with two CFD codes. Case: sweep=10 deg, Ma=0.82, incidence=0.6 deg, nacelle yaw=0 deg. (In the EUVIMB results the missing interpolation of Cp values over the block boundaries is marked with a circle. In the TAU results physical effects of the wing blunt trailing edge are marked with an ellipse.)

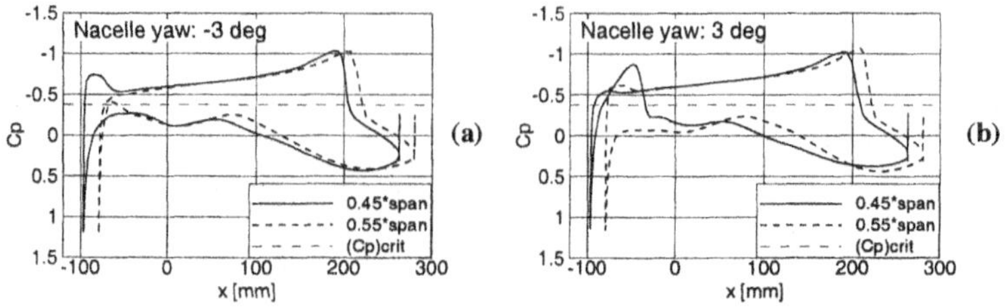

Figure 7 Initial HighPerFlex configuration. Cp distributions on two span-wise sections for two different values of the nacelle yaw angle geometrical parameter. Results obtained with the TAU code. Case: sweep=10 deg, Ma=0.82, incidence=0.6 deg, DX=0.25*chord.

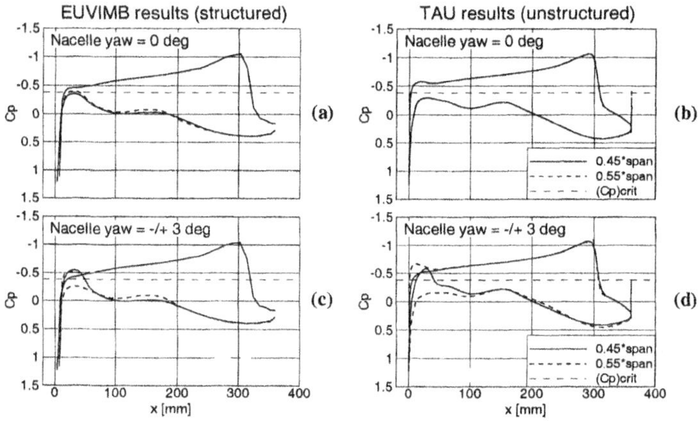

Figure 8 Modified HighPerFlex configuration. Cp distributions on two span-wise sections for three different values of the nacelle yaw angle geometrical parameter. Results obtained with two CFD codes. Case: sweep=0 deg, Ma=0.82, incidence=0.6 deg, DX=0.25*chord.

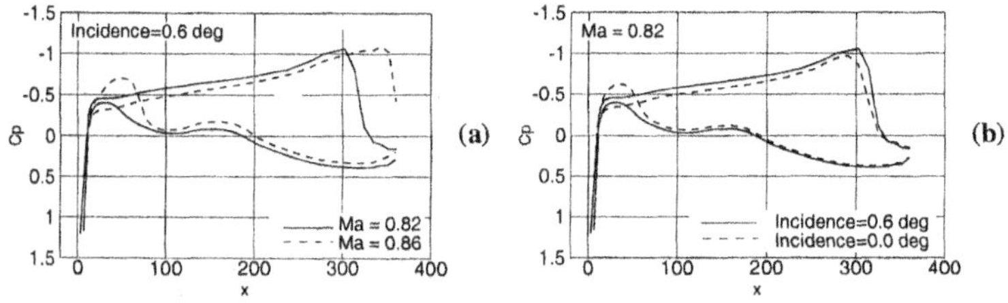

Figure 9 Modified HighPerFlex configuration. Cp distributions on a single 0.55*span section (because of geometrical symmetry the 0.45*span results are identical) for variation of two flow parameters, Ma (a) and incidence (b). Results obtained with the EUVIMB code. Case: sweep=0 deg, DX=0.25*chord, nacelle yaw=0 deg.

Hybrid Navier–Stokes Calculations for Transport Aircraft with Conventional and High–Bypass Engines

O. Brodersen[1], R. Hain[2], and H. Frhr. von Geyr[1]

[1] Research Engineer, German Aerospace Center DLR, Institute of Aerodynamics and Flow Technology, Lilienthalplatz 7, 38108 Braunschweig, Germany

[2] PhD Student, Technical University of Braunschweig, Institute of Fluid Mechanics, Bienroder Weg 3, 38106 Braunschweig, Germany

Summary

A fast and accurate performance prediction including the computation of drag is a major objective for the development of computational fluid dynamics software for aircraft design. It is demonstrated that an unstructured, hybrid method is capable to compute the airframe–engine installation drag for several engine types although drag differences for varying engine positions can be very small. Results are presented for the DLR–F6 and ALVAST configurations with through–flow nacelles and powered engines. Verification by grid refinement as well as validation with wind tunnel data completes this investigation.

1 Introduction

An important objective for the development of new civil transport aircrafts is to increase its efficiency. Ecological aspects such as noise emissions and air pollution will influence the costs for start and landing. The integration of engines with higher bypass ratios is one possibility to fulfill the upcoming requirements because of their reduced specific fuel consumption and lower noise emissions. The tendency to increase the bypass ratio and engines sizes can result in a higher lift loss of the wing and can have a non–negligible impact on the engine installation drag [7, 8].

Another important aspect for the success of a new aircraft design is the design process itself because decisions at the beginning of a project strongly influence the design over its complete life span [15]. A fast and accurate development reduces costs and can improve the efficiency of an aircraft design.

In the past performance predictions have been accomplished generally by experimental investigations because of the long experience with those techniques and the good results. A numerical approach solving the Reynolds–Averaged Navier–Stokes (RANS) equations faces the difficulties that suitable grids, accurate solution algorithms, adequate turbulence modeling, and fast computers are necessary. Structured grids can fulfill the discretisation requirements, but due to a lacking of automatic algorithms the grid generation is very time–consuming. An unstructured, hybrid method promises a very fast semi–automatic grid generation but until now distrust exists whether hybrid techniques can compute drag with a sufficient accuracy.

Since the beginning of 1990 several investigations in the field of airframe-engine integration have been carried out in Europe. Together with ONERA the DLR has used the DLR–F6 configuration with through–flow nacelles to understand the basic transonic flow phenomena [12]. As a follow–on step the DLR–ALVAST model with turbo–powered engine simulators has been used in European projects [2, 7, 11]. Besides wind tunnel tests numerical calculations have been performed solving the Euler and RANS equations using structured techniques. Because of the high effort for structured grid generation only results for selected configurations had been achieved. Effects on the installation drag due to engine position variations could not be computed. Nevertheless it had been clearly shown that the methods are capable to predict the main interference effects and flow phenomena. During the last years DLR has demonstrated that airframe–engine installation effects can be computed with a good accuracy using the MEGAFLOW software [1, 10, 13]. The calculations presented here are focused on installation drag predictions for the F6 configuration with different through–flow nacelles and for ALVAST with different powered engines. The verification of the results will be accomplished by a grid refinement study. Experimental data is available for validation.

2 F6 and ALVAST Configurations

The F6 and the ALVAST configurations represent a twin engine wide body aircraft. The wing aspect ratio is $\Lambda = 9.5$ and the leading edge sweep angle is $\Phi = 27.1°$. For both configurations the cruise flight conditions are $Ma_\infty = 0.75$ and $c_L = 0.5$ which are relevant for the experiments and the simulations considered here. For the F6 model three types of nacelles are available: a conventional CFM56 type with long (CFM-L) and short duct (CFM-S) and a VHBR nacelle. Three different nacelle positions for the CFM-L, one position for the CFM-S, and two positions for the VHBR nacelle have been tested in the wind tunnel. The positions considered here are for CFM-S: $x_{TE}/l = -0.04$, $z_{TE}/l = -0.16$, for VHBR-1: $x_{TE}/l = -0.05$, $z_{TE}/l = -0.12$, and for VHBR-2: $x_{TE}/l = 0.0$, $z_{TE}/l = -0.12$. The variables x_{TE}/l and z_{TE}/l are based on the local chord length and represent the position of the upmost point of the nacelle trailing edge related to the wing leading edge. Figure 1 displays the model with the VHBR nacelles in the S2MA wind tunnel of the ONERA. The ALVAST configuration is similar to F6 but larger [9]. The engine simulators TF, VHBR, and UHBR have bypass ratios of 6.0, 9.2, and 15.7, respectively. Figure 2 shows the half model in the ONERA S1 wind tunnel with the UHBR engine.

3 Experiment

For both configurations several test campaigns have been performed in the 90s. The configurations have been equipped with pressure tabs. The wind tunnel tests for both configurations have been performed for varying Mach numbers and lift coefficients. Here, only the cruise flight conditions (Ma=0.75, $c_L = 0.5$) and engine conditions at the start of cruise flight (SOC) are of interest. The fan pressure and temperature ratios can be found in [11]. Transition strips are used for both models to fix the transition. The Reynolds numbers are 3.0 and 4.3 million for F6 and ALVAST, respectively.

4 Numerical Solution Method

For the solution of the RANS equations the DLR TAU software is used [4, 5]. TAU is a finite-volume vertex-based solver with an edge-based data structure. Together with a dual-grid technique the usage of different element types, e.g. tetrahedrals, prismatic cells, pyramids, and hexahedrals is possible. The discretisation of the convective fluxes has been realized with a second order central difference scheme with scalar dissipation which will be used for these investigations. Time integration to steady state is accomplished with a three-stage Runge-Kutta scheme. For convergence acceleration local time stepping, residual smoothing, multigrid, and MPI based parallelization is available and used here. The Spalart–Allmaras turbulence model with Edwards modification (SAE) is applied for these investigations [14]. The conditions for the powered engines are simulated by prescribing an inlet mass flow and the fan and core pressure and temperature.

The density and quality of grids are of significant importance for solution accuracy. Therefore it is necessary to have a grid generation system that is on the one hand able to generate high quality hybrid grids and on the other hand it should perform this as much automatic as possible. DLR uses the software Centaur from CentaurSoft [3]. 24 prismatic layers have been generated and about 10–16 grid points can be found in the boundary layer. To obtain similar discretisation errors it is of great importance that the initial grids for the different configurations are generated by using exactly the same control parameters. Nevertheless it can not be completely avoided that the grids show some differences. Therefore the grids will be adapted so that initial differences and the overall discretisation errors will be reduced. The final grids have a similar number of points. Table 1 presents the grid densities including the adapted final grids. A solution based grid adaptation is available to refine elements and to move prismatic layers normal to the surfaces to obtain a $y^+ = 1.0$. Here a combination of total pressure and velocity gradients have been used as a refinement sensor. Figure 3 presents exemplarily the adapted surface grid of ALVAST with UHBR engine.

Table 1 F6 and ALVAST (A) grid points [millions]

	F6-WB	F6-CFM-S	F6-VHBR-1	F6-VHBR-2	A-WB	A-TF	A-VHBR	A-UHBR
initial	2.58	4.03	4.70	4.55	3.02	3.80	4.20	5.15
final	5.52	8.72	8.46	8.19	6.16	6.97	7.60	9.31

5 Results

For the F6 configuration calculations for target lift coefficients between 0.2 and 0.5 at Ma=0.75, Re=$3 \cdot 10^6$ have been used to compare experimental and numerical results. The adaptation has only a small influence on the pressure distributions but a slightly larger one on the total drag as demonstrated in Fig. 4 for all F6 configurations. The figure shows a decrease of drag when grids are refined. The change of the drag coefficients is reduced from a coarse to a fine adaptation level. While the discretisation errors can be reduced using finer grids a systematic deviation between the computed and the measured drag values of 5-7% still exists [1]. Possibly this is influenced by various parameters, e.g.: wind tunnel data correction, wing deformation, turbulence modelling, and discretization. This observed systematic deviation nearly cancels out when the installation drag is computed. Therefore the small differences of the drag due to the different engines and their positions can be successfully computed. Figure 5 demonstrates the results for F6 with CFM-S and VHBR nacelles. For CFM-L nacelles the agreement of experiments and computations is even better [1]. The Fig. 6 and 7 demonstrate that besides the coefficients also the flow physics are captured topologically correct looking at near–wall streamlines.

The accomplishment of computations and experiments for powered configurations is more complicated and more sensitive for errors in drag prediction because thrust and drag have to be separated. This is difficult in the experiments because the thrust can not be measured directly and has to be determined from additional calibration tests. Details can be found in the literature with a special emphasis on the correct analysis of experimental data [6].

The quality of the pressure distributions in comparison with the experiments and the structured results from FLOWer can be evaluated in Fig. 8 and 9 [11]. The TAU solution shows a good agreement with the experiments for the ALVAST configurations. Compared to the structured results an improved shock resolution is visible. This is mainly caused by the grid adaptation although the different turbulence models also have a small influence. A discrepancy of the TAU results and the experiments is visible only for UHBR at the engine inboard section on the wing lower surface as plotted in Fig. 10. The outboard section shows again a good result in Fig. 11. The flow acceleration at the engine inboard side on the wing lower surface is overpredicted as it has been detected also for one F6 case [1]. In the rear section the flow tends to separate in the experiments and a difference compared to the computation is visible. A small separation can also be found in the TAU results when looking at the near–wall streamlines in Fig. 12. The thrust and drag book–keeping method is applied to ALVAST, that means that the drag contributions of the engine surfaces that are inside the engine ideal inflow and outflow streamtubes are subtracted from the thrust [6]. The computed drag for UHBR is $c_D = 0.0358$ while $c_D^{Exp} = 0.03465$. The installation drag for all ALVAST configurations is computed higher than measured. Additional grid refinements would reduce the differences between numerical and experimental installation drag as it has been shown for F6 [1].

6 Conclusion

Due to its good handling qualities, turn–around times and solution accuracy the unstructured method in combination with a semi–automatic grid generation is very well suited for performance predictions in an aircraft design process. A confidence has been build up at DLR that the method is capable to compute small drag differences of aircraft design variations while a systematic deviation still exists when results are compared to experiments. It is very important that adequate grids are used and that grid refinements are applied. Special attention has to be given to the determination of thrust and drag components when powered engines are used.

References

[1] Brodersen, O., Stürmer, A.: "*Drag Prediction of Engine–Airframe Interference Effects using Unstructured Navier–Stokes Calculations*". AIAA Paper 2001–2414, (2001)

[2] Burgsmüller, W., Rollin, C., Rossow, C.–C.: "*Engine Integration on Future Transport Aircraft – The European Research Programs DUPRIN/ENIFAIR*". ICAS Paper 98–5.6.2 (1998)

[3] CentaurSoft: "*Centaur*". http://www.centaursoft.com (2002)

[4] Galle, M.: "*Ein Verfahren zur numerischen Simulation kompressibler, reibungsbehafteter Strömungen auf hybriden Netzen*". Promotion Universität Stuttgart, DLR–FB 99–04 (1999)

[5] Gerhold, T., Evans, J.: "*Efficient Computation of 3D–Flows for Complex Configurations with the DLR–Tau Code Using Automatic Adaptation*". Notes on Numerical Fluid Mechanics, Vol. 72, Vieweg Braunschweig, (1998), pp. 178–185

[6] Frhr. v. Geyr, H., Rossow, C.–C., Hoheisel, H.: "*Influence of transonic Flow on the Thrust Determination of TPS during Wind Tunnel Measurements*". to be published STAB (2002)

[7] Hoheisel, H.: "*Aerodynamic Aspects of Engine–Aircraft Integration of Transport Aircraft*". Aerospace Science and Technology, Vol. 1, No. 7 (1997), pp. 475–487

[8] Hurez, A.: "*Recent Progress on Powerplant / Airframe Integration at Aerospatiale Matra Airbus*". Proc. of Workshop on EU–Research on Aerodynamic Engine/Aircraft Integration for Transport Aircraft. DLR Braunschweig, (2000), pp. 2.1–2.10

[9] Kiock, R.: "*The ALVAST Model of DLR*". DLR, IB 129–96 / 22 (1996)

[10] Kroll, N., Rossow, C.–C., Schwamborn, D., Becker, K., Heller, G.: "*MEGAFLOW – A Numerical Flow Simulation Tool for Transport Aircraft Design*". ICAS Paper, (2002)

[11] Laban, M.: "*Aircraft Drag and Thrust Analysis, Project Overview and Key Results*". Proc. of Workshop on EU–Research on Aerodynamic Engine/Aircraft Integration for Transport Aircraft. DLR Braunschweig, (2000), pp. 9.1–9.15

[12] Rossow, C.-C., Godard, J.-L., Hoheisel, H., Schmitt, V.: "*Investigation of Propulsion Integration Interference on a Transport Aircraft Configuration*". AIAA Paper 92–3097 (1992)

[13] Rudnik, R., Rossow, C.-C., Frhr. von Geyr, H.: "*Numerical simulation of engine/airframe integration for high–bypass engines*". Aerospace Science and Technology, Vol. 6, No. 1 (2002), pp. 31–42

[14] Spalart, P., Allmaras, S: "*A One–Equation Turbulence Model for Aerodynamic–Flows*". AIAA Paper 92–0439 (1992)

[15] Szodruch, J., Hilbig, R.: "*Building the Future Aircraft Design for the Next Century*". AIAA Paper 98–0135 (1998)

Figure 1 F6 model wih VHBR nacelles in ONERA S2MA wind tunnel

Figure 2 ALVAST model with UHBR engine in ONERA S1 wind tunnel

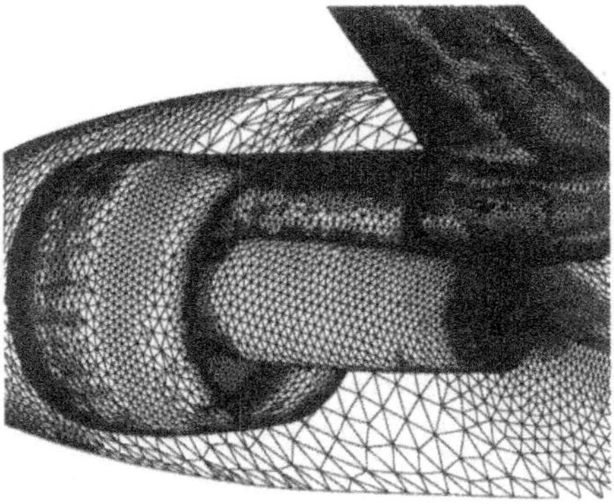

Figure 3 ALVAST–UHBR: adapted surface grid (final)

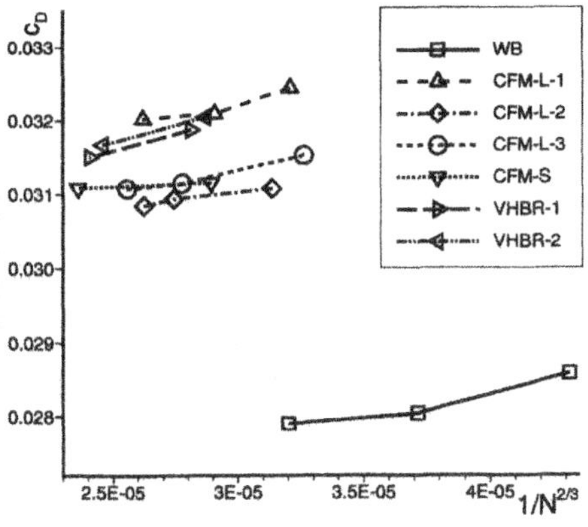

Figure 4 Influence of grid density on the drag coefficient

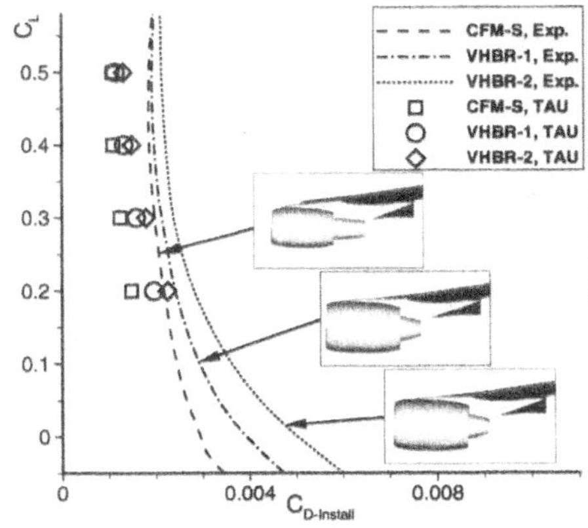

Figure 5 Calculated and measured engine installation drag coefficients

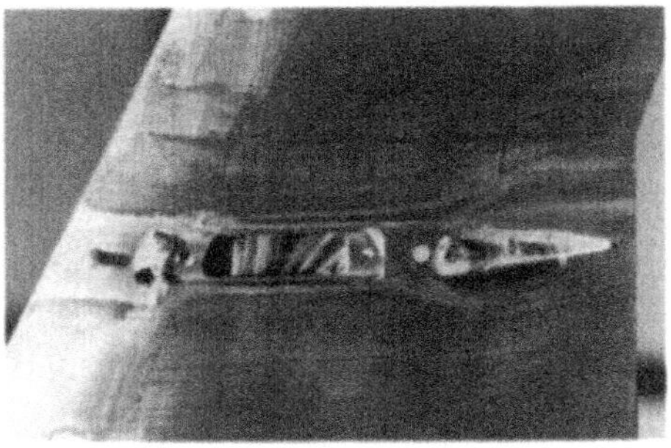

Figure 6 F6 VHBR-2 wind tunnel oil flow pattern, $c_L = 0.5$

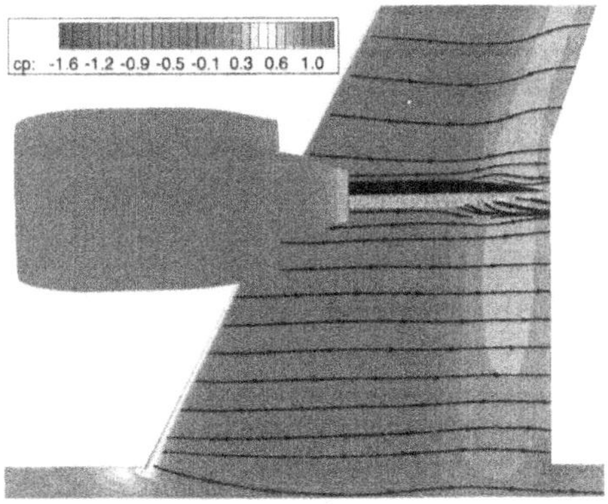

Figure 7 F6 VHBR-2 pressure distribution, near wall streamlines, $c_L = 0.5$

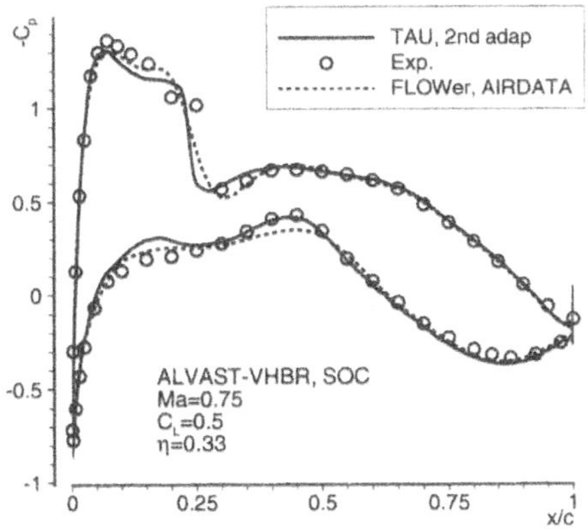

Figure 8 ALVAST-VHBR pressure distribution inboard of engine, Ma=0.75, $c_L = 0.5$

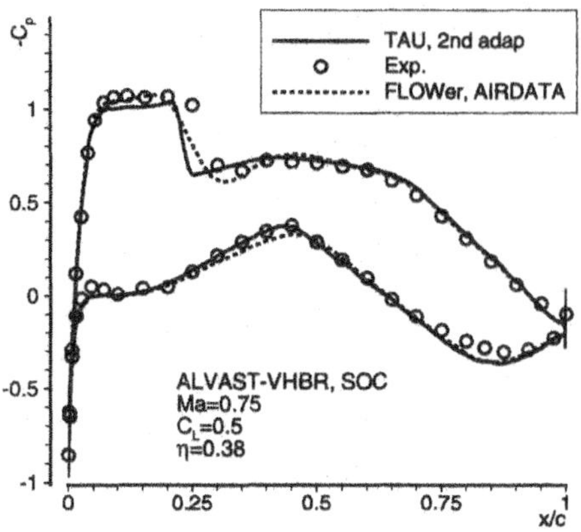

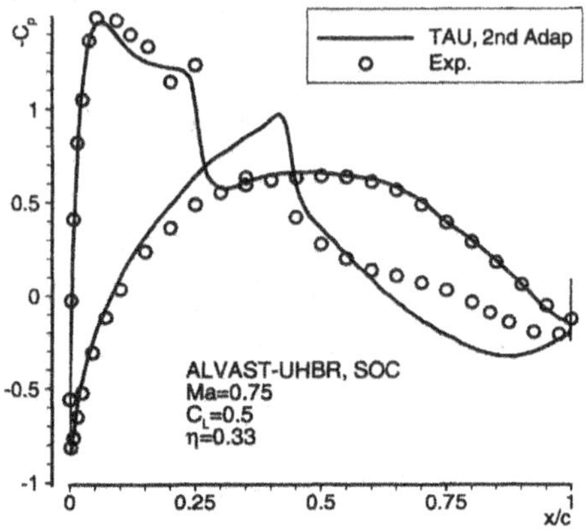

Figure 9 ALVAST-VHBR pressure distribution outboard of engine, Ma=0.75, $c_L = 0.5$

Figure 10 ALVAST-UHBR pressure distribution inboard of engine, Ma=0.75, $c_L = 0.5$

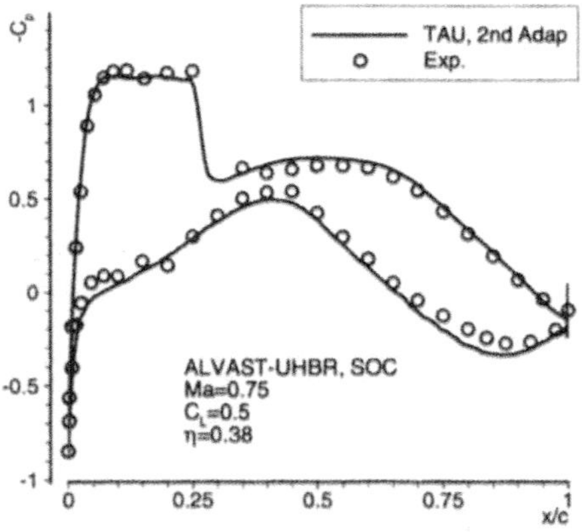

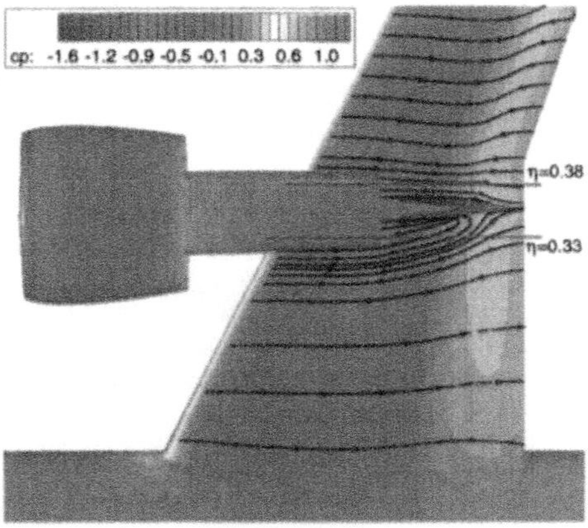

Figure 11 ALVAST-UHBR pressure distribution outboard of engine, Ma=0.75, $c_L = 0.5$

Figure 12 ALVAST-UHBR surface pressure distribution, near–wall streamlines, wing lower side, Ma=0.75, $c_L = 0.5$

DLR τ-Code for High Enthalpy Flows

Bodo Reimann, Ian Johnston and Volker Hannemann

German Aerospace Center
Institute of Aerodynamics and Flow Technology
Bunsenstraße 10
37073 Göttingen
bodo.reimann@dlr.de

Summary

The development of an equilibrium chemistry model for the DLR τ-Code has extended its ability to simulate high enthalpy flows, such as those around reentry vehicles or within high speed test facilities. Details of the development of the chemistry model, and some test cases, are included in this paper. Also presented is an improved axisymmetry treatment for the code, aimed at producing accurate and noise-free solutions for common hypersonic applications such as blunt bodies and shock tunnels.

1 Introduction

The excitation of inner degrees of freedom of molecules and atoms, as well as chemical reactions, play an important role in shaping high enthalpy flows. The limiting case of equilibrium chemistry, where reactions are assumed to occur instantaneously, provides a good approximation for many practical high speed flow cases. Because the complete thermodynamic state of an equilibrium gas is defined by two state variables, there exists the possibility of using a two dimensional interpolation function to approximate its state surfaces. This is a very efficient method for CFD applications, and its development, implementation in the unstructured DLR flow solver τ-Code [3], and test cases are presented in this paper.

Axisymmetric geometries, such as cones and spheres, are frequently used as test cases for basic research in hypersonic flows, and are particularly useful for the validation of CFD against experimental results. Additionally, many hypervelocity test facilities, such as the High Enthalpy Shock Tunnel HEG [5], are axisymmetric in design and produce essentially axisymmetric flow fields. Thus, for modelling hypersonic flows it is of especial importance that the τ-Code is equipped with an accurate, reliable and efficient axisymmetric formulation that works well with upwind-type schemes and with the newly implemented equilibrium chemistry model.

2 Thermochemical Equilibrium Modelling

We consider the equilibrium flow as a mixture of perfect gases consisting of S chemical species, which can be formed from E independent elements according to the following reaction equation,

$$\mathrm{Y}_s \rightleftharpoons \sum_e z_{se}\, \mathrm{X}_e$$

where Y_s and X_e denote the species and elements respectively. The number of elements X_e contained in species Y_s is given by z_{se}. Given that all species are expressed in terms of elements, and the total quantity of each element must remain constant, the S species concentrations ξ_s must satisfy the E element conservation laws

$$\sum_s z_{se}\, \xi_s = \bar{\xi}_e\,, \qquad e = 1, ..., E \tag{1}$$

with a fixed number of elements $\bar{\xi}_e$. Normally the atoms and electrons — if ionisation is taken into account — are used as the independent elements, although it is also possible to use molecules instead.

If we consider the entropy s as a function of density, energy and the species concentrations ξ_s, the change in entropy is given by Liu and Vinokur as [6]

$$T\, ds = de + \frac{p}{\rho^2}\, d\rho - \sum_s g_s\, d\xi_s\,, \tag{2}$$

where g_s denotes the chemical potential of species s, and p is the mixture pressure calculated from the partial pressures of the species using Dalton's law. Thermochemical equilibrium is defined by a maximum of entropy or, equivalently, when no irreversible entropy (the last term in Eq.(2)) is generated by further chemical reaction:

$$\sum_s g_s\, d\xi_s = 0\,. \tag{3}$$

The terms $d\xi_s$ are not all independent. By splitting Eq.(1) and Eq.(3) into E elementary and $N = S - E$ non-elementary species, and substituting the terms $d\xi_s$ in Eq.(3), we can write:

$$g_n - \sum_e z_{se}\, g_e = 0\,, \qquad n = 1, ..., N\,. \tag{4}$$

The N Eqs.(4), the E element conservation Eqs.(1), and the defining equation for the energy $e = \sum_s \xi_s\, e_s$ together determine the S unknown $d\xi_s$ values and the temperature T. To evaluate the required internal energy and entropy distribution for each species we use the partition function described by Esser [2]. The system can then be reduced by substitution to $E + 1$ equations for the unknown ξ_e and T, for a given density and energy. For other combinations of variables, such as enthalpy and entropy, the system is reduced to $E{+}2$ equations. The resulting system can be solved using Gaussian elimination and Newton iteration, for which it is advantageous to work in terms of the logarithms of the unknowns [6].

Data given by Gupta [4] are used for the transport properties of the pure species. To obtain mixture properties, we apply Wilke's rule [7] for viscosity and the Zipperer and Herning [8] approximation for thermal conductivity. The transport of energy by diffusion is taken into account by a constant species Schmidt number of 0.7.

Piecewise Polynomial Interpolation

Iterating towards the equilibrium state for each grid point in a flow simulation at run-time would be computationally prohibitive. To more efficiently calculate the thermodynamic state variables and transport coefficients we use a two dimensional spline fit to approximate each variable f as a function of energy e and density ρ. The thermodynamic range of interest is divided into a number of patch sections; linearly spaced divisions are used for the energy variable whereas a logarithmic scale is applied for density. For each patch, the size of f and its partial derivatives are precomputed and stored in a table.

For the calculation of the conservative fluxes from the primitive variables in the τ-Code, internal energy is required as a function of density and pressure. To achieve this, a starting guess for energy is obtained from a set of additional spline patches expressed in terms of density and pressure. Subsequently, Newton iterations are performed on the original ρ-e patches to ensure consistency is maintained. The use of such an iterative technique requires that each patch is attached to its neighbours continuously and that the approximation is monotone for the state surfaces.

Figure 1 shows a state surface, represented by the sampling points, for temperature T and thermal conductivity κ for a 5 species air mixture (N_2, O_2, NO, N, O) as a function of density and energy with a resolution of 40×20 sampling points.

3 Axisymmetry Axis Treatment

The currently implemented approach for treating axisymmetric flow fields is to model a three-dimensional 'wedge' of small included angle α, which represents a fraction of the full 360 degree flow field. A periodic boundary condition is then applied to corresponding sides of the wedge. A diagram of the concept is shown in Figure 2. Although less efficient than using a two-dimensional formulation with axisymmetry source-terms, this approach requires very little modification to the existing three-dimensional code.

Unfortunately the described technique is subject to a number of problems: (i) On structured grids, a second-order unstructured or one-sided reconstruction of primitive variables is performed near the axisymmetry axis and adds excessive noise to the stagnation region, (ii) The τ-Code uses a dual-grid discretisation scheme which introduces a reconstruction error at the axisymmetry axis and several other boundaries, the magnitude of the error corresponding to a quarter-cell width spatially, (iii) The use of periodic boundary conditions for flow fields not containing whirl needlessly increases the computational cost, (iv) Reconstruction performed in the circumferential direction causes oscillations in the flow field results, and (v) At present the grid adaptation module is not able to properly refine axisymmetric grids consisting of a mixture of hexahedras and prisms. In the course of the present work, all of these issues have been resolved for the case of an upwind scheme. The solution methods are now described.

Shown in Figure 2 is an example of an axisymmetric primary grid for use with τ. Dotted lines indicate the edges of dual-grid control volumes, upon which the finite-volume numerical scheme is applied. A very small separation, of the order of machine precision, has been introduced between neighbouring primary-grid points on the axisymmetry axis. This helps avoid a numerical singularity, and renders cell geometries compatible for use with the τ adaptation module.

The upwind second-order reconstruction scheme used in τ implicitly assumes that the volumetric centroids of dual-grid cells are coincident with primary grid vertices. Although this assumption is generally accurate, the figure shows that cells lying on boundaries are subject to a spatial reconstruction error equal to one-quarter of the corresponding primary-grid cell width. For many boundary conditions such an error has marginal effect on solution quality. However, this error has been observed to significantly disturb symmetry and solution quality in the case of an axisymmetry axis boundary condition.

To work around the issue without major modifications to the τ discretization scheme or code, we implemented a reconstruction and flux-update technique using the idea of a virtual cell. Figure 2 shows a two-dimensional simplification of the primary- and dual-grids near the axis of symmetry, viewed in the circumferential direction. The virtual dual-grid cell with centre p_v is visualised as a reflection about the axisymmetry axis of the internal dual-grid cell with centre p_1. The flow states at p_1 and p_v are constrained to be equal, except for velocity where a mirror transformation is applied:

$$\mathbf{u}_{p_v} = (\mathbf{I} - 2\,\hat{\mathbf{r}}\hat{\mathbf{r}}^{\mathrm{T}})\mathbf{u}_{p_1} \,. \tag{5}$$

The already implemented second-order reconstruction method can then be directly used to calculate the left (L) and right (R) interface states between p_1 and p_2 using the four flow states p_v, p_0, p_1 and p_2 in a symmetric fashion and without a spatial error. An almost zero cell-face area at the axisymmetry axis precludes any fluxes and obviates the need to perform reconstruction there.

Introduction of the virtual cell can be thought of as effectively doubling the size of the dual-grid cell that formerly lay on the axisymmetry axis, such that it may now be visualised as straddling the axis (see the effective dual-grid edge in Figure 2). We can account for this, without having to modify the existing grid properties, by exploiting the symmetry. Simply removing flux components in the radial direction as calculated for the former (half-sized) dual-grid cell, has the same effect as performing a full flux balance over the enlarged dual-grid cell because any radial flux components must cancel. Thus, for example, we set

$$(\rho\mathbf{u})' = \rho\mathbf{u} - (\rho\mathbf{u}\cdot\hat{\mathbf{r}})\hat{\mathbf{r}} \quad , \tag{6}$$

where prime indicates the modified flux and ρ is density.

Finally, radial gradients of all state variables are zeroed on the axisymmetry plane, and velocity gradients on the axisymmetry axis are modified such that:

$$\boldsymbol{\nabla} u'_{x,y,z} = (1 - |r_{x,y,z}|)\boldsymbol{\nabla} u_{x,y,z} + (2|r_{x,y,z}| - 1)(\boldsymbol{\nabla} u_{x,y,z}\cdot\hat{\mathbf{r}})\hat{\mathbf{r}} \,. \tag{7}$$

This ensures that there exists no radial gradient for the axial component of velocity, and that there is likewise no axial gradient for the radial component of velocity along the axisymmetry axis.

Flow state reconstruction is not required in the circumferential direction due to axisymmetry, and so dual-grid faces of this orientation are flagged to save computational effort and to prevent noise being generated by the reconstruction.

Improved Euler Wall Boundary Condition

For axisymmetric flows without whirl, it is more efficient to apply an Euler (slip) boundary condition to the circumferential boundary planes rather than a periodic condition: the reduction in calculation time is approximately 40%.

The default implementation of the τ Euler wall condition assumes that wall pressure is equal to the pressure calculated for the nearest dual-grid cell. Due to the offset of the dual-grid cell centroid, as already noted, this may introduce inaccuracies where there exists a significant component of flow velocity normal to the wall. Such a wall-normal velocity component exists in the axisymmetry treatment by virtue of the angular separation between the slip-wall boundary and neighbouring dual-grid centroids. This is particularly problematic, because the pressure error can induce misalignment of the complete axisymmetric plane.

The pressure error may be reduced by recalculating wall pressure, taking into account the pressure change due to elimination of normal flow velocity. To maintain consistency with the default τ flux solver, we use a second-order AUSM-type pressure splitting to correct the wall pressure,

$$p_w = \begin{cases} p(m_n + 1)^2(2 - m_n)/2 & \text{if } |m_n| \leq 1 \\ p(m_n + |m_n|)/m_n & \text{if } |m_n| > 1 \end{cases}$$

where m_n is the Mach number component normal to the wall, p is the pressure at the dual-grid cell centroid, and p_w is the corrected pressure applied to the wall boundary condition. Momentum fluxes at the wall are subsequently updated using p_w.

4 Results

A test case of hypersonic air flow around a sphere was established to gauge the effectiveness of the new axisymmetry treatment. Results from an Euler calculation, using a perfect gas inflow at Mach 6, are presented in Figure 3. The results show a marked improvement using the new axisymmetry treatment. Noise in the stagnation region has been eliminated, and the downstream-propagating perturbations in the pressure field have also been cured. Pressure at the stagnation point differs from the analytical value by 7.44% for the old axisymmetry treatment, while this error is reduced to 0.04% using the new treatment. Similarly good results have been observed for viscous-flow test cases, and the same reconstruction technique has been successfully applied to the symmetry condition for two-dimensional cylinder flows.

The new equilibrium chemistry model was tested by simulating the hypersonic flow over a cylinder and sphere of radii $R = 0.045$ m. The results are compared with those from CEVCATS-N [1], a structured-grid Navier-Stokes solver, in the contour and stagnation streamline plots of Figures 4 and 5. The calculations were carried out with a free-stream Mach number of $M_\infty = 12$, a free-stream density of $\rho_\infty = 0.0017$ kg/m^3 and a free-stream temperature $T_\infty = 300$ K. The wall temperature T_{wall} was a constant 300 K. An equilibrium database approximated with 400×200 sampling points for density and energy, respectively, was used for the τ calculations. The same grid was used for both codes. The results produced by the two codes are generally in good agreement, with slight differences near the bow shock mainly caused by alternative approximate Riemann solvers used in both codes and small differences in the chemical models. The stagnation point heat transfer and pressure computed by the two codes differ by 4.13% and 0.25% for the cylinder, and 10.65% and 0.26% for the sphere, respectively. The deviation in heat transfer is due to a different modelling of the thermal conductivity.

5 Conclusions

An efficient equilibrium solver has been developed, implemented in the DLR τ-Code, and successfully tested for the case of high-enthalpy flow about a sphere and a cylinder. Results for the new equilibrium model are in good agreement with those produced by the existing CEVCATS-N code. Additionally, an improved treatment for axisymmetric flow simulations has been implemented in the τ-Code. For the simulation of hypersonic flow around a sphere, the improved treatment was shown to produce less noisy and more accurate flow solutions.

References

[1] S. Brück, R. Radespiel, J. M. A. Longo: Comparison of nonequilibrium flows past a simple Space-Shuttle configuration, AIAA 97-0275, 1997.

[2] B. Esser: "Die Zustandsgrößen im Stoßwellenrohr als Ergebnisse eines exakten Riemannlösers". Dissertation, RWTH Aachen, 1991.

[3] T. Gerhold, O. Friedrich, J. Evans, M. Galle: "Calculation of Complex Three-Dimensional Configurations Employing the DLR-TAU-Code", AIAA 97-0167, 1997.

[4] R. N. Gupta, J. M. Yos, R. A. Thompson, K.-P. Lee: "A Rewiev of Reaction Rates and Thermodynamic and Transport Properties for an 11-Species Air Model for Chemical and Thermal Nonequilibrium Calculations to 30000 K". Reference Publication 1232, National Aeronautics & Space Administration, 1990.

[5] K. Hannemann, M. Schnieder, B. Reimann, J. Martinez Schramm: "The Influence and the Delay of Driver Gas Contamination in HEG." AIAA 2000-2593, 2000.

[6] Y. Liu, M. Vinokur: "Equilibrium Gas Flow Computations. I. Accurate and Efficient Calculations of Equilibrium Gas Properties". AIAA 89-1736, 1989.

[7] C. R. Wilke: "A Viscosity Equation for Gas Mixtures". The Journal of Chemical Physics, 18(4), 1950, pp. 517–519.

[8] L. Zipperer, F. Herning: "Beitrag zur Berechnung der Zähigkeit technischer Gasgemische aus den Zähigkeiten der Einzelbestandteile". Das Gas- und Wasserfach, Vol. 4, 49 pp., 1936.

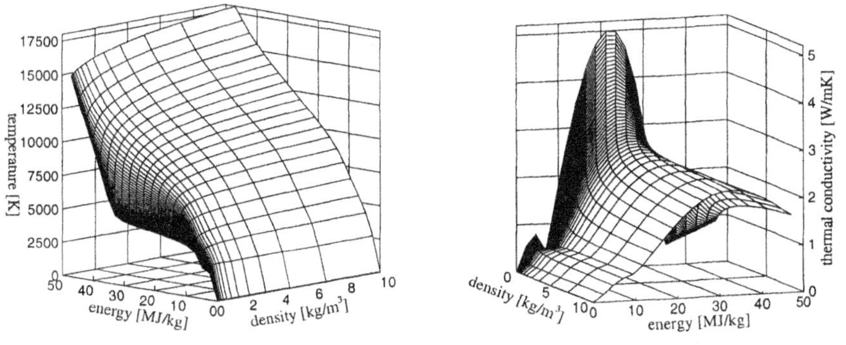

Figure 1 State surface for temperature T (left) and thermal conductivity κ (right) for a 5 species air mixture as a function of density ρ and energy e, using 40×20 sampling points.

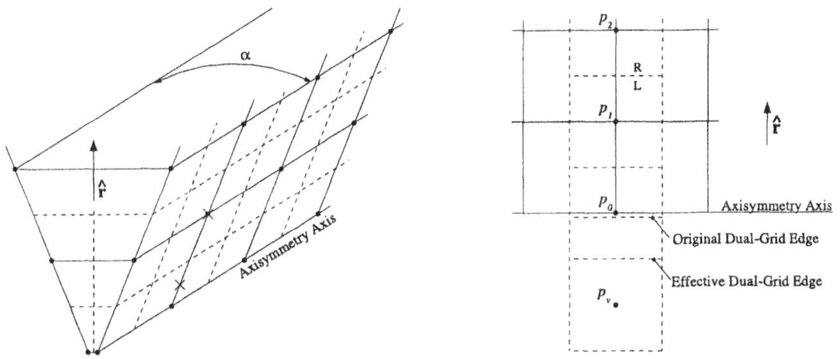

Figure 2 Three-dimensional wedge used to model axisymmetric flows (left), and schematic of the new axisymmetry treatment on the axisymmetry plane, as viewed in the circumferential direction (right). —— Primary grid; - - - Dual grid; • Primary grid vertex; × Projected dual grid centroid.

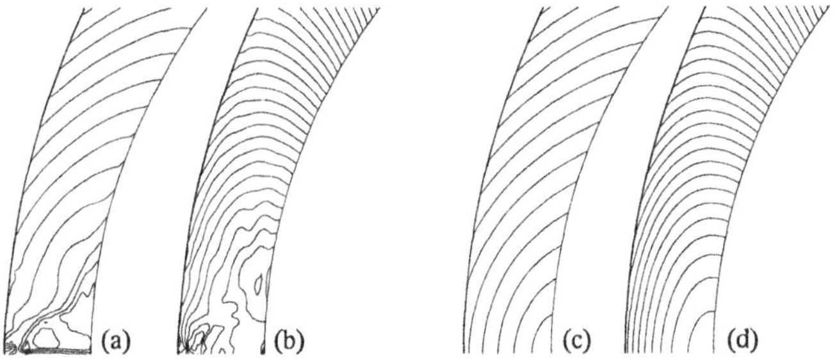

Figure 3 Simulation results for Mach 6 flow past a sphere. (a) Mach number and (b) pressure for the old axisymmetry treatment, compared with (c) Mach number and (d) pressure for the improved treatment.

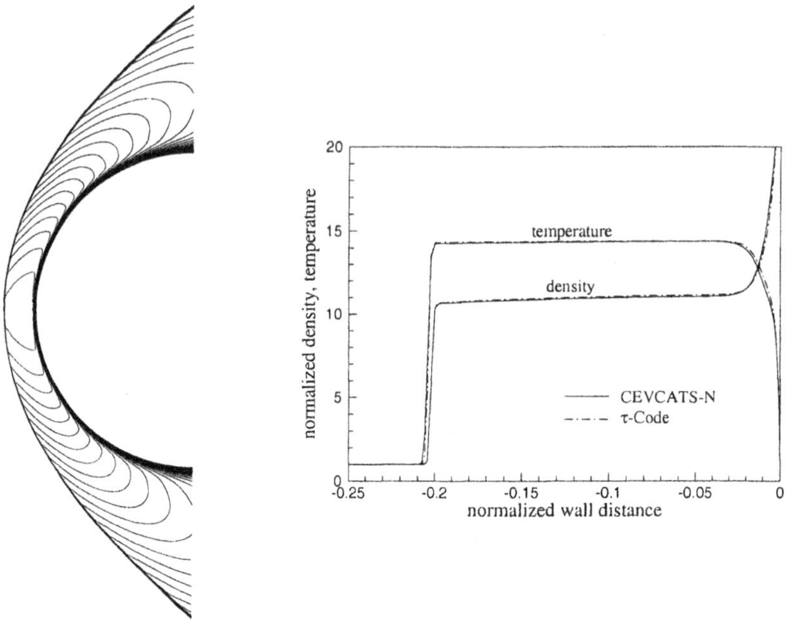

Figure 4 Comparison between DLR τ-Code and CEVCATS-N for a cylinder simulation in thermochemical equilibrium with free-stream properties of $M_\infty = 12$, $\rho_\infty = 0.0017$ kg/m^3, $T_\infty = 300$ K and a wall temperature $T_{wall} = 300$ K. On the left side temperature contours produced by the CEVCATS-N Code (upper half) and the τ-Code (lower half) are shown. Stagnation streamline values of normalized density $\frac{\rho}{\rho_\infty}$ and temperature $\frac{T}{T_\infty}$ versus the normalized wall distance $\frac{x}{R}$ are compared on the right side.

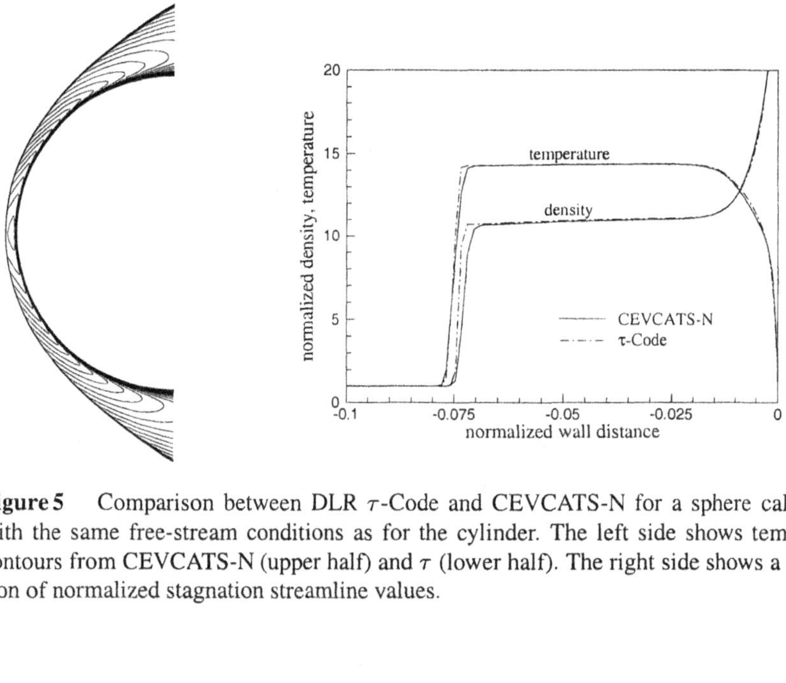

Figure 5 Comparison between DLR τ-Code and CEVCATS-N for a sphere calculation with the same free-stream conditions as for the cylinder. The left side shows temperature contours from CEVCATS-N (upper half) and τ (lower half). The right side shows a comparison of normalized stagnation streamline values.

High Incidence Buffet Flow over Fighter Type Aircraft

Arne Schmid and Christian Breitsamter

Technische Universität München, Lehrstuhl für Fluidmechanik (FLM)
85747 Garching, Germany,
arne@flm.mw.tum.de

Summary

Fin buffet loads of a fighter type delta canard configuration with $50\,^\circ$ swept wing were measured at sideslip angles of $\beta = 0\,^\circ, 5\,^\circ$ and $10\,^\circ$ covering the full angle of attack range up to $\alpha = 90\,^\circ$. The RMS-values of the fin surface pressure fluctuations and related power spectral densities are presented and used to discuss the influence of various flow features. A known relation based on wing sweep angle and angle of attack for the dominant frequency associated with helical vortex breakdown showed good agreement with measured frequencies at $\beta = 0\,^\circ$ and $\beta = 5\,^\circ$ while at $\beta = 10\,^\circ$ holds only for the windward side.

1 Introduction

Modern high agility aircraft reach high angles of attack in the post stall region. The ability of future thrust vectored aircraft to operate at very high incidence conditions and its benefits for air combat are sufficiently demonstrated [6]. Typical fighter aircraft generate, depending on the angle of attack (α), several strong vortices through swept leading edges, strakes and canards which may influence the pressure and velocity distribution near the fin. Especially the wing vortices are known as sources of fin buffet, limiting fin life cycle and flight envelope.

Many investigations on the flowfield and vortex breakdown over moderate to highly swept delta wings cover various aspects e.g.[3],[4],[5], but the fin buffet loads of delta canard configuration with center fin at high α and β are hardly investigated. This configuration is thoroughly investigated up to $\alpha = 31.5\,^\circ$ and $\beta = 5\,^\circ$ regarding flowfield and fin buffet loads, e.g. [1], [2]. As dominating aspect to the fin buffet loads, the helical breakdown of the wing leading edge vortices was identified. The dominant reduced frequency k of the fin buffet loads fulfills the condition $k_{\rm dom} = C\frac{\tan(\varphi_{\rm W})}{\sin(\alpha)}$ where $C = 0.282 \pm 0.025$ at $\alpha < 30\,^\circ$ and $\varphi_{\rm W} = 50\,^\circ \ldots 76\,^\circ$ [1]. Previous studies on similar configurations showed very good agreement up to the α of beginning unsteady vortex shedding. The present work extends the α-range of fin buffet load measurements to $90\,^\circ$ and the sideslip angle range up to $\beta = 10\,^\circ$.

2 Experimental Setup

For the experiments, the steel model of a fighter type aircraft sketched in Fig. 1 with $50°$ swept delta wing, $45°$ swept canard and $54°$ swept central fin was used. The wing span is $2s = 0.74$ m and the mean wing aerodynamic chord $l_\mu = 0.36$ m. To obtain the unsteady fin surface pressures, the fin is equipped with a total of 18 surface mounted piezoresistant pressure transducers located opposite to each other on both sides of the fin as depicted in Fig. 1. The experiments were conducted at the wind tunnel A of the FLM at a freestream velocity of $U_\infty = 40$ m/s, yielding a mean aerodynamic chord based Re-number of $0.97 \cdot 10^6$. The freestream turbulence intensity with open test section was below 0.5 % and maximum geometric blockage reached 6.6 % with $\alpha = 90°$. The 3-axes support was equipped with three interchangeable stings, covering α-ranges of $\alpha = -10° \ldots 30°$, $\alpha = 20° \ldots 60°$ and $\alpha = 50° \ldots 90°$. The unsteady fin surface pressures were low-pass filtered at 256 Hz, sampled at a rate of 2000 Hz, digitized and recorded during the test time of 30 s per α. Transducer accuracy and digitizing account for an achieveable pressure resolution of 3 Pa.

The power spectral density (PSD) was calculated with 2048 frequency intervals through band averaging. The results are shown in terms of reduced frequency $k = f\frac{l_\mu}{U_\infty}$. The results were further averaged to 409 intervals of 2.4 Hz, corresponding to a reduced frequency resolution of $\Delta k = 0.022$.

3 Results

3.1 Fin Buffet Loads

The RMS-values of the fluctuating fin surface pressure coefficients $c_{p,rms}$ were calculated and averaged for each side of the fin ($\bar{c}_{p,rms}$). The results for $\beta = 0°$, $\beta = 5°$ and $\beta = 10°$ are shown in Fig. 2 as a function of α.

Three α-ranges can be distinguished immediately. Up to an $\alpha = 15° \ldots 20°$ the windward $\bar{c}_{p,rms}$ are constant at a low level, followed by increased $\bar{c}_{p,rms}$ up to $\alpha = 56° \ldots 60°$. At $60° < \alpha < 74°$ the values differ slightly from a moderate nearly constant level. At $\alpha > 74°$ these differences vanish at all β. The windward maximum $\bar{c}_{p,rms}$ at the medium α range decrease with increasing β while the overall peak becomes broader and decomposes to several spikes.

The port side $\bar{c}_{p,rms}$ show at $\beta = 0°$ good agreement with the starboard values. At $\alpha < 10°$ the leeward pressure fluctuations increase with increasing β from low to moderate level. Largest difference between windward and leeward curves is a high narrow peak at $\beta = 10°$ and $\alpha = 23°$ of the leeward values. At $\alpha \geq 60°$ the leeward $\bar{c}_{p,rms}$-values become equal to the constant value of the windward side curves. The flowfield has changed to large seperated flow above the wing. Lightsheet visualisation shows formation and shedding of vortices in the trailing edge shear layer at $\alpha < 70°$. At higher α a stable vortical flow from the trailing edge develops and at $\alpha = 90°$ alternating shedding of large vortices from leading and trailing edge at low frequencies can be observed.

The non averaged values allow a more detailed discussion of the fin buffet loads. The leading edge and top stations contribute most to the fin buffet loads. Therefore, these $c_{p,rms}$-values presented in Fig. 3 are used to show the local development of fin buffet loads at three heights on both fin sides. The positions are at 45%, 60% and 80% of the fin span and are marked in Fig. 1.

At $\beta = 0\,°$ the values show approximately the same behaviour as the averaged values. The middle stations show the highest maximum. With increasing α the bursting wing leading edge vortex deflects the canard vortex system to the centerline. While the canard leading edge vortex remains at approximately the same height, the canard trailing edge vortex moves to the fin tip. The corresponding increase of $c_{p,rms}$ near the fin tip is stopped by the decrease of all fin buffet loads above $\alpha = 40\,°$.

At $\beta = 5\,°$ the $c_{p,rms}$-values at the leeward stations show an increase at low α followed by a peak or an inflection point. The lowest station values start at the highest level and reach a low maximum at $\alpha = 10\,°$. With higher α the $c_{p,rms}$ drops to the level of the windward station, following its course to the maximum at $\alpha = 32\,°$. The peak at low α is recognizeable at the middle station at $\alpha = 17.5\,°$ as an inflection point at high $c_{p,rms}$ and at the higher station as a small peak at $\alpha = 20\,°$. With increasing α the windward canard leading edge vortex and canard trailing edge vortex and the leeward canard trailing edge vortex are deflected to the fin. The canard trailing edge vortex moves upward causing peak values of $c_{p,rms}$ at the leeward middle station at $\alpha = 24\,°$ and at the leeward higher station at $\alpha = 29\,°$. The $c_{p,rms}$ -values become maximum at $\alpha = 34\,° \ldots 36\,°$ at all stations except the higher windward station. Within this α-range, the highest values occur at the middle stations followed by the higher stations and the lower stations. At the higher windward station the $c_{p,rms}$-values reach a maximum at $\alpha = 40\,°$. The decrease in $c_{p,rms}$ is interrupted by an α-range of constant values at the lower stations and the middle windward station. That range ends at $\alpha = 48\,°$ at the leeward station and at $\alpha = 52\,°$ at the other stations.

At $\beta = 10\,°$ and $\alpha \leq 10\,°$ the $c_{p,rms}$-values are of the same level as at $\beta = 5\,°$. The windward values remain constant up to $\alpha = 17.5\,°$. The lower and middle station show an increase to a peak at $\alpha = 23\,°$, while at the higher station the values increase up to $\alpha = 30\,°$ with an inflection point at $\alpha = 24\,°$. The leeward values tend to follow those at $\beta = 5\,°$ except at the lower station. Instead of decreasing to the windward side values, they exhibit a peak at $\alpha = 15\,°$ and another peak at $\alpha = 23\,°$. The $c_{p,rms}$-values obtained at the middle leeward station exhibit a very high peak at the same α. At the highest station, the $c_{p,rms}$-values increase up to $\alpha = 24\,°$ and decreases thereafter to the peak value of the higher windward station. Flowfield measurements showed that the expanding region of the burst leeward wing leading edge vortex reaches the fin between $\alpha = 15\,°$ and $30\,°$. The present values indicate that this takes place at $\alpha = 23\,°$. The expansion to the higher stations is delayed by $\Delta\alpha = 1\,°$. Since the leeward surface pressures are still more sensitive to the fluctuations then those at the windward side the resulting $c_{p,rms}$ is higher. Due to further expansion of burst windward wing leading edge vortex, the area of low axial velocity reaches the fin and causes a decrease in $c_{p,rms}$ at all stations except the higher windward side one. All stations exhibit at $\alpha > 30\,°$ two maximums at $\alpha = 35\,° \dots 38\,°$ and $\alpha = 48\,° \dots 50\,°$.

At $60\,° > \alpha > 80\,°$, as the trailing edge shear layer approaches the fin, the windward stations show a local maximum of $c_{p,rms}$ while the leeward values remain constant.

3.2 Power Spectral Densities

The PSDs at the windward middle station are depicted in Fig. 4. The PSDs of the symmetric case exhibit only a wide band increase between $\alpha = 25\,°$ and $\alpha = 50\,°$ with a dominant frequency due to helical mode instability and a maximum at $\alpha = 40\,°$. At $k < 0.012$ the PSDs increase between $\alpha = 24\,°$ and $\alpha = 32\,°$. The following decrease is interrupted by a peak at $\alpha = 35\,°$ and a low one at $\alpha = 50\,° \dots 60\,°$. From higher swept delta wings ($\varphi_W = 60\,°$) the dominant frequency of breakdown location fluctuations due to vortex interaction is determined to $k = 0.02 \dots 0.06$ [4], which agrees well with the observed low frequency values.

The sideslip cases show several wide band PSD concentrations which cause the peak values and inflection points of the $c_{p,rms}$-course. The low frequency band at $\beta = 5\,°$ exhibits peak values at $\alpha = 26\,° \dots 30\,°$ and $\alpha = 35\,° \dots 36\,°$. Comparison to the non-sideslip case indicates that the first peak is caused by breakdown location fluctuations. If the sideslip angle is increased to $\beta = 10\,°$ a smaller peak at $\alpha = 30\,°$ remains at the windward station PSD. The maximum PSD at high α occur at frequencies of $k \leq 0.10$.

Figure 5 shows measured peak values compared with a theoretical relation for the dominant reduced frequency associated with the helical mode instability of burst leading edge vortices, [1]. Hence, the effective wing sweep of the port and starboard side $\hat{\varphi}_W = \varphi_W \pm \beta$ is used in the relation $k_{dom} = (0.282 \pm 0.025)\tan(\hat{\varphi}_W)/\sin(\alpha)$. Without sideslip angle, the values show very good agreement up to $\alpha = 29\,°$. At higher α the measured peak frequencies are slightly below the estimated range. Above $\alpha = 40\,°$ the peaks are mostly within the expected range. The difference may be caused by the occurence of peak values superimposed to the effect of helical mode instability. Although the coincidence with the change of support sting suggests support influence, the comparison to the investigation of Taylor et al. [7] indicates only a minor support interference on vortex breakdown location. There, the investigated sweep angles were higher and the model placed closer to the dummy support than at the present experiment. The results indicate a decreasing influence with decreasing sweep angle and no significant change to the windward wing breakdown location with sideslip angle. The present support is not located within the wake of the burst vortex.

The peak values obtained at sideslip angle $\beta = 5\,°$ exhibit a good agreement with the values estimated for the windward helical mode instability up to $\alpha = 48\,°$, but several peaks at higher frequencies show no agreement with estimated leeward values.

At $\beta = 10\,°$ peaks are observed at the estimated frequencies of leeward and windward helical mode instability at lower α. The higher frequency peaks at $\alpha > 32\,°$ exhibit an increasing deviation with increasing α. Although the course of the measured peaks within $\alpha = 25\,° \ldots 36\,°$ indicates an increasing or at least constant frequency with increasing α, the obtained frequencies accumulate around the estimated values. The influence of the windward wing leading edge vortex on the fin flow is supported by flowfield measurements conducted at $\alpha = 30\,°$. The high frequency peaks are caused rather by a canopy vortex and fin interaction with the burst windward wing leading edge vortex than by distant leeward vortex breakdown. The low frequency peaks at $\alpha \leq 50\,°$ are not related to helical mode instability as Fig. 5 might suggest, since the flow over the low swept windward wing will not sustain a fully developed leading edge vortex. Further investigation is necessary to obtain the α of beginning vortex shedding.

4 Conclusions

While the high fin buffet loads of the investigated delta-canard configuration within $\alpha = 30\,° \ldots 50\,°$ decrease with increasing β, there are more and higher peaks occurring at sideslip within certain frequency ranges and α. They depend predominantly on configuration and flow field details. Wide band high fin buffet loads are found at $\beta = 10\,°$ and $\alpha = 23\,°$ when the burst windward wing leading edge vortex extends to the fin. At $\beta \leq 5\,°$ low frequency peaks due to vortex breakdown location fluctuations could be observed. While the α of the highest fin buffet loads at sideslip is determined by various flow effects, the measured dominant frequencies due to windward helical mode instability agree with the estimated values derived from a formula for $\beta = 5\,°$. At $\beta = 10\,°$ the measured peaks accumulate around the estimated frequencies. The fin buffet loads at $\alpha > 60\,°$ are approximately constant and independent from β. There, the energy concentrates in the low frequency band.

Acknowledgements

The authors would like to thank EADS Deutschland GmbH, especially Dr. J. Becker (MT2) for the continuous support.

References

[1] C. Breitsamter: "Turbulente Strömungsstrukturen an Flugzeugkonfigurationen mit Vorderkantenwirbeln". Dissertation, Herbert Utz Verlag, 1997, ISBN 3-89675-201-4.

[2] C. Breitsamter, B. Laschka: "Fin Buffet Pressure Evaluation Based on Measured Flowfield Velocities". J. Aircraft **35**, 1998, pp. 806-815.

[3] L. E. Ericsson: "Multifaceted Influence of Fuselage Geometry on Delta Wing Aerodynamics". AIAA Paper 2002-0096, 2002.

[4] I. Gursul, W. Xie: "Physics of Buffeting Flows over Delta Wings". AIAA Paper 98-0688, 1998.

[5] J. J. Miau, K. T. Kuo, W. H. Liu, S. J. Hsieh, J. H. Chou and C. K. Lin: "Flow Developments Above 50-Deg Sweep Delta Wings with Different Leading-Edge Profiles". J. Aircraft **32**, 1995, pp. 787-794.
[6] H. Ross: "X-31, The First Aircraft Designed for High Angle of Attack Manoeuvring". RTO-MP-069(**SYA-41**), RTO/AVT Symp. o. Adv. Flow Management, Norway, 2001.
[7] G. Taylor, I. Gursul: "Static Hysteresis of Vortex Breakdown due to Support Interference". AIAA Paper 2001-2452, 2001.

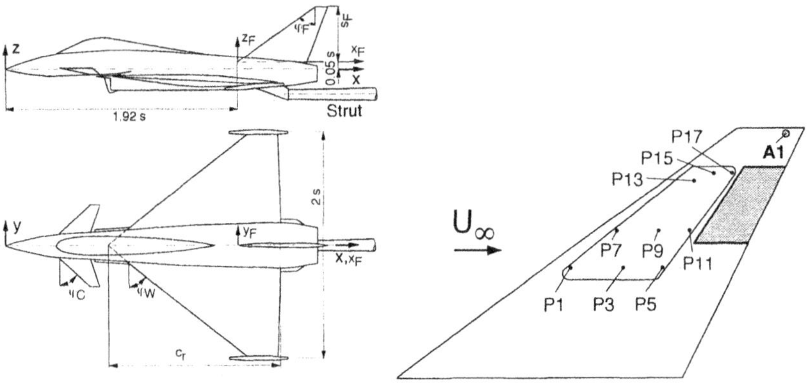

Figure 1 Model geometry and transducer distribution.

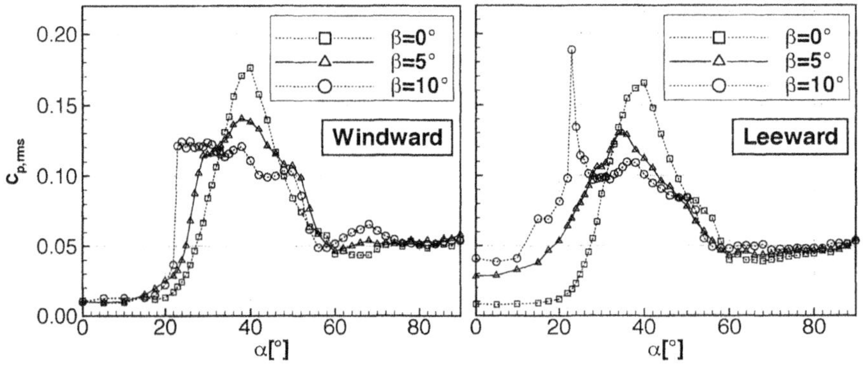

Figure 2 Area averaged fin surface pressure fluctuations as function of α and β.

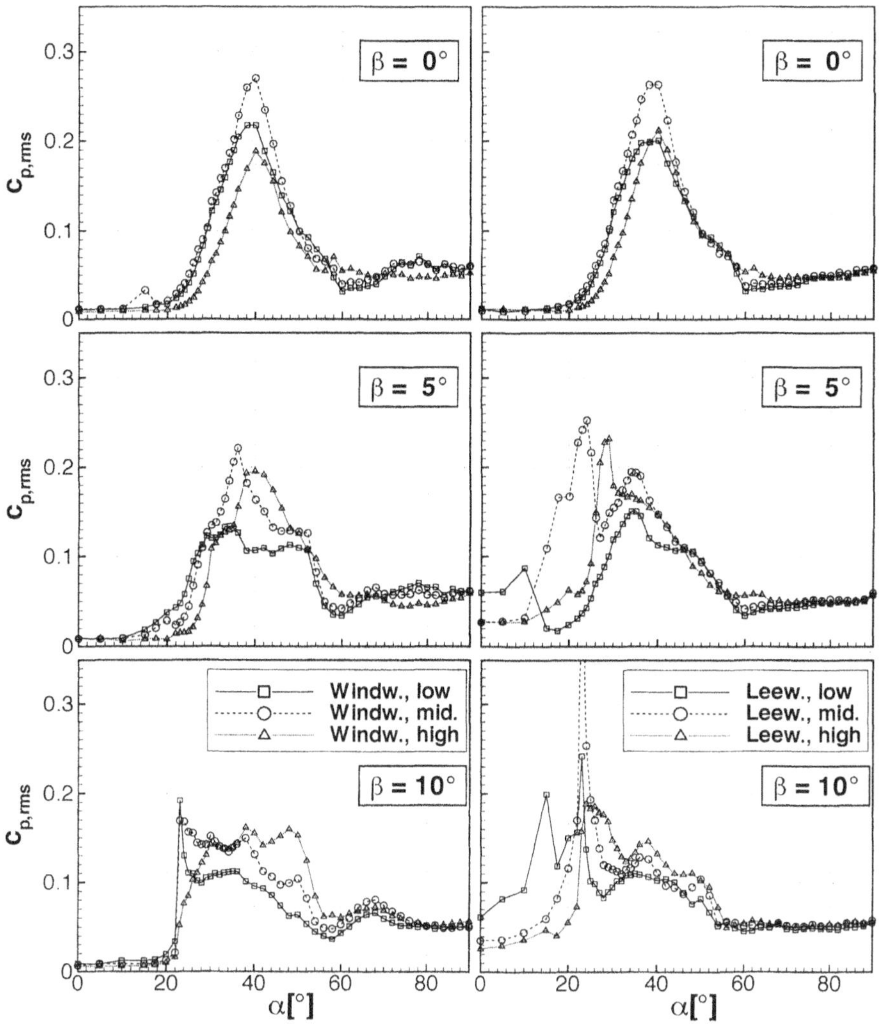

Figure 3 Fin surface pressure fluctuations at two positions along fin leading edge and at one position near to fin tip.

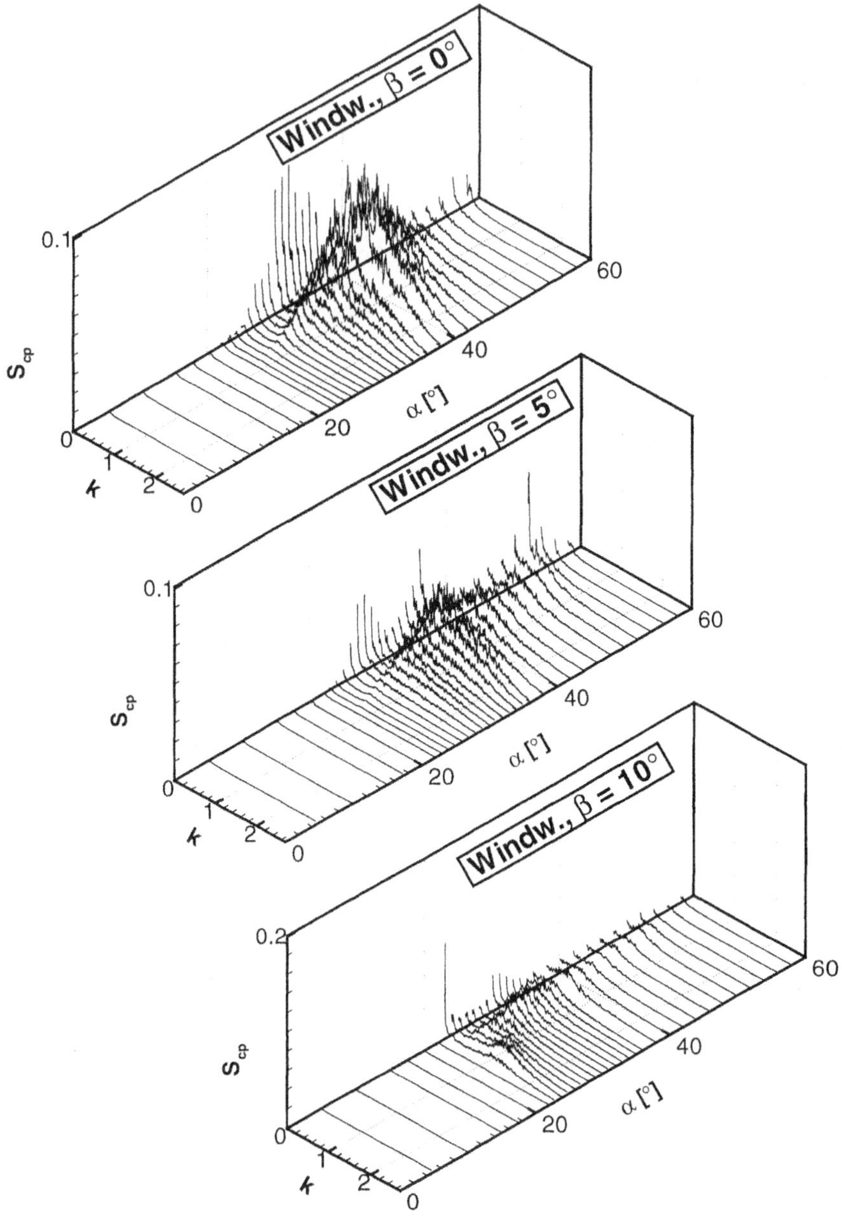

Figure 4 Fin surface pressure PSD at windward middle station, $\Delta k = 0.022$.

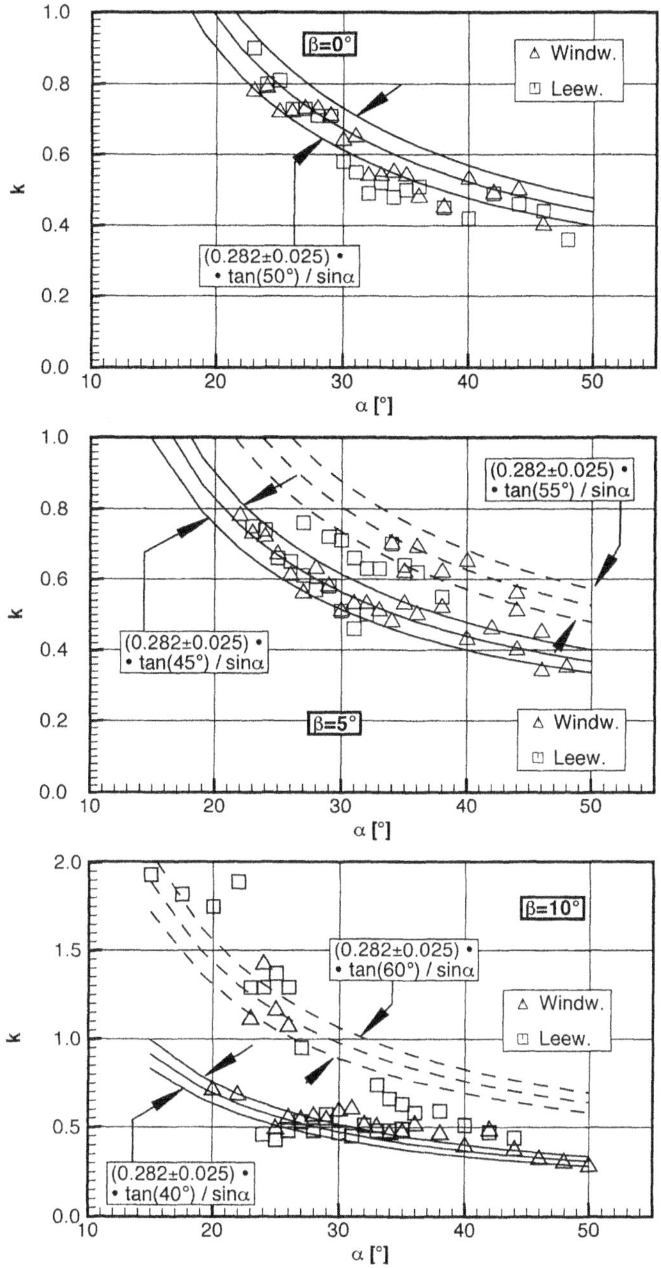

Figure 5 PSD peaks measured at the middle stations compared to calculated frequencies of helical mode instability, $\beta = 0°, \beta = 5°, \beta = 10°$.

Flap efficiency of a delta wing with an external store using an Euler code for small disturbances

Alexander Allen, Caroline Weishäupl, Boris Laschka

Lehrstuhl für Fluidmechanik - Technische Universität München
85747 Garching - Germany

Summary

The results of a numerical investigation on flap efficiency of a delta wing with an external store attached are discussed. The wing has an inboard and an outboard flap. The store is attached to the wing on the lower surface in the area of the outboard flap. For this geometry a structured multiblock mesh has been generated and elliptically smoothed. Numerical investigations are carried out with an Euler code using a small disturbance solver (FLM-SDEu) and an unsteady solver (FLM-Eu). The first assumes that the unsteady disturbances are small in comparison to the overall steady mean flow [1] [2], the latter is only used for comparison to FLM-SDEu. The small disturbance code reduces the calculation time significantly and the results obtained show good conformity with the results from the unsteady solver [3] [4].

1 Introduction

The flap efficiency for a high performance fighter aircraft is of great importance throughout the entire flight envelope. Numerical investigations for a 53° delta wing equipped with an inboard and an outboard flap were conducted at the Lehrstuhl für Fluidmechanik of the Technische Universität München [3] [4]. The results have shown, that the flap efficiency is reduced with increasing angle of attack. Recently an external store has been added in the area of the outboard flap and a similar investigation as before is performed using FLM-SDEu, especially regarding the influence of the additional store on the flap efficiency. The investigations are performed at $M_\infty = 0.8$ and various angles of attack. In order to determine the flap efficiency pitch oscillations of the flaps are performed around the flap hinge line. Here, results for the inboard flap are presented, as the inboard flap is investigated first, because the calculation appeared to cause less problems. The results for the outboard flap are still pending.

2 Geometry and discretization

A delta wing with $\Phi_{LE} = 53^o$ leading edge sweep, $\Phi_{TE} = -3^o$ trailing edge sweep and c_r/s=1.6 (c_r being the root chord length and s the half span) is investigated. The wing has a round leading edge and is equipped with an inboard and an outboard flap,

Fig. 1. An external store is mounted on the lower side of the wing with a pylon, Fig. 2. The center line of the store is at y/s=0.71 and its nose at approx. x/c_r=0.55. The external store has an overall length of $0.396c_r$.
A structured multiblock grid consisting of 34 blocks has been generated with the tool ICEM-CFD. The surface grid consists of 68 cells in chordwise direction on the upper and the lower side, respectively, and 56 cells along the leading and the sharp trailing edge. The external store is discretized with 42 cells in chordwise direction and 62 cells in circumferential direction. The pylon consists of 22 cells in chordwise direction and 8 cells in perpendicular direction. The inboard and outboard flap is discretized with 14 cells chordwise. The inboard flap consists of 12 cells in spanwise direction and the outboard flap of 23 cells, Fig. 3. The grid consists of about 310000 cells.
In the area of the flap hinge line and the flap side edges the grid points are concentrated in order to achieve the optimum resolution of the flow gradients. Between the rigid wing and the flaps a region is introduced to guarantee a smooth change in surface during flap oscillation. The gaps at the flap are not modelled. As a consequence the wing surface is discretized as a continous surface.
Due to the complexity of the geometry a satisfying discretization was accomplished by using several O-Grids, which are positioned in one another like shells around the geometry. Fig. 4 shows a part of the grid in the symmetry plane. The farfield distance is chosen to 10 root chords for x- and z-direction and 5 root chords in spanwise direction. The initial volume grid, which was generated with the tool ICEM-CFD undergoes elliptic smoothing. The body surface is not smoothened and the distance of the first offbody grid line is fixed to 0.5% of the wing root chord c_r.

3 Numerical method

3.1 Nonlinear Euler method FLM-Eu

The Euler equations describe compressible inviscid unsteady flow. The Euler solver FLM-Eu has been developed and extensively validated at the Lehrstuhl für Fluidmechanik. It uses a finite volume approach with Roe's flux difference splitting. An explicit and implicit time integration is available. As an implicit solver the LU-SSOR method is used. Details of FLM-Eu are given in Ref. [2].

3.2 Small Disturbance Euler method FLM-SDEu

In many practical applications the unsteady disturbances are small in comparison to the mean flow, for example in aeroelasticity. Therefore the flow can be interpreted as a superposition of the steady mean flow and small unsteady disturbances in these cases. For FLM-SDEu a harmonic behaviour of the motion is assumed. This is described by the following equation for the x-direction:

$$x(\xi,\eta,\zeta,\tau) = \bar{x}(\xi,\eta,\zeta) + \tilde{x}(\xi,\eta,\zeta)e^{ik\tau} \tag{1}$$

$\bar{x}$ being the steady reference position, $\tilde{x}$ the amplitude of the harmonic grid oscillation and ξ, η and ζ the curvilinear coordinates. The equations for y and z can be derived in the same way. k represents the dimensionsless frequency and τ the dimensionless time. Analogically the unsteady flow variables have to be described by the steady and the perturbation part:

$$\mathbf{q}(\xi, \eta, \zeta, \tau) = \bar{\mathbf{q}}(\xi, \eta, \zeta) + \tilde{\mathbf{q}}(\xi, \eta, \zeta) e^{ik\tau} \tag{2}$$

Introducing the harmonic behaviour into the nonlinear Euler equations the SDEu equations can be derived as follows:

$$\frac{\partial \tilde{\mathbf{Q}}^{(1)}}{\partial \tau} + \frac{\partial \tilde{\mathbf{F}}^{(1)}}{\partial \xi} + \frac{\partial \tilde{\mathbf{G}}^{(1)}}{\partial \eta} + \frac{\partial \tilde{\mathbf{H}}^{(1)}}{\partial \zeta} =$$
$$- \left(\tilde{\mathbf{Q}}^{(1)} ik + \tilde{\mathbf{Q}}^{(2)} ik + \frac{\partial \tilde{\mathbf{F}}^{(2)}}{\partial \xi} + \frac{\partial \tilde{\mathbf{G}}^{(2)}}{\partial \eta} + \frac{\partial \tilde{\mathbf{H}}^{(2)}}{\partial \zeta} \right) \tag{3}$$

with

$$\tilde{\mathbf{Q}}^{(1)} = \bar{J}\tilde{\mathbf{q}}, \quad \tilde{\mathbf{Q}}^{(2)} = \tilde{J}\bar{\mathbf{q}}, \quad \tilde{\mathbf{F}}^{(1)} = \left.\frac{\partial \mathbf{F}}{\partial \mathbf{q}}\right|_{\bar{\mathbf{q}}} \tilde{\mathbf{q}} = \mathbf{A}^{(1)}\tilde{\mathbf{q}} \tag{4}$$

$$\mathbf{Q} = J\mathbf{q} = J \begin{bmatrix} \rho \\ \rho u \\ \rho v \\ \rho w \\ e \end{bmatrix}, \qquad \tilde{\mathbf{F}}^{(2)} = \begin{bmatrix} \bar{\rho}\tilde{U}^* \\ \overline{\rho u}\tilde{U}^* + \widetilde{J\xi_x}\bar{p} \\ \overline{\rho v}\tilde{U}^* + \widetilde{J\xi_y}\bar{p} \\ \overline{\rho w}\tilde{U}^* + \widetilde{J\xi_z}\bar{p} \\ U^{(2)}(\bar{e} + \bar{p}) - \widetilde{J\xi_t}\bar{p} \end{bmatrix}$$

with

$$\tilde{U}^* = \widetilde{J\xi_x}\bar{u} + \widetilde{J\xi_y}\bar{v} + \widetilde{J\xi_z}\bar{w} + \widetilde{J\xi_t} \tag{5}$$

$\mathbf{A}^{(1)}$ is the Jacobian matrix. To solve the above equations the equation of state in linearized form is necessary:

$$\tilde{p} = (\kappa - 1)\left[\tilde{e} - \frac{1}{2}\left(2\bar{u}\widetilde{\rho u} - \bar{u}^2\tilde{\rho} + 2\bar{v}\widetilde{\rho u} - \bar{v}^2\tilde{\rho} + 2\bar{w}\widetilde{\rho u} - \bar{w}^2\tilde{\rho}\right)\right] \tag{6}$$

The numerical solution procedure realized in FLM-SDEu is described in Ref. [2]. The results of FLM-SDEu are compared with the results obtained with FLM-Eu. Therefore a Fourier analysis of the nonlinear results is performed, in order to extract the harmonic parts and split off the unsteady part from the steady reference part.

4 Results

The investigations are performed at $M_\infty = 0.8$ and various angles of attack ($\alpha = 0^o, 2^o, 5^o$ and 8°). Harmonic oscillations are performed for the inboard flap around the zero position with an amplitude of $\Delta\eta = 0.4^o$ and a frequency of $f = 6Hz$ ($k_{red} = 0.7283$). The calculations are performed with the Euler code for small disturbances (FLM-SDEu) as well as the nonlinear Euler code (FLM-Eu). For both methods the density residual which has to be reached is 10^{-5}. The results are compared with the calculation performed on the delta wing alone [3] [4]. The grid of the delta wing alone has a CH topology and consists of 356400 cells in one single block. Fig. 5 shows the steady lift and pitching moment. The pitching moment is calculated around (x,z)=(0,0), compare Fig. 3. A higher angle of attack of 10° is still under investigation at the moment. The external store reduces the lift and also decreases the negative pitching moment slightly. Therefore the absolute value of the pitching moment is reduced, which is due to the reduced lift of which the attack point is behind the moment reference point. The reference area for the delta wing with and without external store are the same.
Fig. 5 also shows, that for the delta wing alone the outboard flap has less efficiency than the inboard flap. This has already been proven by previous investigations [3] [4]. In comparison to these investigations the efficiency of the inboard flap has been decreased. This is due to the store attached to the lower side of the delta wing. The difference between FLM-SDEu and FLM-Eu is larger than the decrease due to the store, as can be seen in Fig. 5. For $c_{l\dot{\eta}}$ and $c_{m\dot{\eta}}$ hardly any changes can be seen.
Fig. 6 shows the pressure distribution in two sections at $y/s = 0.35$ and $y/s = 0.75$. For the nonlinear solver the pressure coefficients are calculated from the Fourier analysis of the unsteady results. Comparing the steady pressure coefficients without store with the ones with store shows the influence of the external store. This influence is very significant at y/s=0.75, but can also be seen at y/s=0.35 in higher negative c_p values.
The first harmonic pressure coefficients for the inboard flap comply very well for the two different solvers. The calculations also show good conformity in the real parts of the first harmonic pressure coefficients compared to the results gathered without external store. In the imaginary part slight differences can be seen, but the overall characteristics are almost identical. Furthermore the imaginary part is one order of magnitude smaller than the real part. The slightly larger difference between the calculations with and without store on the lower side at y/s=0.75 derives from the pylon of the external store. Fig. 6 also shows the typical singularity around the flap hinge line in the real part of the pressure coefficient for the oscillating inboard flap.
The satisfying compliance between the two solvers shows, that FLM-SDEu is very well suited for these kinds of computational tasks. Furthermore FLM-SDEu reduces the calculation time.

5 Conclusion

The solutions obtained with the Euler code for small disturbances (FLM-SDEu) show good correlation with the ones obtained with the nonlinear code (FLM-Eu). The gain considering calculation time is significant.
The calculation of the delta wing with store shows a reduced flap efficiency for the inboard flap than without the external store. The investigation will continue, especially gaining data for higher angles of attack and also for the outboard flap.

Acknowledgements

The authors would like to thank Dr.-Ing. J. Becker (EADS Ottobrunn) for the good cooperation over the past years. The authors would also like to thank Mr. B. Wilke, Mr. M. Lang and Mr. A. Wildschek (all Technische Universität München) for their work towards this project.

References

[1] K. C. Hall, E. F. Crawley: "Calculation of Unsteady Flows in Turbomachinery Using the Linearized Euler Equations". AIAA Journal, Vol. 27, No. 6, pp. 777-787, 1989.
[2] E. Kreiselmaier, B. Laschka: "Small Disturbance Euler Equations: Efficient and Accurate Tool for Unsteady Load Prediction". Journal of Aircraft, Vol. 37, No. 5, pp. 770-778, 2000.
[3] C. Weishäupl, A. Pechloff, U. Sickmüller: "Instationäre Luftkräfte bei Klappenschwingungen mit Hilfe der Euler Gleichungen für kleine Störungen (SDEu)". Technical report, Technische Universität München TUM-FLM-99/26, 1999.
[4] C. Weishäupl, B. Laschka: "Nonlinear and Small-Disturbance Euler (SDEu) Calculations for Delta Wing Flap Efficiency". AIAA Paper 2000-0124, 38th Aerospace Sciences Meeting & Exhibit, Reno, NV, USA, 10-13 January, 2000.

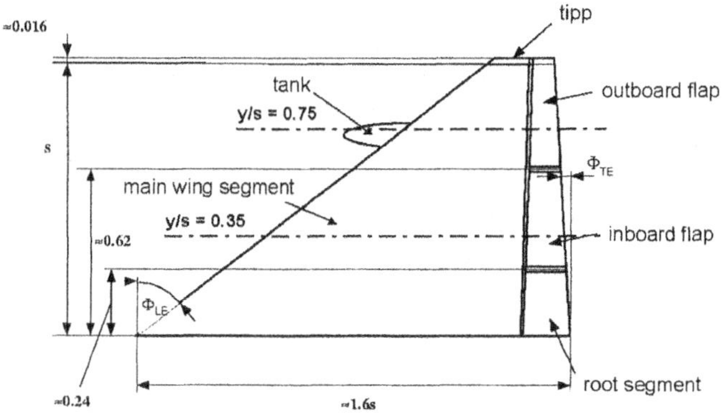

Figure 1 Geometry

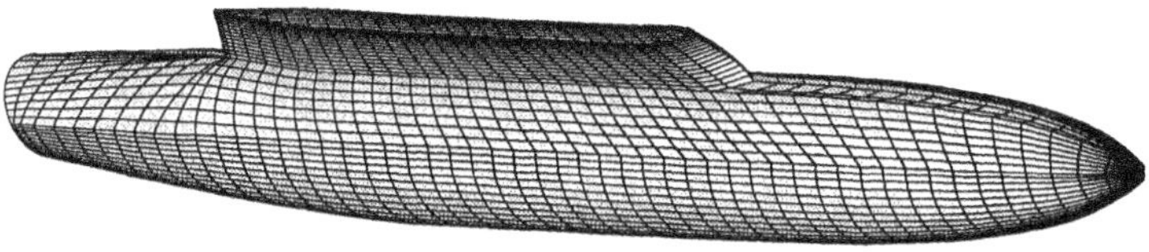

Figure 2 Surface grid of the external store

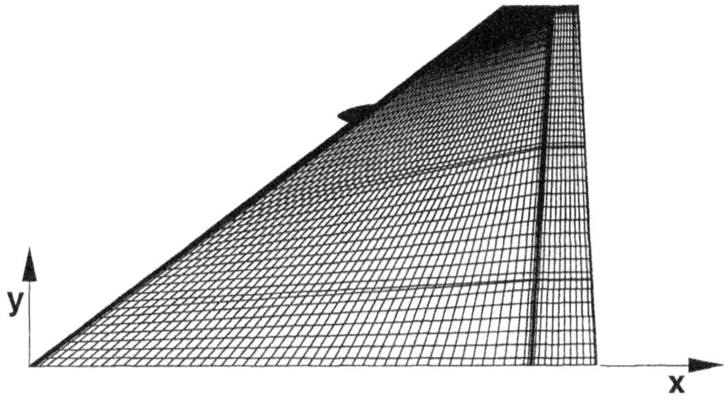

Figure 3 Surface grid of the delta wing

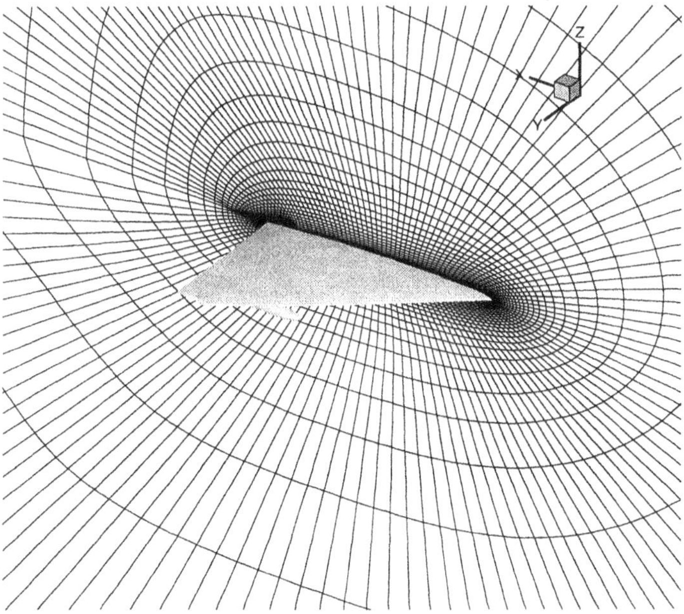

Figure 4 Grid in the symmetry plane

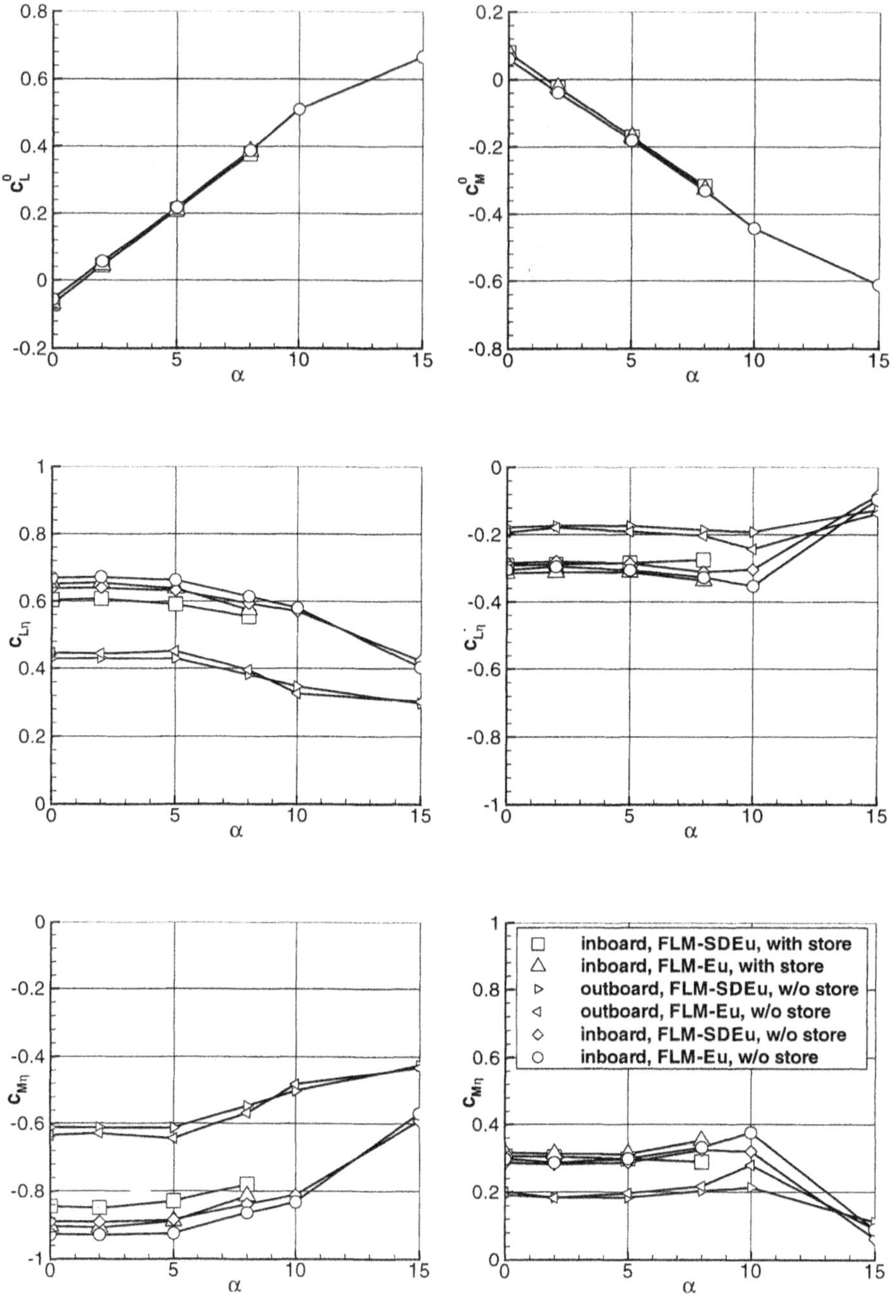

Figure 5 c_L, c_M, $c_{l\eta}$, $c_{l\dot{\eta}}$, $c_{m\eta}$, $c_{m\dot{\eta}}$ at $M_\infty = 0.8$

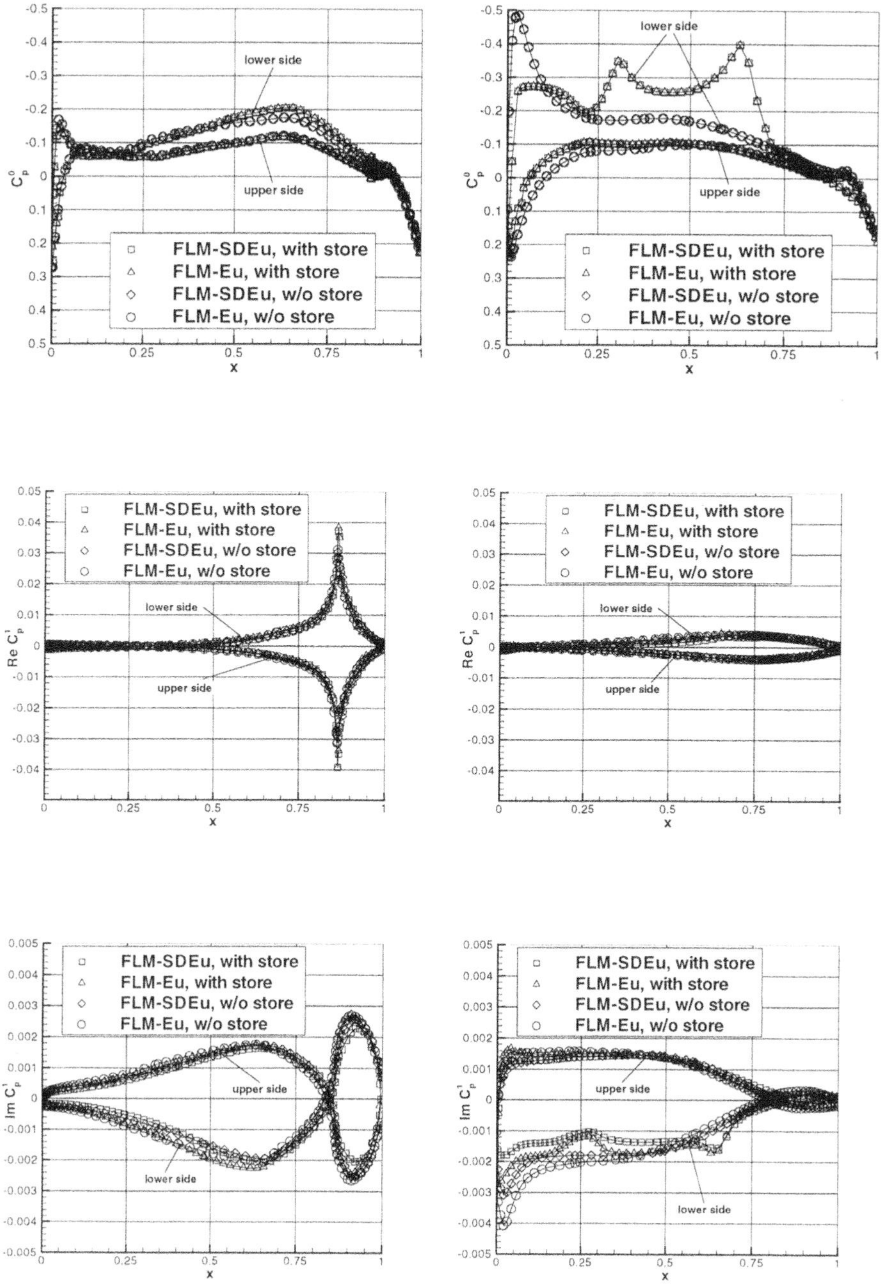

y/s=0.35 y/s=0.75

Figure 6 Pressure coefficient and first harmonic pressure coefficient for the inboard flap at $\alpha = 0^o$ and $M_\infty = 0.8$ for y/s=0.35 and y/s=0.75

Pressure and heat flux measurements at the surface of an asymmetric cone at Mach number 4.6 in ISL's shock tunnel STB

F. SEILER, J. SRULIJES
French-German Research Institute of Saint-Louis (ISL)
5, rue du Général Cassagnou, F-68301 Saint-Louis, France

Summary

New concepts for guiding and controlling the flight of projectiles in the hypersonic speed regime up to Mach number 10 or more, such as the missile's outer contour deformation are under investigation for future applications. Strong lateral forces can be obtained if the deformation takes place at the projectile's nose. For this purpose a flat ramp with ramp angles between 15 deg and 18 deg was machined on conical nose models of 20 deg half angle to create a cone asymmetry. With these models differential interferometry flow visualisations, surface pressure and heat transfer measurements were performed at Mach number 4.6 and 7 km altitude atmospheric conditions at the ISL's shock tunnel STB to examine the resulting flow around the asymmetric cone model.

1 Introduction

The principles of controlling a projectile by deforming its outer shape, including several construction examples are extensively discussed in an ISL patent, see Seiler et al [1]. Fundamental experiments with the prospect of flight trajectory control by deformation of the outer shape of the projectile were carried out. At a conically formed nose of a projectile part of the surface was deformed by flattening to achieve a contour asymmetry, as shown in the sketch of Figure 1. At supersonic flow speed different shock waves develop at the "cone side" and at the "ramp side" as outlined schematically. The flow behaviour at the ramp side is still not well understood. To study this, the aerothermomechanical surface load was determined by pressure and heat flux measurements. The pressure asymmetry, e.g., generates a lateral lift that can be used for projectile guiding. Fig. 1 shows the lift force trend schematically. The experimental results are compared with theoretical estimations of the heat flux development, see Seiler et al [2], and the surface pressure distribution.

2 Experimental procedure

2.1 Shock tunnel test facility

The experiments were carried out at the ISL Shock Tube Laboratory (Figure 2) in which two high-pressure shock tubes, STA and STB, are in operation. These shock tubes have an inner diameter of 100 mm and a total length of about 30 m. For the experiments presented herein the high-pressure shock tube STB was used as a shock tunnel to study the flow field around flattened cylindrical missile nose models during flight in earth atmosphere. In the shock tunnel STB the flow was expanded inside a circular nozzle, driven in the shock reflected mode. A contoured nozzle was used in order to produce a parallel flow. The flow Mach number at the nozzle exit was M = 4.6 with pressure and temperature conditions as present in the earth atmosphere in about 7 km altitude. These conditions are given as mean values, averaged over the experiments performed, as well as their deviations: flow velocity u = 1470 m/s ± 15 m/s, static pressure p = 45 kPa ± 4 kPa, gas density $\rho = 0.60\ g/m^3 \pm 0.03\ kg/m^3$ and static temperature T = 251 K ± 7 K.

The shock tunnel consists of a high-pressure driver tube, a driven tube and the contoured nozzle attached at the end of the shock tube, see Figure 3. The driver gas pressure for the experiments discussed herein is about 30 MPa hydrogen and the initial driven gas pressure is set to 340 kPa nitrogen. The incident shock wave at the nozzle throat is reflected, generating a quasi "reservoir gas". This is expanded and accelerated through the specially designed divergent nozzle contour in such a way that the atmospheric flight conditions mentioned above are reproduced at the nozzle exit. Contrary to the real flight situation, the missile fore-body is fixed in front of the nozzle and the flow is set in motion.

2.2 Ramp flattened cone models

A cone with a diameter of 95 mm at its back is used to model the fore-body of a full scale 95 mm calibre projectile. The total cone angle is 40 deg, i.e., a half angle of 20 deg measured from the line of symmetry. Surface pressure measurements were performed at four positions, two at the cone side and two at the ramp side. Three different ramps with angles of 16 deg, 17 deg and 18 deg were investigated. A picture of the ramp cone model equipped with PCB pressure gauges is shown in Figure 4a. The pressure is measured at each location through a bore hole with 1 mm in diameter and a total length of 1 mm. The measuring ports in Figure 4a are marked with arrows. The pressure sensors are mounted from the back side as shown schematically in Figure 5. The gauges are numbered from 1 to 4 and the positions are given in mm of the measuring ports.

For heat flux measurements cones with 15 deg and 18 deg ramp angles have been used. The cone models are equipped with three heat flux sensors at the cone side (Figure 4b) and three at the ramp side (Figure 4c). The measuring ports are numbered from 1 to 3 at the ramp side and 4 to 6 at the cone side, see Figure 6.

2.3 Temperature gauges

Special temperature sensors both developed and built at ISL were used to determine the surface heat flux. They consist of a small cylindrical glass cylinder (∅ 4 mm), see Figure 7, at the tip of which a 1 mm narrow platinum layer with a thickness of less than 1 µm has been painted and burnt in. The sensor surface is flush mounted with the outer contour of the model as shown in Figures 4b and 4c. When a heat flux from the gas flow reaches the glass support, the glass surface and the platinum layer rise to the same temperature level. The platinum layer temperature change ΔT is determined by measuring the variation ΔR of the resistance R_0 of the platinum layer in time as $\Delta R = \alpha R_0 \Delta T$. Applying a one-dimensional solution of the heat conduction equation, see Oertel [3], the heat flux $\dot{q}$ at $y = 0$ can be calculated as:

$$\dot{q}(0,t) = \sqrt{\frac{\rho c \lambda}{\pi}} \frac{d}{dt} \int_0^t \frac{\Delta T(0,\tau)}{\sqrt{t-\tau}} d\tau .$$

With the calibration factor $E = \sqrt{\pi \rho c \lambda} / 2\pi$, in which the temperature coefficient α, the density ρ and heat conduction λ of the glass support are included, the heat flux $\dot{q}$ is calculated from the recorded temperature change ΔT.

2.4 Model arrangement in STB

The projectile nose model was mounted inside of the measuring section in front of the nozzle as shown on the left hand side in Figure 8. The direct view into the inner section gives more detailed information about the arrangement of the cone model. For visualizing the flow formation around the nose model of a projectile, at both sides of the measuring section optical windows permit taking pictures of the flow development around the wedged cone model. For this purpose the well-known differential interferometry (DI), see Oertel [3], was applied. The DI flow picture in Figure 8 shows the shape of the oblique shock wave around the cone model.

3 Surface heat flux determination

With the thin film gauges described in chapter 2.3 the surface temperature of the gauge (glass) is measured. From that record the heat flux is calculated using the calibration factor E which was determined for each of the sensors used at the measuring ports no. 1 to 3, ramp side, and no. 4 to 6 at the cone side. Three different experiments are presented for each of the following models: 20 /15 deg and 20 /18 deg. An example of the heat flux determined for the 20 deg/18 deg wedged cone is shown in Figure 9. The repeatability of the experimental results can be seen in Figure 10 with three overlaid heat flux records for the 20 deg/18 deg cone. It can be noticed that the heat flux deviations for repeated shock tube experiments are very small and are in the range of some percent, just as the deviations present in the shock tube operation conditions which are given in chapter 2.1. After about the 0.5 ms needed for generating a stable and constant nozzle flow the heat fluxes are nearly constant over the shock tunnel flow duration of about 2 ms. For each of the experiments and for each of the measuring ports, no. 1 up to no. 6, the constant heat flux data have been evaluated as mean values averaged over the electrical noise present in the heat flux records. The results are given in Figure 11, measured at the cone side, and in Figure 12 for the 20 deg/15 deg and the 20 deg/18 deg ramp side angles.

Theoretical estimations done for the boundary layer formation along the surface of a cone as well as along a wedge are compared with the experimentally determined results. The calculated cone side heat flux in Figure 11 is in good agreement with the experiments within the measuring error of about ± 10% shown by the error bars. At measuring port no. 4 the experimentally determined heat flux development becomes smaller than the theoretically calculated value. Maybe this drop is influenced by transition phenomena taking place from laminar to turbulent boundary layer formation downstream near the cone tip. At the wedge side, Figure 12, the experimental situation appears as an interference of the ramp wedge and the cone shaped part. Calculations have been performed for a 15 deg and an 18 deg cone as well as a 15 deg and an18 deg wedge profile. Both results are given in the diagrams in Figure 12 compared with the measured heat flux values. The calculated wedge heat fluxes are slightly higher than the experimental results. They are closer to the results obtained for the cone calculations.

4 Pressure measurements

Pressure measurements were performed for 20 deg/16 deg, 20 deg/17 deg and 20 deg/18 deg wedged cones. For repeatability purposes, similar experiments were done with each of the wedged cones.

Four measuring ports, no. 1 and no. 2 at the wedge side, no. 3 and no. 4 at the cone side have been used, see chapter 2.2. The repeatability of similar experiments is quite good and depends mainly on small shock tube flow uncertainties which occur, e.g., through small differences in the diaphragm burst pressure. For evaluation purposes the surface pressure is deduced from the pressure records taken. The data obtained can be seen in the diagrams of Figures 13 and 14.

The comparison between experimentally recorded surface pressure data and calculated ones is shown in Figure 13 for the cone side and in the diagrams of Figure 14 for the ramp on the 20 deg/16 deg, 20 deg/17 deg and 20 deg/18 deg cone models. The surface pressure measured at the cone side is in good agreement with the calculated pressure level within the measuring error of about ± 10%. This comparison shows that at the cone side the cone flow is practically not influenced by the opposite ramp geometry. The ramp pressure distribution is smaller than that at a pure wedge with the same angle and somewhat higher than that at a cone with also the same angle, i.e., the ramp angle.

The lateral force lift coefficient c_L is an important quantity for the manoeuvrability and control of a missile. An estimate of c_L for the asymmetric cone model, calculated with the measured pressure data as a function of the ramp angle, is presented in the lower right side diagram of Figure 14. Moreover, the pressure drag coefficient is calculated as $c_D = 0.36 \pm 0.04$.

5 Conclusions

Seiler et al. [1], suggested deforming the missiles outer contour to obtain laterally acting forces which can be used for missile guiding purposes. With this idea in mind, investigations to study the action of deforming the fore-body of a supersonically flying projectile on the development of the flow pattern were carried out at the ISL's Shock Tube Laboratory. As shock tube model a conical projectile nose (20 deg half angle) was equipped with a flat ramp, ranging between 15 deg and 18 deg angles, leading to a asymmetric flow. The experiments took place in shock tunnel STB at Mach number 4.6 and for a 7 km altitude flight.

Measurements of wall pressure were carried out at the undisturbed cone surface as well as along the ramp. The predicted pressure asymmetry which can be successfully used to generate lateral forces for missile control was quantified. Future measurements will be carried out with more pressure sensors and with the recently at ISL's Shock Tube Laboratory developed free flight force measuring technique (FFM).

Furthermore, the aerothermal load, especially at the fore-body of missiles flying in the supersonic flow regime plays an important role for their design. The determination of heat fluxes along the cone and the ramp surface showed very high heat fluxes in the range of about 3 MW/m^2. These can stress the surface structure, depending on the flight time, up to surface melting followed by surface ablation. For future high speed missile design the knowledge about the heat transferred onto the projectile surface is of significant importance. Especially the selection of adequate surface materials at those sites where extremely high heat fluxes may be present is of crucial importance.

References

[1] F. Seiler, J. Srulijes, H. Lehr, Verfahren zur Steuerung eines Überschallflugkörpers sowie Überschallflugkörper, Deutsche Patentschrift Nr. 198 39 493, Deutsches Patent- und Markenamt, 1998

[2] F. Seiler, U. Werner, G. Patz, Ground testing facility for Modelling Real Projectile Flight Heating in Earth Atmosphere, Journal of Thermophysics, Vol. 16, Nr. 1, January-March 2002

[3] H. Oertel, Wärmeübergangsmessungen, in "Kurzzeitphysik", Springer Verlag, Wien-New York, 1967

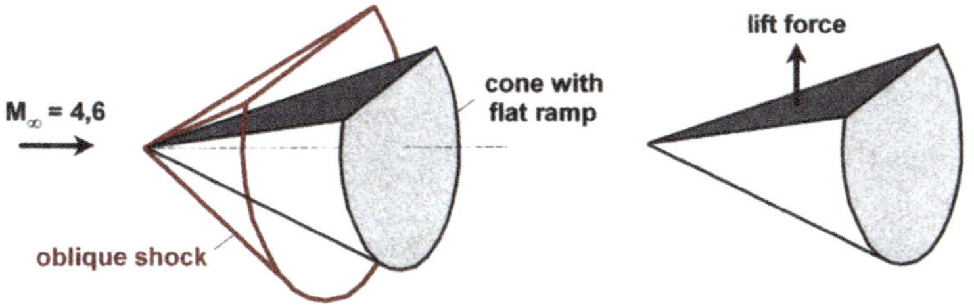

Figure 1 Oblique shock wave in front of the asymmetric cone and lift force trend

Figure 2 ISL's Shock Tube Laboratory

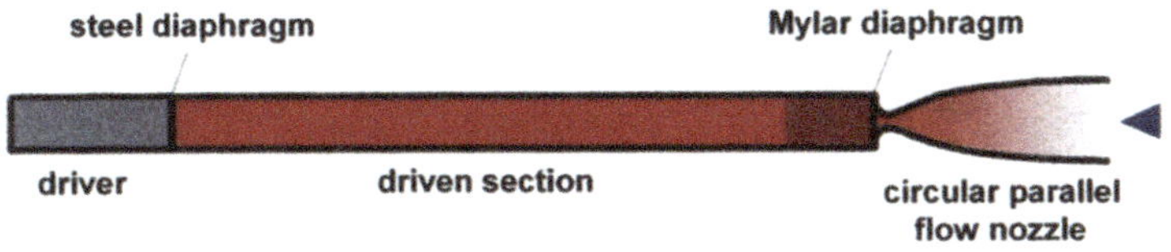

Figure 3 Principle sketch of the shock tunnel facility STB set-up

a) b) c)

Figure 4 Cone models for shock tube applications

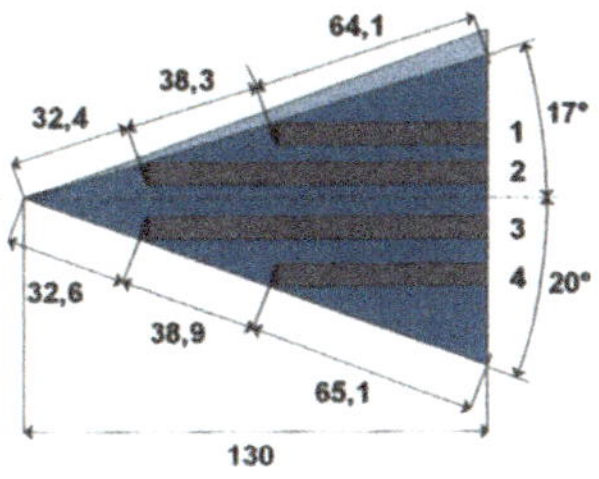

Figure 5 Pressure gauge ports

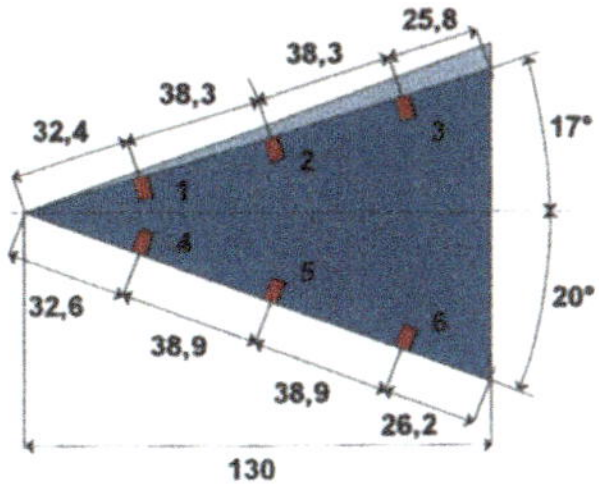

Figure 6 Temperature sensor ports

Figure 7 Set-up of the platinum thin film gauges

Figure 8 Photograph of the measuring section with nozzle and DI flow picture

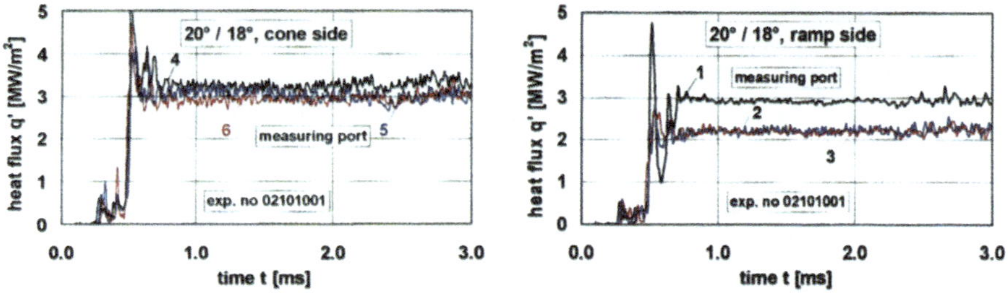

Figure 9 Heat flux for experiment no. 02101001: 20 deg/18 deg cone model

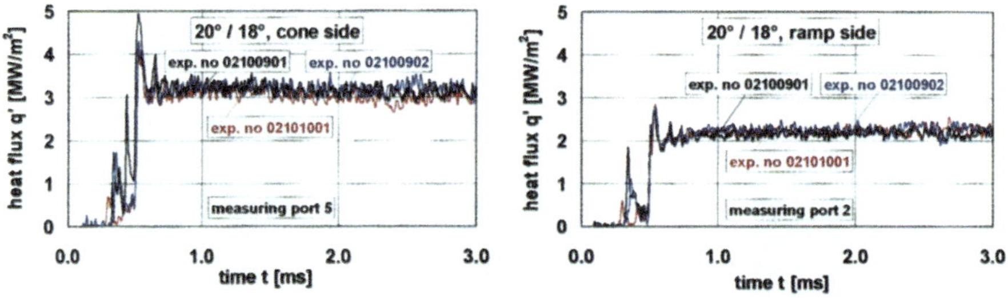

Figure 10 Heat flux rate for the 20 deg/18 deg cone model: repeated experiments for measuring port 5 (left) and port 2 (right)

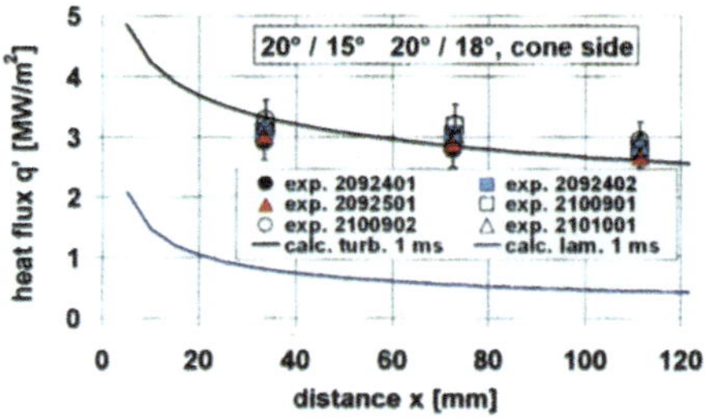

Figure 11 Comparison of theoretical and experimental heat flux rates at the cone side (theoretical estimations: laminar-lower curve, turbulent-upper curve)

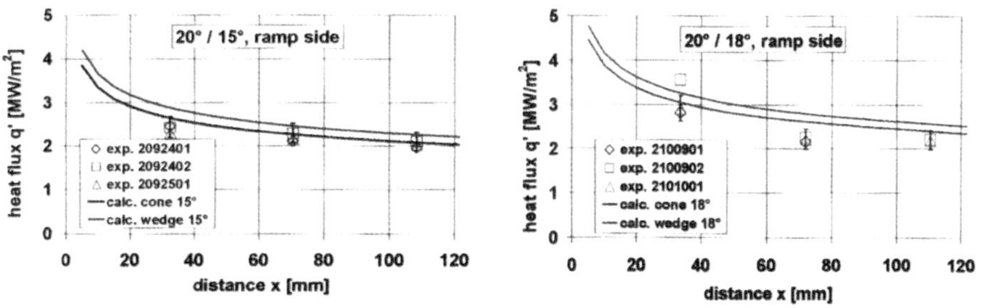

Figure 12 Theoretical and experimental heat flux rates:
20 deg/15 deg, 20 deg/18 deg cone model

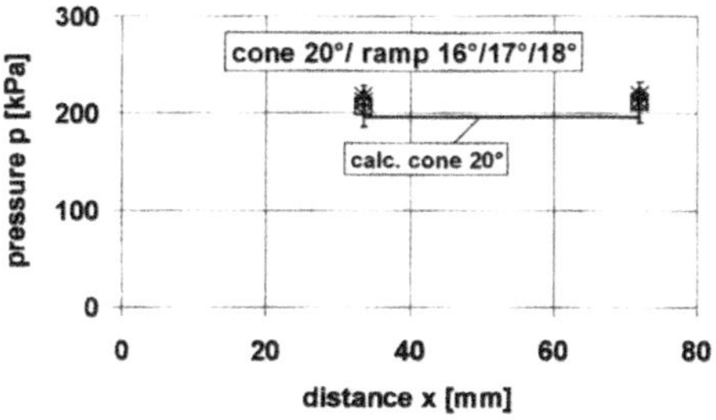

Figure 13 Comparison of experimental and calculated surface pressure: cone side

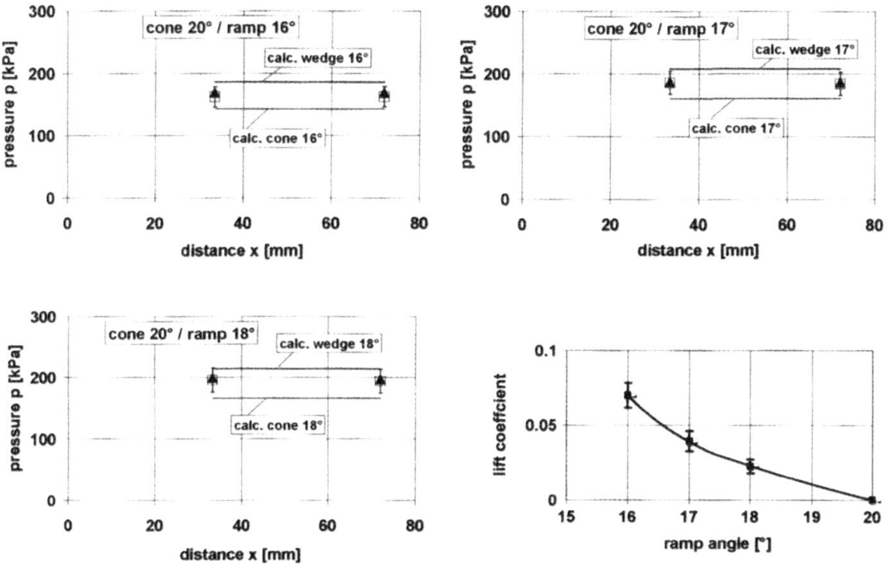

Figure 14 Comparison of experimental and calculated ramp side surface pressure and lift force coefficient

Incorporation of an Actuator Disc for Lattice Wing Modelling in an unstructured Navier-Stokes Solver

Philippe Reynier[1] and Erich Schülein[2]

[1] DLR, Institut für Aerodynamik und Strömungstechnik,
Lilienthalplatz 7, 38108 Braunschweig, Germany, `Philippe.Reynier@dlr.de`,
[2] DLR, Institut für Aerodynamik und Strömungstechnik,
Bunsenstrasse 10, 37073 Göttingen, Germany, `Erich.Schuelein@dlr.de`

Summary

This paper focuses on the modelling of lattice wings for earlier design phases of vehicles equipped with such devices. Due to the complex geometry of such a configuration, the actuator disc approach has been chosen to reduce the computational effort. Using this modelling, the lattice wing is replaced by artificial boundary conditions where the forces due to the grid fin are computed using semi-empirical formulae. The incorporation of this approach in an unstructured Navier-Stokes solver has been achieved. This method allows the prediction of the global axial and normal forces acting on a lattice wing as a function of its geometrical characteristics. The numerical results obtained for a lattice wing alone are compared with experimental data for different Mach numbers and angles of attack in order to validate the method.

1 Introduction

Several investigations [1–3] have shown the interest of lattice wings (also called grid fins) as aerodynamic controls for supersonic applications. The main advantage of grid fins [3, 4] is their lifting capabilities which make them ideal to enhance lift and pitching moment characteristics. They have also other advantages, like small centre of pressure variations over a wide Mach number range [2], a compact storage, and a relatively easy deployment in flight. Another asset is the very small hinge moments reported in several experimental investigations [5, 6], which can reduce the size of control actuator systems and make grid fins very attractive for missile applications. During a long time, the relatively high drag levels were the principal concern [7] which restricted the application of this technology. In the past few years, some studies [6] demonstrated that grid fin drag levels can be reduced considerably with only a minimal impact on lift and other aerodynamic properties.

Different approaches can be used to investigate theoretically the lattice wings. Several studies on grid fins [5, 8, 9] have been performed using linear methods and semi-empirical relations. Comparisons with experimental data have pointed out the inadequacy of linear methods [8, 9] for predicting grid fin aerodynamics especially at angles of attack above 5-8 degrees. Generally, the empirical formulations developed to extend the capabilities of aerodynamic predictions at high angles of attack provide a good agreement [9] with the experimental data for a wide variety of grid fin configurations and Mach numbers. In the last years, different papers have focused on numerical simulations of flows around lattice wings [10, 11] and missiles with lattice wings [4, 3]. These different investigations have shown that generally the Euler predictions underpredicted the experimental results. This has been demonstrated for a missile with lattice wings [3] but also for the flow inside a grid cell [10]. As expected, the Navier-Stokes solutions for a grid fin [11] or a missile with lattice wings [4] show a reasonable agreement with the measurements.

The inconvenience of the Navier-Stokes calculations for a missile with lattice wings is the high computational cost of such a configuration. A previous study [12] has shown that the use of an actuator disc to model the grid fins can reduce by a factor six this cost. Using this approach, the effects of the lattice wing were modelled through forces interpolated from an experimental database. This method has been successfully applied to the numerical simulations around an isolated lattice wing and a missile with grid fins [12]. The results have shown that while the flow conditions were in the database range of validity, the method was able to predict the global forces acting on the vehicle with a margin error which was of the same order that for a body alone. Here, this actuator disc for lattice wings has been incorporated in an unstructured Navier-Stokes solver for reducing the computational effort of mesh generation for future vehicles equipped with this technology. To improve the reliability of the predictions, the forces involved by the presence of the lattice wing are computed using semi-empirical relations [1]. Finally, the tool is applied to an isolated lattice wing and the numerical predictions are compared against experimental data [13] obtained for the same geometry.

2 Modelling

2.1 Actuator disc

An actuator disc consists in replacing a geometry by artificial boundary conditions inside a flow. The main interest of this approach lies in the saving of computational time through the modelling of complex geometries. In this paper, this method is used to model the effects of a lattice wing (see Fig. 1). The grid fin is replaced by boundary conditions where its effects on the flow are modelled through forces computed with semi-empirical relations.

The actuator disc theory is principally based on the application of the conservation laws for mass, momentum and energy. Here, the boundary conditions are briefly described, more details are given in [14]. The actuator disc is split in two sets of boundary conditions: the inflow and outflow conditions. The outflow conditions are located at the upstream side and the inflow conditions at the downstream side of the actuator disc. At the outflow boundary, all the variables are extrapolated if the flow is supersonic while one variable has to be imposed to get a numerically stable condition for a subsonic flow. In the present study, the mass flux is chosen as the variable to be prescribed in subsonic flows. At the downstream side, all the variables are calculated from their value at the upstream side if the flow is supersonic.

When the flow is subsonic one flow variable has to be extrapolated from the flow field. A previous study [14] showed that the extrapolation of the velocity square leads to a more realistic pressure distribution. The forces acting on the lattice wing are computed using the numerical procedure described below and are taken into account at the inflow boundary conditions of the actuator disc.

2.2 Lattice wing theory

At the actuator disc, the forces due to the presence of the lattice wing have to be provided. In a previous study [12], the forces were interpolated from an experimental database. The numerical predictions around a missile with grid fins [12] showed some discrepancies with the experiments [13] originating from the database which was not valid in presence of a vortical flow at high angle of attack. This investigation put in evidence a strong dependence of the result reliability on the database. Here, in order to extend the applicability of the method, the actuator disc is coupled with a numerical procedure based on the semi-empirical lattice wing theory [1]. Using this theory, which is the result of long-time experimental and numerical investigations, the force coefficients of the grid fins are provided by semi-empirical relations. The experimental results obtained at DLR [13], for basic lattice wing configurations and a wide range of Mach numbers, have been used to complete and modify this theory.

The calculation of the lattice wing performances takes advantage of the fact, that at all Mach numbers, the angle of attack corresponding to the maximum lift is considerably higher than for a monoplane wing. The neighbouring lifting planes of the lattice wings induce a more favourable alignment of the flow. This delays the separation to higher angles of attack and causes a smooth separation at supercritical angles. The presence of orthogonal planes avoids the cross flows and gives a good basis for using the linear theory. Each plane of a lattice wing corresponds to a high aspect ratio wing. Another simplification of the theoretical model is the independence of the load capacity of a lattice wing (with given plane spacing and wing sizes) on the internal arrangement of the grid (framework or honeycomb). This hypothesis is valid for all wings with a large number of cells ($n > 10$). Therefore, each cell of the wing is considered to be a square box shaped wing, and its internal flow has an aerodynamic behaviour which is completely independent from the rolling over the longitudinal axis of the wing. These qualities of the grid fins simplify the calculation of the aerodynamic forces at subsonic, transonic and supersonic regimes. A sufficiently exact prediction of the performance of these wings, at any angles of attack and yaw, is performed through a simplified formulation of the flow around the lattice wing.

At subsonic speeds, the calculation model is based on the lifting line scheme. Each individual plane of the grid fins is supposed to be under the same conditions that one of the corresponding polyplanes with infinite span at the same effective angle of attack. Then, the lift of the grid is approximately equal to the lift of the corresponding polyplane of infinite span. The angle of attack of the corresponding polyplane of infinite span is determined by taking into account the average downwash angle from the free vortices of the grid. Afterwards, the induced drag of the wing as well as the friction drag can be calculated.

The model for transonic flows is applied when the sonic speed is reached in the narrowest cross section of the wing. For the calculation of the normal force coefficient, it is supposed that the variation of the normal force with the Mach number is proportional to the variation of the dynamic pressure. The total drag of the wing is increased additionally by the wave drag, which can be calculated using an empirical correlation depending on the wing geometry.

In supersonic regime, the flow between two planes of the lattice wing at moderate incidence angles can be calculated with analytic equations. The pressure distributions on the wing surface and the induced forces (without friction drag) can be determined considering the emerging wave interactions.

Using the semi-empirical lattice wing theory, the aerodynamic forces induced by a lattice wing are calculated as functions of the flow conditions (Mach number, angles of attack and yaw) and grid fin geometrical characteristics (height, chord, span, spacing and plane thickness).

3 Numerical results

3.1 Flow solver

The code TAU [15], developed by DLR, has been used for the numerical simulations. It solves the three-dimensional Navier-Stokes equations using a finite volume approach and can handle structured, unstructured and hybrid meshes built with prisms, pyramids, tetrahedra and hexaedra. The integration in time is carried out using an explicit Runge-Kuttta scheme. The space discretization is done using the AUSM-DV solver and dissipation terms are added to damp high frequency oscillations. The final scheme is accurate to the second order in space.

3.2 Configuration

The flow has been computed around the lattice wing of Fig. 1 without its support. The grid fin is of 36 mm height, 28.82 mm width, with a thickness of 5 mm. The unstructured mesh used for the simulation consists of 155911 tetrahedric cells. From the lattice wing, the grid extends over five lattice-wing heights in the axial direction and four grid-fin widths in the transverse direction. The actuator disc boundary conditions are applied on the lattice wing faces while outflow supersonic conditions are imposed on the sides. On the other boundaries, farfield conditions are set.

3.3 Result discussion

The configuration described above has been computed for several Mach and Reynolds numbers and different angles of attack. The different computed cases are listed in

Table 1. A CFL number of one has been used for the numerical simulations. Converged solutions are obtained after 100 to 500 iterations depending on the Mach number and the angle of attack.

The numerical results have been compared against the experimental data obtained at DLR-Köln [13] for the same lattice wing. In Fig. 3 the experimental, theoretical and numerical values of the drag at different Mach numbers and without angle of attack are plotted. The theoretical values are obtained for a complete lattice wing using the procedure developed from the lattice wing theory described in this paper. Fig. 3 shows that the theoretical and numerical values are nearly similar. The agreement with the experimental data is good at Mach 1.8 and 2 even if the experimental data is underestimated by both theory and simulation. At higher Mach numbers there is a larger difference between the experiments and the predictions. This discrepancy seems to come from the semi-empirical relations used to compute the force coefficients. These relations, based on the lattice wing theory, are valid for rectangular grid fins with uniform cells. The grid fin computed here (see Fig. 1) is octagonal and some cells close to its base differ in shape. Therefore, the semi-empirical relations are not fully adapted to the current lattice wing and their adaptation should improve the agreement with the experiments.

In Fig. 4 the axial and normal force coefficients are plotted as functions of the angle of attack. Several simulations have been performed at Mach 2 for several angles of attack: 0, 5, 10 and 20 degrees. The numerical predictions of both axial and normal force coefficients are in good agreement with the theoretical values. The agreement with the experimental data is generally good and the difference does not exceed 10 %. As for the prediction of the drag without angle of attack (see Fig. 3), the discrepancies are induced by the different shape of the lattice wing used to built the semi-empirical relations. However, the method is able to reproduce the trends of the force coefficient evolution with the Mach number and the angle of attack.

4 Conclusions

A method devoted to the prediction of flows around lattice wings has been incorporated in the unstructured DLR TAU solver. This approach consists in replacing the lattice wing by an actuator disc coupled with a semi-empirical procedure. This procedure provides the force coefficients of the grid fins as functions of the flow conditions and grid fin geometrical characteristics. The main interest of the actuator disc technique is the savings of the computational effort required to solve complex configurations. This approach has been applied to the predictions of the force coefficients of a lattice wing. The force coefficients predicted by the simulations match well with the experimental data and the comparison with the experiments shows that the method is able to reproduce the main trends of the force coefficient evolution as function of Mach number or angle of attack. The remaining discrepancies with the experiments are attributed to the different shape of the lattice wing used to calculate the force coefficients with the lattice wing theory. This demonstrates the potential of this technique to be used for the earlier phases of vehicle design when the geometry and the position of the grid fins has to be determined. In the future this method will be applied to the flow prediction around a vehicle with lattice wings to evaluate its capacity as a tool for system analysis of hypersonic vehicle with grid fins.

Acknowledgements

The authors would like to thank M. Helmut Esch from DLR-Köln for the valuable discussions on his experimental results and Dr. José Longo from DLR-Braunschweig for his comments and remarks on this paper.

References

[1] S. M. Belotserkovsky, L. A. Odnovol, Yu. Z. Safin, A. I. Tylenev, V. P. Frolov and V. A. Shitov: "Reshetchatye krylia". Moskow, Maschinostroenie, 1985, 320 p.

[2] W. D. Washington and M. S. Miller: "Experimental investigations of grid fin aerodynamics: a synopsis of nine wind tunnel and three flight tests". Proceedings of RTO AVT Symposium on Missile Aerodynamics, 11-14 May, Sorrento, Italy, in **RTO MP 5**, 1998, pp. 10.1–10.13.

[3] M. Khalid, Y. Sun and H. Xu: " Computation of flows past grid fin missiles". Proceedings of RTO AVT Symposium on Missile Aerodynamics, 11-14 May, Sorrento, Italy, in **RTO MP 5**, 1998, pp. 12.1–12.11.

[4] J. DeSpirito, H. L. Edge, P. Weinacht, J. Sahu and J. P. G. Dinavahi: "Computational fluid dynamics analysis of a missile with grid fins". Journal of Spacecraft and Rockets, **38(5)**, 2001, pp. 711–718.

[5] R. A. Brooks and J. E. Burkhalter: "Experimental and analytical analysis of grid fin configurations". Journal of Aircraft, **26(9)**, 1989, pp. 885–887.

[6] W. D. Washington and M. S. Miller: "An experimental investigation of grid fin drag reduction techniques". AIAA Paper,**93-0035**, 31st Aerospace Sciences Meeting & Exhibit, Jan. 11-14, Reno, NV, 1993.

[7] E. Y. Fournier: "Wind tunnel investigation of grid fin and conventional planar control surfaces". AIAA Paper,**2001-0256**, 39th Aerospace Sciences Meeting & Exhibit, Jan. 8-11, Reno, NV, 2001.

[8] J. E. Burkhalter, R. J. Hartfield and T. M. Leleux: "Nonlinear Aerodynamic Analysis of Grid Fin Configurations". Journal of Aircraft, **32(3)**, 1995, pp. 547–554.

[9] R. W. Kretzschmar and J. E. Burkhalter: "Aerodynamic Prediction Methodology for Grid Fins". Proceedings of RTO AVT Symposium on Missile Aerodynamics, 11-14 May, Sorrento, Italy, in **RTO MP 5**, 1998, pp. 11.1–11.11.

[10] F. Lesage: "Numerical investigation of the supersonic flow inside a cell". In Proceedings of the 17th International Symposium on Ballistics, BALLISTICS'98, 23-25 March, Midrand, South Africa, 1998, **Vol. 1**, pp. 209–216.

[11] H. Lin, J-C. Huang and C-C. Chieng: "Navier-Stokes computations for body/cruciform gridfin configuration". AIAA Paper, **2002-2722**, 20th AIAA Applied Aerodynamics Conference, June 24-27, St. Louis, Mo, 2002.

[12] Ph. Reynier, U. Reisch , J. Longo and R. Radespiel: "Numerical study of hypersonic missiles with lattice wings using an actuator disk". AIAA Paper, **2002-2719**, 20th AIAA Applied Aerodynamics Conference, June 24-27, St. Louis, Mo, 2002.

[13] H. Esch: "Aerodynamisches Beiwerte der Längsbewegung eines Flugkörpers mit Gitterleitwerk im Überschall". DLR Report, **IB 39113-2000C34**, Köln, 2000.
[14] U. Reisch: "Simulation of lattice wings with the actuator disk concept". DLR Report, **IB 129-2000/13**, Braunschweig, 2000.
[15] T. Gerhold, O. Friedrich, J. Evans and M. Galle: "Calculation of complex three dimensional configurations employing the DLR TAU code". AIAA Paper, **97-0167**, 1997.

Table 1 Angle of attack, Mach and Reynolds numbers of the different computed cases.

Case	Mach number	Reynolds number	Angle of attack
1	1.8	$1.8\ 10^6$	0
2	2	$1.9\ 10^6$	0
3	3	$2.5\ 10^6$	0
4	4	$3.3\ 10^6$	0
5	2	$1.9\ 10^6$	5
6	2	$1.9\ 10^6$	10
7	2	$1.9\ 10^6$	20

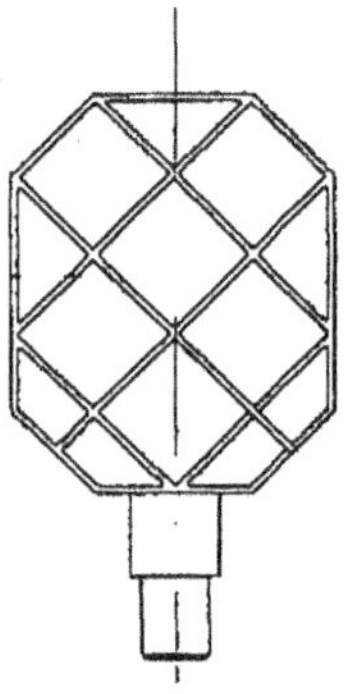

Figure 1 Lattice wing, with its support, experimentally investigated by Esch [13] and numerically studied here.

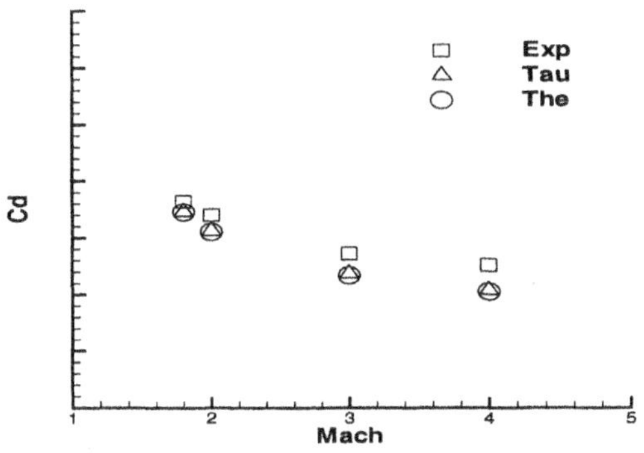

Figure 2 Experimental data, theoretical and computational values of the drag coefficient in function of the Mach number (no angle of attack).

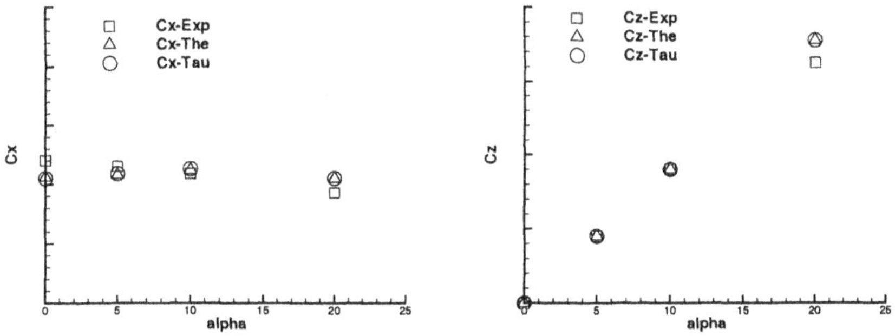

Figure 3 Experimental, theoretical and computational values, at Mach 1.8, of the axial and normal force coefficients in function of the angle of attack. The axial force coefficient is plotted on the left side and the normal force coefficient on the right side.

Control of Flow around Bluff Bodies by Passive Wake Ventilation

F.-R. GROSCHE, M. LU and H.-P. KREPLIN
German Aerospace Center (DLR)
Institute of Aerodynamics and Flow Technology
Bunsenstr. 10, 37073 Göttingen, Germany

Summary

Flow control by passive ventilation of wakes of spheres and elongated bluff bodies of different fineness ratios was investigated in the Göttingen High Pressure Wind Tunnel at Reynolds numbers up to 10^7. Wake ventilation was provided by a central duct along the model axis connecting the stagnation region in front of the body with the wake region in the rear. A very substantial drag reduction was observed at supercritical and transcritical Reynolds numbers, reaching up to 66% for spheres. High drag reductions were also found for elongated bluff bodies at angle of incidence 5 deg and 10 deg. At zero angle of incidence the ventilation effect decreases with increasing fineness ratio. Pitching moment was enhanced by wake ventilation.

1. Introduction

The drag of bluff bodies is determined mainly by the pressure drag caused by flow separation, which leads to a big pressure difference between front and rear surfaces. Additionally, the shear layers of the separated flow tend to be unstable. Therefore the pressure distributions on the body are unsteady and cause vibrations of the body. Passive ventilation of the wake region was proposed by G.E.A. Meier [4] as a possible means to overcome these problems. A bypass channel provides a flow through the bluff body from the stagnation area in front of the body to the near wake regime in the separated flow area at the rear. This duct can simply be realised as a central circular bore through the body. It was assumed that the jet emanating from the rear end of the venting duct might align the free shear layers of the wake to itself and draw them closer to the surface of the bluff body. This would lead to a more streamlined shape of the surrounding flow field resulting in a lower pressure drag. First wind tunnel experiments on the above principle were carried out by G. K. Suryanarayana et al. [4, 5] in the 1m-low speed wind tunnel of DLR Göttingen. Three simple plastic foam spheres of 200 mm diameter with sting support were tested. The inlet diameter of the venting duct was 30 mm. Therefore, the cross section of the straight circular duct was 2.25 % of the cross section of the sphere. Two conical shapes of the duct were also tested, as indicated in Figure 1. The drag of the same spheres with closed ducts (basic spheres) was measured for comparison. The test results of Figure 1 indicate only small effects of wake ventilation on the drag of the spheres at subcritical Reynolds numbers (based on body diameter). But very significant drag reductions around 60 %are found at supercritical Reynolds numbers for both the straight and the conical ducts. Discrepancies of the results in the critical Reynolds number range were caused by slight model inaccuracies of the three individual spheres.

Unfortunately, these drag measurements in a low speed wind tunnel at atmospheric pressure were limited to rather small supercritical Reynolds numbers up to $Re \approx 8.5 \times 10^5$, as shown in Figure 1. From a technical point of view, also much higher Reynolds numbers would be of interest. Therefore, the following question arose: How far extends the Reynolds number range in which passive wake ventilation causes such a strong reduction of aerodynamic drag as realised at low supercritical Reynolds numbers?

1. Wind tunnel tests on the effects of passive wake ventilation at high Reynolds numbers

To answer this question, experiments in the 0.6 m x 0.6 m – High Pressure Wind Tunnel of DLR Göttingen were carried out. By pressurising the air in the tunnel up to 100 bars, very high Reynolds numbers can be obtained at flow velocities up to 35 m/s and model dimensions in the order of 0.1 m to 0.3 m. Also, the Reynolds number of a given model can be changed easily by varying the tunnel pressure and/or velocity.

2.1 Drag measurements on vented spheres at high Reynolds numbers in the High Pressure Wind Tunnel at DLR Göttingen

At first the drag force on a sphere of 200 mm diameter with an asymmetric sting support was measured. Figure 2 illustrates the model. For reference measurements with the basic sphere configuration the inlet and outlet of the venting duct were plugged and smoothed. More details can be found in [2,3].

Figure 3 gives the main results of the tests on the drag of this model. The drag reduction by wake ventilation is close to 60% in the Reynolds number range $4 \times 10^5 < Re \lessdot 6 \times 10^5$. Above that range the wake ventilation effect is somewhat weaker but remains practically constant with a drag reduction of approximately 40 % up to the highest Reynolds number realised in this test.

This proves that the drag of spheres is strongly reduced by passive wake ventilation not only at small supercritical Reynolds numbers but over a large Reynolds number range limited only by technical problems related to excessive vibrations of the model at high flow velocities.

In order to corroborate and extend this finding, more tests were conducted with a smaller sphere – better suited for the test section of the High Pressure Wind Tunnel, which is smaller than the test section of the 1m-wind tunnel. In contrast to the previous sphere models, this sphere was supported symmetrically by a central sting, as illustrated in Figure 4. The inner surface of the ventilating duct of the sphere was connected to the sting by three spokes. The free cross section of the venting duct around the sting is about 2.3 % of the body cross section similar to the previous models. Again the basic sphere configuration was realised by plugging and smoothing the ends of the ventilation duct.

Figure 5 illustrates the test results. With this model the measured drag reduction by passive venting increases with Reynolds number in the supercritical and transcritical region from about 50 % to about 66 % of the drag of the basic sphere. These values correspond well with the former findings. We conclude therefore that the effect of passive wake ventilation on the drag of spheres is not impaired significantly by the disturbances in the rear portion of the model caused by the support sting.

Although the drag reducing effect of the passive wake ventilation is very similar for all spheres investigated, the absolute values of the drag coefficients of both the basic and the vented sphere with central sting support are considerably lower than those of the spheres with asymmetric support. We believe the main reason for this to be that the disturbance of the wake flow by a central sting is weaker than that of an asymmetric sting support, leading to a lower drag. The drag coefficients of the basic sphere in Figure 5 coincide well with results given by Achenbach [1] and others, see [3]. Again, the drag reduction in the subcritical Reynolds number regime is rather small, in the order of 10 %. This is also in agreement with the earlier investigations on the spheres with asymmetric sting support.

2.2 Measurements on elongated bluff bodies at high Reynolds numbers in the High Pressure Wind Tunnel at DLR Göttingen

In order to investigate the effects of passive wake ventilation on bodies which come closer to practical configurations, e.g. to an aircraft fuselage, tests were also made with elongated bluff bodies consisting of a cylindrical mid portion with attached spherical ends, see Figure 6. This configuration was chosen because different fineness ratios L/D could be realised easily by varying the length of the cylindrical portion, which consisted of several cylindrical elements, connected to each other. Fineness ratios L/D between 1.125 and 3 could thus be provided. The model diameter was again D = 120 mm and the diameter of the venting duct was 20 mm, as for the sphere with central sting, see section 2.1. Instead of installing the strain gage balance in the support sting it was mounted inside the model, thus minimising moments on the balance which otherwise might have overloaded the device. For more details on the model see [3]. Measurements were made not only at angle of incidence $\alpha = 0$ deg but also at $\alpha = 5$ deg and $\alpha = 10$ deg in order to examine the effects of cross flow on the passive wake ventilation. Drag, lift and pitching moment were measured in the same Reynolds number range up to $Re \approx 10^7$ (Re again based on body diameter, not length). Only the results on drag and pitching moment will be discussed here but more data are given in [3].

2.2.1 Drag of elongated bluff bodies

As a typical example for the Reynolds number dependence of the drag coefficient, results are presented here only for the elongated body with fineness ratio

L/D = 2.5; for data on the other fineness ratios see [3]. Figure 7 illustrates the effect of passive wake ventilation on the drag at angles of incidence $\alpha = 0$ deg and 10 deg. Rather small drag reductions are observed at zero angle of incidence, with the maximum drop at the highest Reynolds numbers. But at $\alpha = 10$ deg much stronger drag reductions occur and persist over the whole Reynolds number range investigated. The results for $\alpha = 5$ deg – which are not presented here - indicate drag reductions by wake venting similar to the case $\alpha = 10$ deg with the drag coefficients between the values for zero deg and 10 deg.

The main lesson from these tests is that the passive wake ventilation remains effective up to high Reynolds numbers not only for spheres but also for elongated bodies with and without angle of incidence.

As an overview of the results on drag, Figure 8 presents the drag coefficients of the bluff bodies – averaged over the supercritical and transcritical Reynolds number range – as functions of the fineness ratio L/D for all three angles of incidence investigated. L/D = 1 signifies the sphere.

For $\alpha = 0$ deg the drag of the basic model - without wake ventilation – falls off strongly with L/D up to L/D ≈ 1.5 and then stays almost constant. The drag of the vented model is significantly lower in the case of the sphere but is almost independent of L/D over the range investigated. Therefore the drag reduction by wake ventilation declines with L/D from more than 40 % at L/D = 1 to about 10 % at L/D = 3. The drag coefficients of the sphere with symmetric support by a central sting (Figure 4) are lower than with asymmetric support (Figure 2), but the drag reduction ΔC_D by venting is approximately the same.

For $\alpha = 5$ deg the difference in drag between the basic and vented models remains high over the whole range of L/D, in contrast to the case $\alpha = 0$ deg. Both configurations show almost constant drag over L/D. The drag coefficient of the basic model is approximately doubled with respect to the zero angle of incidence case for L/D ≥ 1.5 whereas the drag of the vented models has increased only moderately. Thus the relative drag reduction is more than 40% for all fineness ratios.

At $\alpha = 10$ deg the results are in essence quite similar to the 5 deg case. The drag coefficients of both the basic and the vented models are still higher, as would be expected. The difference – that is the drag reduction by venting – remains about constant over the whole range of fineness ratios L/D. The absolute values of ΔC_D are around 0.07 to 0.085 like in the $\alpha = 5$ deg case. The relative drag reduction ΔC_D / C_D (basic model) ≈ 30% is poorer than for $\alpha = 5$ deg only because of the higher reference drag C_D (basic model) at 10 deg angle of incidence.

In essence these results demonstrate that the flow angle makes a marked difference on the effect of venting. Not only the drag of elongated bluff bodies of the type investigated here increases considerably with angle of incidence, but also the drag reduction by venting is greatly enhanced as compared to the zero incidence case (except for the smallest fineness ratios L/D ≈ 1).

2.2.2 Pitching moment of elongated bluff bodies

The centre of the model was chosen as the reference point for the pitching moment, and the moment is normalised by $q\ D\ \pi\ D^2/4$, defining the pitching moment coefficient C_m. The Reynolds number dependence of the pitching moment is illustrated in Figure 9 for fineness ratio L/D = 2.5, data for other fineness ratios are given in [3]. At zero angle of incidence the pitching moments are not zero – as would be expected from symmetry – but slightly negative. We believe that this is due to the flow distortion by the asymmetric model support. At $\alpha = 5$ deg and 10 deg the pitching moments are positive and increase with angle of incidence.

The effect of wake ventilation extends over the whole Reynolds number range, as for the drag. It leads to a more positive pitching moment of the body. The greater scatter of the measured data at low Reynolds numbers $Re < 4 \times 10^5$ may be due to transitional flow effects,

To give a similar overview as for the results on drag presented in Figure 8, the pitching moment coefficients - also averaged over the supercritical and trans-critical Reynolds number range - are plotted as functions of fineness ratio L/D in Figure 10. As expected, the pitching moment of the basic bodies increases almost linearly with fineness ratio L/D and with angle of incidence α.

The wake ventilation shifts the pitching moments of all configurations to more positive values as compared to the basic models. This means that the pitching moment of the elongated bodies becomes more unstable by wake ventilation. The effect increases with fineness ratio. Interestingly, the few results available on pitching moments of the sphere indicate the inverse influence of venting; the moment becomes more negative at positive angles of incidence. This is true for the model with symmetric (central sting) support as well as with asymmetric support, see Figure 10, but we consider further tests necessary to verify this observation and to study more closely the pitching moment characteristics of the bluff bodies with fineness ratios $L/D \approx 1$.

3. Conclusion

Experiments on passive wake ventilation of bluff bodies in the High Pressure Wind Tunnel of DLR Göttingen proved that the drag reduction remains high or even increases up to the highest transcritical Reynolds numbers $\leq 10^7$ reached in our tests. The drag reduction of elongated bluff bodies depends on fineness ratio L/D and angle of incidence α. At $\alpha = 0$ deg the effect of wake ventilation declines with increasing fineness ratio. At $\alpha \geq 5$ deg the drag reduction by passive wake ventilation remains almost as high as for the sphere for all fineness ratios investigated. Passive wake ventilation increases the pitching moment of the elongated bodies investigated with $L/D \geq 1.125$. Some surface flow visualisations and measurements of pressure distributions on the spheres have been made to examine the flow patterns but we consider more experiments using optical measuring techniques necessary in order to better understand the observed flow effects.

Acknowledgements

The authors acknowledge the very helpful cooperation of H.J. Bendig as wind tunnel operator in performing the tests at the High Pressure Wind Tunnel.

References

[1] Achenbach, E.: Experiments on the flow past spheres at very high Reynolds numbers. J. Fluid Mech., vol. 54, part 3, 1972, pp. 565-575

[2] Lu, M.: Widerstandsverminderung durch Ventilation stumpfer Körper bei höheren Reynoldszahlen. DLR-Forschungsbericht 2001-28, 2001

[3] Lu, M., Grosche, F.-R., Meier, G.E.A: Drag reduction of a sphere at supercritical Reynolds numbers using passive ventilation. Proceedings IUTAM-Symposium on Mechanics of Passive and Active Flow Control, 7.-11.Sept. 1998, Kluwer Academic Publishers, Dordrecht / Boston / London, 1999, pp. 127 – 133

[4] Meier, G.E.A.; Suryanarayana, G.K.; Pauer, H.: Bluff-body drag reduction by passive ventilation. First European Conference on Fluid Mechanics EUROMECH, Cambridge, UK, 16-20 Sept., 1991

[5] Suryanarayana, G.K.; Pauer, H.; Meier, G.E.A.: Bluff-body drag reduction by passive ventilation. Exp. in Fluids 16, 1993, pp. 73-81

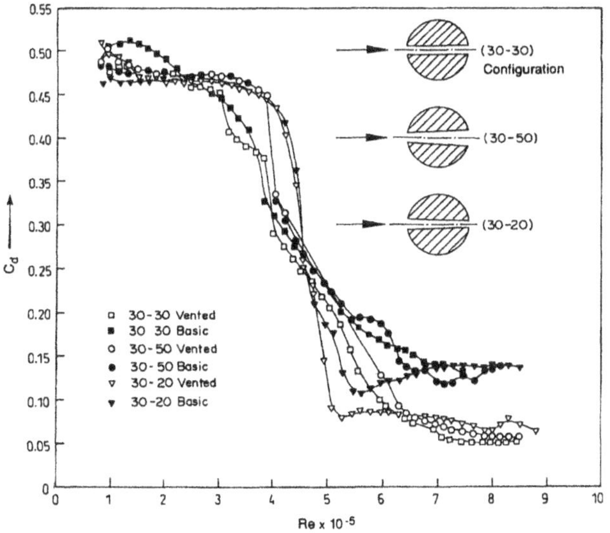

Figure 1: Drag reduction of spheres by passive wake ventilation

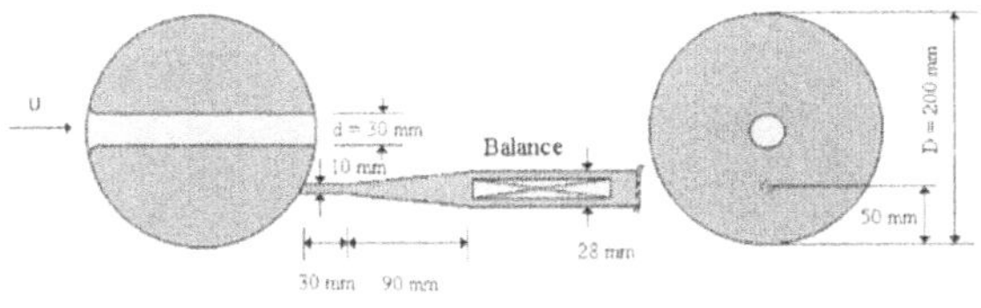

Figure 2: Model sphere with asymmetric sting support

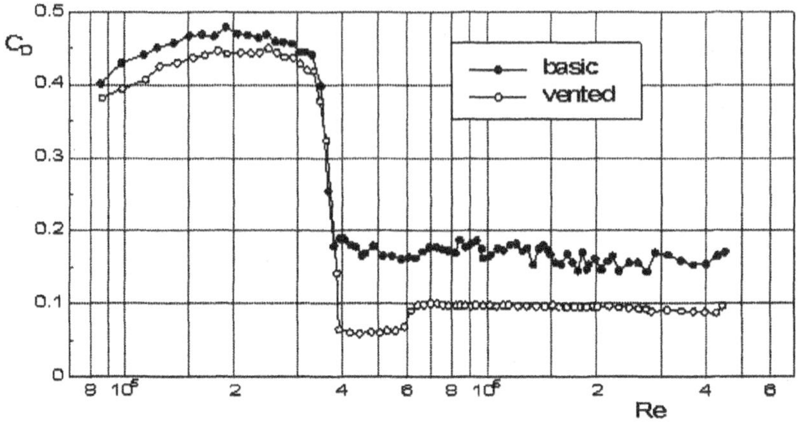

Figure 3: Drag coefficients of sphere with asymmetric sting support

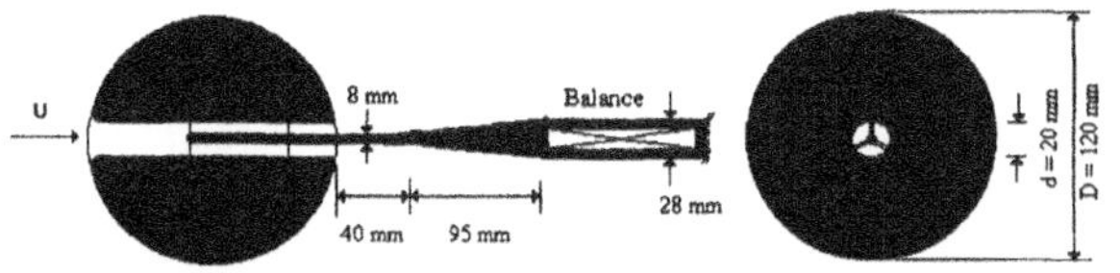

Figure 4: Model sphere with symmetric (central) sting support

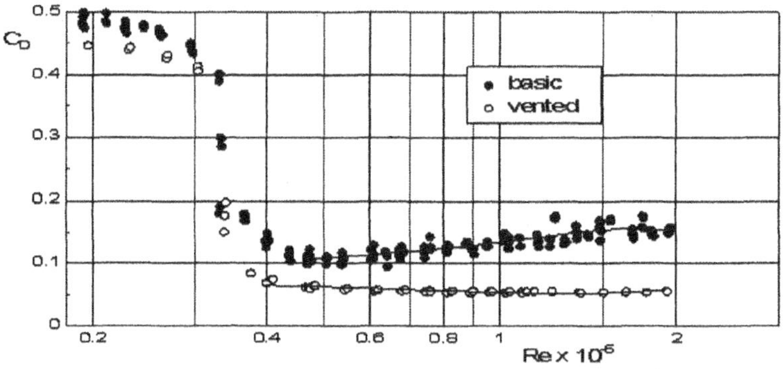

Figure 5: Drag coefficients of sphere with symmetric (central) sting support

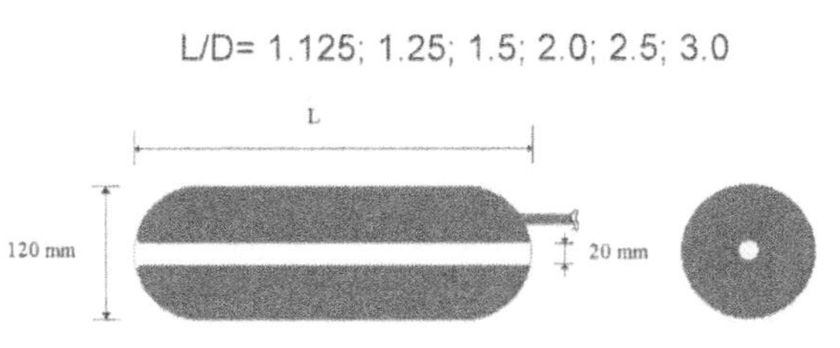

Figure 6: Elongated bluff body with variable fineness ratios L/D

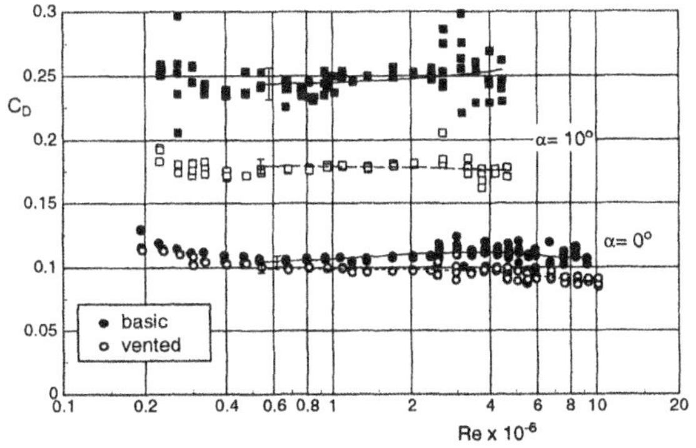

Figure 7:
Drag coefficients of elongated bluff body with fineness ratio L/D = 2.5

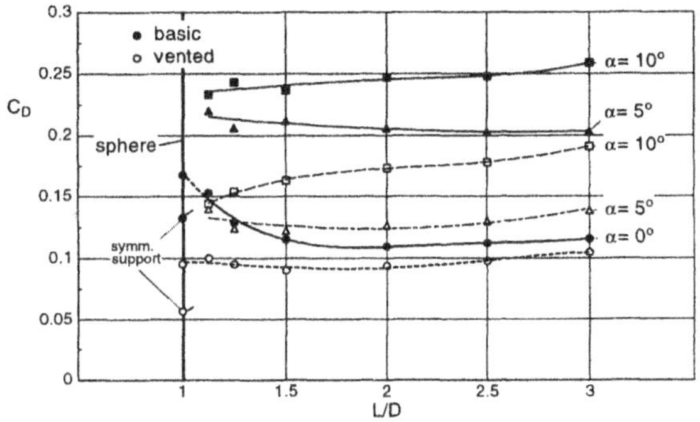

Figure 8:
Drag coefficients of elongated bluff bodies, averaged over Reynolds number, as functions of fineness ratio L/D

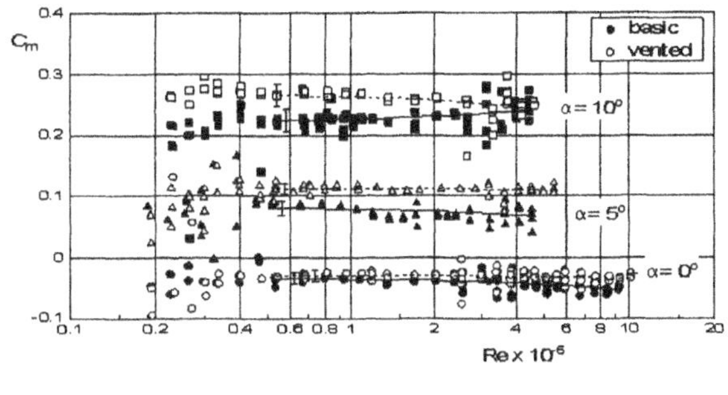

Figure 9:
Pitching moment coefficients of elongated bluff body with fineness ratio L/D = 2.5

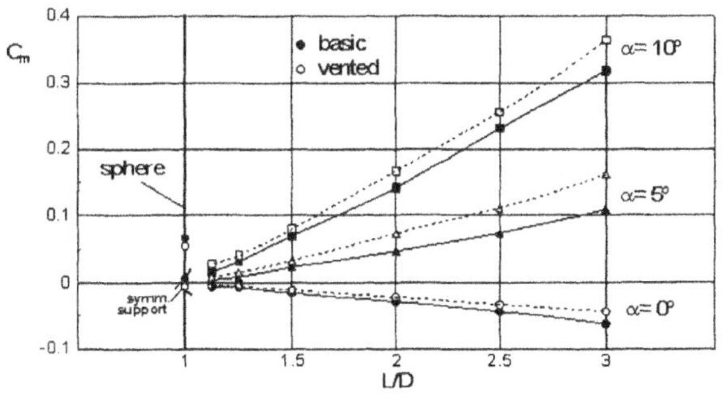

Figure 10:
Pitching moment coefficients of elongated bluff bodies, averaged over Reynolds number, as functions of fineness ratio L/D

Experimental investigation of local vortex roughness receptivity of a boundary layer on an airfoil

S. Herr, W. Würz, S. Wagner

Institut für Aero- und Gasdynamik (IAG), Universität Stuttgart
Pfaffenwaldring 21, D-70550 Stuttgart, Germany
http://www.iag.uni-stuttgart.de/laminarwindkanal/

Summary

In the present paper, the linear local vortex-roughness receptivity of the 2D boundary layer of an airfoil is investigated. The vortex is modeled by the convected wake of a vibrating wire that is oriented perpendicular to the free stream speed vector and parallel to the airfoil surface. Hot-wire measurements were taken downstream of local 2D and 3D roughness elements, phase-locked with the wire vibrations to study the Tollmien-Schlichting (TS) waves developing inside the boundary layer. Experiments with a circular 3D roughness element were performed in order to evaluate the complex 3D vortex-roughness receptivity function. The wave train developing downstream of the roughness was measured in spanwise scans at a constant wall distance. After Fourier decomposition of these scans the development of single wave numbers were compared to linear stability theory which then was used to find the initial Tollmien-Schlichting amplitudes at the position of the roughness. The receptivity function was found by normalizing them to the vortex amplitude measured at the boundary layer edge and the roughness height. The 3D roughness shape was Fourier decomposed so that the results are independent from the actual physical shape.
It was found that the vortex-roughness receptivity is smallest for the 2D case as the acoustic-roughness receptivity [10] [4] and as the vibrational receptivity [4]. The values are about 7 times smaller than for the acoustic-roughness receptivity.

1 Introduction

In the process of boundary layer transition from laminar to turbulent flow the *receptivity* marks the first step by providing the initial amplitudes. These initial amplitudes develop downstream through the linear and non linear stage and finally cause the boundary layer to break down to the turbulent state. The receptivity describes the processes in which external perturbations in the free stream (acoustics or vortices) are converted into instabilities of the boundary layer (TS-waves).
In contrast to the acoustic receptivity the problem of vortex receptivity is studied less so far. The previous investigations in this field are mainly theoretical ones. For *steady* streamwise vortices quantitative experimental data by Kendall is compared to calculations by Bertolotti in a joint work [1]. A very similar problem has

been studied in detail in recent experiments by Boiko [2] for the cases of both 2D (Blasius) and 3D (swept-wing) boundary layer. The mechanism of transformation of wall-normal *steady* free-stream vortices to *steady* boundary-layer perturbations (again streaks) was investigated experimentally and theoretically by Ustinov [5]. It was shown that an interaction of these vortices with a blunt leading edge of the experimental model leads to the formation of very strong streaks (streamwise vortices) inside the boundary layer. The theory predicted very well the experimental observations. The work of Dietz [3] is a quantitative experimental study of the receptivity problem for *unsteady* free-stream vortices. He studied the scattering of a weak vortex street (with 2D spanwise vorticity) on a localized two-dimensional surface roughness in a Blasius flow. The excitation of 2D TS waves has been observed and documented under controlled experimental conditions. This receptivity mechanism was studied theoretically by Wu [7] and a good agreement with experimental data was found.

At the Laminar Wind Tunnel of the University of Stuttgart the acoustic-roughness receptivity problem has been studied in great detail experimentally and with accompanying DNS [10] [4]. Based on this experimental knowledge a first approach to the distributed 3D-vortex receptivity of a boundary layer on an airfoil has been made [9]. In this investigation a $25\mu m$ gilded tungsten wire vibrating with a fixed frequency was used to introduce an antisymmetric vortex street with the vortex vector perpendicular to the free stream and to the leading edge of the airfoil, this led to the development of TS-waves in the boundary layer via the distributed receptivity mechanism.

In the present paper the local vortex-roughness receptivity with the vortex vector perpendicular to the free stream vector and *parallel* to the leading edge of the airfoil is investigated. First the experimental set up is described. In preliminary measurements with 2D roughness elements (strips) the generation of TS-waves in the boundary layer is investigated and the linearity of the problem is shown. Then the TS-wave train developing downstream of a 3D roughness element (circular spot) is studied. The initial amplitudes are found by extrapolation with the help of the linear stability theory and the receptivity function is determined by normalizing them with the causing factors (vortex amplitude and roughness height) analogically to the procedure for the acoustic-roughness receptivity function [10] [4].

2 Experimental set up and generation of unsteady vortex street

The experiments were carried out in the Laminar Wind Tunnel [6] of the *Institut für Aerodynamik und Gasdynamik*. The high contraction ratio of 100:1, as well as five screens and filters, result in a very low turbulence level, less than $Tu = 2 \cdot 10^{-4}$ for a frequency range of $20 - 5000\, Hz$. Hot-wire measurements were performed on a symmetrical airfoil section *XIS40MOD* [8] with 15% thickness. The free-stream velocity was chosen to be 30 $\frac{m}{s}$. This results in a Reynolds number of $Re = 1.2 \cdot 10^6$ based on the arc-length $s_{max} = 0.615\, m$ of the airfoil measured from the leading edge. The velocity distribution and the stability characteristics are shown in figure 1.

The external perturbation was modelled by the unsteady vortex street developing in the wake of a vibrating wire. A gilded tungsten wire with $d = 25\,\mu m$ was used. The wire is positioned $43\,mm$ upstream of the airfoil leading edge and spans the total height of the test section ($\sim 750\,mm$). A piezo actuator installed outside the test section drives the wire with a fixed frequency of $f_{vor} = 1088.9\,Hz$ which is close to the most unstable TS-frequency (figure 1). Due to the mechanical restrictions the wire vibrates in a higher harmonic of its *mechanical* eigenfrequency and a standing wave develops on it (fifth eigen mode with a distance between the knots $\sim 150\,mm$). A maximum amplitude of $3\,mm$ peak to peak is achieved in the anti knot. During the experiment the amplitude is controlled and the mechanical tension of the wire is adjusted by a special mounting. The set up is sketched in figure 2, it is similar to the one used for the investigation of the distributed vortex receptivity [9]. The AC-signal for the vibrating wire and the trigger for the data acquisition were generated strictly phase-locked from the same clock. The time signals at every measurement point were analyzed with a temporal FFT and the amplitude and phase obtained at the excitation frequency ($f = 1088.9\,Hz$) were used for further data processing.

3 Preliminary Measurements

First investigations deal with the disturbance induced by the vibrating wire. The wake has a distance to the model surface of about $8\,mm$. As the Reynolds number of the wire is small ($Re_d = 50$) the influence on the mean flow is weak; the mean velocity deficit is below 1% and the additional turbulence in the wake of the wire without vibration below 0.1% (figure 3). The spanwise distribution of amplitude and phase at the boundary layer edge (marked in figure 3) are shown in figure 4. The amplitudes are depending on the swing of the vibrating wire having a minimum at the knots of the standing wave and a maximum at the maximum swing. The phase is constant along one cycle of the standing wave on the vibrating wire and has a 180° jump in the minimum. The character of the disturbance can be found by evaluating the phase speed. The streamwise development of the phases averaged between the two minima is plotted in figure 5. The concerning phase speed is $a_{vor} = 35.3\frac{m}{s}$ which is 98% of the local free stream speed at the boundary layer edge (U_δ). The associated amplitudes are constant in the region of the roughness position and show regular beatings in the measurement region farther downstream. The dashed line in figure 5 represents the superposition of a disturbance with constant amplitude, travelling with phase speed a_{vor} and a growing TS-wave at the boundary layer edge.
The amplitudes at the spanwise station of the maximum of the wire vibration swing (at $z = 30\,mm$) is plotted in dependence of the averaged velocity in figure 6. It is in good agreement to the eigenfunction of the TS-wave as calculated with linear stability theory. For the following investigation in spanwise direction a constant wall normal distance was chosen to be at the y-position where $\frac{U}{U_\delta} = 0.42$. The amplitude and phase distribution at this height are plotted in figure 7. They show the same characteristics as the distributions at the boundary layer edge. In addition the ratio between the amplitude inside the boundary layer and the one at the boundary

layer edge is plotted, which is nearly constant. The downstream developments of the amplitudes and phases averaged between the two minima are plotted in figure 5. The disturbance inside the boundary layer travels with a speed of $12.2\frac{m}{s}$ this is very close to the phase speed of the TS-wave as it is predicted by linear stability theory with $11.8\frac{m}{s}$.
In the present study the linear receptivity problem was investigated. In figure 8 the TS-amplitudes measured at a position in the measurement region are plotted against the roughness height. The points can be interpolated by a straight line as it is sufficient for linearity. The roughness height used in the experiment ($41\ \mu m$) is marked in figure 8. As the ratio between TS-amplitude and the amplitude at the boundary layer edge is constant in spanwise direction (figure 7) we can assume that the excitation of TS-waves is also linearly depending on the vortex amplitude.

4 Determination of the receptivity function

In a next step measurements were performed with a 3D roughness element (circular spot) in order to evaluate the local complex 3D vortex-roughness receptivity function. Following the definition for the acoustic-roughness receptivity function the definition for the vortex-roughness receptivity function can be expressed as

$$\widetilde{G}_{vr}(\alpha,\beta) = \frac{\widetilde{B}_{inTS}(\alpha,\beta)}{\widetilde{A}_{vor} \cdot \widetilde{A}_r(\alpha,\beta)} \tag{1}$$

with the complex initial TS-amplitudes $\widetilde{B}_{inTS}(\alpha,\beta)$, the complex amplitude of the vortex street at the position of the roughness $\widetilde{A}_{vor}$, measured at the boundary layer edge, and the complex amplitude of the roughness itself $\widetilde{A}_r(\alpha,\beta)$. The formulation in wave number space has the advantage that the resulting values are independent from the actual physical shape especially of the roughness.
As the initial TS-amplitudes cannot be measured directly due to continuous-spectrum modes and forced oscillations in the vicinity of the surface roughness the boundary layer serves as a selective amplifier of the unstable (discrete-spectrum) modes and, simultaneously, as an attenuator of the continuous-spectrum modes and the forced oscillations. The downstream development of the wave train was measured in 5 spanwise scans, that covered the whole width of the TS-wave train developing downstream of the roughness, at the same constant wall distance as in section 3, which is close to the maximum amplitude of 2D and inclined TS-waves in their eigenfunctions. This position was adjusted during the experiment by keeping $\frac{U}{U_\delta} = 0.42$ constant. The time signals at every measurement point were analyzed with a temporal FFT and the amplitude and phase forming the complex value for the investigated TS-frequency ($f_{TS} = 1088.9Hz$) were used for further processing. The wave train at this frequency as it develops downstream of the roughness is plotted in figures 9 and 10. The spanwise scans were mapped by the complex Fourier transform

$$\widetilde{B}_{TS}(s_i,\beta) = \int_{-\infty}^{\infty} \widetilde{A}_{TS}(s_i,z) \cdot e^{-i\beta z} dz\ .$$

to the wave number space. After this decomposition, the downstream development of waves with different spanwise wave numbers can be followed separately. In figure 11 the downstream development of amplitude and phase for the TS-wave with wave number $\beta\delta_1 = \pm 0.096$ is plotted. The amplitude development predicted by the linear stability theory is added and due to the good agreement in the measurement region it can be used to extrapolate the initial amplitude at the position of the roughness. The phases can be extrapolated by a straight line (figure 11). The gradient of this line corresponds to the streamwise wave number $\alpha_{r,TS}$ of the TS-wave. As a result, the dependence of $\alpha_{r,TS}$ on β (dispersion relationship) is obtained and the propagation angle Θ of the corresponding TS-wave is found as $\Theta = \arctan \frac{\beta}{\alpha_{r,TS}}$. As the resonance condition demands [3]

$$\alpha_{rough.} = \alpha_{r,TS} - \alpha_{vor}$$

we found the amplitudes that generate TS-eigenmodes from the double Fourier decomposition of the shape of the roughness by shifting the dispersion relation of the TS-wave by the wave number α_{vor} of the vortex street. For the values along this curve the receptivity coefficients G_{vr} were evaluated according to equation 1. The receptivity functions G_{vr} are presented in dependence of the propagation angle Θ of the excited TS-wave in figure 12, the values obtained for a single 2D measurement are added. Basic properties of the vortex-roughness receptivity can be seen from these results. The scattering of the vortical disturbances on a local roughness is more effective for inclined TS-waves rather than for 2D ones. This behavior is similar to the vibrational and the acoustic-roughness receptivity [10] [4], but in contrast to the *distributed* vortex receptivity studied in [9]. The phase-delays of the generated TS-waves depend only weakly on the wave propagation angle Θ. The total value of the amplitude part of the vortex-roughness receptivity function is about 7 times smaller than the corresponding value for acoustic-roughness receptivity.

5 Conclusions and remarks

The linear vortex-roughness receptivity of the 2D boundary layer on an airfoil was investigated. The unsteady vortex street in the wake of a vibrating wire excites TS-waves of the same frequency on a roughness inside the boundary layer. The wave train behind a 3D roughness element was measured and the comparison to the linear stability theory was used to find the initial TS-amplitudes. Normalizing them with the causing factors (vortex amplitude and roughness height) leads to the vortex-roughness receptivity function. It is found that free stream vortex perturbations interacting with surface non-uniformities excite more effectively oblique (3D) TS-waves rather than 2D ones. The dependence of the phase part of the receptivity function on the TS-wave obliqueness (the spanwise wave number) is weak. The course of the vortex-roughness receptivity function with the propagation angle is analogous to the vibrational and the acoustic-roughness receptivity function [10] [4], but different from the distributed vortex receptivity as studied in [9]. Due to the total Fourier decomposition of the roughness shape the receptivity coefficients obtained in the present investigation are independent from the actual physical shape of the roughness. We note here that the presented values for the receptivity function are preliminary results because the post processing of the data is still in progress.

Acknowledgment

This work was performed under grant of the German Research Council (DFG) and is a contribution to the research program 'Transition'.

References

[1] Bertolotti F.P., Kendall J.M.: "Response of the Blasius boundary layer to controlled free-stream vortices of axial form". AIAA Paper 97-2018, 1997.

[2] Boiko A.V.: "Receptivity of boundary layers to free-stream axial vortices". Report IB 223-2000 A10, Institut für Strömungsmechanik, DLR Göttingen, 2000.

[3] Dietz A.J.: "Local boundary-layer receptivity to a convected free-stream disturbance". J. Fluid Mech. 378, pp. 291-317, 1999.

[4] Herr S., Würz W., Wörner A., Rist U., Wagner S., Ivanov A., Kachanov Y.: "Systematic investigations of 3D acoustic receptivity with respect to steady and unsteady disturbances. Experiment and DNS". Abschlusskolloquium DFG-Schwerpunktprogramm Transition, April 2002.

[5] Ustinov M.V.: "Streaks production by free-stream nonuniformity interaction with blunt leading edge". In 20th Int. Congress of Theoretical and Applied Mechanics (ICTAM 2000), Abstract Book. Tech. Rep. No. 950. DTAM Univ. Illinois at Urbana-Champaign, Abstract JA4, 2000.

[6] Wortmann F.X., Althaus D.: "Der Laminarwindkanal des Instituts für Aerodynamik und Gasdynamik der Technischen Hochschule Stuttgart". Zeitschrift für Flugwissenschaften Nr.12 Heft 4, 1964.

[7] Wu X.: "Receptivity of boundary layers with distributed roughness to vortical disturbances". In 20th Int. Congress of Theoretical and Applied Mechanics (ICTAM 2000), Abstract Book. Tech. Rep. No. 950. DTAM Univ. Illinois at Urbana-Champaign, Abstract GB06, 2000.

[8] Würz W.: "Hitzdrahtmessungen zum laminar-turbulenten Strömungsumschlag in anliegenden Grenzschichten und Ablöseblasen sowie Vergleich mit der linearen Stabilitätstheorie und empirischen Umschlagskriterien". Dissertation, Universität Stuttgart, 1995.

[9] Würz W., Herr S., Wagner S., Kachanov Y.S.: "A first experimental approach to the distributed 3D-vortex receptivity of a boundary layer on an airfoil". ICMAR 2002, Proceedings Part. II - Novosibirsk: ITAM, pp. 173-178, 2002

[10] Würz W., Herr S., Wörner A., Rist U., Wagner S., Kachanov Y.S.: "Three-dimensional roughness-acoustic receptivity of a boundary layer on an airfoil. Experiment and DNS". accepted for publication in J. Fluid Mech.

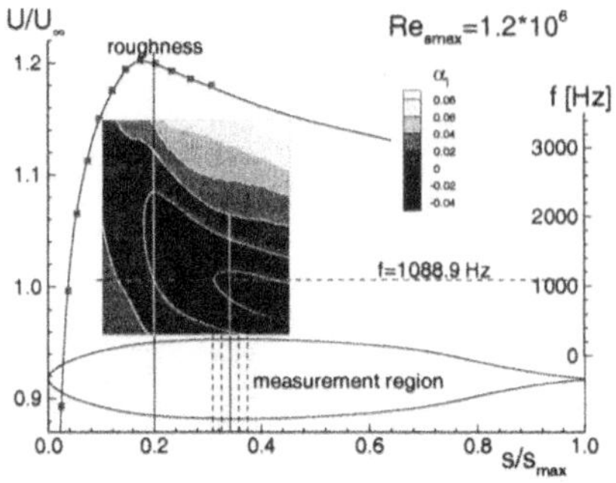

Figure 1 Velocity distribution and position of the roughness in the stability diagram

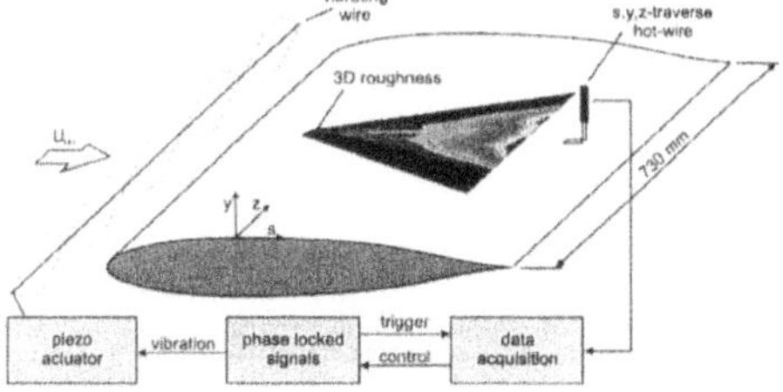

Figure 2 Experimental set up

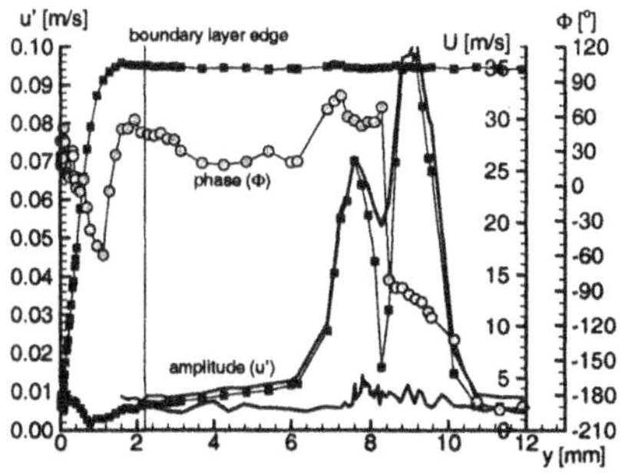

Figure 3 Wall normal distributions of phase and amplitude ($f = 1088.9Hz$) and total fluctuation amplitude u'_{RMS}

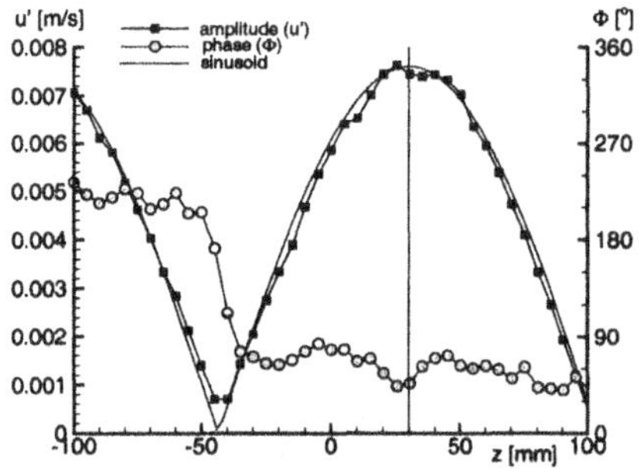

Figure 4 Spanwise distribution of amplitude and phase at the boundary layer edge

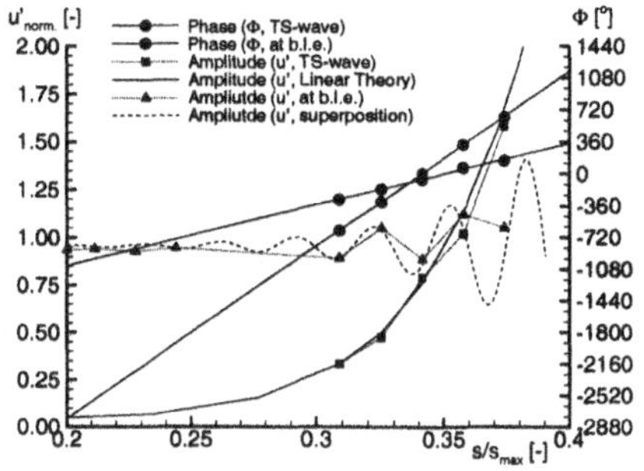

Figure 5 Streamwise development of amplitude and phase inside the boundary layer and at the boundary layer edge

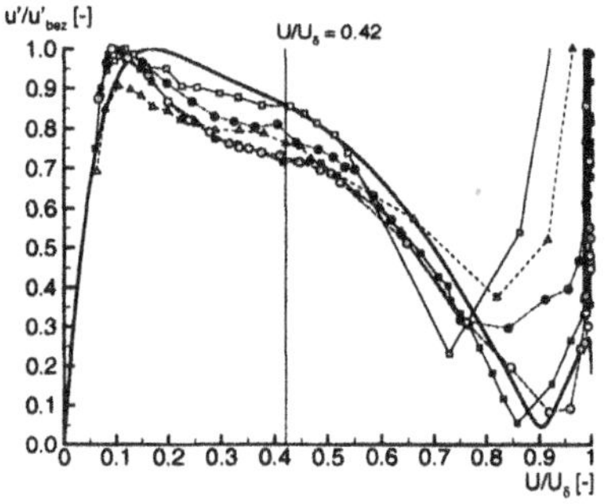

Figure 6 Wall normal distribution of amplitude inside the boundary layer in comparison to the eigenfunction for $f_{TS} = 1088.9Hz$

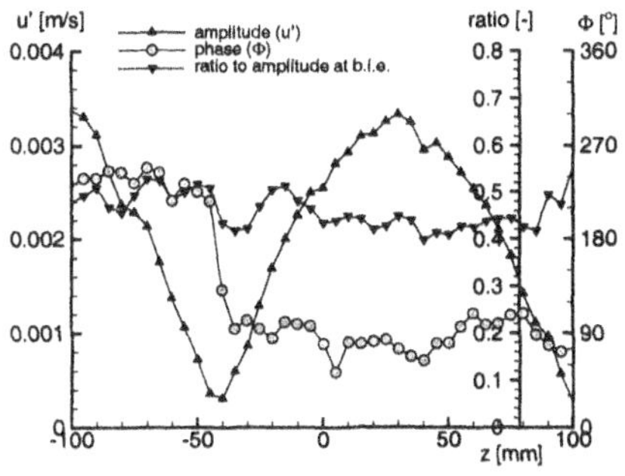

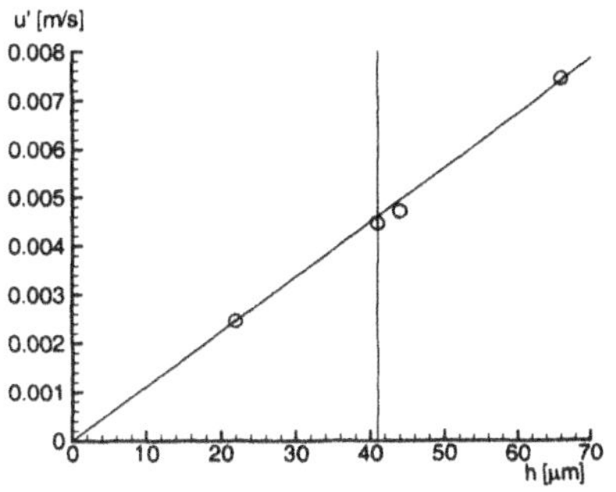

Figure 7 Spanwise distribution of amplitude and phase inside the boundary layer

Figure 8 TS-amplitude in dependence of roughness height

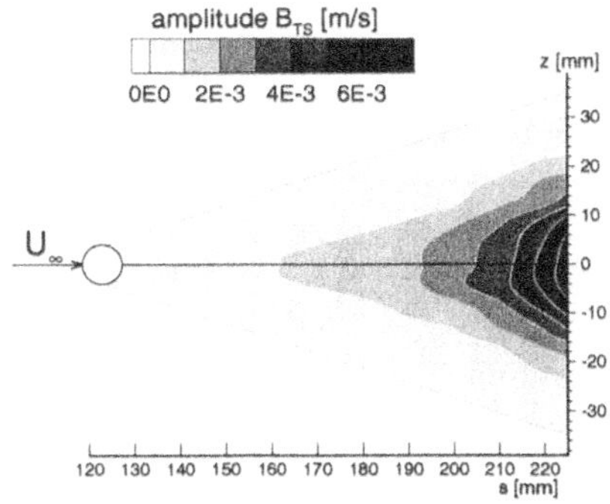

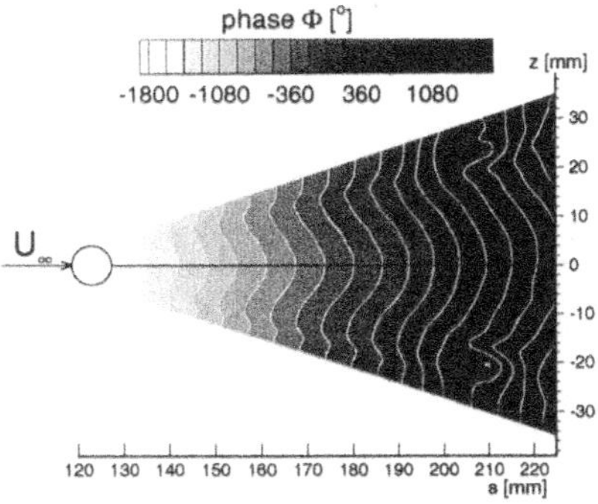

Figure 9 Amplitude part of wave train downstream of the 3D roughness element

Figure 10 Phase part of wave train downstream of the 3D roughness element

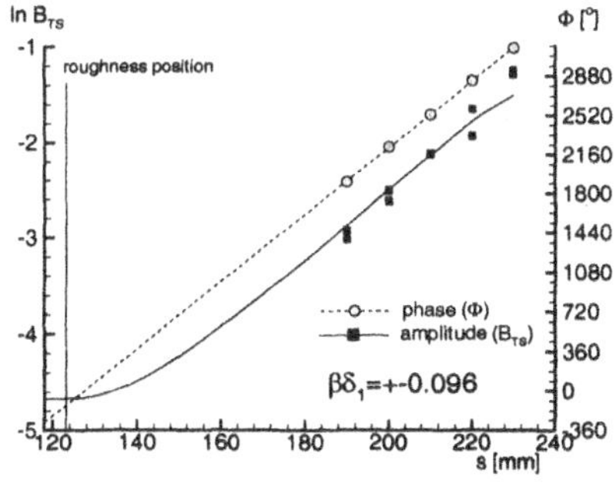

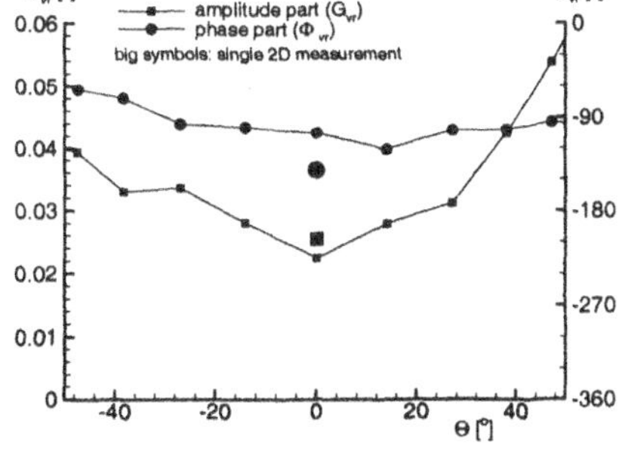

Figure 11 Downstream development of TS-amplitude and phase ($\beta \cdot \delta_1 = \pm 0.096$)

Figure 12 Amplitude- and phase part of vortex-roughness receptivity function

Adaptive Transonic Wing Section with Natural Transition

Alexander Meijering and Wolfgang Schröder

Aerodynamisches Institut, RWTH Aachen,
Wüllnerstr. zw. 5 u. 7, 52062 Aachen, Germany

Summary

To investigate the potential to increase the aerodynamic performance of aircraft by incorporating the adaptive wing technology, a demonstrator model with an adaptive upper surface over the entire chord length has been developed and tested in the wind tunnel at transonic speeds with natural transition. The airfoil shapes tested in the experiment have been obtained by a numerical optimization method. To accurately investigate the state of the boundary layer with respect to the transition location and the occurrence of separation bubbles, the hot film technique and the shear sensitive liquid crystal coating technique have been used. The geometry of the shock wave has been captured using the shadowgraph technique. The experimental results showed the good quality and the adjusment accuracy of the surface of the adaptive wing section model, which allowed the investigation of natural transition of the TOLLMIEN-SCHLICHTING waves at transonic speeds at REYNOLDS numbers of approximately 2.7 million. The shape modification resulted in the elimination of an existing separation bubble and in the improvement of the aerodynamic performance. By combining a numerical airfoil shape optimization and an adaptive wing section wind-tunnel model with integrated sensor technology to experimentally acquire the current flow conditions and boundary-layer state information a first step towards a closed-loop adaptive wing technology is provided.

1 Introduction

During the long distance flight of modern commercial aircraft at transonic speeds the flight conditions change substantially. Additionally, the strong demands for the reduction of fuel consumption from the airlines require the total drag of transonic wings to be reduced. To increase the aerodynamic performance and exploit the aerodynamic potential of the wing at each operation point of the flight envelope the adaptive wing technology is considered. This concept implies the geometrical modification of the airfoil shape by an arbitrary technique to obtain the desired wing contour that is considered the optimum choice for the current flow condition with the option to further modify the airfoil shape as the flight conditions change and thereby to maximize the aerodynamic benefits during flight.

At high cruise speeds the viscous drag of the wings that depends on the state of the boundary layer accounts for approximately 25% of the total amount of friction drag of an aircraft [1]. Compared to the laminar boundary layer the turbulent boundary layer is characterized by a substantially higher wall shear stress. Therefore, the laminar flow concept which aims at the delay of the transition far downstream of the leading edge is a promising method to reduce the friction drag and thus the total drag [2].

Another goal of the aerodynamic design is to delay the increase of total drag caused by the wave drag to higher MACH numbers by reducing the shock strength. The sudden pressure increase across the shock has a large impact on the state of the boundary layer and can cause separation. If the transition is delayed by the laminarization concepts up to the shock, the laminar boundary layer is likely to separate upstream of the shock forming a laminar separation bubble if the REYNOLDS number is small enough. These phenomena yield a substantial change of the aerodynamic performance, especially an increase of drag. Hence, the benefits of the laminarization techniques will be reduced or even annihilated.

The potential of the adaptive wing technology at transonic speeds is investigated numerically and experimentally for a wing section without sweep with natural transition. Hence, of the three boundary-layer instability mechanisms of attachment line transition, cross-flow instability and TOLLMIEN-SCHLICHTING instability, only the latter mechanism is significant for this investigation [2]. For the experimental analysis a wind-tunnel model with a rectangular top-view with an adaptive upper surface over the entire chord length has been developed. The model is tested at transonic speeds to verify the aerodynamic performance of the numerically optimized airfoil shapes.

2 Aerodynamic Design

The optimization of the airfoil shape is based on the Evolution Strategy (ES) [3], which is a stochastical search method that drives a population of individuals, each representing an airfoil shape, to find the global extremum through the use of evolution operators. It is a very robust method for the optimization of highly nonlinear multi-modal objective functions like the design of a laminar-type airfoil subject to shock boundary-layer interaction [4, 6].

The main objective of the present aerodynamic design of an airfoil shape is to increase the lift-to-drag ratio over a wide range of lift coefficients by modifying the airfoil upper surface. The occurrence of separation bubbles decreases the aerodynamic performance and hence should be prevented. Additionally, the strength of the wing structure requires a maximum thickness of at least 12% chord. These constraints have been taken into account by a penalty function imposed on the objective function. The laminar-type airfoil

DA LVA-1A will be used as initial geometry for the aerodynamic design and the experimental investigation. In the ES each individual of the population represents an airfoil shape, which has been parameterized by eight geometric modes. For the evaluation of the objective function of each individual one calculation of the flow field with the EULER-boundary-layer method MSES by DRELA [5] has to be performed. The transition location is predicted by the e^N-criterion. The limiting N-factor N=7, which depends on the turbulence level of the wind tunnel, has been correlated with previous transition measurements on the DA LVA-1A in the trisonic wind tunnel [6].

3 Experimental Techniques

3.1 Adaptive Wing Model

The two major requirements for an aerodynamic configuration with adaptive features of the surface geometry contradict each other. On the one hand, the shape has to be flexible to allow adaptation of the entire upper side of the airfoil and on the other hand, the structure has to possess sufficient stiffness to maintain the shape under the high aerodynamic loads. Additionally, the slender bodies of supercritical airfoils leave little space for the integration of shape adjustment kinematics and measurement equipment inside the model with 200 mm chord. To prevent premature transition, the surface must be smooth without curvature discontinuities. This imposes major constraints on the adjustment accuracy of the wing contour.

The present adaptive wing section model with a projected area of 200 × 400 mm^2 consists of three modules. The initial airfoil geometry corresponds to the laminar-type airfoil LVA-1A designed by DASA for flight conditions of Ma_∞=0.73, C_L=0.4 and $Re_\infty = 20 \cdot 10^6$. The main module absorbs the aerodynamic loads and introduces them to the wind tunnel supports (figure 1). It is equipped with nine drive mechanisms that are distributed equidistantly from 12.5% to 72.5% of the chord length each of which supports and adjusts the carbon fiber composite upper surface shell at five positions in spanwise direction. The adjustment kinematics, that transforms a rotating movement into a translation, allow a continuous adjustment over 2 mm. This is sufficient for the aerodynamic optimization. The main advantage of this type of adjustment kinematics is that energy is needed only to adjust the airfoil shape, whereas no additional energy is required to retain the position of the surface. The flexible carbon fiber composite shell represents the second module and extends over the leading edge towards the lower surface to avoid premature transition due to edges. For spanwise stiffness the shell is equipped with nine steel pipes that are hinge connected to the drive kinematics to prevent curvature discontinuities. The third module completes the lower side of the test model. Through a T-slot mechanism this module can be removed without the use of bolts, which would damage the surface. This mechanics enables the simple model assembly and provides access to the drive adjustment kinematics. For a more detailled description of the model refer to [6].

The measurements have been performed in the trisonic wind tunnel of the AERODYNAMISCHE INSTITUT AACHEN. It is an intermittent working suction type wind tunnel that is equipped with an adaptive-wall test section of 40×40 cm^2 cross section to reduce wall interference [4].

3.2 Detection of Transition and Separation Bubbles

To obtain the lift the model is equipped with 47 pressure orifices distributed in a staggered pattern at an angle of 15° to the flow direction. The drag is obtained by a wake rake with 60 total pressure taps and 4 static pressure taps positioned off the center line of the test section outside the range of the surface pressure orifices at one chord length behind the trailing edge. The arrangement of the different measurement techniques is depicted in figure 2. To identify the transition location the multi-sensor hot-film technique is applied. 20 flush mounted sensors, which operate in the constant-temperature mode (CTA), cover an area from 50% to 80% chord in a staggered pattern. The determination of the location of transition and of the location and extent of the separation bubble based on the unsteady hot-film sensor data will be supplemented by the shear sensitive liquid-crystal coating technique, that is applied on the remaining spanwise part of the airfoil's upper surface. The combination of the different measurement techniques is indispensable for a detailed analysis on the state of the boundary layer and the overall flow field over the airfoil. For example, to avoid the misinterpretation of the hot-film data in case of the occurrence of a turbulent wedge covering the sensors, that can be interpreted as a fully turbulent boundary layer the liquid crystal coating clarifies the picture. On the other hand, the hot-film technique is able to identify transition when a separation bubble is present. This is not possible with the liquid-crystal coating. To ensure a correct analysis of the surface applied measurements techniques, the shadowgraph technique is utilized to visualize the shock configuration on the upper surface of the airfoil.

4 Results and Discussion

4.1 Numerical Results of the Aerodynamic Design

The numerical optimization pursues two objectives. On the one hand, the potential of the increase of the airfoil performance of a mission-adaptive wing with an adjustable upper surface is investigated. Therefore, one-point designs for 20 different transonic freestream conditions have been performed.

On the other hand, the numerical optimization should provide a moderate number of airfoil shapes for the experimental investigation that not only show enhanced performance at the design point, but also perform well over a wide range of flight conditions compared to the initial airfoil shape. Hence, the aerodynamic performance of the optimized one-point designs is analyzed at off-design conditions, to identify wing geometries with good overall aerodynamic characteristics. The REYNOLDS number for all designs corresponds to that achievable in the wind-tunnel tests and is determined to $\mathrm{Re}_\infty = 2.7 \cdot 10^6$ based on the chord length

At this REYNOLDS number the flow field over the initial airfoil DA LVA-1A at transonic speeds is characterized by a favorable pressure distribution that delays the transition far downstream from the leading edge. However, the shock causes laminar separation. Due to the penalty on separation, this implies, that the optimization procedure starts in the infeasible domain. Hence, the major effort of the optimization algorithm is to search for the feasible domain of airfoil geometries for which no separation occurs, even if this means a decrease of the lift-to-drag ratio. Subsequently, the optimizer aims to increase the lift-to-drag ratio within the feasible domain. The optimization algorithm was able to remove the separation bubble, if it occurred, and also to increase the lift-to-drag ratio for all 20 one-point designs in the MACH-C_L range investigated. At subsonic speeds where no separation bubble is present, the transition location is shifted even further downstream to increase the aerodynamic performance [6]. At transonic flows without separation the shock strength, and hence the wave drag, has been reduced [6].

For the experimental investigation of the aerodynamic characteristics of the adaptive wing two airfoil shapes have been selected. The performance of these airfoil shapes with respect to drag polars for different MACH numbers compared to the DA LVA-1A initial airfoil is shown in figures 3 and 4.

From the figures it is obvious that the CS05NT airfoil, design point of which is $\mathrm{Ma}_\infty = 0.74$ and $C_L = 0.4$, increases the performance at the higher MACH number, whereas the CS18NT airfoil, design point $\mathrm{Ma}_\infty = 0.70$ and $C_L = 0.6$, shows better aerodynamic characteristics than the DA LVA-1A initial airfoil at the lower MACH number. This result is especially valuable since it introduces the possibility to reduce the number of shape adaptations to changing flight conditions and to remain still in the neighborhood of optimal performance.

4.2 Experimental Investigation of the Adaptive Airfoil

The experimental investigation aims at validating the aerodynamic design by determining the important aforementioned flow phenomena and at demonstrating the potential of the adaptive-wing concept on a wind-tunnel model.

Figure 5 displays the pressure distribution on the airfoil of the CS05NT compared with the DA LVA-1A airfoil at $\mathrm{Ma}_\infty = 0.74$ and $C_L = 0.67$. From the surface pressure distribution it is obvious that the shock strength is reduced by a slight smearing of the shock. The reddish color of liquid-crystal coating on the DA LVA-1A airfoil surface shown in figure 6 characterizes a long extent of the laminar boundary layer. The following area of bright red color depicts a region of low shear stress that indicates the occurrence of a separation bubble. This observation is verified by the linear correlation coefficient of neighboring hot-film sensors in figure 8. This coefficient is recognized to be more effective to indicate separation in the vicinity of shocks and turbulent boundary layers than the 180° phase shift criterion [6, 7]. The two distinct peaks of strong negative values of the linear correlation coefficient evidence the location and extent of a separation bubble within 57% - 70% chord, which corresponds with the observations of the liquid-crystal coating image.

Comparing these observations with the liquid-crystal coating image of the CS05NT airfoil at the same freestream conditions in figure 7, also exhibits a long extent of the laminar boundary layer, which in this case is terminated directly by the blue color at 62% chord, which is characteristic for laminar-turbulent boundary-layer transition without separation. This corresponds to the linear correlation coefficient results in figure 9. Thus, with respect to the DA LVA-1A airfoil the CS05NT airfoil shape smoothes the shock, and hence reduces the shock strength. The laminar boundary layer remains attached to the surface which reduces the total drag. Further analysis of the wake rake data is required to determine the development of the airfoil drag.

These observations show the feasibility to develop airfoil shapes that incorporate the concept of natural laminarization at the investigated REYNOLDS numbers in the sense that transition is delayed almost up to the location of the shock. The mechanism is combined with a reduction of the shock-boundary layer interaction to prevent separation bubbles.

5 Conclusion

To demonstrate the possible increase of aerodynamic performance at transonic speeds the adaptive wing technology has been investigated numerically and experimentally using the DA LVA-1A as reference airfoil shape. The optimization algorithm is based on the evolution strategy to design airfoils, which incorporate the natural laminar flow concept in conjunction with shock-boundary layer control. The aerodynamic design provides two airfoil shapes for the experimental investigation. A wind tunnel model with an adaptive upper surface over the entire chord length equipped with a wide range of sensor technology for a detailed acquisition of the flow field has been built to successfully verify the numerical results. The combination of numerical, experimental and structure enhancing techniques presented here, provide the first step towards a closed-loop adaptive wing system.

References

[1] Robert, J.P.: Drag Reduction: An Industrial Challenge. In: AGARD Special Coarse on Skin Friction Drag Reduction. AGARD-R 786 (1992) 2-1 - 2-15

[2] Redeker, G., Horstmann, K.H., Köster, H., Quast, A.: Investigations on High Reynolds Number Laminar Flow Airfoils. J. Aircraft **25** (1988) 583 – 590
[3] Rechenberg, I.: Evolutionsstrategie: Optimierung technischer Systeme nach Prinzipien der biologischen Evolution. Frommann-Holzboog, Stuttgart (1973)
[4] Meijering, A., Limberg, W., Schröder, W.: Aerodynamic Design of an Adaptive Airfoil at Transonic Speeds. NNFM **84**, Springer-Verlag (2003) 85 – 103
[5] Drela, M., Giles, M.B.: Viscous-Inviscid Analysis of Transonic and Low Reynolds Number Airfoils. AIAA J. **25** (1987) 1347–1355
[6] Meijering, A.: Design of Adaptive Wing Sections with Natural Transition. PhD thesis, Aerodyn. Inst. RWTH-Aachen (2003)
[7] Abstiens, R., Fühling, S., Schröder, W.: Boundary-layer Measurements on the ELAC Configuration at $Re = 20 \times 10^6$. In: ICTAM 2000 Chicago. (2000)

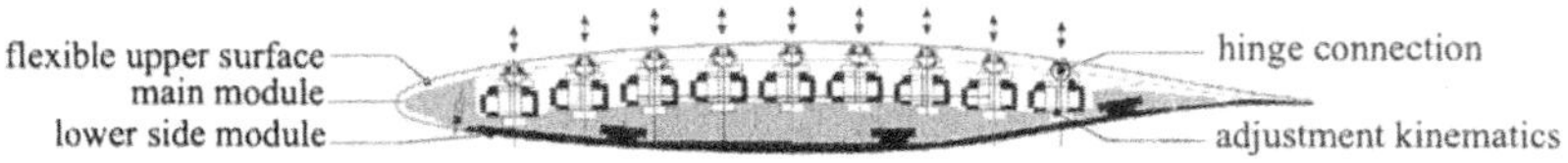

Figure 1 Sketch of the adaptive wing model cross section

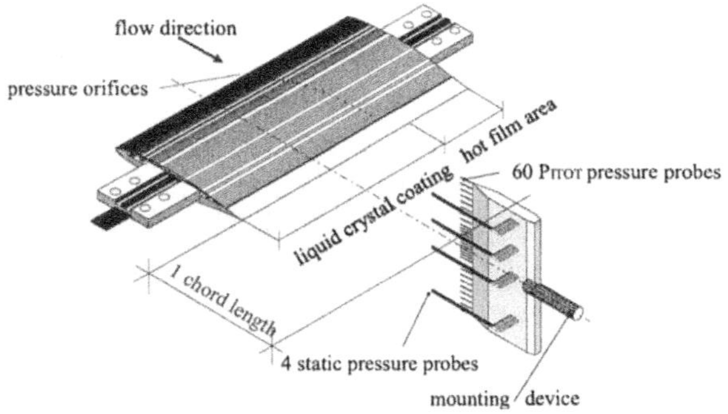

Figure 2 Arrangement of the measurement techniques on the model

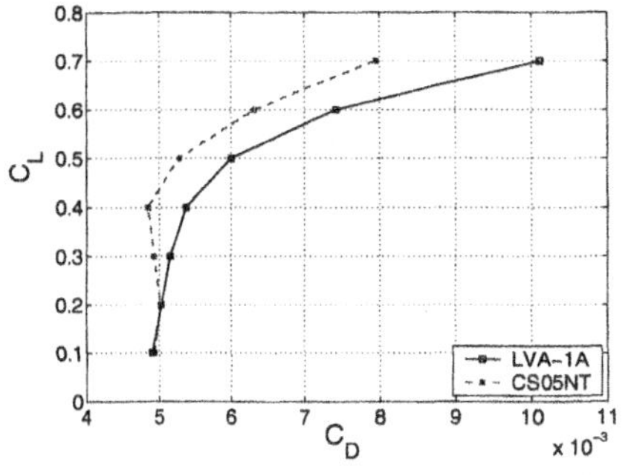

Figure 3 Numerically determined drag polar at $Ma_\infty = 0.74$ for the DA LVA-1A and the CS05NT designed for $Ma_\infty = 0.74$ and $C_L = 0.4$

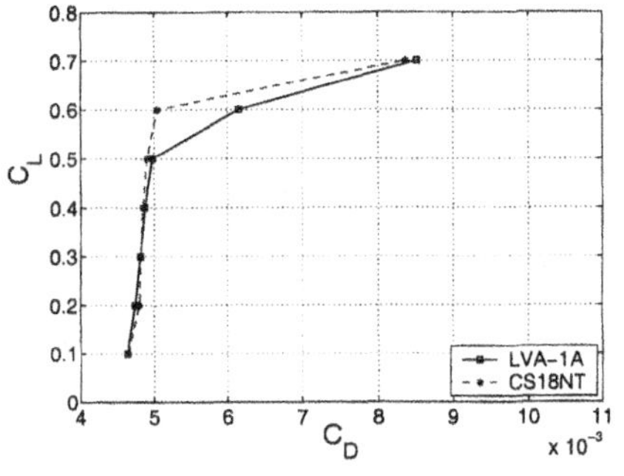

Figure 4 Numerically determined drag polar at $Ma_\infty = 0.70$ for the DA LVA-1A and the CS18NT designed for $Ma_\infty = 0.70$ and $C_L = 0.6$

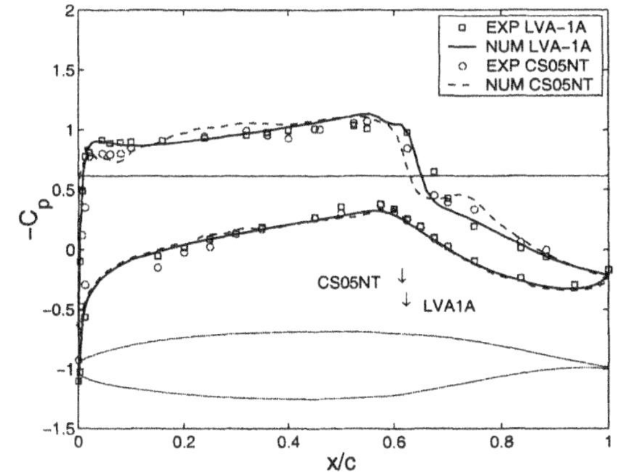

Figure 5 Pressure distribution on the airfoil of the DA LVA-1A and the CS05NT airfoils at $Ma_\infty = 0.74$, $Re_\infty = 2.7 \cdot 10^6$ and $C_L = 0.67$. The arrows denote the transition location.

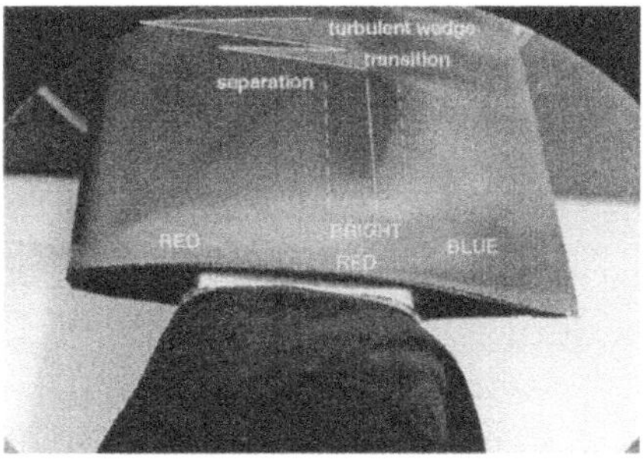

Figure 6 Liquid-crystal image of the DA LVA-1A at $Ma_\infty = 0.74$ and $C_L = 0.67$

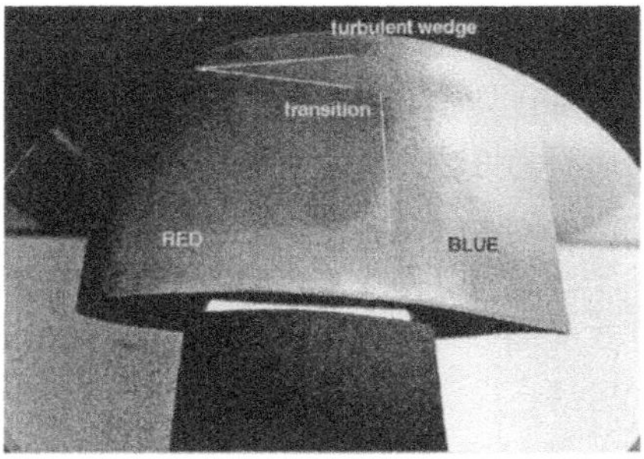

Figure 7 Liquid-crystal image of the CS05NT at $Ma_\infty = 0.74$ and $C_L = 0.67$

To make it easier to understand the gray-scale images of the figures 6 and 7 the various areas are directly indicated.

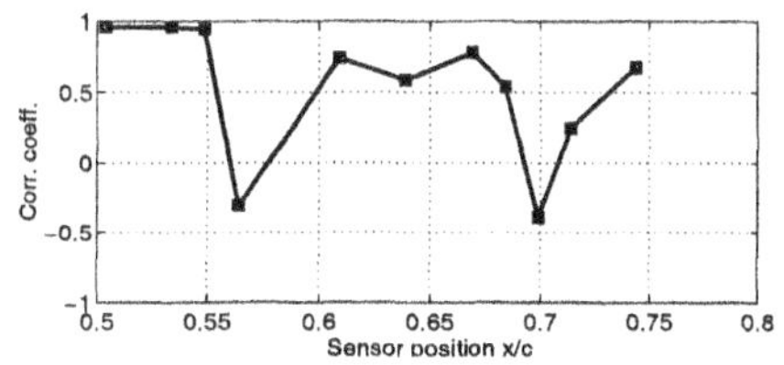

Figure 8 Linear correlation coefficient of neighboring hot-film sensors of the DA LVA-1A airfoil

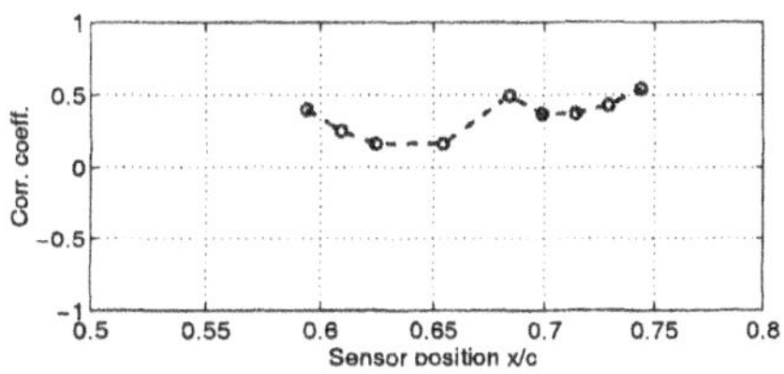

Figure 9 Linear correlation coefficient of neighboring hot-film sensors of the CS05NT airfoil

Secondary instability of superposed steady and unsteady crossflow vortices

G. Bonfigli and M. Kloker

Institut für Aerodynamik und Gasdynamik,
Universität Stuttgart, Pfaffenwaldring 21, D-70550, Germany

Summary

Laminar-turbulent transition in a 3-D boundary layer under controlled simultaneous forcing of unstable steady and unsteady crossflow waves is investigated by means of direct numerical simulations. The flow configuration reproduces the accelerated 3-D flat-plate boundary layer of the "Querströmungsprinzipexperiment" of the DLR-Göttingen [1]. The superposition of high-amplitude steady and unsteady waves induces a spanwise and streamwise modulation of the flow, which is shown to trigger the laminar breakdown also in the absence of additional background disurbances. The mechanism has some similarities to the one documented for the case of purely steady vortices [6] [8]. However the interaction of the primarily perturbed waves provides both the preconditions and the forcing for such mechanism.

1 Introduction

Relevant progress has been achieved recently with respect to the final stages of transition in crossflow (CF) unstable 3-D boundary layers. First experimental results [1] [10], then theoretical works [4] [6], and finally direct numerical simulations (DNS) [8] [9] [11] showed that a strong secondary instability leads to the growth of high-frequency perturbations, which in the end trigger the onset of the turbulent regime. Furthermore, the secondary instability could be put in relation with the modulation induced by the the primary vortices in the distribution of the velocity components over crosscut-planes orthogonal to their axes. However, the major part of these results is confined to fundamental scenarios where the primarily unstable modes are steady in a proper, eventually translating, galilean reference system (this is the case when the primary state is the consequence of the excitation of a single unsteady wave). Exceptions are the experimental work carried out at the DLR-Göttingen [1], where steady and unsteady unstable modes were perturbed simultaneously, and temporal DNS [11]. In the experiment high-frequency perturbations were observed but the flow was too complex to provide a conclusive picture of the breakdown mechanism, also because the theoretical model [4] [6] treating the fundamental cases might fail since its assumptions of steadiness and invariance in downstream direction are violated. Temporal DNS [11] suffer from a basic flaw in the model since initially steady modes can not be kept steady (rather the spatial wavevector is fixed).

All results presented hereafter are computed solving the vorticity-velocity formulation of the unsteady incompressible three-dimensional Navier-Stokes equations. The spatial discretization is based on a well-tested combination of 6th-order compact finite-differences (in downstream and wall-normal direction) and spanwise Fourier-spectral expansions. A 4th-order Runge-Kutta scheme is used for marching in time. Lengths are made nondimensional with $L^* = 0.1m$, velocities with $u^*_\infty = 14m/s$. The nondimensional grid steps in wall-normal and streamwise direction are $1.1 \cdot 10^{-4} \leq \Delta y \leq 4.6 \cdot 10^{-4}$ (varying depending on the wall distance) and $\Delta x = 1.3 \cdot 10^{-3}$, respectively. The number of Fourier harmonics used in spanwise direction amounts to 82. A description of the numerical procedure can be found in [8].

1.1 Baseflow and transition scenario

The baseflow reproduces the 3-D accelerated flat-plate boundary layer of the "Querströmungsprinzipexperiment" of the DLR-Göttingen [1] for $Q^*_\infty = 19m/s$. The downstream development of the most relevant boundary-layer parameters is shown in figure 1. Velocity profiles at different x-positions are also given and show everywhere the classical CF shape. Experimental data for $x = 3.0$ [1] are provided for comparison. Here and in the following, x, y and z are the downstream, the wall-normal and the spanwise coordinates, respectively, in a plate-oriented system with z parallel to the plate leading edge.

Over the whole integration domain amplification rates are maximal for CF modes with a spanwise wavelength approximately equal to $\Lambda^*_z = 12mm \approx 4 \cdot \delta^*_s$ and with frequencies in the range $0Hz \leq f^* \leq 250Hz$. Correspondingly, the nondimensional fundamental wavenumber of the spanwise Fourier expansion was set equal to $\gamma = 2\pi/\Lambda_z = 52.4$, and the frequency $f^* = 133Hz$ ($\omega_0 = 6.0$) was chosen for the unsteady disturbance. Indicating by (h, k) modes with frequency $\omega = h \cdot \omega_0$ and spanwise wavenumber $\gamma = k \cdot \gamma_0$, the modes $(0, 1)$ and $(1, 1)$ were perturbed at $x = 1.56$ by suction and blowing within a thin wall stripe parallel to the plate leading edge. The perturbation amplitudes were $A_{(0,1)} = 1.7 \cdot 10^{-3}$ and $A_{(1,1)} = 1.5 \cdot 10^{-4}$.

Figure 2 documents the downstream amplitude development for the perturbed modes. It has been shown [2] that the transition scenario varies strongly depending on the relative amplitudes of the primary modes before and after the onset of saturation. In the present simulation (hereafter called case $(0, 1) + (1, 1)$, corresponding to case R5 in [2] and to case $(0, 1) + (1, 1)$ in [5]) the CF vortex $(0, 1)$ is perturbed with larger amplitude than the CF wave $(1, 1)$ and dominates, despite weaker growth, the transitional region. At $x = 2.6$ its amplitude is nearly 10% of the local free-stream velocity u_{bse}, and further downstream it strongly influences $(1, 1)$. The amplification of both primary modes is damped by nonlinear effects. The influence of mode $(1, 1)$ on the onset of saturation, if any, is marginal. However its amplitude is not negligible ($8\% \cdot u_{bse}$) and indeed it plays a determinant role in the further development of the flow.

1.2 Transition mechanism

Up to saturation of the primary modes ($x \leq 3.4$) the flow field is still determined by the quasi-linear superposition of the disturbance fields associated with the primary modes $(0,1)$ and $(1,1)$ and their superharmonics. Reference simulations have been carried out where modes $(0,1)$ or $(1,1)$ were perturbed alone (case $(0,1)$ and case $(1,1)$, respectively). The resulting flow fields and the one obtained by superposing them linearly were then compared with case $(0,1) + (1,1)$. As figures 3a and 3b show for ω'_ξ, i. e. for the component of the perturbation vorticity parallel to the ξ-axis and nearly parallel to the axes of the most relevant vortical structures, both primary modes tend to generate stripe patterns in the vorticity distribution over every wall-parallel plane cutting the cores of the corresponding main vortices. Regions of negative ω'_ξ corresponding to the neighboring CF main vortices are separated by regions of positive shear ω'_ξ, which in case $(0,1)$ develop into small counterrotating streamwise secondary vortices. When $(0,1)$ and $(1,1)$ are superimposed (fig. 3d) a checkered unsteady pattern emerges, where modulations are evident both in the direction normal and parallel to the axis of the dominating vortices $(0,1)$. The similarity between 3c and 3d for $\xi \leq 1.0$ ($x \leq 2.85$) confirms that the vorticity modulation in case $(0,1) + (1,1)$ is primarily originated by linear effects.

Concentrating on case $(0,1) + (1,1)$, secondary instability mechanisms in the sense of the secondary linear instability theory (SLST) [4] [6] [8] [10], based on the instantaneous local crosscut flow fields, determine the flow development in the saturation region. Figure 4 shows secondary vortical structures connected to high-frequency modes I in the left column and low-frequency modes III in the right column: i) for flow fields obtained by superposing the primary state from case $(0,1)$ with properly normed secondary modes from own SLST-computations (in conjunction with [7] and [3], figs. 4a-b); ii) for snapshots from case $(0,1)$ under additional unsteady excitation of secondary modes in the saturation region (DNS, figs. 4c-d); iii) for the superposition of instantaneous primary states from case $(0,1)+(1,1)$ and the corresponding SLST secondary modes (figs. 4e-f); and iv) for snapshots from case $(0,1) + (1,1)$ (DNS, figs. 4g-h). The SLST-analysis for iii) has been carried out for crosscut planes of instantaneous flow fields from case $(0,1) + (1,1)$. Time steps and plane positions were chosen so that the perturbations finally developing into the structures of iv) crossed the considered planes at the considered time steps. The comparison of i) and ii) shows that in case $(0,1)$ the eigenfunctions of the SLST problem provide a good approximation of the flow field also at stages where secondary vortical structures become visible and perturbation amplitudes are virtually too large for the linearity assumption to hold. It is then possible to identify instability mechanisms by considering the shape of the resulting secondary vortical structures. This can be done for case $(0,1) + (1,1)$ comparing iii) and iv) and one may conclude that both instability mechanisms of type I (compare figs. 4e and 4g, capital letter indicate corresponding structures) and type III (compare figs. 4f and 4h) are active in case $(0,1) + (1,1)$. The frequency ω and the streamwise wavenumber β_ψ of the instabilities of the secondary modes connected to the structures in iv) may be assumed equal to those of the modes generating the structures in iii) (the spatial extension of the structures is inversely proportional to β_ψ). This gives $\omega \approx 30.0$, $\beta_\psi \approx 40.0$, for mode I and $\omega \approx 20.0$, $\beta_\psi \approx 40.0$, for mode III.

Caution is needed in applying the SLST to case $(0, 1) + (1, 1)$ since as shown in fig. 3c the flow field is strongly modulated in downstream direction, not steady in any reference system, and consequently violates the basic assumptions of the theoretical approach. However neither the generation of isolated vortical structures (see figs. 4g-h) contradicts the connection between SLST and the mechanism governing the present flow, nor does the results of fig. 5, according to which the mode-I structure of fig. 4g develops where the instantaneous velocity profiles are not particularly unstable (2/5T curve) and is associated to weakly unstable frequencies ($\omega \approx 30$). On the contrary, both aspects may be explained considering the localized character of the forcing initiating the instability mechanism. Particularly relevant is the concentration of positive ω'_ξ marked as $M1$ in figure 3c and originating from the modulation induced by the superposition of the primary modes $(0, 1)$ and $(1, 1)$. Both mode I and mode III are associated to local concentrations of positive vorticity on the updraft side of the primary vortex and may be expected to be efficiently excited by vorticity concentrations in that position. Following the flow development it becomes clear that the structures of figs. 4g-h both originate from the maximum $M1$, whose positions after one and two periods of the CF wave $(1, 1)$ are marked in figure 3c by $M2$ and $M3$, respectively. The development of the two distinct ω'_ξ-maxima $M3I$ and $M3III$ (fig. 3c) from the original concentration $M1$ is the first step toward the generation of the two distinct structures and is in agreement with the known characteristics of the instability modes involved in the process. The group velocity of mode I is indeed significantly higher than that of modes III. The convection velocity of the associated structures, and before that of the local vorticity concentrations $M3I$ (connected to mode I and developing into the structure of fig. 4g) and $M3III$ (connected to mode III and developing into the structure of fig. 4h), are correspondingly different.

In conclusion it can be stated that the instability mechanism connected to modes I and III of the SLST is robust and quite insensitive to the violation of the periodicity assumption made in the theory for the downstream direction. It is driven by the interaction between the flow field generated by the primary vortex and the region of positive ω'_ξ on its updraft side and is initiated by downstream, periodical or local, modulations of the flow. The selection of the dominant instability is influenced by the character of the excitation, with the result that the observable really dominating mode is not necessarily the most amplified one. Rather the effective initial amplitudes of the various modes are important, and in the present case they decrease with increasing frequency since all unsteady perturbations are nonlinearly generated by the fundamental wave. The interaction of steady and unsteady primary modes determines the initial streamwise flow modulation and consequently the development of the transition process.

It has been verified in other DNS that, when a flat spectrum background pulse is provided, more unstable higher-frequency type-I modes prevail and generate the smaller secondary vortices documented in [8]. In case $(0, 1) + (1, 1)$ the laminar breakdown is triggered by the mode-I structure of fig. 4g and in no case the breakdown was initiated by mode-III structures. Even in the simulation of figure 4d, where mode III was explicitly excited, the turbulent regime is initiated by the mode connected to the first superharmonic of the forcing, which is already of type I.

1.3 Comparison with the experiment

A comparison between numerical and experimental results for case $(0,1) + (1,1)$ is provided in figure 6 on the basis of a visualization by Lerche [5] showing the isosurfaces of the wall-normal gradient and of the second time derivative of the streamwise velocity component. The good agreement for the wall-normal gradient shows that the DNS closely reproduces the transition scenario of the experiment, while differences in the position of the high frequency perturbations are probably due to uncontrolled background unsteady perturbations in the experiment. Indeed, background noise might be there the reason for the observed relevant growth of high frequency secondary modes (the spectrum in fig. 66 of [5] shows a local maximum for $f^* \approx 2000Hz$), while in the DNS, where unsteady excitation is provided exclusively through the low frequency primary mode $(1,1)$, modal amplitudes fall monotonously for growing frequencies.

The comparison of the temporal visualization of fig. 6 (data originate from a fixed (y,z)-plane for different time steps) and the spatial snapshot of figure 7 explains the origin of the climbing isosurfaces visible in fig. 6 for the wall-normal gradient of the downstream velocity. They follow from the profile deformation induced by the primary vortex, the climbing shape being a consequence of the streamwise flow modulation induced by the superposition of steady and unsteady primary perturbations. Finer structures, like the ones induced by the secondary instability mechanisms discussed above, are smoothed out.

The financial support by the Landesgraduiertenförderung (LGF) under grant 7631 is gratefully acknowledged.

References

[1] H. Bippes. *Progress in Aerospace Sciences*, 35:363–412, 1999.
[2] G. Bonfigli and M. Kloker.
[3] W. Koch. Private comunications on spectral SLST-solver and eigenvalue tracking, 2001.
[4] W. Koch, F. P. Bertolotti, A. Stolte, and S. Hein. *J. Fluid Mech.*, 406:131–174, 2000.
[5] T. Lerche. Dissertation, Universität Göttingen, 1997.
[6] M. R. Malik and C. L. Chang. *J. Fluid Mech.*, 268:1–36, 1994.
[7] R. Messing. Private comunications on finite-difference SLST-solver, 2002.
[8] P. Wassermann and M. Kloker. *J. Fluid Mech.*, 456:49–84, 2002.
[9] P. Wassermann and M. Kloker. Transition mechanisms induced by travelling crossflow vortices in a three-dimensional boundary layer. *Accepted for publication in J. Fluid Mech.*, 2002.
[10] E. B. White, W. S. Saric, R. D. Gladden, and P. M. Gabet. AIAA 2001-0271, 2001.
[11] T. Wintergerste and L. Kleiser. In J. Chollet, P. R. Voke, and L. Kleiser, editors, *Direct and large-eddy simulation*, volume II. Kluwer academic press, 1997.

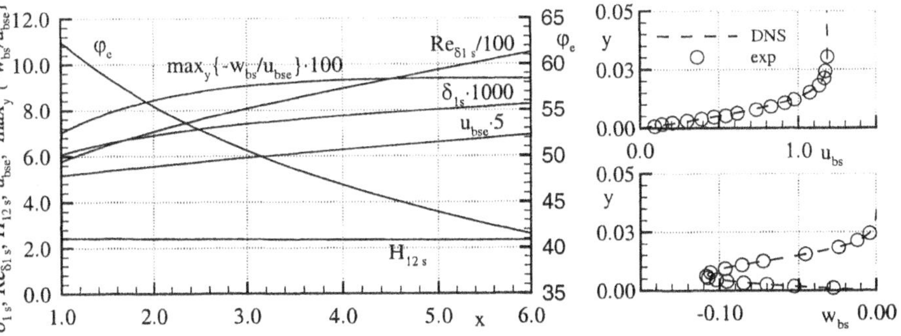

Figure 1 Parameters of the baseflow in the streamline-oriented coordinate system (local sweep angle with respect to the fixed system φ_e, displacement thickness $\delta_{1\,s}$, local Reynolds number $Re_{\delta 1\,s}$, shape factor $H_{12\,s}$, maximum local CF velocity $\max_y\{-w_{b\,e\,s}\}$) and velocity profiles at $x = 3.0$ for DNS and experiment.

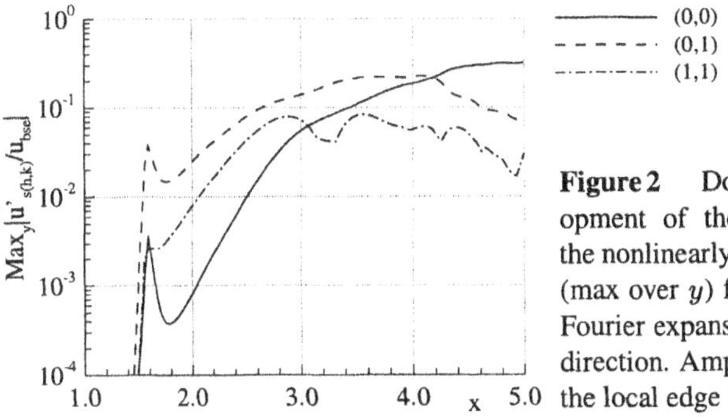

Figure 2 Downstream amplitude development of the perturbed modes and of the nonlinearly induced baseflow distortion (max over y) for case $(0,1) + (1,1)$ after Fourier expansion in time and in spanwise direction. Amplitudes are normalized with the local edge velocity.

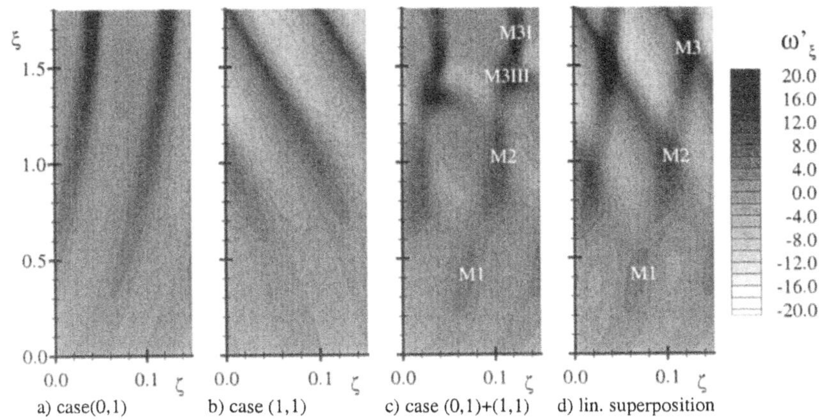

Figure 3 Component ω'_ξ of the perturbation vorticity in the plane $y = 0.85\cdot 10^{-2} \approx 0.25\cdot\delta$ for the different simulation cases and for the linear superposition of the perturbation fields from cases $(0,1)$ and $(1,1)$ (fig. d). The primary main vortices rotate clockwise when looking downstream and have negative ω_ξ values. The ξ-axis spans an angle of 45 deg. with the x-axis and is nearly parallel to the local edge velocity. The origin of the rotated system lies in $x = 2.13$, $z = 0.07$.

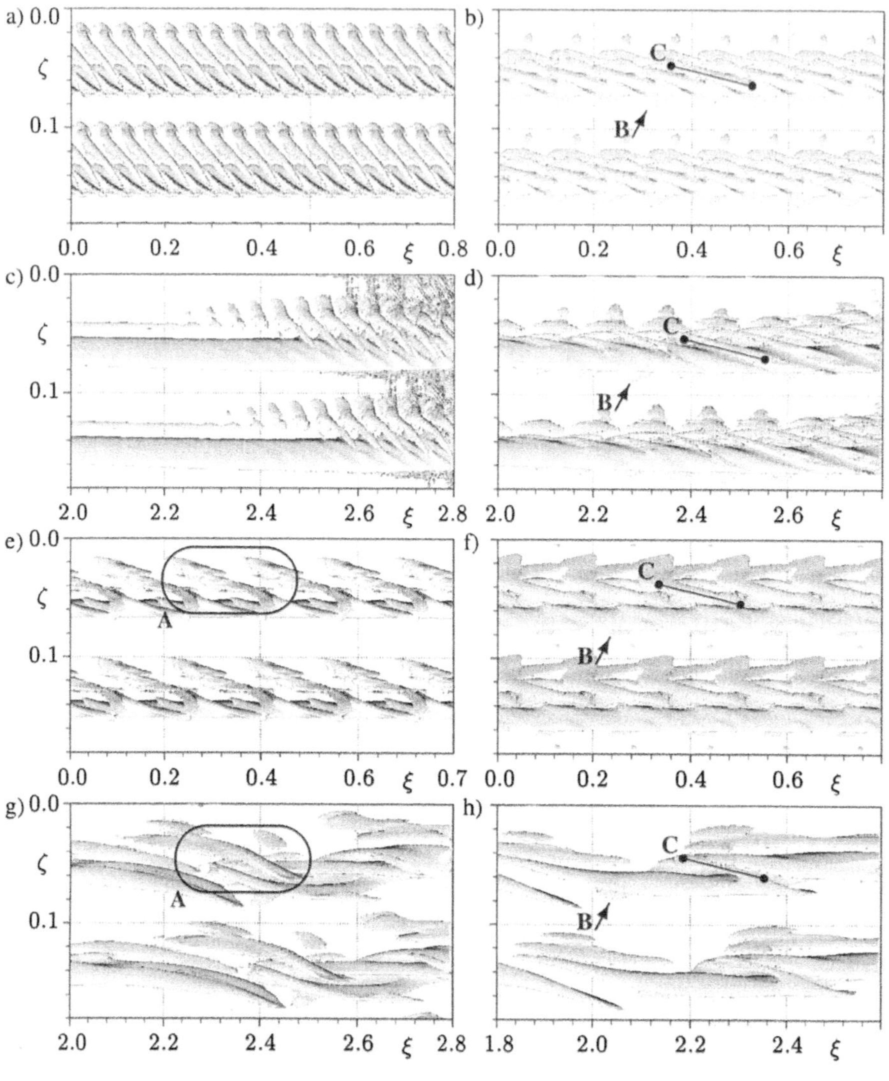

Figure 4 Visualization (top view, two spanwise wavelengths) of secondary structures by means of λ_2-isosurfaces: a) superposition of the primary state from case $(0,1)$ ($x = 3.32$, $\xi \approx 1.58$) and eigenfunctions for mode I ($\beta_\psi = 134$, $\omega \approx 120$, maximal amplitude of the u_ξ component of the eigenfunction $A'' = 0.1 \cdot u_{bse}$); b) as a) for mode III ($\beta_\psi = 62$, $\omega \approx 30$, $A'' = 0.1 \cdot u_{bse}$); c) DNS for case $(0,1)$ under additional excitation of mode I ($\omega = 120$) in the saturation region; d) as c) for mode III ($\omega = 30$); e) superposition of primary state from case $(0,1) + (1,1)$ ($x = 3.42$, $\xi \approx 1.58$, $t = 0.4 \cdot T$) and eigenfunctions for mode I ($\beta_\psi = 40$, $\omega \approx 30$, $A'' = 0.25 \cdot u_{bse}$); f) superposition of primary state from case $(0,1) + (1,1)$ ($x = 3.61$, $\xi \approx 2.11$ $t = 0.44 \cdot T$) and mode III ($\beta_\psi = 40$, $\omega \approx 20$, $A'' = 0.25 \cdot u_{bse}$); g) mode-I structure from DNS for case $(0,1) + (1,1)$ ($t = 0.48 \cdot T$); h) mode-III structure from DNS for case $(0,1) + (1,1)$ ($t = 0.26 \cdot T$). Same letters indicate corresponding structures. See caption of fig. 3 for definition of the reference system.

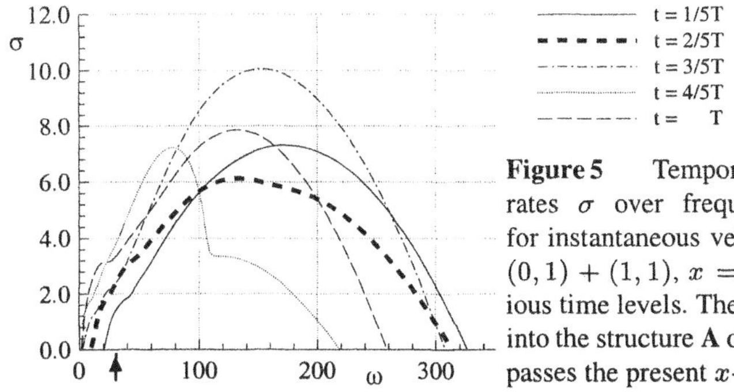

Figure 5 Temporal SLST amplification rates σ over frequency for mode I and for instantaneous velocity profiles from case $(0,1)+(1,1)$, $x=3.42$, $\xi\approx1.58$, at various time levels. The perturbation developing into the structure **A** of fig. 4g ($\omega\approx30$, arrow) passes the present x-position for $t=2/5T$.

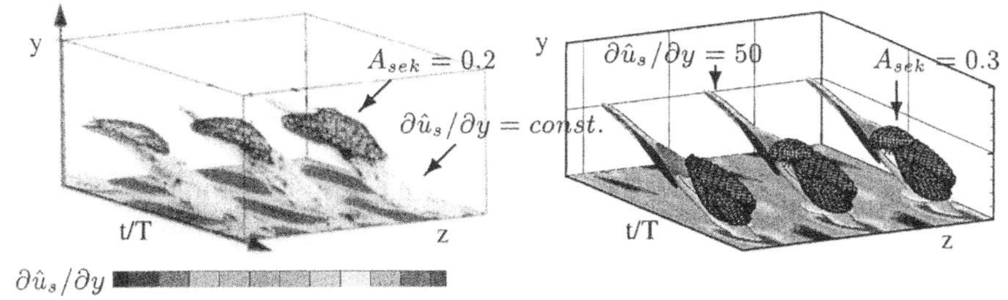

Figure 6 Comparison of experimental [5] and DNS results for case $(0,1)+(1,1)$: a) experiment [5]; b) DNS for $x=3.63$. $\hat{u}_s$ is the time-dependent component of the downstream velocity: $\hat{u}_s=u_s-<u_s>$; $\partial\hat{u}_s/\partial y$ is the wall-normal shear of the unsteady part of the flow. According to [5] A_{sek} is a normalization of the second time derivative of u_s, and is large in regions presenting high frequency perturbations.

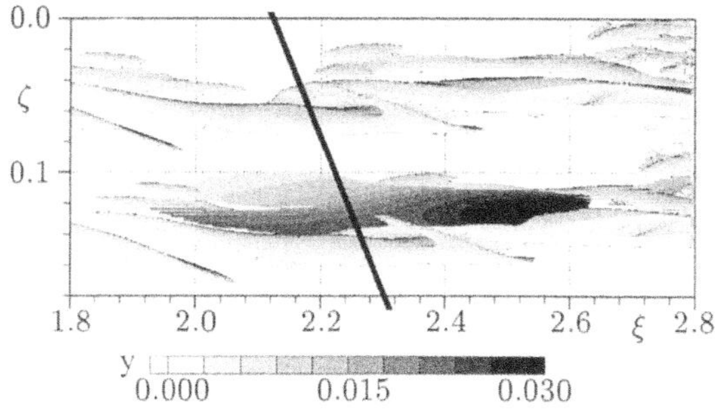

Figure 7 Spatial visualization for case $(0,1)+(1,1)$, $t=0.52\cdot T$, of the isosurfaces $\partial\hat{u}_s/\partial y=50.0$ from fig. 6b (shaded, the color shows the distance of the isosurface points from the wall) and of λ_2-isosurfaces (not shaded). The thick line shows the section plane, data for fig. 6b originate from.

Inviscid Flow Computation for the Shuttle-Like Configuration PHOENIX

M. Hesse, B. U. Reinartz, J. Ballmann
Lehr- und Forschungsgebiet für Mechanik
RWTH Aachen, Templergraben 64, 52062 Aachen, Germany

Summary

The development of the shuttle-like technology demonstrator PHOENIX is intended to prove the feasibility of a future European reusable space transportation system. The present work focuses on aerodynamic solutions for the clean-configuration of PHOENIX by solving the Euler and Navier-Stokes equations for three-dimensional flow. Details of the structured grid generation process are discussed and a converged solution for supersonic flow is presented. The agreement of the obtained aerodynamic coefficients with wind tunnel and design data is favorable. Computations are in progress for viscous hypersonic flow about PHOENIX for different flight conditions.

1 Introduction

Within the frame of the Collaborative Research Center SFB 253 "Fundamentals of the Design of Aerospace Planes", compressible hypersonic flow problems are studied numerically using the RANS flow solver FLOWer, version 116.0+ [8]. This code was originally developed, employing finite volume technique and central differencing throughout, for the simulation of flows around airfoils in subsonic and transonic regimes. Over the last years different upwind schemes and turbulence models have been implemented for the simulation of hypersonic flows and extensively validated against measurements as well as other numerical schemes [1, 2, 6, 8]. A further extension presently under work concerns non-equilibrium chemistry [7].

Currently, the SFB 253 is involved in the German research and development program ASTRA. One of the foci of ASTRA is the development and testing of the technology demonstrator PHOENIX, a small scale model of HOPPER, which is a winged lifting body with blunt leading edges and similar in appearance to the US Space Shuttle [11]. Its reusable main stage is intended to take-off horizontally from a skid and to land horizontally on an airfield. The contribution of the SFB concentrates on numerical simulations of the flow around the clean configuration of PHOENIX by 3D Euler and Navier-Stokes computations.

Substantial work was performed to generate a multi-block grid for numerical applications starting from the CAD surface data of PHOENIX. Additionally, inviscid flow computations are performed for different angles of attack to tailor the grid to the flow field and to determine the aerodynamic coefficients for specified reference flow conditions. Currently, the grid is being adapted to the requirements of viscous, turbulent flow. Future investigations will include the simulation of flow according to the full Navier-Stokes equations at high-enthalpy conditions for different flight Mach numbers and different angles of attack.

2 Physical Model

The governing equations for hypersonic, turbulent flow are the Reynolds-averaged Navier-Stokes equations for unsteady, compressible fluid flow in integral form,

$$\frac{\partial}{\partial t}\int_V \mathbf{U}\,\mathrm{d}V + \oint_{\partial V} \left(\mathbf{F}^{\mathrm{c}} - \mathbf{F}^{\mathrm{d}}\right)\,\mathbf{n}\,\mathrm{d}S = \mathbf{0} \quad , \tag{1}$$

with $\mathbf{U} = (\rho,\ \rho u,\ \rho v,\ \rho w,\ \rho e)^{\mathrm{T}}$ being the vector of conserved quantities: mass density, the three Cartesian components of momentum density, and total energy density. The quantity V denotes an arbitrary control volume with the boundary ∂V and the outer normal $\mathbf{n}$. The flux array is divided into two parts, $\mathbf{F}^{\mathrm{c}}$ designating the convective terms including pressure and $\mathbf{F}^{\mathrm{d}}$ denoting the diffusive part,

$$\mathbf{F}^c = \begin{bmatrix} \rho\,\mathbf{v} \\ \rho\,\mathbf{v}\circ\mathbf{v} + p\,\mathbf{E} \\ \rho\,e\,\mathbf{v} + p\,\mathbf{v} \end{bmatrix} \quad \text{and} \quad \mathbf{F}^d = \begin{bmatrix} 0 \\ \boldsymbol{\tau} \\ \boldsymbol{\psi} \end{bmatrix} ,$$

where $\mathbf{E}$ is the identity tensor and $\circ$ denotes the dyadic product[1]. $\mathbf{v}$ represents the velocity vector of fluid motion, p is the static pressure, $\boldsymbol{\tau}$ the viscous stress tensor and $\boldsymbol{\psi}$ the energy supply due to viscous stresses and heat conduction. A detailed description of the implemented diffusive part including the formulation of the symmetrical stress tensor components with Stokes' hypothesis of vanishing pressure viscosity can be found in [3]. The molecular viscosity is approximated in case of laminar flow by Sutherland's formula as a function of the statical temperature only. The vector of heat conduction is modeled by Fourier's law with the heat conductivity k assuming a constant Prandtl number $Pr = 0.72$. For turbulent flows, the laminar viscosity μ is replaced by $\mu + \mu_t$ with the turbulent part provided by the applied turbulence model. Finally, since the air is assumed to behave as a calorically and thermally perfect gas, the pressure is calculated by the equation of state

$$p = (\kappa - 1)\rho(e - \frac{1}{2}\mathbf{v}^2)$$

with κ denoting the ratio of specific heats (air as perfect gas: $\kappa = 1.4$).

[1]Scalar Products of dyadics formed by two vectors $\mathbf{a}$ and $\mathbf{b}$ with a vector $\mathbf{c}$ are defined as usual, i.e., $\mathbf{a}\circ\mathbf{bc} = \mathbf{a}(\mathbf{bc})$, $\mathbf{ca}\circ\mathbf{b} = (\mathbf{ca})\mathbf{b}$.

3 Numerical Method

The computations are performed with the extended DLR FLOWer code. The computational domain is divided into non-overlapping hexahedra in general curvilinear coordinates ξ, η, ζ, and the integral formulation (1) is then applied to each cell (i, j, k) separately. Central discretization of the system of equation (1) turned out to be applicable for Mach numbers up to 2.5. It results in a set of ordinary differential equations for the time rates of change of the volume-averaged conserved quantities $\mathbf{W}_{i,j,k}$ which are in balance with the sum of the corresponding area-averaged fluxes, $\mathbf{R}^{\mathrm{c}}_{i,j,k}$ and $\mathbf{R}^{\mathrm{d}}_{i,j,k}$, across the cell faces and, for stability reasons, the artificial dissipation $\mathbf{D}_{i,j,k}$:

$$\frac{\mathrm{d}\mathbf{W}_{i,j,k}}{\mathrm{d}t} = -\frac{1}{V_{i,j,k}}\left(\mathbf{R}^{\mathrm{c}}_{i,j,k} - \mathbf{R}^{\mathrm{d}}_{i,j,k} + \mathbf{D}_{i,j,k}\right) = \mathbf{Res}_{i,j,k} \quad . \tag{2}$$

In addition, with emphasis on hypersonic flow, the directed propagation of information in the inviscid part of the equations, the AUSM (Advection Upstream Splitting Method) scheme is used for the approximation of the convective flux functions [4]. Higher-order accuracy and consistency with the central differences used for the diffusive terms is achieved by MUSCL (Monotonic Upstream Scheme for Conservation Laws) Extrapolation, and TVD (Total Variation Diminishing) property of the scheme is ensured by a modified van Albada limiter function.

The system of ordinary differential equations (2) is solved by an explicit five-stage Runge-Kutta time-stepping scheme in combination with different convergence acceleration techniques like multigrid and local time stepping for asymptotically steady-state solutions [3, 9]. Since the total enthalpy is a constant throughout the flow field for inviscid flows, its numerical deviation is applied as forcing function to accelerate convergence. For turbulent flow, the time integration of the turbulence equations is decoupled from the mean equations and the turbulence equations are solved using a Diagonal Dominant Alternating Direction Implicit (DDADI) scheme. The implicit scheme increases the numerical stability of turbulent flow simulations, which is especially important since the low Reynolds number damping terms as well as the high grid cell aspect ratios near the wall make the system of turbulence equations stiff. Due to the CFL-condition for explicit schemes, the CFL number of the Runge-Kutta scheme has an upper limit of 4. Implicit residual smoothing allows to increase the explicit stability limit by a factor of 2 to 3 [9] and, thus, a CFL number of 7.5 is realized for PHOENIX.

4 Results

4.1 Grid Generation

The grid generation is an integral part of the numerical solution process and, for complex geometries such as PHOENIX, it cannot be performed independently. Therefore, a preliminary solution of the governing equations on the current grid is computed which is then followed by a redistribution of points to improve the grid where necessary until a satisfactory solution is achieved.

The applied grid generation software MegaCads has been developed by the DLR as part of the MEGAFLOW project from which the FLOWer code originated, too [8, 5]. MegaCads is an interactive grid generation system which generates structured, three-dimensional, multi-block grids for flow computations. Its characterizing feature is the ability to process block-structured grids harmonizing with the flow solver. Furthermore, where appropriate, neighbouring blocks can be merged to reduce the overall block number and improve handling. Following this approach, the currently used three-dimensional Euler grid contains finally 20 blocks for the half configuration of PHOENIX and consists of approximately 2.15 million grid points. Fig. 1 presents a schematic overview of the distribution of blocks assembled into a single block-structured grid. In Fig. 2, the node distribution along with the block boundaries in the symmetry plane and in one additional grid plane (with one constant curvilinear coordinate) projected onto the y-z-plane are shown.

4.2 Inviscid Flow Computation

Euler calculations for the complete demonstrator PHOENIX at a freestream Mach number of $M_\infty = 2.21$ and different angles of attack have been performed. The calculations were conducted switching between 2-level W-cycle multigrid and single grid computations to damp high wavelength oscillations due to unfavorable grid spacing and to stabilize the inviscid flow solution around the vertical tail, respectively. No steady state solution has been found for the base and wake flow behind the vehicle, thus the density residual shows no consistent convergence. Therefore, the convergence behavior of the lift was used instead. A perspective view on the flow field in the symmetry plane is shown in Fig. 3. Due to the angle of attack, the bow shock is weaker on the leeward side of PHOENIX than on its windward side. The flow accelerates over the leeward side until it meets the shock in front of the tail fin and, thereafter, accelerates again. The flow along PHOENIX's windward side is followed by a strong expansion around the vehicle base.

The aerodynamic coefficients at different flight conditions are important design criteria for determining the shape and the trajectory of PHOENIX. For inviscid flow, only the pressure force acting on the surface has to be considered. Contour lines of the pressure coefficient distribution are shown in Fig. 4, where the pressure coefficient c_p is defined as deviation from the undisturbed static pressure normalized by the dynamic pressure $\frac{1}{2}\rho_\infty v_\infty^2$. As the CAD data of the body's surface had to be converted to generate a block structured grid with MegaCads [5] and no smoothing has been applied afterwards, the curvature of the surface of PHOENIX is not entirely smooth, which causes small oscillations in the pressure distribution, as can be seen at block boundaries and in the aft part of the fuselage. The lift coefficient is calculated by projecting the force coefficients in the direction perpendicular to the undisturbed inflow. Similarly, the drag coefficient is a projection in the inflow direction. Since PHOENIX has a symmetry plane and the yaw angle is zero for all computations, the side force and the yaw moment are equal to zero. The computed aerodynamic coefficients of the Euler solutions are presented in Table 1. Juxtaposition with aerodynamic design data provided by Astrium is shown in Fig. 5.

4.3 Future Investigations

Additional grid layers for the resolution of boundary layers have to be inserted into the Euler grid before the computations of the viscous high-speed flow are started. These investigations include a full Navier-Stokes analysis of PHOENIX for high-enthalpy conditions. A first Navier-Stokes computation assuming laminar flow and employing the upwind scheme has been performed on the Euler grid with angle of attack $\alpha = 20^o$ and a Mach number of $M_\infty = 4.5$, see Fig. 6.

5 Conclusions

Inviscid flow computations for the technology demonstrator PHOENIX are presented. The applied 3D grid consists of 20 blocks with a total of 2.15 million grid points and uses structured, curvilinear coordinates. An interaction between multigrid and single grid computations has been necessary to stabilize the solution process of the flow solver FLOWer. Converged solutions are presented for a flight Mach number of 2.21. The obtained aerodynamic coefficients are in good agreement with design and wind tunnel data provided by Astrium. A first result for viscous flow simulations has been obtained.

Acknowledgements

The financial support from the Deutsche Forschungsgemeinschaft (DFG) and the funding of Astrium Space Company is gratefully acknowledged.

References

[1] Coratekin T., van Keuk J., Ballmann J. "On the Performance of Upwind Schemes and Turbulence Models in Hypersonic Flows", AIAA Paper 2001–1752 (2001)

[2] Coratekin T. "Numercal Simulation of Turbulent Shock/Boundary-Layer Interactions in Hypersonic Flows", Diss., VDI Verlag 2000, ISBN 3-18-338107-9

[3] Heinrich R., Pahlke K., Bleecke H. "A three dimensional dual-time stepping method for the solution of the unsteady Navier-Stokes Equations", Proc. 'Unsteady Aerodynamics' Conference in London, pp. 5.1-5.12, 1996

[4] Kroll N., Radespiel R. "An Improved Flux Vector Split Discretization Scheme for Viscous Flows", DLR-Forschungsbericht **93/53** (1993)

[5] "Das MegaCads-Handbuch", DLR Braunschweig, 1996

[6] v. Keuk J. "Numerische Analyse und Bewertung von Upwind-Verfahren für die Anwendung auf Mehrdimensionale Über- und Hyperschallströmungen von Gasen", Diss., Shaker Verlag 2000, ISBN 3-8265-8080-X

[7] v. Keuk J., Ballmann J., Sanderson S. R., Hornung H. G. "Numerical simulation of experiments on shock wave interactions in hypervelocity flows with chemical reactions", AIAA Paper 03-0960 (2003)

[8] Kroll N., Rossow C.-C., Becker K., Thiele F. "The MEGAFLOW Project", Aerospace Science and Technology **4** (2000) 223–237

[9] Radespiel R., Rossow C., Swanson R. "Efficient Cell-Vertex Multigrid Scheme for the 3D Navier-Stokes Equations", AIAA J. **28/8** (1990) 1464–1472

[10] Reinartz B., v. Keuk J., Coratekin T., Ballmann J. "Computation of Wall Heat Fluxes in Hypersonic Inlet Flows", AIAA Paper 2002-0506 (2002)

[11] Somsok, A. "Wiederverwendbare Raumtransportsysteme – Optionen für die Zukunft", Luft und Raumfahrt 3/2001 32–34

Table 1 Aerodynamic coefficients for inviscid flow computation ($M_\infty = 2.21$).

Angle of attack α	Lift coeff. c_l	Drag coeff. c_d	Moment coeff. c_m
10.0^o	0.19258	0.0982	$-0.2582 \cdot 10^{-2}$
11.0^o	0.21133	0.1055	$-0.3616 \cdot 10^{-2}$
12.0^o	0.23010	0.1135	$-0.4625 \cdot 10^{-2}$
12.5^o	0.23952	0.1177	$-0.5169 \cdot 10^{-2}$
13.0^o	0.24900	0.1223	$-0.5723 \cdot 10^{-2}$
15.0^o	0.28762	0.1423	$-0.7915 \cdot 10^{-2}$

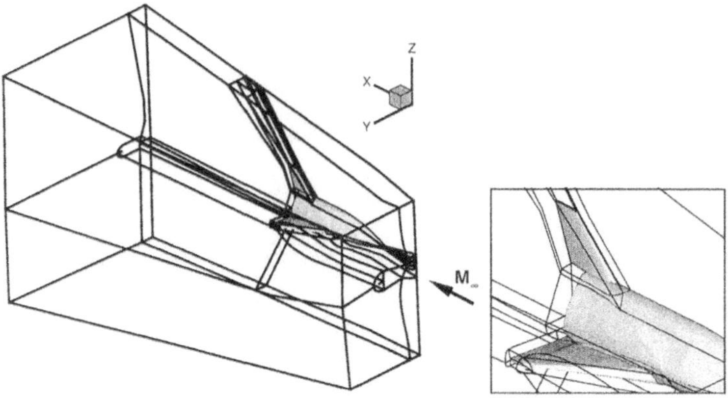

Figure 1 Multi-block grid with 20 blocks and 2.15 million grid nodes.

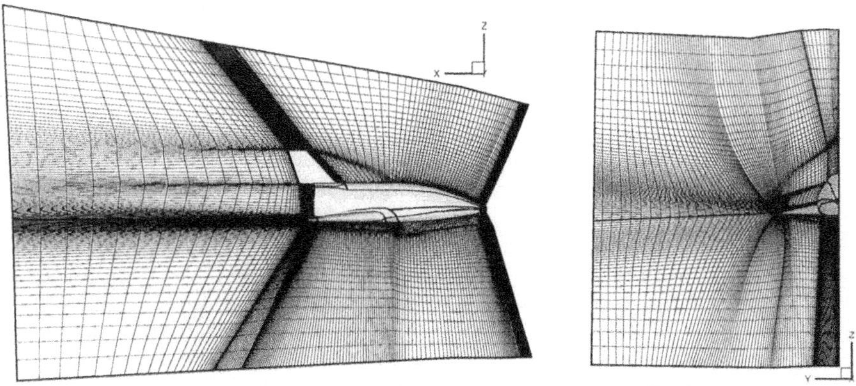

Figure 2 Grid point distribution along symmetry plane (left) and in one grid plane along the wing center projected on the y-z-plane (right).

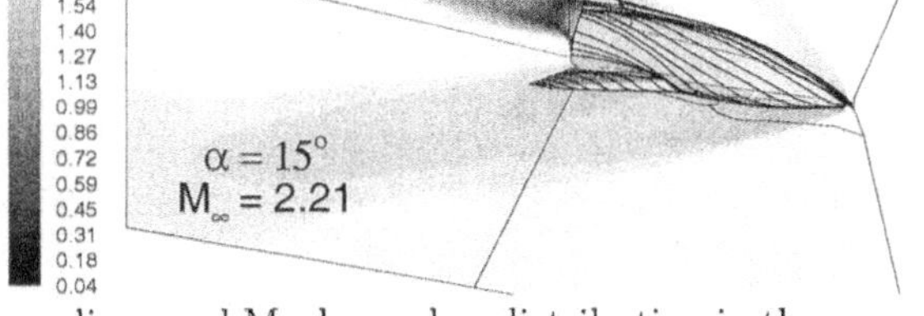

Figure 3 Streamlines and Mach number distribution in the symmetry plane with additional streamlines on the surface of PHOENIX .

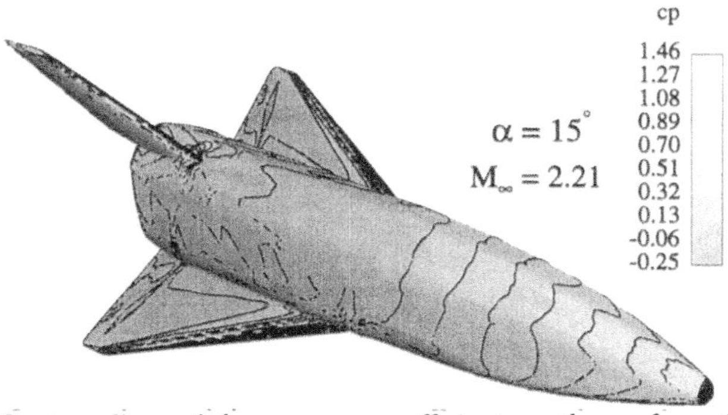

Figure 4 Contour lines of the pressure coefficient on the surface of PHOENIX

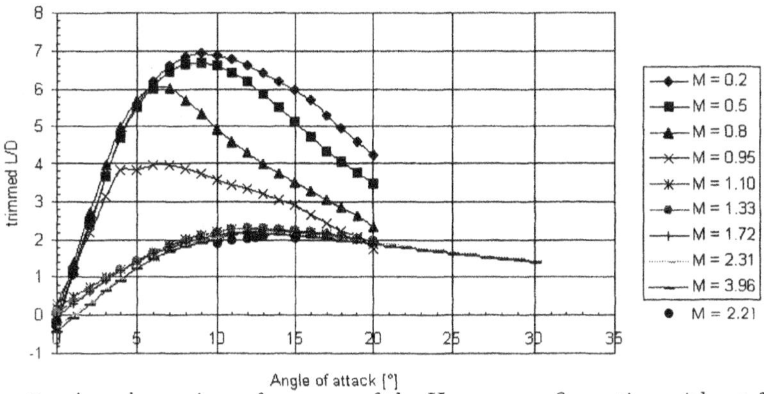

Figure 5 Aerodynamic performance of the Hopper configuration without friction (provided by Astrium); extra data from inviscid flow computations ($M_\infty = 2.21$) added

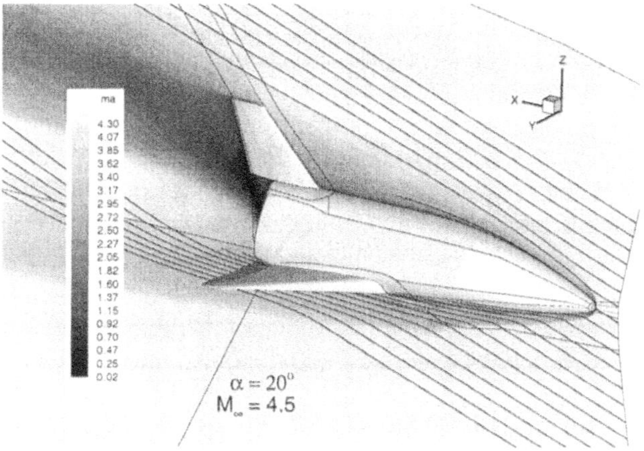

Figure 6 Streamlines and Mach number distribution in the symmetry plane for viscous flow calculation

Aerodynamic Design of an Ariane 5 Liquid Fly-Back Booster

Thino Eggers

DLR, Institute of Aerodynamics and Flow Technology, Lilienthalplatz 7, 38108 Braunschweig, Germany

Summary

Within the German future space launcher technology research program ASTRA two reusable first stage designs are under investigation. The one dedicated for near term application with an existing expendable core stage is called a winged liquid fly-back booster (LFBB). The main goal of the present study is the refinement of an existing LFBB design. The regarded partially reusable space tranportation system consists of two booster stages, which are attached to the expendable Ariane 5 core stage at an upgraded future technology level. The area of interest discussed here is the descent after the staging procedure which dominates the aerodynamic design of the LFBB. The aerodynamic parameter studies pointed out that a LFBB may not be trimmed or stabilised without the introduction of canards as control surface. By introduction of canards a more robust configurations is obtained.

1. Introduction

The aerodynamic and flight dynamic simulation applied in the second design loop of the LFBB requires trimmed aerodynamic datasets for the complete return flight trajectory ($0.27 < M < 7$ and angles of attack of $5° < \alpha < 35°$). Within these studies tight margin concerning longitudinal stability and trim have to be taken into account for the isolated LFBB as well as for the complete system (Fig. 1). Due to the structural design of the Ariane 5 core stage the position of the LFBB with regard to the core stage is fixed. The aerodynamic behaviour of the booster has to be robust over the complete Mach number range. This also provides sufficient safety with view to the uncertainties coming from the applied simplifications of the aerodynamic approach. The aerodynamic studies are based on Euler calculations applying unstructured grids with the DLR TAU-Code for $M < 2$. For $M > 2$ the DLR surface inclination method HOTSOSE is applied. As shown in [1] and [2] both codes are well established for these kind of applications. The investigations show that robust configurations are obtained if canards are considered as control surfaces. The behaviour of the final configuration is discussed in a more detailed analysis of the aerodynamic coefficients and the flow fields.

2. Aerodynamic Design and Analysis of the LFBB

During the LFBB design several configurations were considered. The initial LFBB-design, based on the results of [3], is the configuration "Y6" (Fig. 2). The vehicle has a length of l_{ref}=42m, a span of b=21m and a planform area of A_{ref}=360m^2. With view to the hypersonic region of the descent trajectory, where the aerodynamic coefficients of the vehicle are dominated by the lower surface, the trapezoidal wing has an airfoil with flat lower surface. In order to allow an indifferent behaviour with acceptable flap deflections, the goal with view to the longitudinal stability is to shift the neutral point into the region of the center of gravity, $x_S/l \approx 0.62$. Nevertheless, the results indicated severe problems concerning logitudinal stability and trim for configuration "Y6". Figure 3 shows that the trim of this vehicle is only possible for M < 5.6 and assuming an actually not achievable far backward position of the center of gravity, $x_S/l=0.75$. The required flap deflections in hypersonic flow are still in the order of $\eta=30°$ and more than $\eta=-10°$ for M < 2. These results are obtained with very large wing flaps and a body flap which were deflected simoultaneously. Additionally, the configuration is highly unstable (Fig. 3). Based on these results it is decided to investigate LFBB configurations with canards. During the design cycle of the canard configuration a large number of wing planforms and canards is investigated. The numerical results for all the configurations are obtained with the surface inclination method HOTSOSE [4]. For the subsonic part of the descent trajectory handbook methods like [5] are applied. They allow to avoid the time consuming and complex grid generation process of Euler or Panel methods.

Planform views of the initial ("C60") and the final configuration ("Y7") of this study are given in Fig. 2. All investigated LFBB variants have canards with unswept trailing edges and a leading edge sweep of $\varphi=60°$. With view to the canard efficiency and controllability in subsonic flow, this sweep angle is chosen in order to establish leading edge vortices up to angles of attack for the canards of $\alpha \approx 15°$. Even for bursted vortices in the trailing edge region the maximum lift of the canards is not expected for $\alpha<25°$. The strategy for the application of the control surfaces is to consider a clean wing, a body flap with constant deflection and variable canard deflection for trim. In order to allow a direct comparison to the LFBB without canard ("Y6") the initial canard configuration "C60" has an unchanged planform area of A_{ref}=360m^2. Therefore, also the wing load is constant. The final configuration "Y7" has a slighty reduced planform area of (A_{ref}=350m^2) obtained by minor changes of canard and wing size. This is required because the masses of wing and canard have to be reduced [6]. The body itself is unchanged due to the tight margins concerning the connection with the Ariane 5 core stage. The trim results show that the configurations „C60" and „Y7" are very robust even for a realistic position of center of gravity at $x_S/l=0.62$ (Fig. 4). The obtained canard deflections are always smaller than $\eta_{can}=5°$. The configurations are stable for M > 5.5 and only slightly unstable for 2 < M < 5.5. Furthermore, estimations from handbook methods indicate that the required canard deflections for subsonic conditions are always smaller than $\eta_{can}=10°$. An additional assessment of configuration "Y7" with a deflected body flap ($\eta_{BF}=5°$) again reduces the flap deflections and results in a slightly increased longitudinal stability. Using these results configuration "Y7" is chosen as final LFBB for this design cycle. Additionally, a wind tunnel model of this configuration is projected. This model is going to be used for force measurements in the DLR wind tunnels TMK and H2K in November 2002. The measurements will cover the Mach number range $0.5 < M < 7.5$ for conditions without and with sideslip. The results are going to be used to validate the aerodynamic approach and to obtain a trimmed dataset for longitudinal as well as sideslip motion.

For the analysis of configuration "Y7" in subsonic, transonic and supersonic flow, Euler grids around the configuration with a fixed body flap deflection of $\eta_{BF}=5°$ are generated for canard deflections of $\eta_{can}=0°$ and $\eta_{can}=10°$. In order to be able to simulate the influence of the leading edge vortices on the canard efficiency, also based on the Euler calculations, the original NACA 0008 was slightly sharpened along the leading edge. This approach of course does not allow to obtain the real vortex structure, but it gives a good approximation of the vortex strength and it also enables to predict a possible vortex bursting. The goal is to generate a trimmed aerodynamic dataset for the Mach number range $0.27 < M < 2.0$. The basis for this dataset are calculations for 15 Mach numbers, two canard deflections and four angles of attack for each (Ma, η_{can})-combination (Fig. 5). For the trim calculations and the definition of the final aerodynamic dataset the lift and pitching moment coefficients are obtained by linear interpolation. For the interpolation of the pressure drag coefficients, quadratic polars are fitted into the Euler results. The skin friction drag is estimated based on the assumption of a turbulent flat plate as $C_{D,f} = c_f \cdot A_{wet}/A_{ref}$ [7]. The trim results of this approach are summarised in Fig. 6. A first comparison with the HOTSOSE results for $M > 2$ shows a good agreement of both codes. For M=2, the obtained angles of attack are nearly identical and the canard deflections show an uncertainty of $\eta \approx 1°$. Due to the fact that the calculation of the neutral point is very sensitive concerning small changes in pitching moment, the largest differences between both codes appear during the analysis of the static margin. The Euler results indicate a slightly unstable behaviour for M=2. The HOTSOSE results are more optimistic and show an indifferent behaviour and therefore, also a slightly reduced canard deflection. Due to the fact that surface inclination methods are known to be sufficiently accurate for $M > 4$, a static margin of $(x_N-x_S)/l=-0.02$ can be expected in the range $2<M<4$ (dotted line, Fig. 6).

Considering the flap deflections for subsonic and transonic flight conditions it turns out that, with the exception of M=0.95, all canard deflections are smaller than $\eta_{can}=10°$. This shows that the careful application of handbook methods for estimation of the vehicle behaviour in subsonic flow proved oneself. A comparison of the flow fields for M=0.7 and M=0.95 (Fig. 7) explains the increased canard deflections for $0.9 < M < 1$. For M=0.7 a vortex is established along the leading edge of the canard. It is responsible for the relatively small canard deflections in subsonic flow because it increases the canard efficiency. For transonic conditions it is visible that vortex bursting appears at the trailing edge of the canard. This behaviour results in the situation, that the lift coefficient in transonic flow is nearly independent of the canard deflection (Fig. 5). Therefore, larger canard deflections are

required.
A second interesting effect in transonic flow is the sudden tailward shift of the neutral point for $0.75 < M < 0.95$ and the following frontward movement for $0.95 < M < 1.025$. This behaviour may be explained with the help of Fig. 8. For the explanation it has to be recalled that the aerodynamic forces resulting from a change in angle of attack are acting in the neutral point. Therefore, a comparison of the pressure distributions for two angles of attack allows an assessment of the influence of the Mach number on the neutral point position. First, the upper surface pressure distributions for M=0.7 are considered. The comparison of $\alpha=0°$ and $\alpha=5°$ shows that the leading edge vortex of the canard is responsible for the increase of the lift force. The vortex strength along the canard leading edge increases and this vortex also induces a low pressure region over the outer part of the main wing. If now the results for M=0.9 are considered it is visible that the overall upper surface pressure levels of the canard do not differ significantly for the shown angles of attack. Along the inner part of the leading edge the vortex strength increases with the angle of attack but in the trailing edge region vortex bursting appears. This effect partly revokes the gain in lift obtained in the inner part of the canard. Additionally, the effect of the canard vortex on the upper surface of the main wing is significantly reduced. A second important effect on the increase of lift results from the influence of the trailing edge shock on the upper surface of the wing. In subcritical transonic flow, a shock moves rearward with increasing Mach number and for angle of attack. This is visible when the results for M=0.975 are considered in addition. Here the shock already reached the trailing edge. For $M < 0.975$ the shift of the shock position is still possible and therefore, an increase on angle of attack results in a rearward extension of the low pressure region on the upper surface.
In summary, regarding the results for the Mach numbers M=0.7 and M=0.9, the additional lift resulting from an increased angle of attack at M=0.9 is mainly produced in the wing region and it is not dominated by the canard vortex. This explains the sudden neutral point shift to a more backward position for $0.7 < M < 0.9$. For M=0.975 the same situation appears except that the shock on the upper surface of the wing already reaches the trailing edge. This effect is independent of the angle of attack and thus, the resulting lift force acts slightly more frontward. The lack of additional suction force along the trailing edge then leads to the noseward jump of the neutral point for $0.9 < M < 1.025$.
Based on the Euler results the aforesaid discussion about the neutral point position also explains the lack of longitudinal stability ($(x_N-x_S)/l=-0.05$) for the subsonic cruise condition (Fig. 6). The pressure distributions in Fig. 8 show that the strong interaction of the leading edge vortex of the canard and main wing are responsible for the given unstable behaviour of "Y7". The results indicate, that the problem can be solved if the increase of the wing lift with angle of attack is higher than the one of the canard. Additionally, it would help to shift the center of gravity in noseward direction. For a final discussion of this problem at first the wind tunnel experiments have to be analysed concerning the existence of a vortex along the leading edge of the main wing which could increase the slope of the wing lift curve. Due to the fact

that the leading edge has a large bluntness the Euler calculations are not able to predict these phenomena.

3. Conclusions

The second aerodynamic design cycle of an Ariane 5 liquid fly-back booster required detailed aerodynamic datasets which allow the trim and balance of the vehicle along the complete return flight trajectory. In the range $0.27 < M < 2$ these were obtained based on Euler calculations using unstructured grids. For $M > 2$ surface inclination methods were applied. With view to the trim of the vehicle the results show that there is an essential need to consider a configuration with canards. The canard deflections may be limited to $\eta_{can} \approx 8°$ for subsonic flow and they are always smaller than $\eta_{can} = -5°$ for super- and hypersonic flow conditions. One remaining problem of the final configuration is the lack of longitudinal stability during the subsonic cruise flight. The Euler results showed that the reason for this problem is the strong interaction between the leading edge vortex of the canard and the upper surface of the main wing. The problem could be solved if the slope of the lift curve of the wing would be larger than the one of the canard. Additionally the center of gravity should be shifted in frontward direction. However, a final discussion of this problem will be possible when the wind tunnel experiments for the final LFBB configurations are analysed concerning the existence of wing leading edge vortices which might increase the slope of wing lift curve.

References

[1] A. Mack; V. Hannemann: 'Validation of the Unstructured DLR-TAU-Code for Hypersonic Flows". AIAA Paper 2002-3111, 2002.

[2] Th. Eggers, Ph. Novelli; M. Haupt: „Design Studies of the JAPHAR Experimental Vehicle for Dual Mode Ramjet Demonstration". AIAA Paper 2001-1921, 2001.

[3] M. Sippel; U. Atanassow; J. Klevanski; V. Schmid: „First Stage Design Variations of Partially Reusable Launch Vehicles". J. of Spacecraft, Vol. 25, No. 4, July-August 2002.

[4] U. Reisch; Y. Anseaume: „Validation of the Approximate Calculation Procedure HOTSOSE for Aerodynamic and Thermal Loads in Hypersonic Flow with Existing Experimental and Numerical Results". DLR-FB 98-23, 1998.

[5] J. Klevanski; M. Sippel: „Beschreibung des Programms CAC zur aerodynamischen Voranalyse, Version 1.0". SART TN-003/2001, DLR-IB 645-2001/02,2001.

[6] M. Sippel; J. Klevanski; H. Burkhardt; Th. Eggers; O. Bozic; Ph. Langholf; A. Rittweger: „Progress in the Design of a Reusable Launch Vehicle Stage". AIAA Paper 2002-5220, 2002.

[7] Th. Eggers: „Studies for the Design of a Dual Mode Ramjet Flight Test Vehicle". DLR-IB 129-2000/1, 2000.

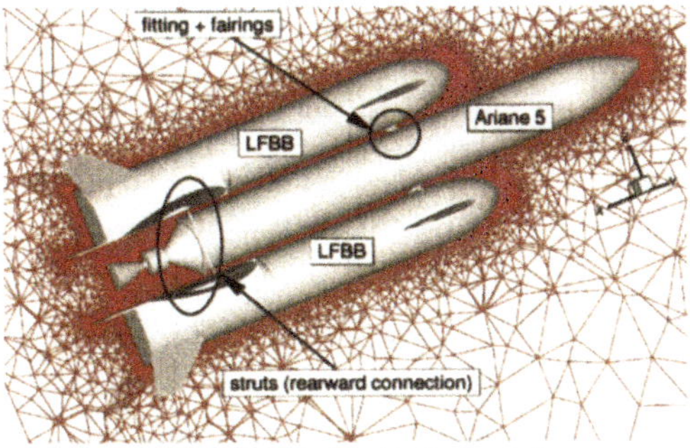

Fig. 1: Ariane 5 with liquid fly-back Boosters

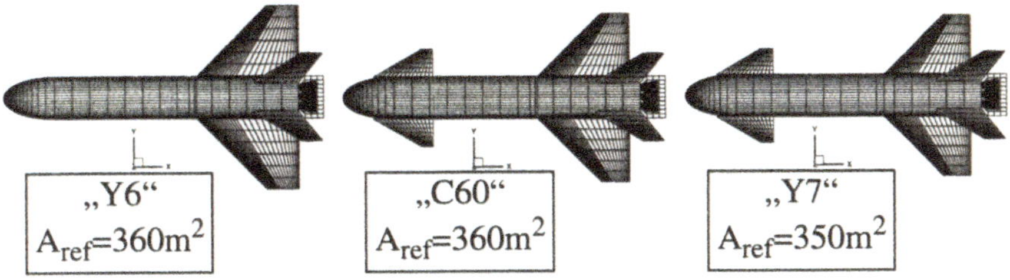

Fig. 2: Planform views of the discussed configurations (l_{ref}=42m)

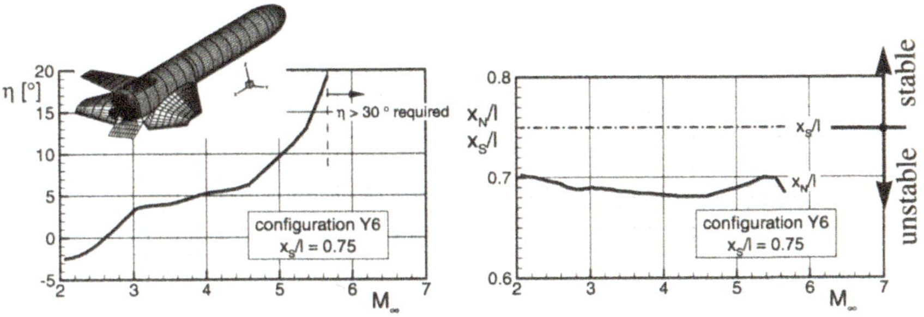

Fig. 3: Trim results for configuration "Y6"

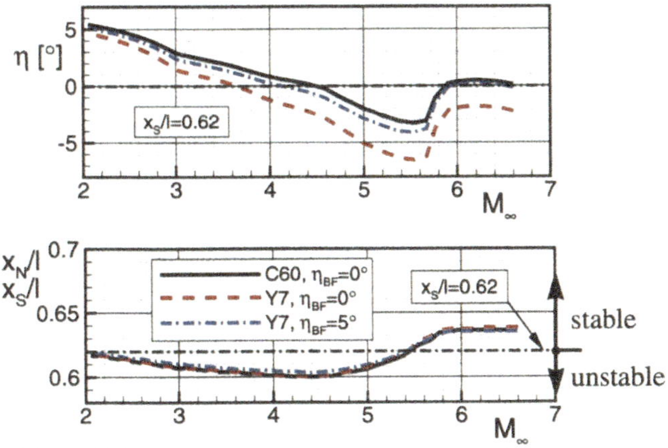

Fig. 4: Trim results for the main canard configurations

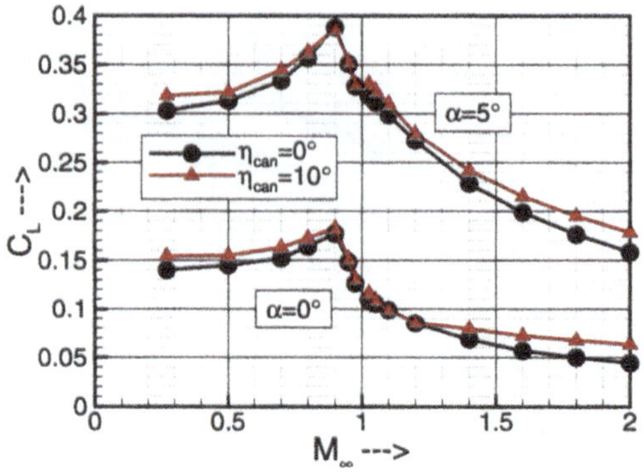

Fig. 5: Euler lift coefficients of configuration "Y7"

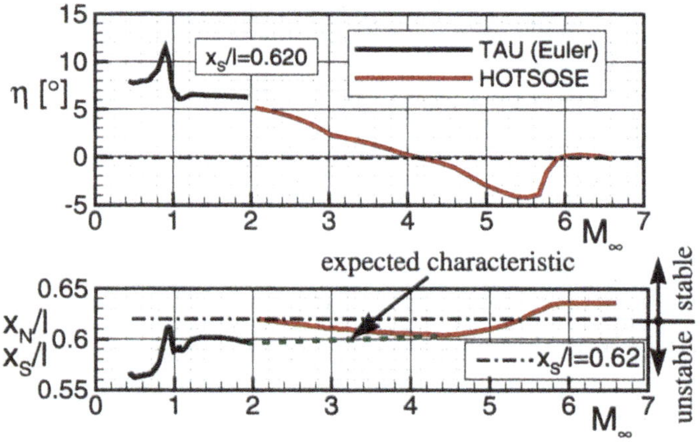

Fig. 6: Trim results along the complete trajectory (configuration "Y7")

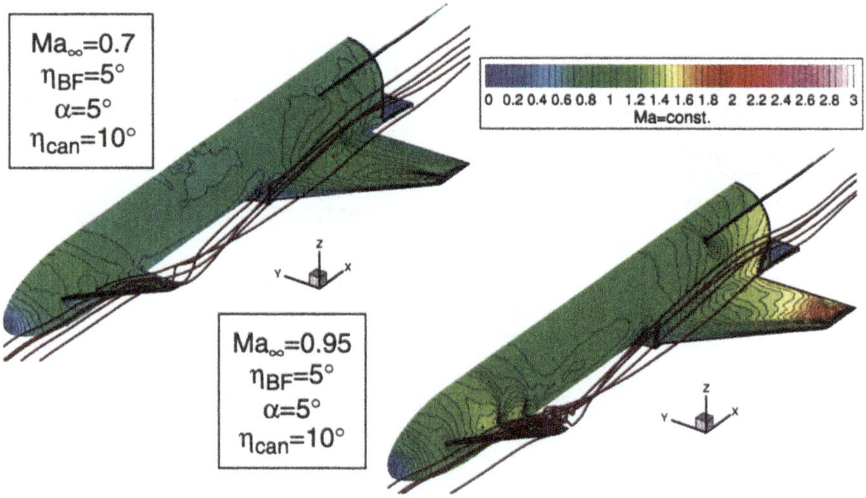

Fig. 7: Flow fields around configuration "Y7"

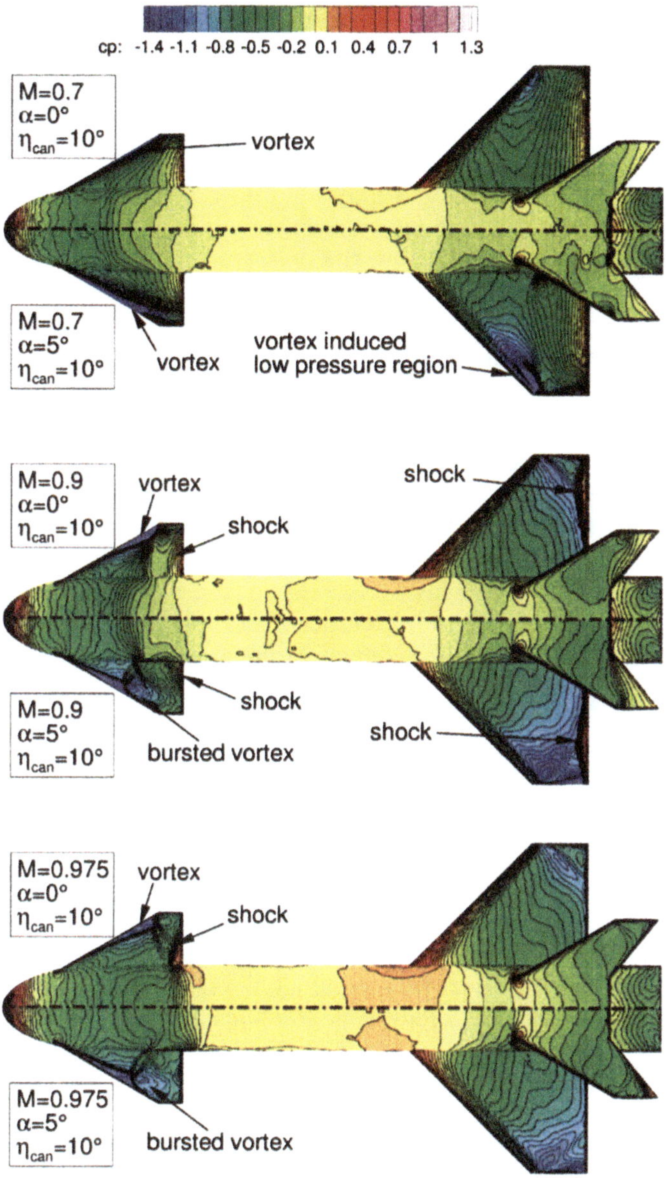

	$C_{L\alpha}$ [1/rad]
canard	0.355
wing	1.031
Y71	2.076

	$C_{L\alpha}$ [1/rad]
canard	0.309
wing	1.217
Y71	2.351

	$C_{L\alpha}$ [1/rad]
canard	0.309
wing	1.238
Y71	2.292

Fig. 8: Upper surface pressure distribution of configuration „"Y7"

Vortical type flow with respect to jet-external airflow interaction

Wilke Zeiss, Roland Behr

Astrium GmbH, 81663 München, Germany
wilke.zeiss@astrium-space.com

Summary

Basic phenomena of jet-airflow interaction are studied at several test cases. Simplified cases consisting of a flat plate at zero inclination and a vertically blowing thruster as well as two prospective thrusters on the X-38 re-entry vehicle are investigated. The corresponding flow topology is discussed and a general dependency on the ratio of onset flow and thruster jet strength is found. Within this study two particular vortex systems are observed. The first is a horseshoe vortex with a spanwise extension upstream of the thruster and two downstream legs. The second system is a closed ring vortex located around the perimeter of the thruster exit. In some cases this ring vortex breaks up into two counter rotating horseshoe vortices with spanwise filaments up- and downstream of the jet.

1 Introduction

Future reusable launch vehicles (RLV) follow the concept of lifting bodies. In the upper part of the trajectory, where the density is small, the flight control of the vehicle is conducted by a thruster-based reaction control system (RCS), while in the lower part of the trajectory, where the aerodynamic forces increase, aerodynamic control surfaces (flaps, rudders, etc.) are used. A RCS system typically consists of small thrusters in different places on the vehicle surface. They are fired to perform an impulse on the vehicle. The efficiency of the RCS thrusters may be raised by the complex interaction with the outer flow field. In addition, on the one hand, severe heating and loading problems associated with the exposure of the control surface (i.e. flaps) in supersonic stream may be avoided, but on the other hand firing thrusters will lead to flow separation and reattachment with the consequence of producing hot spots. In [1] an in detail investigation of these problems for the US space shuttle is given, with the analysis tools availbale at that time. Moreover, the vehicle surface can be loaded by an impinging thruster jet. As the flowfield of the jet/external airflow interaction is expected to be sophisticated, first studies are performed for simplified test cases. In a second step the flow around the X-38 vehicle with thruster 'on' and 'off' is computed. The results of both simulations will be compared.

2 Numerical method

2.1 Flow solver

The Navier-Stokes flow solver DAVIS-VOL [3] [4] has been developed at Astrium. The code solves the compressible Favre-averaged Navier-Stokes equations, using a cell-centered finite volume discretization technique and structured meshes. The convective fluxes are disctretized using Yee's [5] second-order upwind Total Variation Diminishing (TVD) scheme. The viscous terms are second-order centrally discretized. In order to increase robustness and to accelerate convergence, an implicit time-stepping procedure is used. The resulting system of linear equations is solved by a line Gauss-Seidel relaxation method. Since the goal of this study is to identify the basic phenomena of this kind of flow, like flow separation and reattachement as well as vortex formation, the physical modelling was in a first step simplified with the assumption of laminar flow for all computational cases, perfect gas for the flat plate test cases and equilibrium real gas for the X-38 cases. Of course, turbulence and to some extent the local thermodynamic state of air (non-equilibrium effects) will have an influence on the flow field but the experience has revealed that a complete change of the topology has not to be expected.

2.2 Modeling and boundary conditions

Test cases: At the inflow plane and at the thruster exit Mach number, pressure and temperature are prescribed (see chapter 3). At the supersonic outflow boundary all flow variables are extrapolated. At the walls no-slip condition is applied and radiation adiabatic boundary condition with constant emissivity ($\varepsilon = 0.8$) is used.

X-38 cases: The modelling of the air in thermodynamic equilibrium is based on analytically defined functions developed by Mundt et al. [2]. The boundary conditions applied are the same as for the case described above.

3 Flow parameters

The free stream conditions of all cases are taken from a X-38 trajectory provided by NASA in the frame of Germany's technology program TETRA. To vary the flow conditions upstream of the thrusters, two different angles of attack are considered. The thrusters applied are authentic products manufactured by NASA resp. Astrium.

"Test case 1" is set up identical to the X-38 thruster2.

The inflow conditions at the thruster exit plane of "test case 2" are gained by increasing the dynamic part of the thrust of "test case 1". This is realized by increasing the pressure accordingly.

"Test case 3" produces the same dynamic part of the thrust as "test case 2". For this case the pressure of "test case 1" remains unchanged, and the Mach number is increased.

4 Computational grid

Test case: A flat plate with an embedded thruster is chosen. The surface mesh is refined towards the thruster (see fig. 1). The whole mesh is more refined than in case of the X-38. In total the mesh consists of 300,806 cells.

X-38: For the jet/airflow interaction problem only the front part of the vehicle is taken into account. As the side-slip angle is set to be zero, only the half configuration is regarded. Anyhow, the total number of cells is approx. 1.2 mio. The position of the thrusters is given in fig. 2. This picture represents both thrusters simultaneously. The detail screens show the meshes around the thrusters. Thruster1 is the one next to the vehicle nose, thruster2 is positioned on the rudimentary canopy. The exit plane of the thruster lies flush with the surface of the vehicle. The thruster jet is injected without any velocity profile resp. boundary layer.

5 Computational results

"Test case 1": Fig. 3 shows the distribution of pressure coefficient on the surface and the symmetry plane with streamlines of "test case 1". The thruster jet is refracted by the oncoming free stream. The free stream itself is redirected by a shock upstream of the thruster jet. Two major vortex systems can be identified. A closed ring vortex lies around the thruster exit and a horseshoe vortex in front of the ring vortex. This ring vortex can be particularly seen behind the thruster, whereas upstream of the thruster it is lying much more closely at the surface. Here the horseshoe vortex is better observable. The separation and reattachement lines of these vortices are clearly visible. The pressure coefficient increases close to the reattachement line of the horseshoe vortex.

"Test case 2": A view of the situation in this case is presented in fig. 4. Increasing the pressure of the thruster affects the thruster jet as well as the flow around the thruster. The strengthened jet leaves the surface more perpendicularly. Upstream of the jet the ring vortex and the horseshoe vortex increase in size. Downstream of the thruster the ring vortex is reduced. The separation line of the horseshoe vortex in front of the thruster moves upstream. The reattachement zone between the ring vortex and the horseshoe vortex is amplified.

"Test case 3": The solution of the flow field is shown in fig. 5. Both, the ring vortex around the thruster and the horseshoe vortex also exist here. Compared to "test case 1" the vortices upstream of the jet increase in size, but less than in "test case 2". The reattachement between the ring vortex and the horseshoe vortex is weakened compared to "test case 2". The separation line of the horseshoe vortex is lying less upstream compared to "test case 2". As the dynamic part of the thrust is the same for "test case 2" and "test case 3", the position of the separation line is related to the static pressure of the thruster. The higher the pressure the more the separation line moves upstream.

X-38 thruster2 $\alpha = 0°$: The flow phenomena at the X-38 vehicle with thruster2 $\alpha = 0°$ are given in fig. 6. In front of the X-38 vehicle a bow shock exists. The thruster is positioned beneath the bow shock. Therefore, the oncoming flow is not identical to the free stream condition as it was for the test cases. The flow topology is different to the one of the test cases. Neither the horseshoe vortex upstream of the thruster nor the ring vortex around the thruster can clearly be found. Instead of both systems there seems to exist a narrow horseshoe vortex downstream of the thruster near the symmetry plane only.

X-38 thruster2 $\alpha = 40°$: Fig. 7 shows the flow topology next to thruster2 for $\alpha = 40°$. Again the oncoming flow for the thruster changed due to the finite angle of attack. The flow topology looks similar the one of $\alpha = 0°$. In principal the vortex pattern induced by the thruster show the same characteristics with different spatial extension. The separation line beside the thruster can be found also in the results without thruster as it is the global separation line of the flow about the X-38 vehicle bulged by the thruster jet.

X-38 thruster1 $\alpha = 0°$: The streamlines on the surface of the vehicle with thruster1 and $\alpha = 0°$ are plotted in fig. 8. In consequence of the different positions of the X-38 thruster1 and 2 the flow next to thruster1 differs to the one in case of thruster2. Here we have a flow topology very similar to what we have observed for "test case 2" and "test case 3" (fig. 4, 5). Upstream of the thruster there are consecutively existing two horseshoe vortices. The one more away from the thruster also exists in the test cases. The one close to the thruster is a burst ring vortex. As in the other X-38 cases there is a narrow horseshoe vortex downstream of the thruster near the symmetry plane.

6 Conclusion

In the frame of this study mainly two vortex systems are observed for the jet-external airflow interaction problem. A horseshoe vortex located upstream of the thruster jet and a ring vortex lying around the thruster exit appear. Depending on the pressure and the kinetic energy ratio between free stream and thruster jet the ring vortex exists as a closed ring or it breaks up into a pair of horseshoe vortices. One infront of the thruster and the other – here always appearing as very narrow – horseshoe vortex is formed downstream of the thruster.

The simplified configuration chosen for the test cases allows a comprehensive analysis of the flow phenomena and the corresponding flow field topology. Of course, further investigations with variation of thruster and/or freestream data will be helpful and necessary to completely understand the intricate flow occurring in case of the jet-external airflow interaction in applications on re-entry vehicles like the X-38 with strong 3-dimensional effects in the onset flow.

References

[1] W. J. Monta, J. R. Rausch: "Effects of reaction control system jet-flow field interactions on a 0.015 scale model space shuttle orbiter aerodynamic characteristics", NASA-CR-134074, 1973

[2] Ch. Mundt, R. Keraus, J. Fischer: "New accurate vectorized approximations of state surfaces for the thermodynamic and transport properties of equilibrium air", ZWF, 15(6),1991,pp.179-184

[3] C. Weber: "Improvement of numerical simulation methods", DLR-TETRA Programmm, TET-DASA-21-TN-2407, Mai 2001

[4] C. Weber, R. Behr, C. Weiland: "Investigation of hypersonic turbulent flow over the X-38 crew return vehicle", AIAA paper No. 2000-2601, Denver, 2000

[5] H. Yee: "Upwind and symmetric shock-capturing schemes", Technical memorandum 89464, NASA, 1997.

Table 1 Free stream conditions for test cases (left) and X-38 flows (right)

test case free stream			X-38 free stream		
γ	=	1.4			
Ma_∞	=	17.5	Ma_∞	=	17.5
Re_∞	=	$1.02 \cdot 10^6$	Re_∞	=	$1.02 \cdot 10^6$
p_∞	=	26.2 Pa	p_∞	=	26.2 Pa
T_∞	=	258 K	T_∞	=	258 K
angle of attack	=	$0°$	angle of attack	=	$0°, 40°$

Table 2 thruster conditions for the flow cases computed

	"test case 1"	"test case 2"	"test case 3"	X-38 thruster1	X-38 thruster2
p	192 Pa	1921 Pa	192 Pa	16210 Pa	192 Pa
T	728 K	728 K	728 K	57 K	728 K
Ma	5.99	5.99	18.95	4.25	5.99

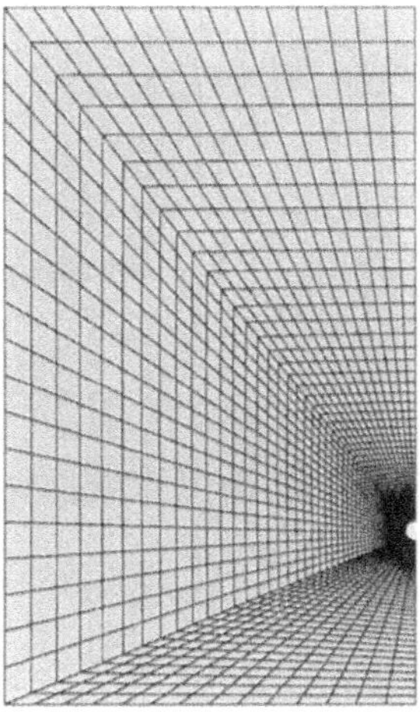

Figure 1 mesh on surface of test case with thruster

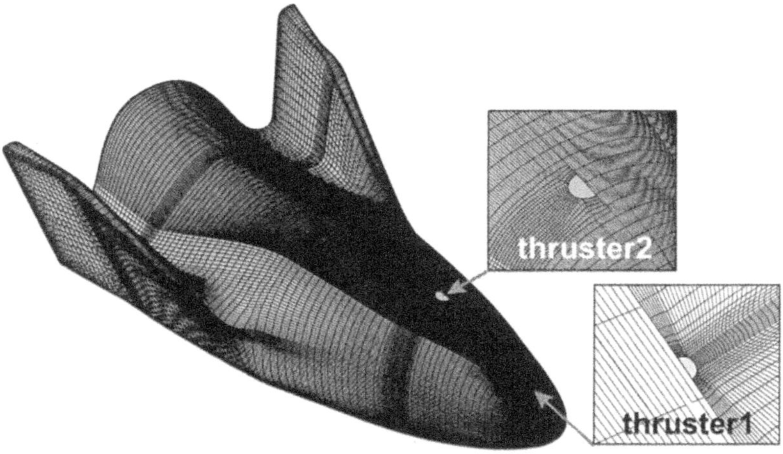

Figure 2 mesh on surface of X-38 with thrusters, upper detail thruster2, lower detail thruster1

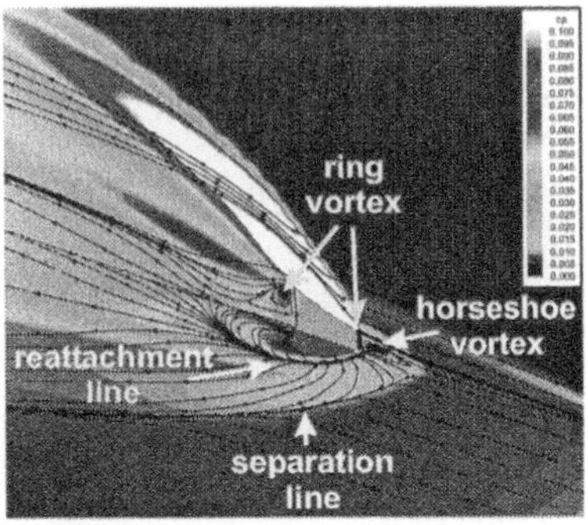

Figure 3 "test case 1" pressure coefficient isosurfaces and streamlines in symmetry plane and on surface

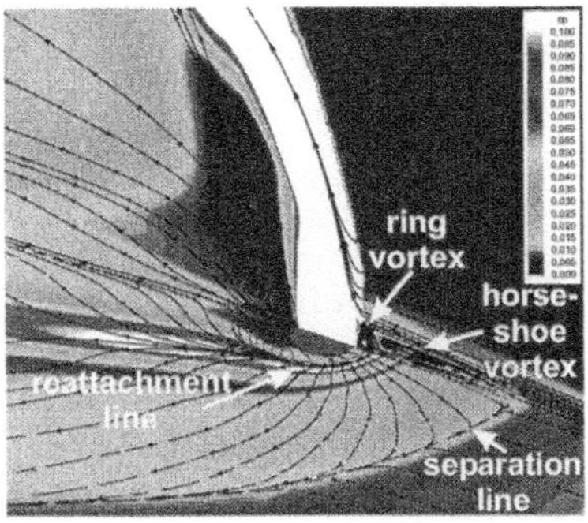

Figure 4 "test case 2" pressure coefficient isosurfaces and streamlines in symmetry plane and on surface

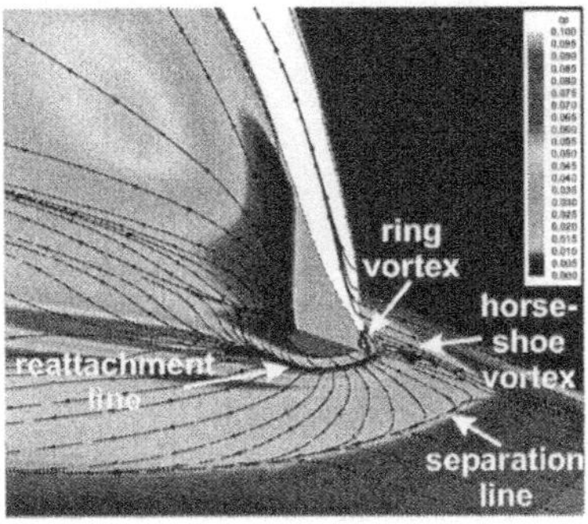

Figure 5 "test case 3" pressure coefficient isosurfaces and streamlines in symmetry plane and on surface

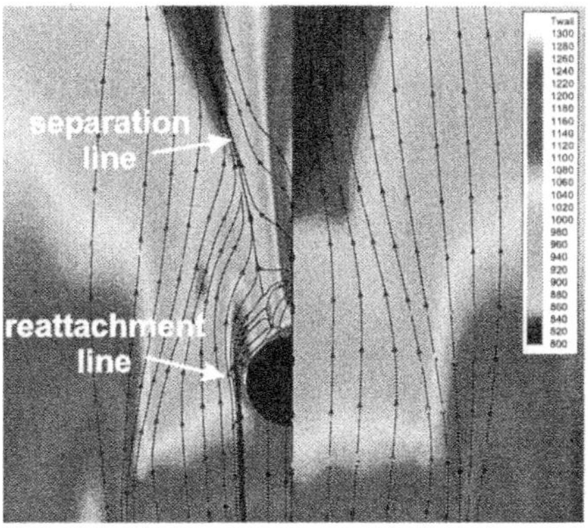

Figure 6 temperature isosurfaces and streamlines on the surface X-38 $\alpha = 0°$ left: with thruster2 right: without thruster

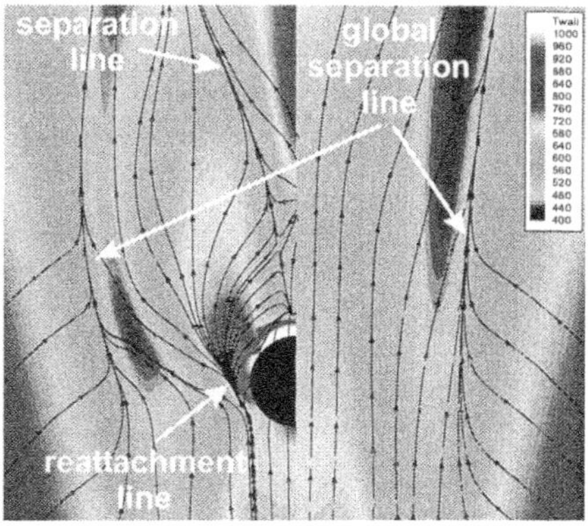

Figure 7 temperature isosurfaces and streamlines on the surface X-38 $\alpha = 40°$ left: with thruster2 right: without thruster

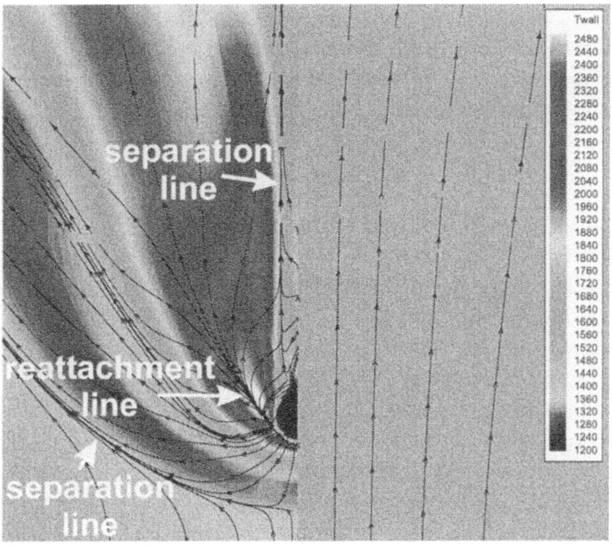

Figure 8 temperature isosurfaces and streamlines on the surface X-38 $\alpha = 0°$ left: with thruster1 right: without thruster

Flowfield Topology Changes due to Fluid-Structure Interaction in Hypersonic Flow using ANSYS and TAU

A. MACK[1], R. SCHÄFER[2], A. GÜLHAN[3] and B. ESSER[3]
[1]DLR, Institute of Aerodynamics and Flow Technology,
Lilienthalplatz 7, 38108 Braunschweig, Germany
[2]DLR, Institute of Structures and Design,
Pfaffenwaldring 38-40, 70569 Stuttgart, Germany
[3]DLR, Institute of Aerodynamics and Flow Technology,
Linder Höhe, 51170 Köln, Germany

Summary

The present investigation is aimed to evaluate the improvement of heat flux predictions for hypersonic vehicles by coupled simulation of fluid mechanics and structure in stationary flow. Radiation equilibrium simulations with an unstructured fluid solver (TAU) as well as thermally coupled solutions with a structural solver (ANSYS) have been performed for a wedge compression corner both for 2D and 3D. It turned out that there is a very strong 3D effect in the flow field that changes the flow topology completely. In addition, heat conduction inside the structure leads to a different temperature distribution reducing heat peaks at exposed parts significantly. It can be shown that the physics is qualitatively well predicted by the numerical tool and the results show good agreement with the experimental data, gained in the arc heated facility L3K.

1. Introduction

Until today, the design of thermal protection systems (TPS) for space vehicles is done very conservatively due to the lack of accurate predictions of heat flux, resulting in expensive, heavy and overdesigned parts. Heat loads are usually estimated using radiation equilibrium wall conditions without taking into account the heat conduction inside the structure. This leads to an inaccurate prediction of the boundary layer and heat fluxes. To solve this problem, a coupled analysis of fluid and structure is necessary. Taking thermal coupling into account, a higher accuracy for heat load predictions of problems like thermally driven deformations on TPS systems including steps and gaps is possible. But also the treatment of problems with large changes in the flow topology induced by heat conduction inside the structure is possible.

The objective of the joint DLR-university-industry project IMENS (**I**ntegrated **M**ulti-disciplinary d**E**sig**N** of hot **S**tructures) is to develop a tool for industrial purposes to deal with coupled problems between fluid and structure for space vehicles. This includes also the prediction of deformations. The IMENS tool allows to use different flow solvers and structural solvers. The interpolations between the different surface grids of the fluid and the structure codes are done by using the

MPCCI interpolation routine, integrated by the TENT system [1],[2]. The validation of the system is done by different experiments performed in DLR arc-heated facilities, but also using results obtained during the TETRA program [3].

Particularly, in the TETRA program a generic flap model with included hinge line gap, that either can be opened or closed, was investigated numerically and experimentally. Until now no numerical solution [4],[5] could match those results, which is the motivation of the present investigation for the stationary coupled problem. The numerical solutions obtained here are performed for the fluid part with the unstructured DLR TAU code [6], whereas for the thermal analysis the ANSYS code is used [7].

2. CFD Investigations (2D)

In a first step 2D perfect gas calculations for laminar flow have been performed around the considered wedge compression corner to get an idea of the flow field topology with open and closed gap.

For both configurations (open and closed gap) the fluid solution was obtained using radiation equilibrium wall conditions. Insulation parts have a factor of emissivity of 0.95, the C/SiC-plates respectively 0.8. Degraded radiation cooling in the gap is approximated by using an emissivity of 0.3. The meshes for the fluid solution are shown in Fig. 1-left. They have been adapted following large flow gradients and also sufficient boundary layer resolution. For the closed gap (Fig. 1-left) a large separation bubble lies over the hinge line between the front part of the model and the ramp. The shocks induced by the separation, reattachment and the ramp interact with the bow shock. In Fig. 2-left a detailed view on the gap entry is given. The reverse flow leads to a saddle point on the gap forward wall. Also, a vortex is located at the gap entry that insulates the gap flow from the separation bubble. Surface temperature is displayed in Fig. 3-left. The wall temperature on the front plate decreases from 1350 K in front of the separation to 800 K. Therefore heat flux is reduced by almost 80%. The saddle point (attachment of flow) on the gap forward wall is marked by a peak in temperature and heat flux. Inside the gap the heat flux is approximately zero, the temperature drops to 370 K.

The configuration with open gap leads to a completely different flow field. The fluid exits the open gap at sonic speed. In Fig. 1-right the strong ramp shock can be clearly seen interacting with the bow shock. The large separation bubble and the associated shocks of the closed gap case have disappeared. Here only a small separation bubble, well embedded in the boundary layer, is present at the foot of the ramp shock. A detailed view of the gap entry in Fig. 2-right shows a main separation vortex reattaching at the gap forward wall. The figure displays also the streamlines of the flow exiting the gap coming from the front of the model and from the reverse flow of the ramp. A small separation appears on the gap rearward wall due to pressure increase at the curved wall. Surface values of temperature distribution are shown in Fig. 3-right. The reattachment point of the main separation leads to the highest overall temperature of the configuration, 1650 K. Also, the temperature at the gap exit reaches 1240 K.

3. Structural Model (2D)

To obtain a coupled fluid-structure solution a structural model has been formulated. The model basically consists of two curved C/SiC plates, a C/SiC rear wall and a Al_2O_3 insulation inside. In the gap region, the radiation is calculated by taking view-factors into account to simulate the radiation exchange between the neighbouring surfaces. This procedure calculates a view-factor for every surface from every other surface. For the radiation of the outer surfaces, an ambient space node formulation is used where all energy not radiated to other elements is absorbed by this node. On the surfaces of the model affected by the coupling, finer meshing compared to the other region has been applied. Furthermore, non-linear formulations for the material properties, like conductivity and emissivity, are taken into account. At each node, thermal 2D-elements with four nodes and a single degree of freedom, the temperature, are used (Fig. 4-left). The thermal analysis calculates the temperature distribution within the structural model based on the CFD heat flux load (Fig. 4-right). The obtained temperature distribution on the coupling surface is then used as a boundary condition for the next fluid solution.

4. Fluid-Structure Interaction (2D)

For this model strong effects of fluid-structure interaction are expected due to the fact that the heat conductivity in the C/SiC plates is high. Strong heat transportation is expected in wall parallel direction since the conductivity of the C/SiC plate is much higher in this direction than in the wall normal direction. This leads to an equalisation of the temperature inside the model parallel to the surface and on its surface.

Indeed, for the closed gap configuration the main coupling effect is the heat conduction inside the structural model affecting the uncoupled surface temperature (Fig. 5-left). Regions which previously show high temperature, cool down whereas the cold ones heat up compared to the radiation equilibrium solution. The heat flux (not shown here) in cooled regions increases, whereas it decreases in regions with increasing temperature. The most significant part for the interaction of fluid and structure is the gap, where temperature is increased from 400 K to almost 900 K. The related thermal energy derives from the front plate and the ramp, where the temperatures decrease by approximately 100 K. Temperature at the rear plate also decreases significantly due to the increased heat conductivity in this plate. However, this plate shows no significant heat load due to the expanding flow field, leading to strong radiation cooling in this area. Also, in contrast to the uncoupled solution, the local heat and temperature peak at the front wall of the gap and the drop inside the separation on the front plate are smeared out and lead to a monotonous decrease of temperature on the front plate. The continuously adapted (34000 to 72000 points) and therefore partially refined CFD mesh leads to an increased accuracy of the flow field. Separation length increases significantly. A converged solution is obtained after 5 coupling iterations. However, as in previous investigations, the results do not agree with the experimental data, neither in quantity nor in quality due to a 3D effect discussed later.

The open gap configuration shows the same main coupling effects in heat conduction as described above. Major effect is the local decrease of the temperature at the reattachment point of the main separation on the gap forward wall (Fig. 5-right), where the temperature is reduced from 1650 to 1350 K. Heat is conducted mainly forward into the separation region and into the front plate. The gap exit temperature increases from 1200 to 1280 K. Inside the gap the coupling effects are not as strong as with the closed gap configuration because the radiation equilibrium wall assumption reproduces the real conditions quite well. Due to the fact that the flow topology does not change significantly, the heat flux (not shown here) almost remains constant during the coupling iterations. The results differ from the experiments due to a 3D effect shown later, the gradients are not well predicted.

5. Configurations in 3D

To study the impact on the flow topology investigations with a 3D configuration are performed on a hybrid CFD mesh (Fig. 6-left) for open and closed gap. The structural model is built with the same formulations for material properties and radiation used in 2D. Apart from the side plate it also includes the same components investigated in the 2D model. A small part of insulation between the side plate and the other C/SiC plates has been inserted. For the meshing, thermal 3D elements with eight nodes and one degree of freedom at each node are applied, as shown in Fig. 6-right.

As can be seen from the surface streamlines and temperature in Fig. 7-left, the flow topology of the closed gap configuration reveals a considerable 3D effect. On the front plate the separation has almost disappeared, remaining only on the ramp. Therefore, the reattachment line moved upstream in this case. Although the gap is closed, there is flow through the gap, entering the gap from the separation on the ramp and exiting by the side. This can be seen on the side plate of the model by a cold exit flow leading to diverging streamlines. At the gap entry, only a small separation occurs leading to a local temperature maximum at the reattachment line. Temperature peaks occur on the side edges of the model. This means that even in the gap region of the symmetry plane, there exists a strong 3D effect that dominates the whole flow field and changes the topology completely, leading to an increase in temperature of 100 K on the ramp and 400 K at the end of the front plate compared with 2D (Fig. 8-left). It turns out that the 3D solution agrees better with the experiments except at the gap itself.

Four coupling iterations are performed to get a converged solution. In comparison to the uncoupled solution it is remarkable that the heating at the side plate and edges is completely removed due to the large heat conduction down to the cooled model holder in these plates. The temperature distributions at the symmetry plane are plotted in Fig. 8-left; main effects are the reduction of the reattachment temperature peak on the gap forward wall, the increase of temperature in the gap bottom from 500 to 1000 K and the reduced temperature on the front plate and the ramp. The coupled solution is much closer to the experiments now, particularly in the gap area, predicting there the gradients correctly (Fig. 8-right).

For the open gap, the 3D flow topology also differs from the previous 2D solution. The reattachment line on the ramp moves significantly upstream and there is also flow in spanwise direction leading to high temperatures at the side edge of the model and at the gap (Fig. 7-right). The temperature peak at the main separation is also reduced. To predict the effects of the fluid-structure interaction four coupling iterations are necessary. The surface temperature distribution at the symmetry plane is displayed in Fig. 9-left. The heat peaks at the gap entry (forward and backward wall) are significantly weakened. The conditions inside the gap (Fig. 9-right) are also better predicted. For both open and closed gap, remaining quantitative discrepancies between experimental and computational data may appear due to the differences in flow assumptions. While in the experiments the flow was in non-equilibrium, the CFD simulations are carried out for perfect gas. Furthermore, the experimental results are not completely steady state. Although, the surface temperatures during the experiments were close to steady state, the gap temperatures were not since gap heating is a slow process.

6. Conclusions

In this study a hypersonic wedge compression corner with a closed or open gap has been simulated numerically and compared with available experimental data. Two approaches have been used: 2D and 3D simulations, with and without consideration of thermal conduction inside the model structure. The study has shown that in the 3D simulation, also in the gap region of the symmetry plane, there is a strong cross flow which changes the flow topology completely. Indeed, for the 3D simulations a smaller separation bubble appears in agreement with the experimental data. Strong fluid-structure interaction is present, particularly at the surface of the model with small geometrical radius. There, the coupled solutions provide a smoother surface temperature than the uncoupled ones, also in agreement with the experimental data. Minor differences due to the perfect gas model still occur.

Acknowledgement

The authors would like to thank Dr. J.M.A. Longo from DLR Braunschweig for many valuable discussions during the study and remarks on the paper.

References

[1] http://www.mpcci.org

[2] http://www.dlr.de/sc/produkte/TENT

[3] Gülhan, A.: Investigation of Gap heating on a Flap Model in the Arc Heated Facility L3K, TET-DLR-21-TN-3101, Feb. 1999

[4] Frühauf, H.-H., Infeld, F., Fertig, F., Olawsky, F.: Thermal Loads for the Experimental Vehicle X-38, ESA SP-487, 2002

[5] Behr, R.: CFD Analysis of gap flow phenomena, TETRA-report TET-DASA-21-TN-2403, 2000, pp. 529-536

[6] Mack, A., Hannemann, V.: Validation of the Unstructured DLR-TAU-Code for Hypersonic Flows, AIAA 2002-3111

[7] ANSYS, Inc.: ANSYS User's Manual, Revision 5.2, August 31st, 1995

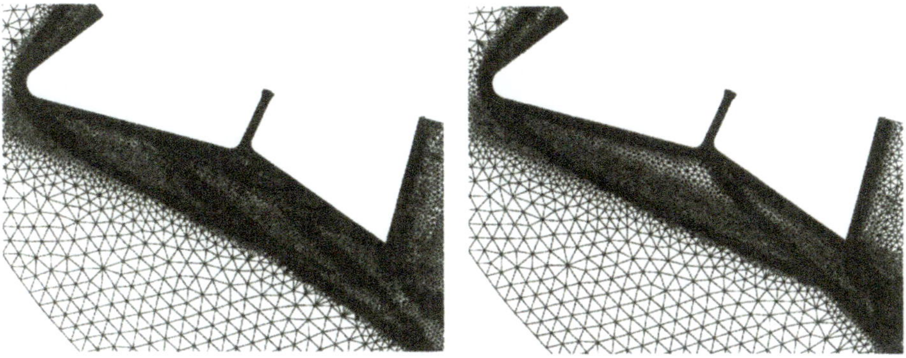

Fig. 1: Adapted grids for 2D simulation; (left) closed gap, (right) open gap

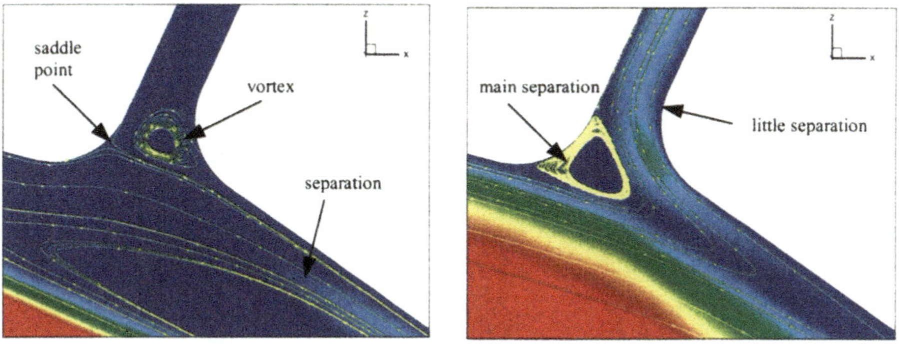

Fig. 2: Flowfield topology at gap; (left) closed gap, (right) open gap

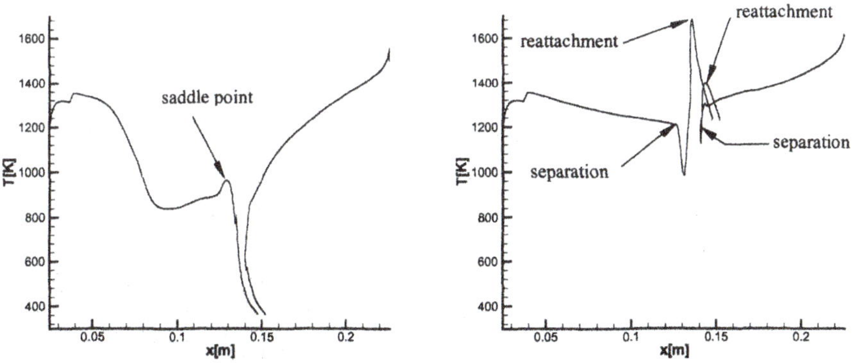

Fig. 3: Surface temperatures (2D simulations); (left) closed gap, (right) open gap

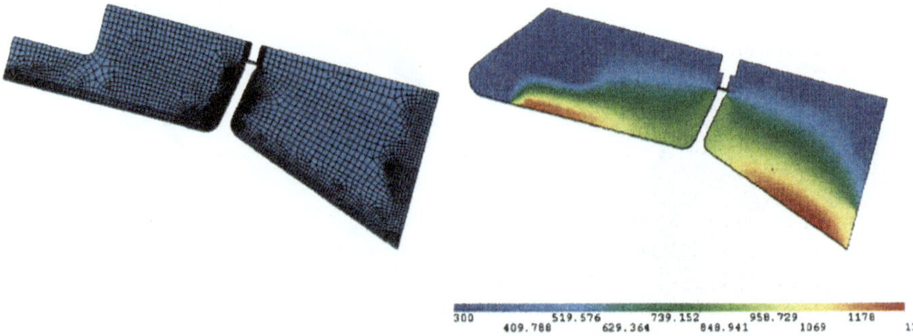

Fig. 4: ANSYS 2D; (left) structural grid (right) thermal solution

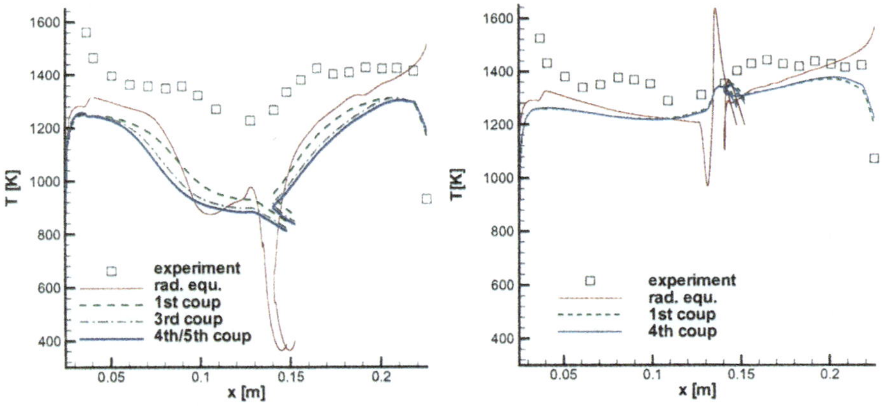

Fig. 5: Coupled 2D simulation; (left) closed gap, (right) open gap

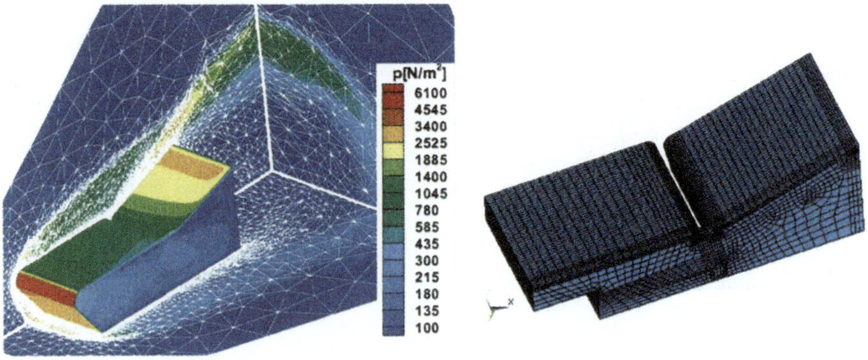

Fig. 6: 3D models; (left) pressure distribution TAU, (right) structure ANSYS

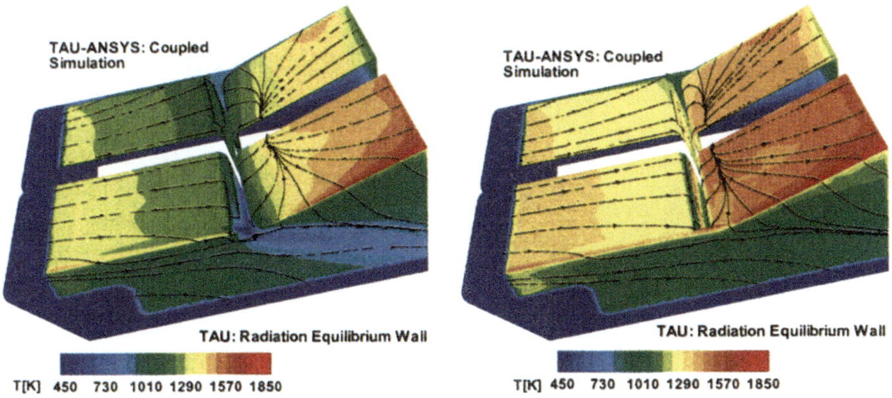

Fig. 7: Coupled 3D solution; streamlines and temperature distribution (left) closed gap, (right) open gap

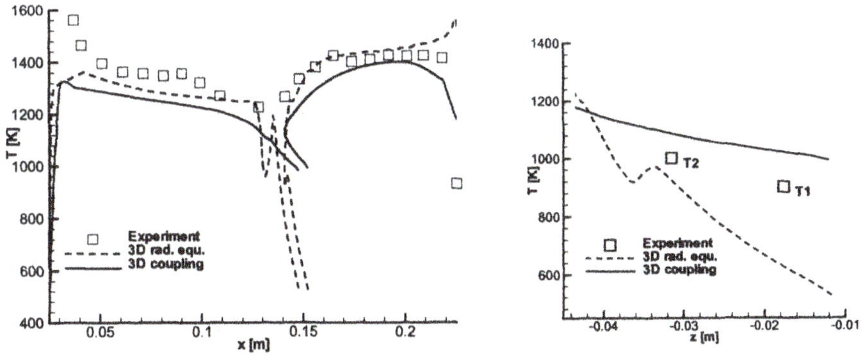

Fig. 8: Surface temperature at symmetry plane for a closed gap. (left) in x direction, (right) on the backward gap side, in z direction

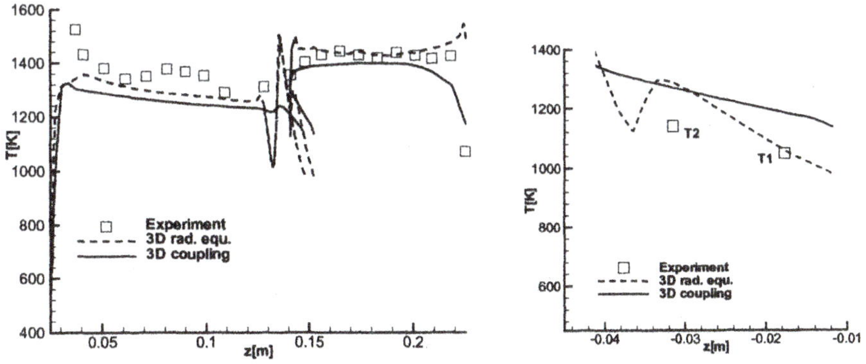

Fig. 9: Surface temperatures at symmetry plane for an opened gap. (left) in x direction, (right) on the backward gap side, in z direction

Generation of a Pre-Design Dataset for an Airbreathing Upper Stage of a Reusable Space Transportation System

M. Hornung, W. Staudacher
University of the German Armed Forces Munich
Institute for Aeronautics
Werner-Heisenberg-Weg 39
D-85577 Neubiberg, Germany

Summary

The generation of a basic dataset as the key issue in the pre-design process of an airbreathing upper stage is presented. The integration of the airbreathing propulsion system into the upper stage of a Two-Stage-to-Orbit (TSTO) showed a strong coupling between the aerodynamics and the propulsion system. By sticking to the bookkeeping conventions a complete dataset for all flight conditions has been established, which allowed an optimisation of the configuration and a sensitivity analysis as the basis for a design evaluation.

1. Introduction

Evaluation of existing TSTO-concepts leads to a pure accelerator configuration with stage separation at Ma $\approx$ 4.0. In this concept a "simple" lower stage can be used with a high potential of spin-off to other applications. In contrast to the lower stage, the upper stage increases in complexity and weight, as a higher share of the total amount of kinetic energy has to be delivered by the upper stage. This tendency leads to the idea of using an airbreathing engine in the upper stage, too. Especially the Scramjet introduces the demand of a high degree of integration into the configuration. The consequences are a strong interaction between the aerodynamics, structure and the propulsion system.

2. Dataset Generation

2.1 Bookkeeping

For a strongly coupled configuration like an airbreathing upper stage a stringent bookkeeping system is mandatory to account for the aerothermodynamical effects at the lower side of the fuselage. All moments and forces in the up- and down-stream influence of the airbreathing engine were accounted for in the propulsion dataset during airbreathing operations (Figure 1).

2.2 Propulsion

The thrust of the scramjet is decreasing with increasing Mach number and relies on the existence of atmospheric oxygen. During the ascent at high hypersonic Mach numbers and the in orbit operations additional rocket engines are necessary. Rocket engine performance data were used for state of the art high pressure engines. Modifications of the nozzles were necessary due to the integration into the Semi-Expansion-Ramp-Nozzle (SERN). The shortened 2d nozzles have been embedded between the body-flaps of the orbiter.

The scramjet is subject to investigations worldwide as it might increase the specific impulse for a hypersonic propulsion system by the factor of 4 – 6. Available performance data are still theoretical validated by wind tunnel tests and vary in a wide range. Additionally the performance of a scramjet is strongly dependent on the configuration in which it is integrated. This makes it difficult to adapt existing data to a configurational study. In this study the geometry and some performance parameters of the scramjet were taken from [1] and scaled in size. In order to establish the thrust, fuel consumption and the aerodynamics data a recalculation including the configuration had to be performed. In contrast to [2,3] a "black box" approach was used, where the combustion chamber conditions were calculated externally and transferred into the CFD calculation (Figure 2).

In the first step a 2d CFD calculation has been performed on the lower side including the scramjet with non-running combustion chamber. The results are shaping the "cold" dataset for return flight investigations and delivered the boundary conditions at combustion chamber entry. Based on those results and the burner geometry, the combustion chamber exit conditions for the core flow could be calculated by a 1d-analysis integrated in the HAP-Burner Software [4]. After transforming those results to boundary conditions at the combustion chamber exit a "hot gas" 2d CFD calculation has been performed (Figure 3). The characteristics of the burned air – hydrogen mixture has been corrected by adapting the specific heat to the burner exit conditions in the ideal gas law. The Spalart-Allmaras model was used to describe the turbulence.

Integration of the wall conditions provided the forces and moments of the propulsion system. Figure 4 shows the net-thrust at different Mach numbers and AoA in comparison with scaled FESTIP (Future European Space Transportation Investigation Programme) data.

2.3 Aerodynamics

The aerodynamic shape of the configuration excluding the lower side of the fuselage (= Scramjet) has been designed to minimize drag along the trajectory and establish a flyable configuration (also in regard to pitching moment characteristics). Especially the latter are difficult to handle with a scramjet operating through a wide Mach number range. By means of a body flap it was not possible to compensate the pitching moments of the scramjet throughout the complete speed range. Additional flaps at the scramjet nacelle are necessary to vector the thrust.

The wide range of operations from low speed during descent and landing and the high-speed accelerated ascent required a broad variety of methods to establish the aerodynamic dataset. For the orbiter CFD calculations (Euler) in combination with panelmethods (impact theory for hypersonics) and handbook calculations have been used. The analysis of the different results allowed a synthesis into one dataset valid for the complete Mach number range.

Three different conditions along the trajectory had to be investigated. From Ma = 4 to 12 a Scramjet propelled configuration is needed. Above Ma = 12 the Scramjet will be closed to protect the engine during rocket ascent and reentry. During approach and landing the scramjet will be opened again to reduce the base drag in the SERN-nozzle. According to those phases a splitted aerodynamic data set is generated to do the performance calculation. Figure 5 shows the zero lift drag during the above-mentioned phases of flight.

2.4 Thermal Protection System / Masses

The primary objective of a space transportation system is the acceleration of the payload to orbital velocity. To accomplish this task high accelerations are necessary. In a configurational sense this means a high net thrust combined with low drag and a low mass of the upper stage. Especially in terms of thrust and masses the scramjet is reducing the performance of a upper stage. The thrust of a scramjet rapidly decreases with Mach number. In order to establish the highest possible thrust in hypersonics a trajectory with high dynamic pressures has to be chosen (~ 45 kPa). Therefore the thermal and structural loads are high even in ascent conditions. This requires a specially tailored thermal protection system (TPS) for ascent and reentry. To enable the estimation of the thermal loads for a variety of configurations a quick calculation of the surface temperatures is necessary. To accomplish this task an existing aerodynamics tool, based on impact theories, has been modified to determine the thermal loads on the windward side. The implemented theories are based upon Tauber and Hankey. Both theories depend on the local Reynolds number [5,6].

The investigations on the surface temperatures showed that large areas of the configuration are dominated by the ascent loads at low angles of attack. Only the rear part of the upper side had the highest loads during "upside-down" reentry. Figure 6 shows the surface temperatures at Ma = 12 and 6° AoA.

Due to the low volumetric efficiency of the slender low-drag configuration the structure increases in weight in comparison with a conventional upper stage utilizing pure rocket propulsion (Horus 40 – based on the Horus upper stage of the "Sänger" programme with a separation Mach number of Ma_{Sep} = 4.0). For the same payload the structural weight increases from 28 tons (Horus 40) to 47.5 tons

(Scram 5). The higher thermal loads lead to an increase in TPS-mass fraction from 17.8 % to 23.2 % of the empty weight.

3. Sensitivity Analysis

One key element in the performance and risk analysis of a design is the sensitivity analysis. The growth factors of the configuration indicate the areas where high quality datasets are needed. In comparison to the rocket-propelled Horus 40 upper stage the airbreathing Scram 5 is less sensible in means of weight and rocket efficiency (Figure 7). The low thrust of the scramjet at high Mach numbers leads to high growth factors in zerolift drag and scramjet thrust. Especially the high scramjet performance sensitivity in combination with the wide uncertainties in the available datasets result in a very high design risk. Further improvements in performance estimations of scramjet engines have to be made, before a viable scramjet configuration can be designed.

4. Conclusion

The airbreathing engines isolated from the surrounding configuration show a high specific impulse compared to a rocket engine. Integrated into an accelerator the airbreather lacks the high thrust to weight ratio and the independency of dynamic pressure of the rocket engine. Therefore a highly integrated low-drag configuration has to be designed, which has to withstand extreme thermal loads during the scramjet-propelled ascent. Higher structural and TPS masses for the orbiter are the consequence. Even with optimistic performance data for the scramjet a slight increase in separation mass is encountered.

In terms of costs this means a higher development and production effort according to the increase in empty mass and complexity. Additionally the risks by changing from pure rocket propulsion to a combined propulsion system are drastically increased. A major contributor to this is the high uncertainty in the performance estimations of the airbreathing propulsion system.

Utilizing an airbreathing propulsion system in a hypersonic accelerator vehicle like a Space Transportation System does **not** increase the performance and does **not** decrease the risks/costs of such a vehicle. The high specific impulse of the airbreather might be advantageous for long range cruise vehicles, but even in this case a higher accuracy in the prediction of the engine performance is mandatory.

References

[1] Kurth, G.: *Prozeßanalyse und Leistungsvorhersage von Staustrahltriebwerken mit Überschallverbrennung für einstufige Raumtransportersysteme*, Dissertation, Fakultät für Maschinenwesen der Rheinisch-Westfälischen Technischen Hochschule Aachen, 1997

[2] Perrier, P., Rostand, P.: *Hypersonic Airbreathing Aircraft Integration through CFD: Global simulations for global thinking*, AIAA 94-3090, 30th AIAA/ASME/SAE/ASEE Joint Propulsion Conference, Indianapolis, 1994

[3] Perrier, P., Rostand, P.: *Integration of an Air Breathing Hypersonic Airplane*, AIAA 93-5032, AIAA/DGLR fifth International Aerospace Planes and Hypersonics Technologies Conference, Munich, 1993

[4] Heiser, W.H., Pratt, D.T.: *Hypersonic Airbreathing Propulsion*, AIAA Education Series, 1994

[5] Hankey, W.K.: Re-Entry Aerodynamics, AIAA Education Series, 1988

[6] Schöttle, U., Frühauf, H.-H.: Wiedereintrittsprobleme von Raumflugsystemen, Teil I – III, Institut für Raumfahrtsysteme der Universität Stuttgart, 1996

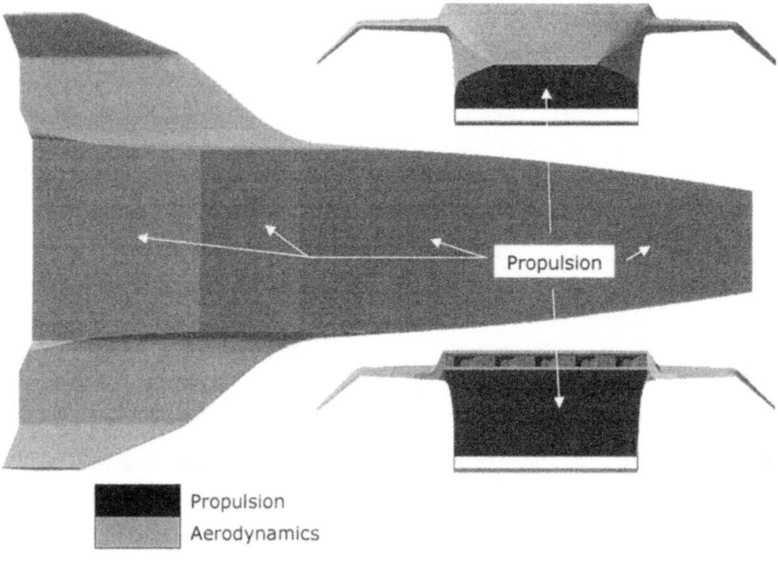

Figure 1 Bookkeeping Scram 5

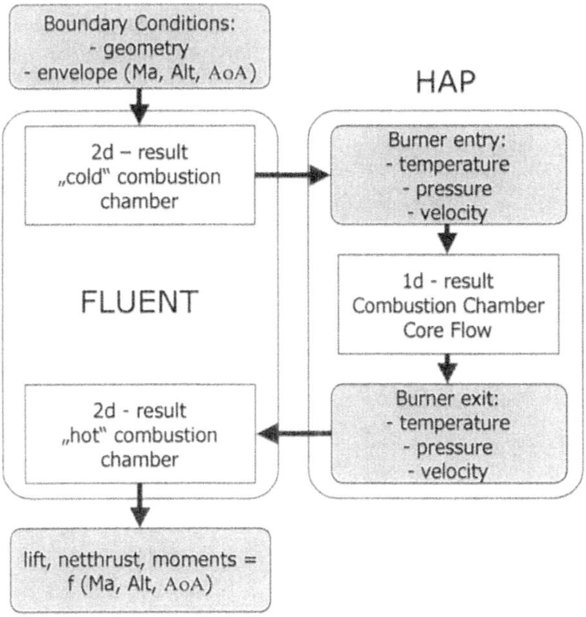

Figure 2 Scramjet performance calculation

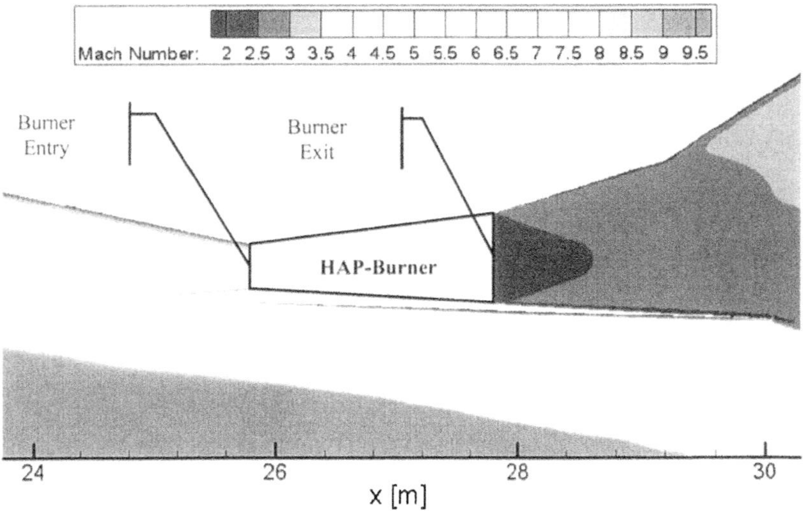

Figure 3 Zoom of the Burner Section with "hot" combustion chamber

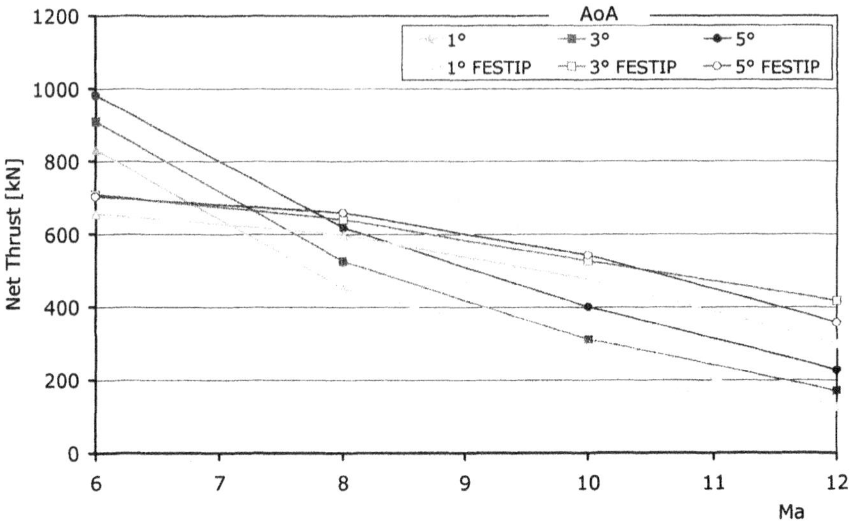

Figure 4 Net-Thrust of the Scramjet

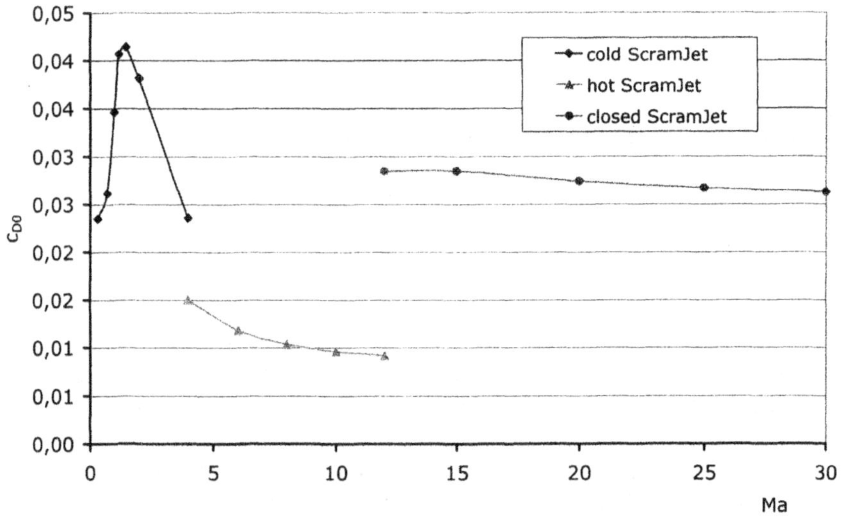

Figure 5 Zero lift drag at different phases of flight

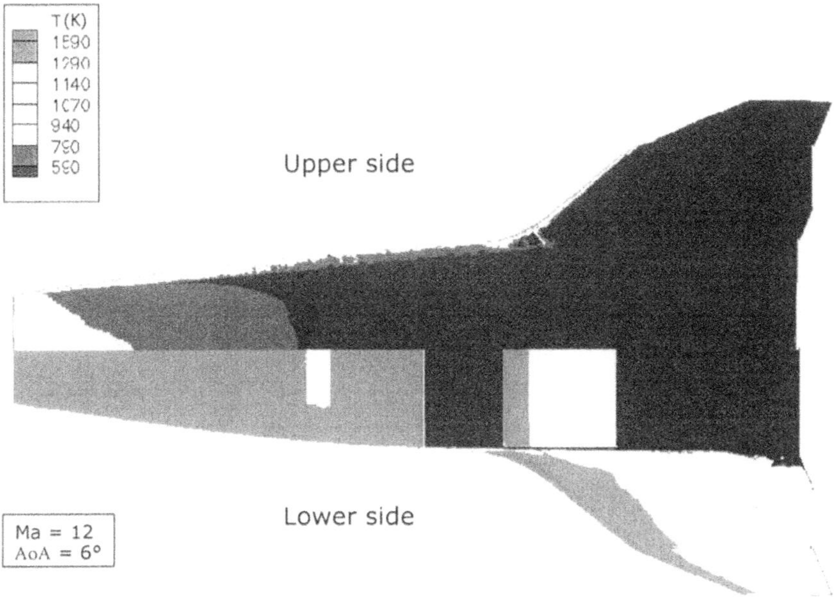

Figure 6 Surface temperatures during ascent

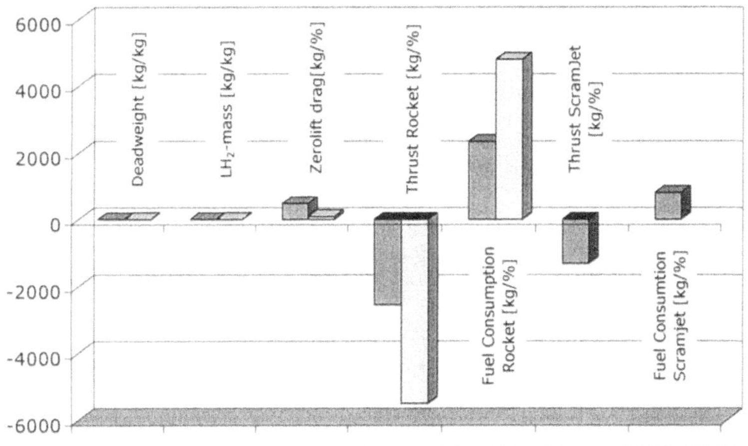

Scram 5	5,2	5,5	480	-2540	2340	-1310	810
Horus 40	10,8	11,7	94	-5500	4800		

Figure 7 Growth Factor Comparison

High enthalpy shock tunnel flow past a cylinder : A basis for CFD validation

Sebastian Karl, Jan Martinez Schramm and Klaus Hannemann

German Aerospace Center DLR
Institute of Aerodynamics and Flow Technology
Bunsenstraße 10
37073 Göttingen
Sebastian.Karl@dlr.de

Summary

Basic aerodynamic configurations are well suited to study fundamental aspects of high enthalpy flows. A test campaign of the flow around a cylinder has been performed in the High Enthalpy Shock Tunnel Göttingen HEG. Density distributions were measured in the shock layer as well as surface pressure and heat flux. The flow in the shock layer is subjected to various relaxation phenomena and therefore the presented measurements provide a good basis for the validation of physico-chemical models used in CFD-codes. This paper presents results of the test campaign and comparisons of measurements to numerical computations. The sensitivity of the CFD results to different parameters of physico-chemical models will be discussed.

1 Introduction

One important step during the development of a CFD code for high enthalpy flow computations is the validation of the models used to describe high temperature effects. The strategy that is generally pursued is to validate these physico-chemical models with data obtained in ground testing facilities. The validated code can then be used for ground to flight extrapolation and for the computation of the flow past a re-entry vehicle at free flight conditions. A test campaign of the flow around a cylinder has been performed in HEG. The cylinder, placed with its axis transverse to the flow has a diameter of 90mm and a span of 380mm. The resulting large shock standoff distance simplifies the use of optical measurement techniques to investigate flow properties in the shock layer. The HEG is a large scale shock tunnel operating in reflected shock mode. The overall length of the facility is 60m and it consists of three main sections: the driver (air buffer and compression tube), the shock tube and a subsequent nozzle. The compression of the driver gas is performed by a free-flying piston. The exit radius of the conical nozzle is 0.44m. Experiments at HEG condition III with air and nitrogen as test gas were performed. For this condition the total enthalpy of the flow amounts to about 14 MJ/kg. Further discussion of the performance of the HEG is given by Hannemann et al. [5]. The experimental results presented in this paper include surface pressure and surface heat flux measurements as well as phase shift and density distributions in the shock layer obtained by holographic interferometry. The considered HEG experiments are run 627 (test gas: air) and run 615 (nitrogen) at condition III. The respective free stream conditions are summarized in table 3.

2 Measurements

The cylinder model was equipped with 17 pressure transducers and the same number of thermocouples to measure surface pressure and heat flux distributions. These transducers are distributed along six rows located close to the plane of symmetry at mid-span location (see also Hannemann et al. [6]). The model was mounted on the nozzle centerline. The main advantage of holographic interferometry is the ability to extract the phase information from the interference patterns using the phase step technique. The evaluated phase shifts are directly comparable to numerical simulations. Furthermore, the application of 3D CFD analysis allows to derive quantitative density information from the phase shift field. However, the use of interferometric techniques in a large shock tunnel such as HEG is challenging. The size of the HEG and the test section results in a field of view of 300 mm diameter and an optical path length of almost 12 m. Therefore, the system must be stable and well isolated from the vibrations of the tunnel. A detailed description of the experimental setup is given by Hannemann et al.[6].

3 DLR CEVCATS-N Code

For the present investigations the DLR CEVCATS-N code [2] was used to compute the flow past the cylinder model and also for the numerical rebuilding of the HEG nozzle flow to determine the free stream conditions in the test section. This block-structured, three dimensional finite-volume scheme solves the Navier-Stokes equations and uses multigrid strategies with residual averaging and local time-stepping to accelerate convergence to a steady state. The test gas, air, is modeled as a chemically reacting mixture of perfect gases and is assumed to consist of five species, namely molecular and atomic nitrogen and oxygen and nitric oxide. The viscous fluxes such as diffusion, shear stress and heat flux include the assumptions of Fick, Stokes and Fourier, respectively. The species viscosities are provided by curve fits published by Blottner [1] while the species thermal conductivities are given by the Eucken correction and the mixture viscosity and thermal conductivity are calculated by Wilke's mixture rule. All computations were performed with non-equilibrium chemistry. Reaction rate models by Park[11], Dunn and Kang[3] and Gupta[4] were used. The pure nitrogen flow was assumed to be in thermal nonequilibrium. The exchange between vibrational and translational/rotational energy was modeled with the Landau-Teller approach and the vibrational relaxation times were obtained with the semi-empirical curve fit by Millikan and White [10]. The algebraic turbulence model by Baldwin and Lomax was applied for the computation of the nozzle flow.

4 Determination of the Free Stream Conditions

The evaluation of the free stream conditions was based on numerical analysis using a suitable set of measured input parameters. First, the reservoir conditions were computed with ESTC[9] using the measured initial shock tube filling pressure, shock speed and nozzle reservoir pressure. Based on these reservoir conditions the free stream is subsequently determined by numerical rebuilding of the nozzle flow using the CEVCATS-N code. The transition point from laminar to turbulent nozzle wall boundary layer is set according to the location that is determined by heat transfer measurements along the nozzle wall (see also Hannemann et al. [5]). The HEG permanent probe measures the static pressure, Pitot pressure and stagnation point heat flux in the free stream close to the nozzle exit. These measurements are used to judge the accuracy of the CFD analysis of the nozzle flow. For that, additional numerical simulations of the flow around the Pitot pressure and stagnation heat flux probes are performed.

5 Numerical computation of the flow past the cylinder model

Two and three-dimensional computations of the flow past the cylinder model were performed assuming laminar flow, chemical and thermal nonequilibrium and a fully catalytic cylinder wall. The computational grid used for the 3D simulations is shown in the left part of Figure 1. It consists of 33x81x101 grid points in streamwise, normal and spanwise direction. The symmetry plane of the 3D grid was used for the 2D computations to avoid discrepancies caused by different resolutions. A comparison between 2D and 3D results for surface pressure and heat flux along the stagnation line in spanwise direction is shown in the right part of Figure 1 for run 627. It is clear from this plot that the 2D results are sufficient to evaluate surface pressure and heat flux distributions at the midspan location. Also the comparison of the 2D and 3D CFD density distributions along the stagnation streamline in figure 5 is a proof of the two-dimensional nature of the flow in the symmetry plane.

Numerical phase shift distributions were obtained from the 3D CFD solutions by integration along a set of lines of sight as described by Hannemann et al. [6]. For the 2D case this integration was replaced by a multiplication of the local phase shifts in the symmetry plane with the span of the cylinder.

6 Parameter Studies and Results

Different reaction rate models were used for the nozzle and cylinder flow computations. The resulting shock stand-off distances, stagnation pressures and heat fluxes of the cylinder flow as well as the static pressure and density in the free stream were compared to measurements. The results are presented in Table 1. The stagnation values vary only slightly and are therefore of little use for the validation of chemical rate constants.

The air flow of HEG run 627 was subjected to significant chemical nonequilibrium effects in both the nozzle and in the cylinder shock layer. The main mechanisms for this condition are oxygen dissociation and recombination and the formation of nitric oxide. The static pressure in the free stream strongly depends on the chemical kinetics in the nozzle and varies about 15% for different reaction rate models. The shock stand-off distance being a function of the nonequilibrium situation in the free stream and the reaction velocities in the shock layer varies by an amount of about 20%. The reaction rate model by Gupta was chosen for the CFD rebuilding of the flow because of the best agreement in the above values. The air flow was assumed to be in thermal equilibrium because of the presence of oxygen atoms which act as an effective catalyst for vibrational relaxation.

The nitrogen flow of HEG run 615 showed significant dependence on chemistry models in the shock layer region only because the amount of dissociation in the nozzle was very low. The variation in shock stand-off distance was less than for the air flow and amounted to 10%. This is because nitrogen dissociation is less important than oxygen based reactions at the considered conditions. The chemistry model by Park led to the best results and was chosen for the numerical rebuilding of HEG run 615. The pure nitrogen flow is subjected to strong vibrational nonequilibrium. There are indications in the literature (e.g. Hurle [8]) that the relaxation in expanding flows is not accurately predicted by models that are mainly tuned to shock layer problems. For that reason the vibrational relaxation time was altered for the nozzle flow computation. The Millikan and White relaxation times were simply multiplied with a "Vibrational relaxation factor" and the results are given in Table 2. It can be seen in that table that there is a significant impact of different vibrational relaxation times in the nozzle for the conditions considered here. However, this result is not a sufficient basis for a final conclusion and further work is in progress to more accurately assess vibrational relaxation processes in the HEG nozzle flow. For the final evaluation of the free stream conditions the original Millikan and White model was chosen.

Surface pressure and heat flux distributions obtained with the free stream conditions from Table 3 are presented in Figures 2 and 3 . Good agreement was achieved for both the air and the nitrogen flow. Figure 4 shows the computed and measured phase shift distributions along the stagnation line. It is clear from those Figures that it is necessary to consider and compute the entire 3D flow field past the cylinder model in order to correctly assess the results of a line-of-sight measurement technique such as holographic interferometry. The deviation between numerical and experimental phase shifts in this figure are mainly due to uncertainties in the prediction of the absolute density in the free stream (see section 4). This deviation is still not resolved and further work is in progress to clarify this problem. Phase shift measurements can be used to evaluate the free stream density as described by Hornung [7]. The phase shift is used to compute the density distribution along the stagnation line in the shock layer. The free stream density can subsequentely be derived. The ability to simulate the line-of-sight experiment with a 3D CFD flow field solution allows the results to be corrected for the 3D effects. The ratio of $\rho_{\infty}^{corrected}/\rho_{\infty}^{measurement}$

was 1.072 for run 615 and 1.076 for run 627 . A comparison of the obtained normalized density distributions of the experiment and simulated experiment is given in Figure 5. The good agreement of the density ratio distributions between CFD experiment in Figure 5 shows that the kinetics of the considered relaxation phenomena in the shock layer are correctly modeled by the applied CFD tools.

7 Conclusion

Experiments in HEG were performed using a cylinder as a standard configuration for CFD validation. This configuration has the advantage of a large shock stand-off distance that permits optical measurement techniques to investigate gas properties in the shock layer. The experiments at the considered HEG conditions are well suited to investigate chemical and thermal relaxation phenomena. The static pressure measurements proved to be a sensible indicator of the correctness of relaxation models. The interpretation of optical line of sight measurement methods such as the holographic interferometry requires the knowledge of the entire flow field to quantify 3D effects. Three-dimensional CFD flow field analysis allowed in this case the application of the measurements to both the validation of CFD modeling of the flow in the shock layer and the experimental evaluation of the free stream density in the HEG test section.

References

[1] Blottner, F.G., Johnson, M., Ellis, M., Chemically reacting viscous flow program for multi-component gas mixtures, Sandia Laboratories, SC-RR-70-745, 1971

[2] Brück S., Radespiel R., Longo J. M. A., Comparison of nonequilibrium flows past a simplified Space-Shuttle configuration, AIAA Paper 97-0275, 1997

[3] Dunn M.G., Kang S. W., Theoretical and experimental studies of reentry plasmas, NASA CR-2232, 1973

[4] Gupta R. N., Yoss J. M., Thompson R. A., Lee K.P., A review of reaction rates and thermodynamic transport properties for an 11-species air model for chemical and thermal nonequilibrium calculations to 30 000K, NASA-RP-1232, 1990

[5] Hannemann K.,Schnieder M.,Reimann B.,Martinez-Schramm J.: The Influence and the Delay of Driver Gas Contamination in HEG. AIAA Paper 2000-2593, 2000

[6] Hannemann K., Martinez-Schramm J., Karl S., Beck W.: Cylinder Shock Layer Density Profiles Measured in High Enthalpy Flows in HEG. AIAA Paper 2002-2913, 2002

[7] Hornung H. G.: Non-equilibrium dissociating nitrogen flow over spheres and circular cylinders. J. Fluid Mech. vol 53, part 1, pp. 149-176, 1972

[8] Hurle I.R.: Nonequilibrium flows with special reference to the nozzle-flow problem. Proceedings of the 8th International Shock Tube Symposium, 3:1-37, 1971

[9] McIntosh M. K. : Computer Program for the Numerical Calculation of Frozen and Equilibrium Conditions in Shock Tunnels. Department of Physics, Australian National University, Canberra A.C.T., 1968.

[10] Millikan R. C.,White D. R.: Systematics of Vibrational Relaxation. Journal of Chem. Physics, vol. 39, no. 12, 1963

[11] Park C., On convergence of computation of chemically reacting flows, AIAA Paper 85-0247, 1985

Table 1 Influence of the reaction rate model on the shock stand-off distance Δ, the static free stream pressure p_{stat}, the cylinder stagnation pressure p_{stag}, the cylinder stagnation heat flux q_{stag} and the free stream density ρ_∞ for HEG shot 615 (nitrogen) and 627 (air)

HEG run 615	total enthalpy H_0 = 14.4 MJ/kg				
Reaction rate model	Δ [mm]	p_{stat} [Pa]	q_{stag} [MW/m^2]	p_{stag} [kPa]	ρ_∞ [g/m^3]
Park	12.30	522	4.58	66.7	2.755
Dunn and Kang	10.95	526	4.67	67.1	2.744
Gupta	11.12	521	4.66	67.9	2.753
Measured	12.4	500	4.8	68.0	2.298
HEG run 627	total enthalpy H_0 = 13.4 MJ/kg				
Park	9.9	835	4.77	70.66	3.103
Dunn and Kang	11.1	785	4.78	70.56	3.149
Gupta	12.3	713	4.79	70.98	3.255
Measured	11.9	726	4.83	68.0	3.541

Table 2 Influence of the vibrational relaxation time on the shock stand-off distance Δ, the static free stream pressure p_{stat}, the cylinder stagnation pressure p_{stag}, the cylinder stagnation heat flux q_{stag} and the free stream density ρ_∞ for HEG shot 615 (nitrogen)

Vibrational relaxation factor	Δ [mm]	p_{stat} [Pa]	q_{stag} [MW/m^2]	p_{stag} [kPa]	ρ_∞ [g/m^3]
0.25	12.19	545	4.57	67.3	2.728
0.5	12.24	534	4.60	67.4	2.741
1.0	12.30	522	4.58	67.4	2.755
2.0	12.35	511	4.60	67.5	2.770
4.0	12.43	500	4.66	67.5	2.785
Measured	12.4	500	4.8	68.0	2.298

Table 3 Calculated free stream conditions for HEG run 615 (nitrogen, Park rates) and 627 (air, Gupta rates) (ξ_i are species mass fractions)

	run 627	run 615		run 627	run 615
p_∞ [Pa]	687	500	ξ_{N_2}	0.73555	0.9891
T_∞ [K]	694	610	ξ_{O_2}	0.1340	—
u_∞ [m/s]	4776	5090	ξ_{NO}	0.05090	—
ρ_∞ [kg/m^3]	3.255	2.755	ξ_N	1.0E-09	0.0109
M	8.78	10.0	ξ_O	0.07955	—
T_{vib,N_2} [K]	—	3188			

Table 4 Numerical results versus permanent probe measurements for run 615 and 627

	run 627		run 615	
Free stream probe	CFD	Measurement	CFD	Measurement
Static pressure [Pa]	713	726	522	500
Pitot pressure [kPa]	70.75	72.0	67.4	68.0
Stagn. heat transfer [MW/m^2]	15.13	14.5	15.5	14.2

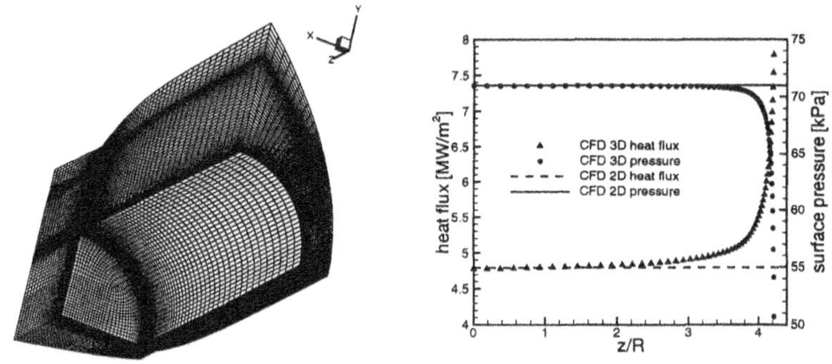

Figure 1 Computational grid (left) and CFD solution for run 627 (right). Shown are the surface pressure and heat flux along the stagnation line in spanwise direction (z/R=0 represents the symmetry plane at midspan location).

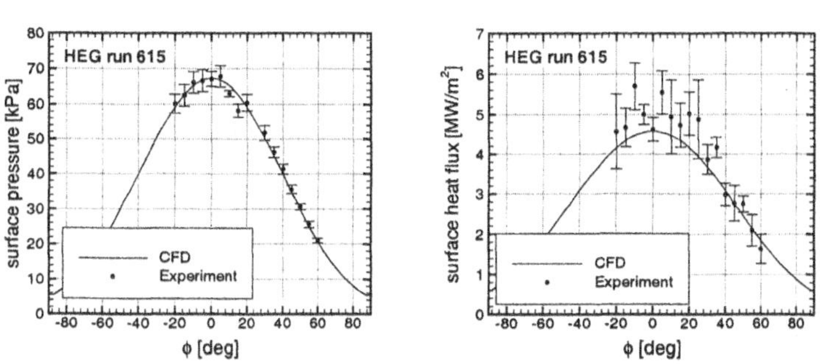

Figure 2 Measured and calculated surface pressure and surface heat flux for HEG run 615

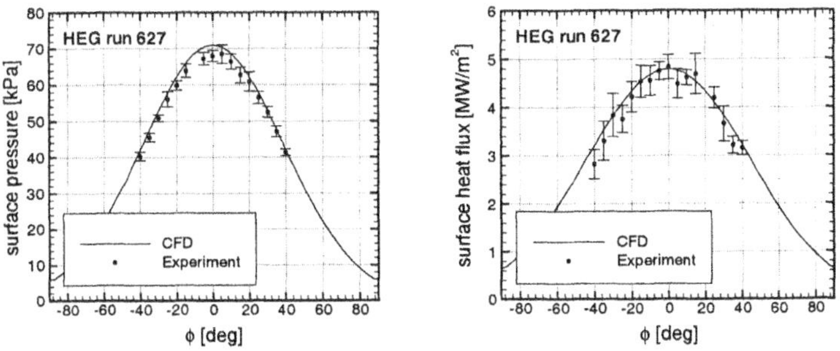

Figure 3 Measured and calculated surface pressure and surface heat flux for HEG run 627

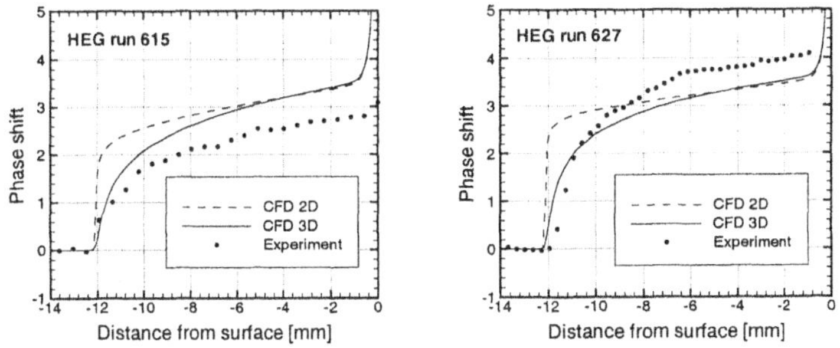

Figure 4 Comparison of computed and measured phase shift distributions along the stagnation line for HEG run 615 (left) and 627(right). CFD 2D: purely two-dimensional flow assumption, CFD 3D: simulated experiment using a raytracing algorithm)

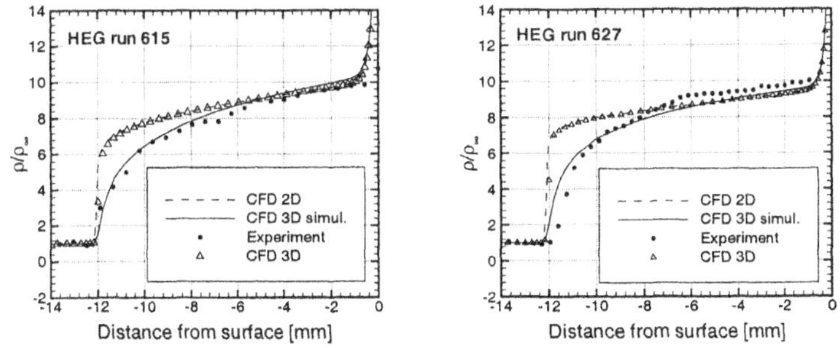

Figure 5 Comparison of computed and measured normalized density distributions along the stagnation line in the symmetry plane. The solid line "CFD 3D simul." represents the simulated experiment using the 3D flow field solution and a raytracing algorithm

End-to-end modelling and design of a new operating condition for HEG

A.D. Gardner, P.A. Jacobs, and K. Hannemann

Institute of Aerodynamics and Flow Technology, German Aerospace Centre (DLR), Bunsenstr 10, 37073, Göttingen, Germany.

Summary

A new operating condition has been designed for the High Enthalpy Shock Tunnel Göttingen (HEG). It corresponds to the free-flight conditions expected during the SHEFEX and HYSHOT free-flight experiments and duplicates Mach 8 flight at 28 km altitude. The new operating condition was designed using a semi-analytic approach and then verified with an end-to-end facility simulation code.

1 Introduction

Free piston driven shock tunnels [2] can generate test flows with stagnation enthalpies up to 23 MJ/kg [3], and as low as 2 MJ/kg [4]. Currently, the HEG has verified operating conditions with stagnation enthalpies between 10 MJ/kg and 23 MJ/kg [5]. To match the flight conditions encountered during the SHEFEX flight, a new operating condition has been designed to give a stagnation pressure of 20 MPa at a stagnation enthalpy of 3.1 MJ/kg. This allows duplication of flight conditions for testing of small hypersonic airbreathing flight experiments such as HYSHOT [6], and SHEFEX [1]. Another benefit of the HEG is that, because its operating envelope permits an increase in the test-section density by a factor of eight beyond this new operating condition, full Reynolds similarity testing of larger flight experiments is also possible. Experiments can now be performed over a large range of enthalpies in the one facility.

The extreme conditions encountered in free piston shock tunnels mean that Computational Fluid Dynamic (CFD) methods are heavily utilised in the calculation of generated conditions in the freestream. The potential for damage due to the fast-moving piston, means that considerable care must be invested into the design of new conditions. The two- or three-dimensional unsteady modelling of a full facility is still prohibitively expensive computationally, so analytical tools and, increasingly, one-dimensional CFD are used.

2 The free piston reflected shock tunnel HEG

The HEG facility [3], uses a high pressure air buffer driving a free piston to compress a driver gas mixture of helium and argon in a transient process, allowing generation of high driver temperatures and pressures (Figure 1). The driver pressure is regulated by the bursting of a scored steel diaphragm. After diaphragm burst, a strong shock wave propagates down the shock tube and is reflected from the end wall, creating a region of high pressure, shock-heated test gas. The shock tube is usually operated in tailored mode which means that, as the test gas is expanded through a nozzle to provide the freestream flow, constant conditions are maintained in the shock-reflection region.

While the test flow is occurring at the test section, the piston is being brought to rest by the fraction of driver gas that remains in the compression tube. The orifice plate, located between the compression and shock tubes, regulates the flow of driver gas into the shock tube and thus maintains the fraction of driver gas that decelerates the piston. At low enthalpy in HEG (3 MJ/kg), bringing the piston to a soft landing solely by throttling the driver gas is not possible. Hence, a new piston with an additional braking mechanism to prevent reverse motion of the piston after diaphragm burst is required.

3 One-dimensional simulation with L1D

Shock tunnels have a strong reliance on computation to define the freestream conditions in the test section. It has been standard practice at HEG to use the program ESTC [11] to compute the nozzle supply conditions using values measured in the shock tube. This program uses the shock speed, shock tube filling pressure and measured nozzle supply pressure to analytically calculate the temperature behind the reverse shock. The expansion of the gas through the nozzle is calculated in two-dimensions using the DLR code CEVCATS-N [9].

The analytic code PaCT [12] has achieved some success in the prediction of new operating conditions. Although several new conditions have been predicted using it, PaCT suffers from reliability and accuracy problems, and does not model the shock tube, requiring extensive (and therefore expensive) experimental tinkering to achieve a new condition. An improvement in accuracy, without excessive computational expense is offered by a combination of analytic methods to predict the new condition and a one-dimensional transient-flow simulation code such as L1D [8] to verify it. This code has lower calculation times than two- or three-dimensional CFD tools. Typical calculation times can be between 5 minutes and two hours on an AMD Athlon 850MHz processor, depending on options chosen.

L1D is a one-dimensional discrete-element time-explicit code for the simulation of impulse facilities including shock tubes, shock tunnels [8], gun tunnels, and expansion tubes. Models of physical processes such as:

- Piston motion (with seal friction),
- Diaphragm rupture at a predetermined pressure,
- One dimensional wave processes within the gas slugs,
- Wall shear stress,
- Heat transfer and
- High temperature effects (such as dissociation and ionisation)

all appear explicitly in the simulation code. The definition of tube geometry, diaphragm behaviour and piston mechanics is specific to each facility.

Although most of the code components are standard one-dimensional gas flow models, two correction models are also included. The first permits the definition of regions in which a fraction of the momentum of the gas flow is removed. This correction allows compensation for two-dimensional effects, like the losses incurred in the breaking of the diaphragm and the complex oblique shock train that may exist in the entrance of the shock tube. The losses are specified once for each class of operating condition of each facility (e.g. high enthalpy, low enthalpy, high pressure), using experimental data. The loss regions are constrained in space, and the loss in each region is specified by a traditional engineering K-factor for head loss. The difference between appropriate loss factors for different conditions is small compared with the difference between facilities.

The second model is that of a distributed contact surface, which may be manually specified to better capture the interaction between the reflected shock and the contact surface. The blending length is somewhat arbitrary (we used 1.5m), and serves only to remove some non-physical noise from the nozzle supply region pressure signal. With these correction models specified, L1D calculations can produce pressure histories which agree with experiment simultaneously at key points in the compression and shock tubes.

4 Calibration of L1D for HEG

In order to ensure that L1D worked correctly for HEG, and to find appropriate loss factors, an existing calibrated operating condition, Condition III [9], (Table 1) was used as a test case (A more complete study of HEG operating conditions I-VI is available in Reference [10]) .

The pressure at the downstream end of the compression tube compares well with experiment until t=6ms (Figure 2). At this time, a shock, propagating upstream from a (non-physical) loss region located near the upstream-end of the shock tube reaches the orifice plate and affects the flow entering the shock tube. This loss region is placed such that the shock does not intersect any regions of interest until after the test time is finished. The propagation of this shock is clearly visible in the x-t diagram (starting from x=36) for this condition (Figure 5). With the loss region in place, the nozzle reservoir pressure is well predicted (Figure 3). The prediction of the nozzle temperature is in good agreement with the prediction from ESTC (Figure 4), but is time resolved, rather than a single value, and shows a decrease with time. This is an improved accuracy, as the value predicted by ESTC is 5% too high, at the test time. The shock speed calculated using the same method as is used experimentally (comparing the time of the shock induced pressure rise at four discrete locations in the shock tube) is 3680 m/s, 1.1% higher than the experimental value.

5 Prediction of a new operating condition

The new SHEFEX condition is of lower enthalpy and pressure than existing conditions, requiring a different main diaphragm burst pressure and a different mixture of driver gases in the compression tube.

Calculation of the new operating condition was first done semi-analytically using ideal gas models and an adiabatic, ideal compression by the piston. The expansions were assumed to be isenthalpic for computation of tailored shock tube conditions. The standard equations [14] for predicting the reflected shock properties in a shock tube from the filling conditions were reversed to calculate the filling conditions required to produce known nozzle reservoir conditions [13]. The input data is in Table 2, and the calculated conditions are in Table 3 (column 1). Due to the fact that the analytical method does not adequately model the wave processes in the facility, L1D is used to verify the new operating condition.

These filling conditions obtained by the ideal, analytic method were inserted into L1D. Multiple iterations in L1D were used to find an optimal filling pressure for the air buffer.The filling pressures decided on after this step are noted in Table 2, column 2. L1D was used because the analytic method does not adequately model the wave processes in the air buffer gas during expansion.

For this calculation, the maximum piston weight that can be physically handled in HEG (810 kg) was used, with the assumption that the piston had brakes preventing reverse motion after diaphragm burst. It was found that the analytically predicted air buffer pressure overestimated the pressure required by L1D by about 20%. A holding time of approximately 5 ms was achieved in the compression tube, without significant overpressure, however the nozzle stagnation pressure did not hold steady for the required 5 ms (Figure 6).

To overcome this problem, the decision was made to run this condition slightly overtailored, to increase the test time. Experimental investigation will be used to verify whether this has a detrimental effect on the driver gas contamination of the test flow. The shock tube pressure was reduced until an acceptable pressure profile was reached, and then the diaphragm burst pressure was increased to compensate for the lowered nozzle reservoir pressure [13]. The results of these changes on the HEG filling conditions are summarised in column 3 of Table 3. A holding time of approximately 5 ms was again achieved in the compression tube and the nozzle stagnation pressure holds for approximately 5 ms as desired (Figure 7).

6 Summary

The design of a new HEG operating condition for an enthalpy of 3.1 MJ/kg based on the use of an analytical predictive method, and simulation by the one dimensional Lagrangian code, L1D was performed. The intention of implementing this technique to design new operating conditions for HEG is to reduce the experimental component of the design process. L1D was tailored for the HEG and validated with a calibrated operating condition of HEG.

Acknowledgements

Peter Jacobs was supported in 2002 by the Alexander von Humboldt Stiftung.

References

[1] H. Weihs, J. Longo, A. Gülhan, "The sharp edge flight experiment SHEFEX", Proc. of the 4th European Workshop on Hot Structure and Thermal Protection Systems for Space Vehicles. Palermo, Italy, November 2002

[2] R.J. Stalker: "Development of a hypervelocity wind tunnel", Journal of the Royal Aeronautical Society, **76**, pp. 374-384, 1972.

[3] G. Eitelberg: "First results of the calibration and use of the HEG", AIAA 1994-2525, 18th AIAA Aerospace Ground Testing Conference, Colorado Springs, June 20-23, 1994.

[4] A. Paull: "Hypersonic airbreathing propulsion". Paper 1620 presented at the 22nd International Symposium on Shock Waves, Imperial College, London, UK, July 18-23, 1999.

[5] K. Hannemann, M. Schneider, B. Reimann, J. Martinez Schramm, "The influence and the delay of driver gas contamination in HEG" AIAA 2000-2595, 21st AIAA Aerodynamic Measurement Technology and Ground Testing Conference, Denver, 19-22 June, 2000.

[6] A. Paull, H. Alesi, S. Anderson: "HYSHOT flight program and how it was developed". AIAA 2002-4939. AIAA/AAAF 11th International Space Planes and Hypersonic Systems and Technologies Conference, Orleans, France, September 29th- October 4th, 2002.

[7] P.A. Jacobs: "L1D: a computer program for the simulation of transient-flow facilities". Dept. of Mechanical Engineering Report 1/99, University of Queensland, 1999.

[8] P.A. Jacobs: "Quasi-one-dimensional modelling of a free-piston shock tunnel", A.I.A.A. Journal **32**(1), pp. 137-145, 1994.

[9] K. Hannemann, J. Martinez Schramm, S. Karl, W.H. Beck, " Cylinder shock layer profiles measured in high enthalpy flows in HEG", AIAA 2002-2913, 22nd AIAA Aerodynamic Measurement Technology and Ground Testing Conference, St Louis, June 24-28, 2002.

[10] P.A. Jacobs, A.D. Gardner, K. Hannemann, "Gas-dynamic modelling of the HEG shock tunnel", DLR-IB 224-2003A02.

[11] M.K. McIntosh, "Computer program for the numerical calculation of frozen and equilibrium conditions in shock tunnels." Department of Physics Unpublished Technical Report, Australian National University, Canberra, 1968.

[12] T.J. McIntyre, C. Atcitty, "Pistons and Compression tubes- piston motions in the high-enthalpy shock tunnel (HEG)" DLR Internal report, DLR-IB 222-90A20, Institute of Experimental Flow Mechanics, DLR, Göttingen, 1991.

[13] A.D. Gardner, "Calculation of low enthalpy conditions for the HEG", DLR-IB 224-2003-A03.

[14] J.D. Anderson Jr, "Hypersonic and high temperature gas dynamics", McGraw-Hill Inc, USA, 1989.

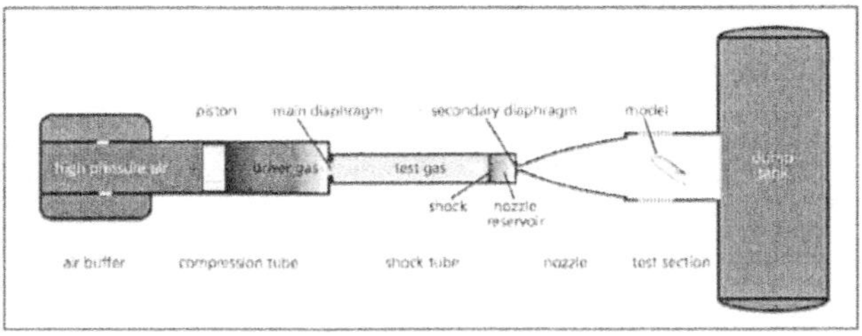

Figure 1 Free piston shock tunnel HEG

Table 1 Data for HEG Condition III.

Fill pressure: Air buffer	4.91 MPa
Fill temperature: Air buffer	310.5 K
Piston mass	294 kg
Fill pressure: Compression tube	72.46 kPa
Compression tube argon (by pressure)	10 %
Fill pressure: Shock tube	50.20 kPa Nitrogen
Diaphragm burst pressure	52.6 MPa
Shock speed	3700 m/s
Nozzle stagnation pressure	52.9 MPa
Nozzle stagnation temperature	8364 K
Nozzle stagnation enthalpy	14.17 MJ/kg

Table 2 Input data for the analytical calculation of the new tailored condition in HEG.

Shock tube length	17.060 m
Compression tube length	34.070
Shock tube diameter	0.150 m
Compression tube diameter	0.550
Nozzle orifice diameter	0.140 m
Diaphragm orifice diameter	0.070
Room temperature	300 K
Air buffer fill temperature	310 K
Compression ratio	40
Air buffer volume	5.5 m^3
Target nozzle reservoir enthalpy (H_5)	3.1 MJ/kg
Target nozzle reservoir pressure (P_5)	20 MPa
Target nozzle reservoir temperature (T_5) (Ideal)	3097 K
Target hold time	5 ms

Table 3 Filling conditions for the SHEFEX condition in HEG.

	Analytical Prediction	**L1D optimised Air buffer**	**L1D overtailored**
Fill pressure: Shock tube	109.9 kPa	110 kPa	84.19 kPa
Fill pressure: Compression tube	37.73 kPa	37.73 kPa	45.44 kPa
Shock speed	1607 m/s	1607 m/s	1607 m/s
Diaphragm burst pressure (P_4)	17.8 MPa	17.8 MPa	21.43 MPa
Compression tube argon (by pressure)	62.5 %	62.5 %	62.5 %
Air buffer filling pressure	2.024 MPa	1.7 MPa	2.0 MPa

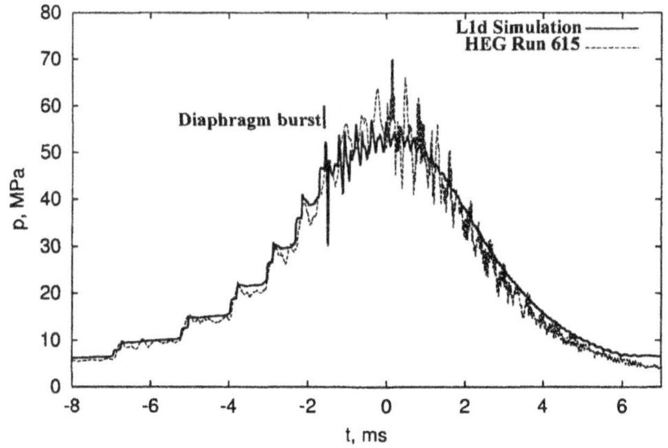

Figure 2 Comparison between calculated and measured pressure at the downstream end of the compression tube; HEG operating condition III. t=0 to match an arbitrary experimental zero.

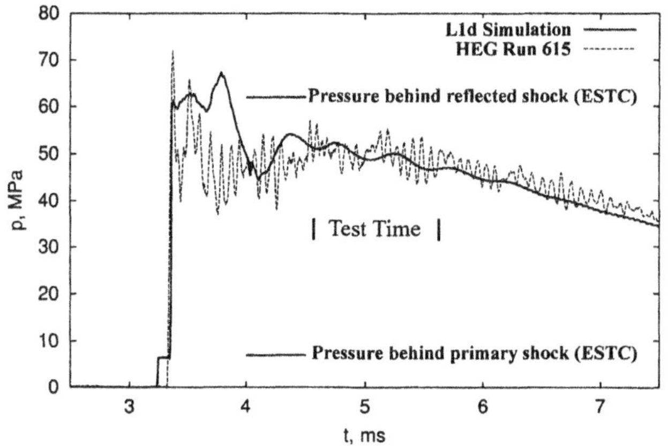

Figure 3 Comparison between calculated and measured pressure at the downstream end of the compression tube; HEG operating condition III. t=0 to match an arbitrary experimental zero.

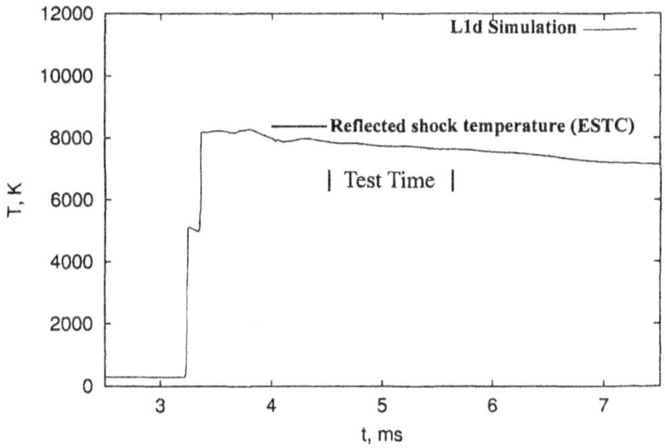

Figure 4 Comparison between nozzle reservoir temperatures calculated by L1D and ESTC. t=0 to match experimental zero.

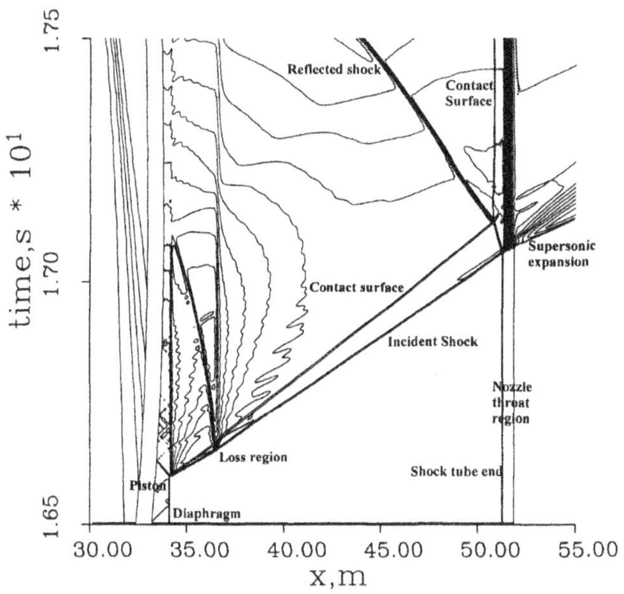

Figure 5 x-t diagram based on L1D in the shock tube for Condition III. t=0 at piston release. Shown are pressure contours and the meeting points between blocks of cells; the latter making the contact surface and the piston visible.

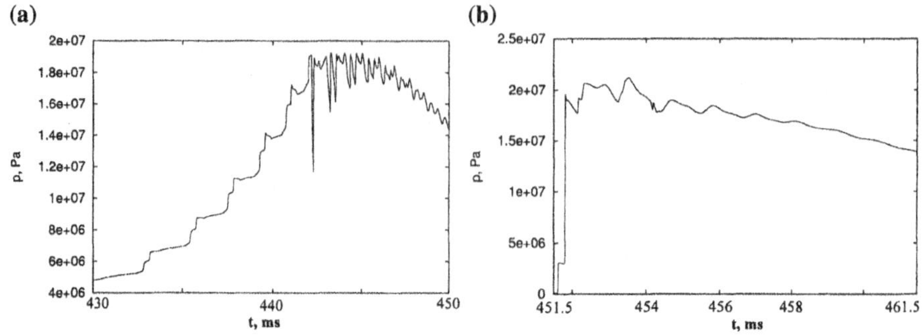

Figure 6 SHEFEX condition: L1D calculations for a tailored condition a) Compression tube pressure b) Nozzle reservoir pressure. t=0 at piston release.

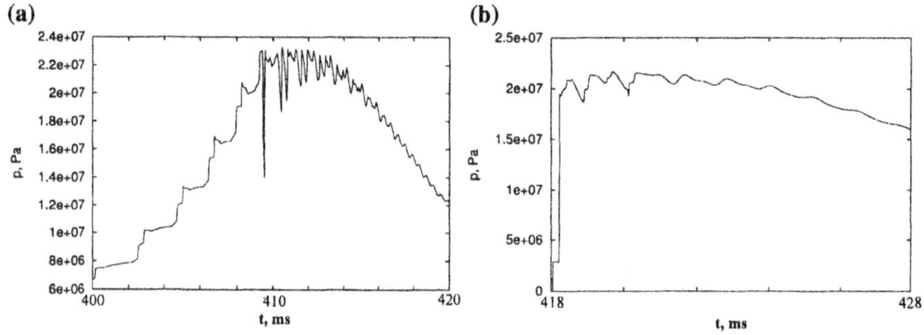

Figure 7 SHEFEX condition: L1D calculations for slightly overtailored condition a) Compression tube pressure b) Nozzle reservoir pressure. t=0 at piston release.

High Speed Flow Visualization at HEG

Jan Martinez Schramm, Sebastian Karl and Klaus Hannemann

German Aerospace Center
Institute of Aerodynamics and Flow Technology
Bunsenstraße 10
37073 Göttingen
Germany
Jan.Martinez@dlr.de, Sebastian.Karl@dlr.de,
Klaus.Hannemann@dlr.de

Summary

The High Enthalpy Shock Tunnel Göttingen (HEG) of the German Aerospace Center (DLR) is one of the largest free piston driven shock tunnels capable of simulating high temperature effects in re-entry flows. Full utilization of such a ground based high enthalpy impulse facility, with typical test times in the order of a few milliseconds, requires detailed knowledge of the test section flow, and its starting process. Further, the temporal development of the model flow must be known in order to assure that a steady flow field has developed during the test time window. Therefore, in addition of time resolved model surface property measurements such as pressure and heat transfer, a high speed shadowgraph and schlieren system was implemented at HEG. For the first tests with the new system a cylinder flow was chosen. The large shock stand-off distance of the cylinder flow allows high spatially resolved optical measurements. Due to the sensitivity of the shock stand-off distance to the gas composition, the measurements are also a valuable aid in detecting driver gas contamination in the test section. The results presented in this paper are related to HEG operating condition III (h_0= 13.4 MJ/kg, p_0= 48.3 MPa, Ma=8.7, u_∞=4776 m/s, T_∞=694 K, p_∞=687 Pa, $\rho_\infty = 3.255$ g/m^3).

1 Introduction

The flight velocity of a vehicle re-entering the atmosphere from low earth orbit is approximately 7 km/s, and causes the vehicle to encounter flow conditions with stagnation temperatures reaching up to 10 000 K. These temperatures give rise to dissociation of the molecular oxygen and nitrogen and can have an influence on the aerodynamics of the vehicle. The primary requirement for ground based hypervelocity testing on reduced-scale re-entry models is the duplication of the flight total enthalpy to ensure that the stream energies are sufficient to generate high temperature effects of appropiate strength. One of the most successful facility types for hypervelocity and high enthalpy ground testing in a range of total enthalpies up to approximately 24 MJ/kg is the free piston driven shock tunnel. The free flying

by Hannemann et al. [2]. Due to the nature of the free piston driven shock tunnel, the useful test time for this type of impulse facility is in the order of a few milliseconds. Therefore, one important aspect of the flow characterization process is to obtain the knowledge of the flow establishment time, especially when the flow or parts of the flow are dominated by viscous effects.

2 Experimental Set-up

High speed flow visualization at HEG is performed using shadowgraph and schlieren techniques. The fundamentals for both types of flow visualization may be found in Merzkirch [4], Schardin [6] or Holder and North [7]. The high speed flow visualization at HEG is realized by using the existing optical components installed for the holographic interferometry system. Specific details about this system are described by Kastell et al. [5]. The setup uses a z-path layout for the object light path with spherical mirrors (Halle SDH4300) having a diameter of 300 mm and a focal length of 1500 mm. The length of the light path is approximately 12 m. Figure 1 shows a schematic of the object light path. The light source is expanded with a telescope to a diameter of 30 mm because a reference beam of this diameter is separated from the object light path, which is needed for interferometric measurements. A f = -50 mm lens (L3) overexpands the beam on to the spherical mirror P2, which aligns the beam to a parallel light-bundle of 300 mm diameter through the test section. On the other side of the test section the beam is collimated by the second spherical mirror P1. The beam is then magnified and focused by a lens or lens system, depending on the portion of the test section which is chosen to be visualized. The high speed flow visualization uses the object light path only and may be used as a shadowgraph, where the test section plane is defocused on the film plane, or as a schlieren system, where the test section is focussed onto the film plane, and a razorblade is used at the focal point after L4. The laser used is a diode pulsed Nd:YAG from Lightwave Electronics (Model 612). It has the capability to operate at pulse rates up to 50 kHz either in continuous or burst mode. The laser emits at 532 nm. The pulse width varies with frequency and ranges from 43 ns to 155 ns. The images are recorded with a Cordin rotating drum camera (Model 318) on black and white film (Kodak TMAX 100). The camera consists of a rotating drum, which is equipped with 35 mm film strips. The drum has a circumference of 1 m. Figure 2 shows a schematic of the drum camera. A relay lens system with a 90° mirror forms an image onto the inner side of the drum. The drum resides in an evacuated housing, to minimize the air friction. Due to evacuation rotation speeds of the imaging plane up to 300 m/s are possible. The drum camera is a streak camera. Therefore, the shutter remains open during the experiment; the narrow pulse width of the laser performs a pseudo-framing function. The system relies on the short duration of the laser pulse to record the image onto the moving film.

The cylinder model selected for testing the performance of the high speed flow visualization system was simultaneously used to obtain validation data for high temperature models used in CFD tools by holographic interferometry [3]. It has a diameter of 90 mm and a span of 380 mm and is placed with its axis transverse to the flow. The model is equipped with 17 pressure transducers and 17 thermocouples, distributed along six rows located close to the plane of symmetry at the mid-span location (see also [3]). The cylinder was mounted on the centerline of the flow.

3 Results

In addition to the application of the high speed flow visualization system in order to obtain information of the temporal development of the shock shape and the shock stand-off distance the combined test campaign also included surface pressure, surface heat flux measurements and the determination of two dimensional shock layer phase shift distributions by holographic interferometry [3]. Further, 3D flow field computations were performed using the Navier-Stokes solver CEVCATS-N including non-equlibrium chemistry. Details of the comparison of the data obtained by holographic interferometry and CFD are discussed by Hannemann et al. [3].

Figure 3 shows the comparison between the schlieren picture at t = 2.266 ms obtained for HEG operating condition III with air as a test gas and a line of sight shock shape extracted from a 3D CFD computation for the defined test time. In the CFD computation free stream properties are used which result from the simulation of the flow in the conical HEG nozzle using the same CFD code as for the cylinder computations. As seen in Figure 3 the shock stand-off distance and the shock shape are well reproduced by the CFD computation. It should be emphazised here, that the numerical shock stand-off distance is very sensitive to the chemical reaction rate model used in the CFD code (see [3]). Therefore, it is a useful quantity for the validation of physico-chemical models used in CFD codes. Figure 4 shows a sequence consisting of 7 frames out of 54 showing the flow start-up around the cylinder. The last frame in the figure is taken during the test time window. From this sequence, the shock stand-off distance was extracted and is shown in the upper plot of Figure 5. Here the measured shock stand-off distance resulting from all frames is plotted versus time. Included in the figure are the development of the stagnation pressure measurement of the cylinder (middle plot) and the stagnation heat flux (lower plot). In the lower plot the passage of the first and secondary shock, which are formed during the starting process in the nozzle, can be clearly identified by the two strong initial peaks in the heat flux trace. After the passage of this shock system the unsteady expansion and boundary layer development in the nozzle may be identified by the slow rise of the stagnation pressure signal in the middle plot of Figure 5. After a constant expansion is developed in the nozzle, the stagnation pressure as well as the stagnation point heat flux reach a constant level. In the upper plot of Figure 5 it can be seen that approximately 2.3 ms after shock reflection, the shock stand-off distance reaches a constant level, which is also reproduced by the CFD computation. The test time window for this specific run, was determined to be between 2.3 ms and 2.8 ms. At approximately 3 ms after shock reflection the reservoir pressure drops due to the arrival of the expansion waves reflected off the piston face. For operating condition III this process limits the period of constant test flow. The sudden rise of the shock stand-off distance at 3.7 ms is caused by the arrival of driver gas in the test section. Further details concerning the driver gas contamination in HEG can be found in [3].

4 Conclusion

A high speed flow visualization system consisting of a high speed drum camera and a diode pulsed Nd:YAG laser, was succesfully implemented at HEG. It has been shown that the resulting temporally and spatially resolved flow visualization is a very valuable tool for the determination of test time windows and the investigation of flow establishment times in impulse ground testing facilities such as HEG. The experimental results could also be verified by Computational Fluid Mechanics investigations. Future work will include the possible extension of the system to obtain time resolved interferometric measurements for the evaluation of the temporal development of density fields in HEG.

Acknowledgements

The generous support of the Arnold Engineering Development Center (AEDC) which provided a high speed flow visualization system during the evaluation phase of this technique is gratefully acknowledged.

References

[1] R.J. Stalker, A Study of the free-piston shock tunnel, AIAA Journal, 5(12):2160-2165, 1967

[2] K. Hannemann, M. Schnieder, B. Reimann, J. Martinez Schramm, The Influence and the Delay of Driver Gas Contamination in HEG, 21st AIAA Aerodynamic Measurement Technology and Ground Testing Conference, June 19-22, Denver CO, 2000.

[3] K. Hannemann, J. Martinez Schramm, S. Karl, W.H. Beck, Cylinder Shock Layer Density Profiles Measured in High Enthalpy Flows in HEG, 22nd AIAA Aerodynamic Measurement Technology and Ground Testing Conference, June 24-28, 2002 / St. Louis, MO

[4] W. Merzkirch, Flow Visualization, Academic Press, 1974.

[5] D. Kastell, M. Carl, G. Eitelberg, A combined holographic interferometer and laser-schlieren system applied to high temperature high velocity cylinder flow, Experiments in Fluids, 22(1);57-66, 1996.

[6] H. Shardin, Das Toeplersche Schlierenverfahren, VDI (Verein Deutscher Ingenieure), Forschungsheft No. 367, 1934.

[7] D.W. Holder, R.J. North, and G.P. Wood, Optical methods for examining the flow in high-speed wind tunnel, AGARDograph No. 23, November 1956.

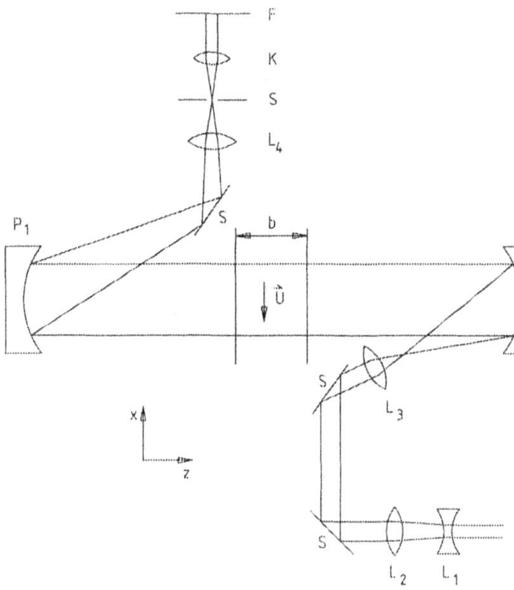

Figure 1 Schematic drawing of the z-path optic system at HEG. P are spherical mirrors of 300 mm diameter. L1 and L2 are forming a telescope for widening of the laserbeam, which enters from the bottom right. F is the film plane of the drum camera.

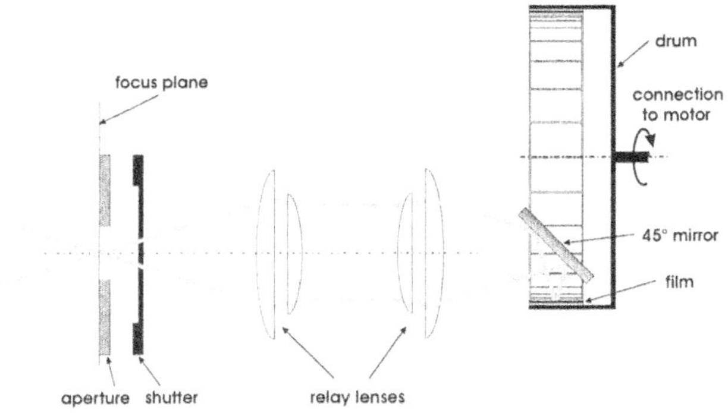

Figure 2 Schematic drawing of the drum camera.

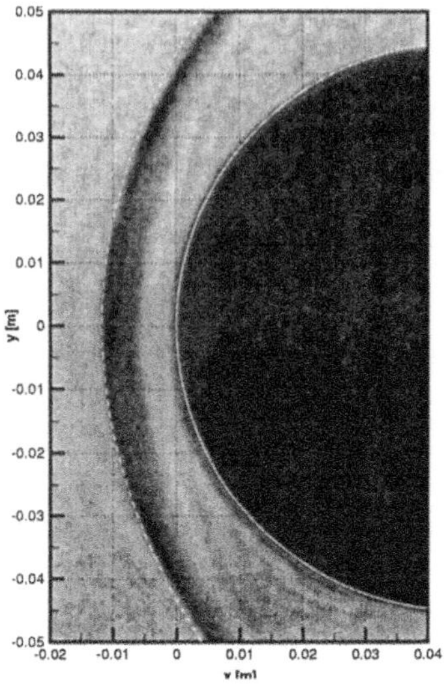

Figure 3 Comparison of the schlieren picture with CFD simulation of cylinder fow. The dashed white line show the numerical shock position. This enlarged picture is the last picture from the sequence in the figure below.

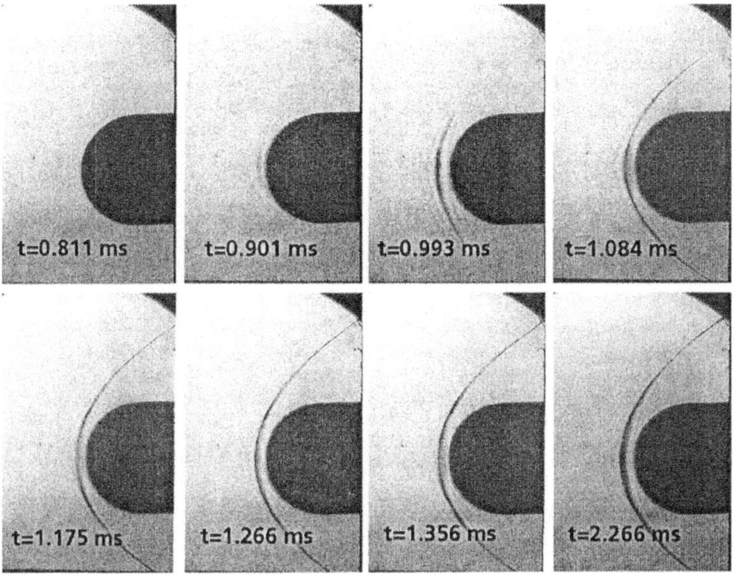

Figure 4 Visualization of the start-up of the cylinder flow for HEG operating condition III.

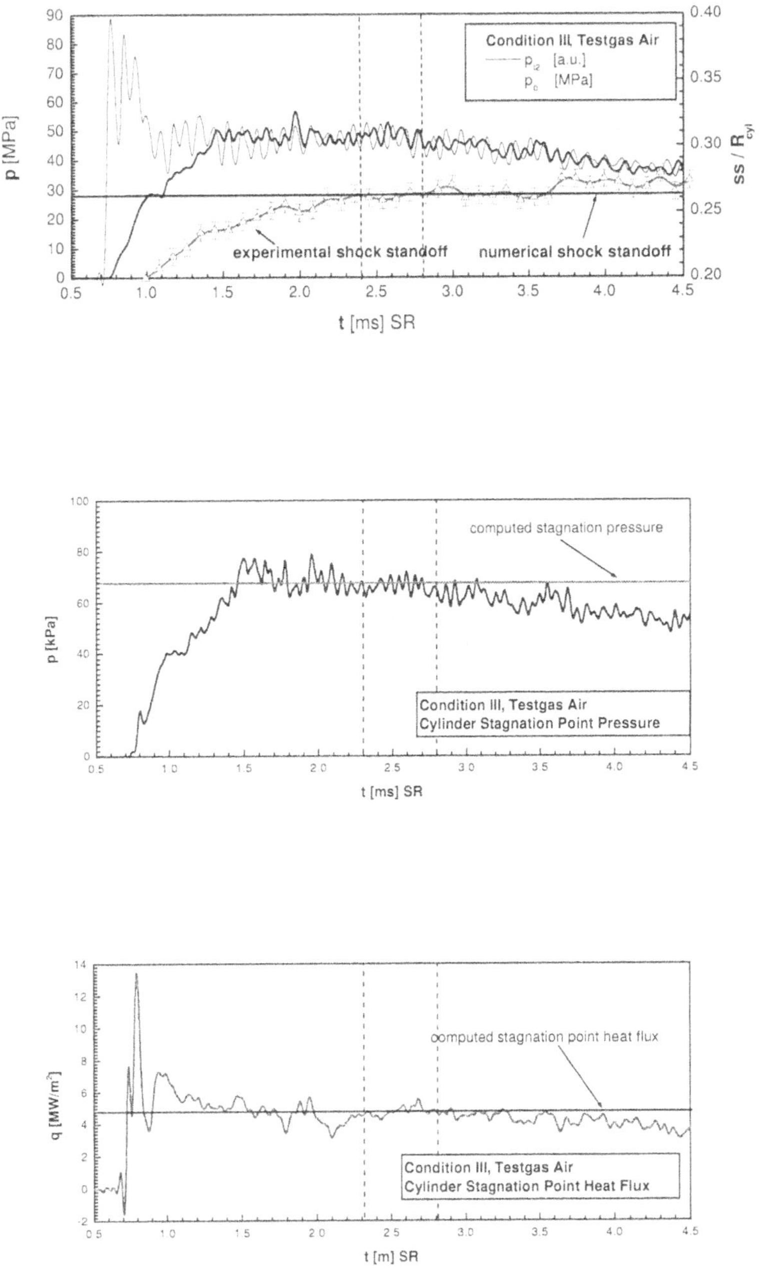

Figure 5 Temporal development of the shock stand-off distance, reservoir pressure and Pitot pressure(upper). Cylinder stagnation point pressure (middle). Stagnation point heatflux (lower). HEG Condition III, The test time window is marked by the dashed lines.

Simulation of a SERN/Afterbody Flow Field Regarding Heated Nozzle Flow in the Hypersonic Facility H2K

P. GRUHN and A. HENCKELS
German Aerospace Center (DLR),
Institute of Aerodynamics and Flow Technology,
Wind tunnel division Köln, Linder Höhe, D-51147 Köln, Germany

Summary

An experimental study with respect to the influence of temperature changes on the interaction of a SERN nozzle flow with the external flow around a nozzle model has been performed in the hypersonic wind tunnel H2K. The static pressure along the expansion ramp and the nozzle flap was measured at varying total temperatures of the nozzle flow between 300 K and 1000 K. Additional information about the flow was gained by Schlieren optics. It was found that the axial thrust was increased by up to 1.5 per cent with increasing air temperature. Considering the impact of the nozzle thrust on the net thrust of airbreathing engines at hypersonic speeds, this demonstrates the importance of simulating the "hot" nozzle flow for the investigation of interacting nozzle flow fields.

1 Introduction

An essential task for the design of future space transportation systems is the integration of the propulsion unit into the air frame. Particularly the integration of the nozzle is of interest, because of the strong interactions of the nozzle flow with the free stream. Here, Single Expansion Ramp Nozzles (SERN) are often considered as an adequate solution [7].

Several studies were performed to understand the mechanisms of the flow field interactions governed by the free stream and the SERN nozzle flow, aiming at the improvement of nozzle gross thrust and a reduction of the thrust vector rotation at transsonic speeds [2, 10, 11]. As one result, it was found that the nozzle performance is strongly influenced by the flow at the flap cowl [1, 8], and hence can be improved significantly by an aerodynamic optimisation of the flap cowl geometry [1, 3, 4]. Although the total temperature of the real nozzle flow field of space transportation vehicles exceeds 3000 K, the nozzle flow was simulated by pressurised cold gas in most of the experimental studies because of technological limits of conventional ground testing facilities. Studies on hot nozzle flows concentrated rather on real gas effects [12] or the overall engine performance [5, 6] than on the interaction of the flow. The impact of temperature effects, like, for example, the change of dynamic viscosity or the density due to the hot nozzle flow, was neglected by experimental studies so far.

The aim of the present activity is to study the interaction between the nozzle and the external flow around the model with respect to changes of the nozzle flow temperature. A new electrical heater was implemented into the hypersonic blow down facility H2K to provide hot compressed air to simulate the nozzle flow. Although the high temperatures of real nozzles cannot be reached, achievable total temperatures up to 1100 K allowed to capture tendencies in the change of the flow field due to increased nozzle flow temperature. Thus, the H2K facility with the new heater offers an excellent opportunity to study interacting flow fields of nozzle flows or other jet plumes regarding temperature effects.

To compare the nozzle performance between "cold" and "hot" nozzle flow and to provide data for the validation of numerical tools, the pressure was measured at 13 positions along the expansion ramp and the nozzle flap. Additional information about the flow topology was gained by Schlieren optics.

2 Experimental Setup

The study has been performed at the hypersonic blow down facility H2K at DLR Cologne using a SERN nozzle model. Depending on the flow condition, the facility offers a test duration up to 30 seconds. To avoid condensation effects the flow upstream of the wind tunnel nozzle was heated by electrical heaters up to 700 K. The runs were performed at a Mach number of 6 using an axisymmetric contoured Laval nozzle with 600 mm exit diameter. During tests the unit Reynolds number was adjusted by the stagnation pressure while keeping the stagnation temperature constant. The unit Reynolds number ranged from $2.7 \cdot 10^6\ m^{-1}$ to $17.2 \cdot 10^6\ m^{-1}$, which corresponds to Reynolds numbers from $1.2 \cdot 10^6$ to $7.4 \cdot 10^6$ for an overall model length of 430 mm of the SERN nozzle model.

The SERN nozzle flow was heated by a recently installed heater shown in Figure 1. Its total length is about 2.5 m. Inside the heating tube, the flow is electrically heated by three segments with a total of twelve Kanthal-spirals of 24 mm diameter. The power rating of 400 kW is sufficient to provide a dried air mass flow of up to $0.3\ kg \cdot s^{-1}$ at pressures up to 60 bar and temperatures up to 1100 K. During the heating phase the exhaust air is diverted into the ambient atmosphere by a hot gas valve until the desired steady state temperature is achieved. The left side of Figure 2 shows the additional heater together with the insulated air supply pipe connected to the H2K test chamber. On the right side of this figure the SERN nozzle model installed inside the test chamber is shown.

Figure 3 shows a sketch of the SERN nozzle model (left side). The total pressure and the temperature of the nozzle flow were measured inside the nozzle chamber of the model. Additionally the static pressure was measured at 13 positions, one located inside the nozzle chamber, five along the expansion ramp and seven at the nozzle flap, i.e. two at the upper and five at the lower flap side (Figure 3, right side). Depending on the expected maximum pressures, different pressure modules from PSI™, located outside the flow to minimize errors due to the jet temperature, were utilized. Table 1 gives the accuracy of measurement of these modules.

For the present study, the total pressure inside the model was limited to 10 bar, leading to nozzle pressure ratios (the ratio between the nozzle flow total pressure and the ambient pressure) of up to $\Pi = 650$, a typical design pressure ratio for TSTO-configurations at $Ma = 6$. The total temperature of the nozzle flow was varied between 300 K (without heating) and 1000 K. The information gained by pressure measurements was supported by additional Schlieren diagnostics.

3 Results

Figure 4 shows a Schlieren photograph taken from an experiment at $Re = 7.4 \cdot 10^6$, a nozzle flow temperature of $T_D = 300$ K and a nozzle pressure ratio of $\Pi = 98$. The typical structure of a SERN nozzle flow field is clearly identified, i.e. the shear layer that separates the nozzle flow and the exterior flow, and the interior shock and the plume shock which adjust the pressure of both parts of the flow. Furthermore the separation shock can be identified. It is caused by the separation of the exterior flow at the nozzle flap, which on the other hand occurs due to the blockage of the exterior flow by the exhaust plume [9].

The comparison of Schlieren images at the same nozzle pressure ratios shows, that the size of the nozzle jet grows by increasing the temperature. As example, the flow fields at $Re = 7.3 \cdot 10^6$, $\Pi = 615$ and total temperatures of the nozzle flow of $T_D = 298$ K and $T_D = 986$ K respectively are compared in Figure 5. Here, one reference value was defined as vertical span between the end of the expansion ramp and the interior shock, the other one was defined as horizontal span between the flap tip and the interior shock.

A more quantitative comparison of the impact of different nozzle flow temperatures shows Figure 6. Here, the ratio between the static pressure along the expansion ramp and the total pressure inside the nozzle chamber is given at three different levels of the nozzle total temperature ($T_D = 298$ K; 594 K and 986 K) at a fixed Reynolds number of $Re = 7.3 \cdot 10^6$ and a nozzle pressure ratio of about $\Pi \approx 615$. As illustrated, the pressure ratio grows with the total temperature, which results in a higher nozzle thrust at increased temperatures for a given nozzle total pressure.

Since the static pressure is only measured at seven positions in the area that accounts for the nozzle thrust, the experimental pressure data on its own is not sufficient to calculate the nozzle thrust. The pressure and the momentum at the nozzle throat were calculated from the absolute values inside the nozzle chamber by one-dimensional gasdynamic equations. The pressures at the end of the expansion ramp and at the flap tip were extrapolated from measured pressures. Since the pressure gradient changes significantly in the area between locations 2 and 3 and between locations 7 and 8 (see Figure 3), so that a just linear interpolation would lead to overestimated pressures, the static pressures in that area were approximated by a fitting Bezier curve. As example, Figure 7 presents the obtained curve for the pressure ratio at the expansion ramp and the upper flap side at $Re = 7.3 \cdot 10^6$, $T_D = 297$ K and $\Pi = 615$.

With the gained pressure distribution, the nozzle thrust and the axial thrust coefficient $c_{fg,x}$, which is defined as the ratio between the axial thrust and the ideal thrust, were calculated for a comparison of the nozzle performance. Figure 8 gives the axial thrust coefficient at $Re \approx 7.3 \cdot 10^6$ and at two different nozzle pressure ratios of $\Pi \approx 100$ and $\Pi \approx 600$ for a variation of the nozzle total temperature between $T_D = 296$ K and $T_D = 986$ K. The axial thrust coefficient rises with increasing temperatures, by about one percent at the pressure ratio $\Pi \approx 600$ and by about one and a half percent at $\Pi \approx 100$. Since the force at the inlet of an airbreathing engine is of the same magnitude as the nozzle thrust at hypersonic speeds, this leads to a significant impact of several percent on the engines net thrust [7].

4 Conclusion

Experimental results in the hypersonic wind tunnel H2K confirmed that the variation of the nozzle flow temperature influences the interaction between the nozzle flow with the external flow around the model. The comparison of Schlieren images of the nozzle flow field showed, that the extent of the nozzle flow grows with increasing temperature, thus leading to a displacement of the typical shocks of the interacting flow field.

Furthermore, the comparison of the pressure curves and the calculated axial thrust coefficients showed, that the nozzle thrust at a typical design nozzle pressure ratio of $\Pi = 600$ increases by about one percent when the total temperature of the nozzle flow is raised from $T_D = 300$ K to about $T_D = 1000$ K. Considering the severe impact of nozzle thrust on the net thrust of airbreathing engines at hypersonic speeds, this study demonstrates the necessity of taking into account temperature effects for wind tunnel simulation of nozzle flow fields.

Acknowledgements

The presented study was carried out at the Institute of Aerodynamics and Flow Technology, Wind Tunnel Division Cologne, of the German Aerospace Center (DLR). It is part of the DFG Collaborative Research Center 253 (SFB 253) at the Aachen University of Technology. The authors gratefully acknowledge the support by DFG and DLR's space program directorate.

References

[1] Th. Esch, A. Bauer and H. Rick. "Simulation and nozzle/afterbody integration of hypersonic propulsion systems". Z. Flugwiss. Weltraumforsch. 19, pp. 19-28, 1995.

[2] T.A. Gronland, J.L. Cambier and S. Wallin. "Nozzle/Afterbody performance for Hypersonic Airbreathing Vehicles". 33rd AIAA/ASME/SAE/ASEE Joint Propulsion Conference & Exhibit, AIAA-97-3166, 1997.

[3] P. Gruhn, A. Henckels and S. Kirschstein. "Flap contour optimization for highly integrated SERN nozzles". Aerospace Science and Technology, pp. 555-565, Vol. 4, 2000.

[4] P. Gruhn, A. Henckels and G. Sieberger. "Improvement of the SERN nozzle performance by aerodynamic flap design". Aerospace Science and Technology, pp. 395-405, Vol. 6, 2002.

[5] L.D. Huebner, K.E. Rock, D.W. Witte and E.G. Ruf. "Hyper-X Engine Testing in the NASA Langley 8-Foot High Temperature Tunnel". 36th AIAA/ASME/SAE/ASEE Joint Propulsion Conference, AIAA-2000-3605, 2000.

[6] R. Lederer. "Testing the actively cooled, fully variable hypersonic demonstrator nozzle". AIAA-96-4550-CP, 1996.

[7] R. Lederer and W. Krüger. "Nozzle Development as a key element for Hypersonics". AIAA/DGLR 5th International Aerospace Planes and Hypersonics Conference, AIAA-93-5058, 1993.

[8] J. Niezgodka. "Druckverteilungsmessungen zur Simulation der Interferenz des Triebwerksstrahls auf das Heck der Sänger-Unterstufe (MTU-Modell)". DLR internal report, DLR IB-39113-93C15, 1993.

[9] C.W. Palko and J.C. Dutton. "Velocity Measurements in a Shock-Separated Free Shear Layer". 36th Aerospace Sciences Meeting & Exhibit, AIAA-98-0698, 1998.

[10] F.W. Spaid, E.R. Keener and F.C.L. Hui. "Experimental Results for a Hypersonic Nozzle/Afterbody Flow Field". NASA Technical Memorandum 4638, 1995.

[11] C. Trefny. "On the use of External Burning to reduce Aerospace Vehicle Transonic Drag". 26th AIAA/ASME/SAE/ASEE Joint Propulsion Conference, AIAA-90-1935, 1990.

[12] H. Weissgerber, M. Fischer, F. Magens, A. Winandy, W. Foerster, M. Beversdorff. "Experimental analysis of the flow of exhaust gas in a hypersonic nozzle". AIAA-98-1600, 1998.

Table 1 Accuracy of measurement of PSI™-modules

PSI™-Module	Accuracy of measurement [%]	[10^{-3} psi]	[10^{-3} bar]
1 psi	0.20	2.0	0.14
5 psi	0.15	7.5	0.52
15 psi	0.15	22.5	1.55
100 psi	0.15	150.0	10.34

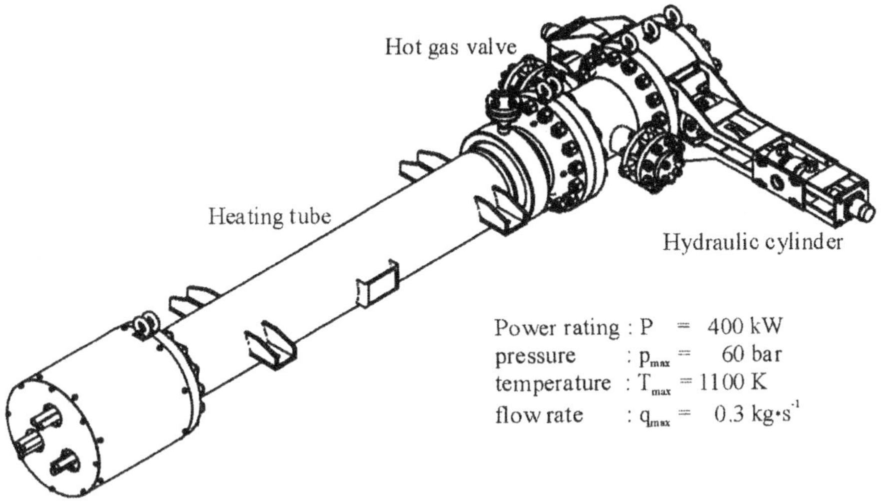

Figure 1 Sketch of additional heater for the nozzle flow

Figure 2 Additional heater connected to the H2K test chamber (left) and SERN nozzle model mounted inside the test chamber (right)

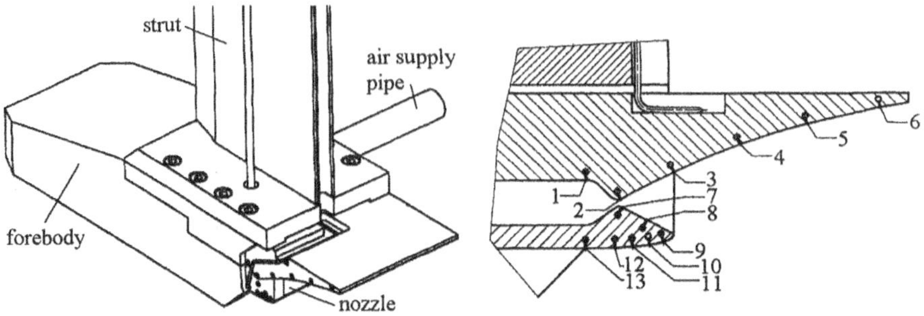

Figure 3 3D drawing of the SERN nozzle model (left) and nozzle cross section with static pressure test positions (right)

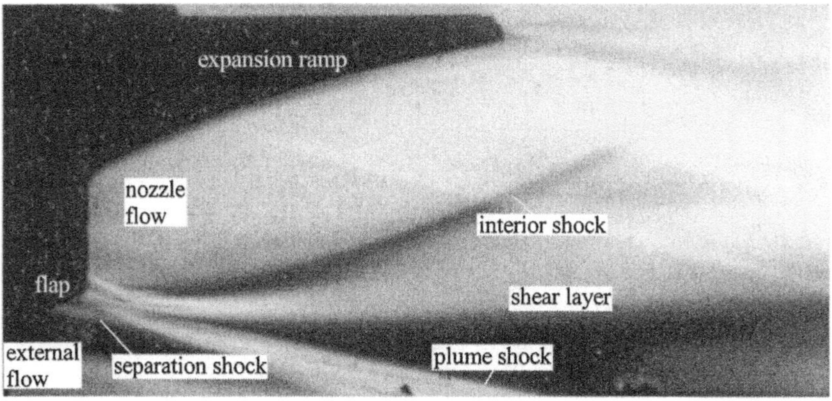

Figure 4 SERN nozzle flow field (Re = $7.4 \cdot 10^6$; T_D = 300 K; Π = 98)

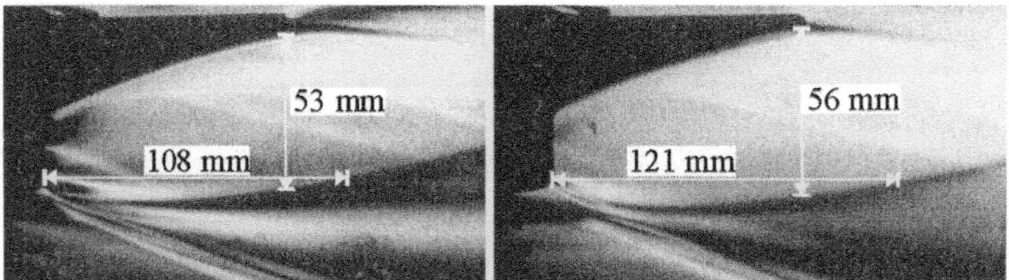

Figure 5 Extension of the nozzle flow field at T_D = 298 K (left) and T_D = 986 K (right; Re = $7.3 \cdot 10^6$; Π = 615)

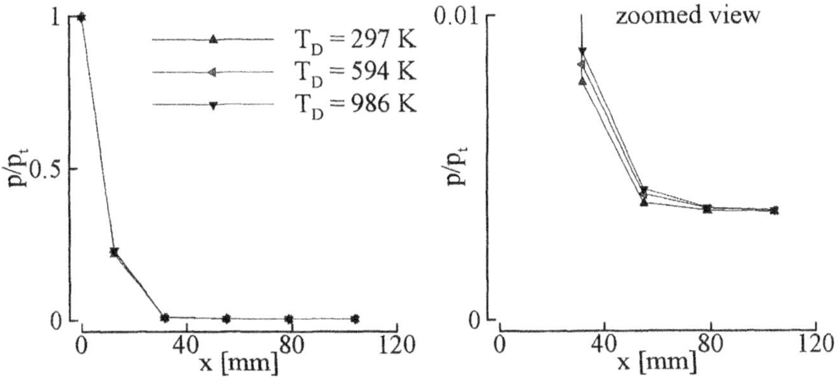

Figure 6 Ratio of static ramp pressure and total pressure inside the nozzle chamber at different total temperatures of the nozzle flow ($Re = 7.3 \cdot 10^6$; $\Pi \approx 615$)

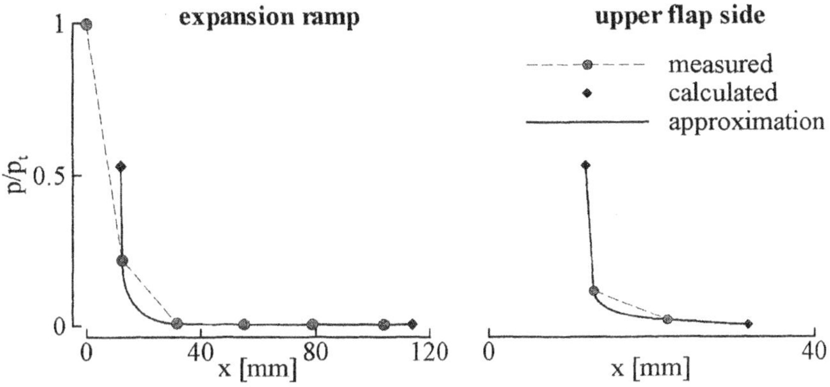

Figure 7 Approximated pressure curves ($Re = 7.3 \cdot 10^6$; $T_D = 297$ K; $\Pi = 615$)

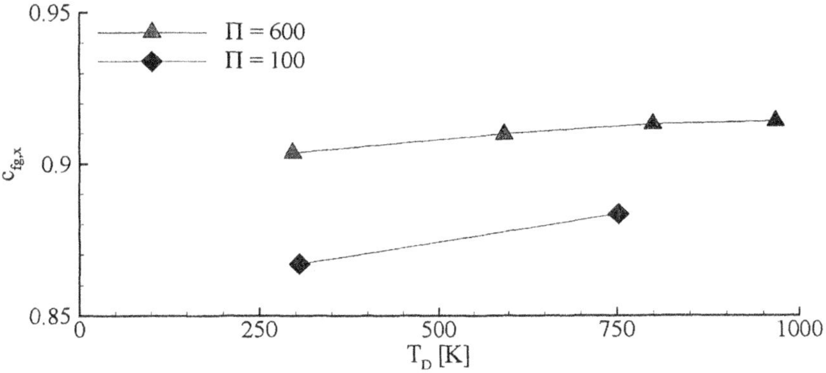

Figure 8 Axial thrust coefficient $c_{fg,x}$ at different nozzle pressure ratios Π ($Re = 7.3 \cdot 10^6$)

Heat Transfer Measurements and Flow Visualizations in Supersonic Flow

C.-C. Ting and A. Henze

Aerodynamisches Institut, RWTH Aachen
Wüllnerstr. zw. 5 u. 7, 52062 Aachen, Germany

Summary

Experimental investigations of supersonic flow around the hypersonic configuration ELAC1c emphasizing the heat transfer rate on its upper surface using the liquid crystal display technique are presented. The flow visualizations show that at large angles of attack ($\alpha \geq 5°$) the flow separates above the leading edge and forms conical vortices on the delta wing. These vortices interact with the bow shock of the upper stage and are distorted by the interaction. For $Ma \geq 1.5$ and $\alpha \geq 12°$ the velocity in the streamwise cross sections becomes supersonic and forms embedded shocks above the delta wing, which are shown by the oil flow pattern. At small angles of attack (about $\alpha \leq 5$) the bow shock of ELAC's upper stage impinges on the upper surface of ELAC and causes the large heat transfer in the region of the shock.

1 Introduction

A hypersonic flight is characterized by large heat loads [8, 16]. During flight the flow undergoes a dramatic deceleration across the shock layer upstream of the vehicle, which converts a great amount of kinetic energy into internal and thermal energy. Considering the development of a fully reusable vehicle this immediately leads to the question of materials and structures, which can stand such heat loads. In this study the reusable geometry considered is based on the ELAC configuration. ELAC [9] is a two-stage-to-orbit space transportation system, which is meant as a reference geometry for a fully reusable airplane-like space transportation system for horizontal take-off.

A reliable prediction of the surface temperature distribution is a must to develop cooling and heat protection systems such as surface radiation cooling or protective cooling material [8, 11]. The numerical method uses the full Navier-Stokes equations to calculate the surface temperature [7]. The heat transfer measurements are conducted using the optical liquid crystal display (LCD) technique [10, 16]. In this study we compare numerical and experimental data to analyze the temperature distribution on the upper surface of the ELAC1c model.

2 Experimental Setup

2.1 ELAC1c Model

ELAC1c is the configuration, which combines the lower stage (ELAC1a) and the orbital stage (EOS), where ELAC1a is a delta wing model with different semi-elliptic cross-sections for the lower and the upper sides. EOS is a shuttle-like model that separates at an altitude of about $30km$ and $Ma \approx 7$. Figure 1 shows ELAC1c at front, side, and top views [4]. The scale of the model, which is used in the trisonic wind tunnel, is 1:240. The model is made out of UREOL, which has a low thermal conductivity and behaves as a thermal insulator [10].

2.2 Trisonic Wind Tunnel

The experiments are carried out in the trisonic wind tunnel, which is sketched in Figure 2. In the vacuum tunnel, which possesses a cross section of 0.4 m × 0.4 m, Mach numbers ranging from subsonic to supersonic ($0.2 \leq Ma \leq 4.0$) can be reached. It operates intermittent at atmospheric pressure and temperature, allowing test periods of 2 to 5 seconds. At transonic Mach numbers Reynolds numbers of 10 to 15 million per meter are achieved.

For the heat transfer measurements the heat flux ($\dot{q}_s$) on the model surface possesses a negative value during the cooling process. The reservoir fluid flows under constant enthalpy conditions through the test section into the vacuum tanks, such that the temperature in the test section decreases during the test runs. This means the cold flow in the vicinity of the body leads to a large heat transfer on the model surface. Due to its definition [10] the heat transfer q_s is negative which also holds for the modified Stanton number, that is defined in equation (2). This fact has to be kept in mind when the heat flux on the ELAC1c surface is discussed.

$$\dot{q}_s = \sqrt{\frac{\rho c \lambda}{\pi}} \left[\frac{T(t) - T(0)}{\sqrt{t}} \right] \tag{1}$$

where ρ is the density, c the specific heat of the model material, and λ the thermal conductivity.

2.3 LCD Technique

The LCD technique is a measurement method, which uses the temperature sensitive liquid crystal (TLC) as material for temperature registration and the full spectrum white light as light source. The TLC is an organic liquid, which can be applied to the model surface with an air brush. Since the derivative of the crystal pitch (P) of TLC, where P is a variable geometric quantity, is a monotone decreasing function of the temperature it can be used for temperature measurements in aerodynamics [3].

Bragg's scattering [12] states that during test only one wave length among the white light spectrum can be scattered by a constant pitch length of TLC at a constant angle, since the rate of change of the TLC pitch length and the pitch itself depend only on the temperature. From the "Hue - Saturation - Luminance" (HSL) a "hue" value of the color space system is determined that corresponds to the wave length [5]. This means that the surface temperature on the model can be determined by the recorded "hue" value.

For this purpose a three RGB chip CCD camera of Hitachi HV-C20A and a Matrox Meteor II/MC automatic capture system with no compression factor TIFF format are used to take photos of the model during test runs with 10Hz recording frequency. From the recorded scattered wave length the RGB data are converted into HSL color space coordinates as described in [15, 16].

The quantitative reconstruction of the model surface temperature by calibration is the most important but also the most difficult work when the LCD technique is used [2]. The optimum setup for the LCD technique is realized when the calibration and the test arrangements are exactly the same. To minimize the test error the same arrangement with different models for calibration and the test runs are used in this study [16].

3 Results

The research focuses on the separated flow field on the upper side of ELAC which occurs at angles of attack $\alpha \geq 5°$. Moreover, the intriguing phenomena due to the interaction between the lower and upper stage flow field will be considered.

3.1 Conical Vortex / Bow Shock Interaction

The shock-vortex interaction on the upper side of ELAC1c is a real three-dimensional problem. Figure 3 shows a schematic of the flow field around the ELAC1c model, indicating the main shocks and the shock-vortex interaction region. The vortex is generated at the leading edge of the lower stage. When it moves downstream it crosses the bow shock of the upper stage (Figure 4) whose pressure gradient changes its initially circular shape. This is evidenced by the vapor-screen image in figure 5(a) which illustrates the flow structure in the cross section x/L=0.7 at Ma=2.0 and an angle of attack $\alpha = 10°$. In figure 5(a) the moist fluid in the vortex core is evaporated due to the low pressure and accelerated outwards. The shape of the core is evidenced by the black area whose contour is determined by the difference in the laser light reflection inside and outside the vortex core. The vortex phenomena encountered in flows separating at the leading edge of a delta wing for large angles of attack can be described in detail using topology rules discussed e.g. in [14]. At large angles of attack the separated subsonic or weakly supersonic flow forms a conical flow on the lower stage's upper surface. The strongly deformed vortex (figure 5(a)) is located on the wing of the upper stage. It possesses a kidney-like shape which is determined by the spanwise pressure field generated by the shock-vortex interaction as sketched in figure 5(b). As discussed in [6] the larger pressure downstream of the bow shock leads to a compression of the flow that alters the circular shape of the vortex core, which is observed in an unperturbed flow field, into a kidney form.

3.2 Embedded Shock

A vortex induced shock appears on the upper surface of ELAC1a. Such a phenomenon is well-known for delta wings with sharp leading edges as sketched in figure 6 [13]. The shock impingement lines of the embedded shocks on ELAC1a are visualized by the oil flow pattern in figure 7. Note that the induced pressure gradient is not strong enough for the flow to separate.

3.3 Heat Transfer and Wall Temperature

The heat transfer is discussed using equation (1) and the modified Stanton number (St'), which is a rewritten version of the original definition with the difference $h_\infty - h_w$ neglected [1].

$$St = \frac{\dot{q}_s}{\rho_\infty V_\infty (h_O - h_W)} \quad \Rightarrow \quad St' = \frac{\dot{q}_s}{\frac{1}{2}\rho_\infty V_\infty^3} \tag{2}$$

This means due to the definition of $\dot{q}_s$ in equation (1) that a larger heat flux corresponds to a larger modified Stanton number with negative sign. Since in contrast to the real flow conditions the temperature of the body surface of the model is larger than that of the freestream flow field the temperature distributions within the boundary layer differ for the real and the model problem. That is, no quantitative analysis is possible, however, considering the main flow phenomena such as vortices, shocks etc., which are determined by the interaction of inertia, pressure, and viscous forces, the results show qualitatively the impact of the angle of attack. Figure 8(a) shows the measurement at Ma=2 and $\alpha = 5°$. Two lines of large heat transfer rate occur near the leading edge of the delta wing. At $\alpha = 10°$ these lines, which move closer to the symmetry plane, are even more pronounced. Figure 8(c) also indicates the secondary separation and reattachment above the delta wing. The large heat transfer rate in the measurements corresponds to the low temperature due to the cooling process in the trisonic wind tunnel. The numerical result for the wall temperature in figure 8(b) shows the strong influence of the shock of the upper stage on the temperature distribution. At large angles of attack the shock strength caused by the mounted upper stage is reduced, which results in smaller gradients in the wall temperature distribution (figure 8(d)).

4 Conclusion

The heat loads and temperature distributions on the upper surface of ELAC1c model are experimentally and numerically investigated at supersonic flow conditions and angles of attack in the range $0° \leq \alpha \leq 10°$, while the experimental flow visualization was carried out in the range $-9° \leq \alpha \leq 15°$. The joint discussion of the measurements and the numerical data shows that at angles of attack larger than 5° the vortical flow field on the upper surface is strongly influenced by the mounted second stage (EOS). The bow shock of EOS results in a pronounced modification of the shape of the vortices evolving from the leading edge of the stage.

Acknowledgements

The investigations were supported by the Graduiertenkolleg "Transportvorgänge in Hyperschallströmungen" and the Sonderforschungsbereich 253 "Grundlagen des Entwurfs von Raumflugzeugen" of the DFG.

References

[1] Anderson, J. D. *Introduction to Flight.* McGraw-Hill, inc, fourth edition, 2000.

[2] Behle, M., Schulz, K., Leiner, W., and Fiebig, M. *Color-Based Image Processing to Measure Local Temperature Distributions by Wide-Band Liquid Crystal Thermography.* Applied Scientific Research, **56**:113–143, 1996.

[3] DE Gennes, P. G. and Prost, J. *The Physics of Liquid Crystals.* Clarendon Press, Oxford, second edition, 1998.

[4] Engler, V. *Geometrie der Oberstufenkonfiguration EOS und Integration in die Unterstufe ELAC 1.* IB A-97-1. März 1997. Sonderforschungsbereich 253, Institut für Luft- und Raumfahrt, Rheinisch-Westfälische Technische Hochschule Aachen.

[5] Ernst, H. *Einführung in die digitale Bildverarbeitung.* Franzis-Verlag GmbH, München, 1991.

[6] Grasso, F. and Pirozzoli, S. *Shock-Wave-Vortex Interactions: Shock and Vortex Deformations, and Sound Production.* Theoretical and Computational Fluid Dynamics, Springer-Verlag, **2000(13)**:421–456, 1999.

[7] Henze, A., Schröder, W., and Meinke, M. *Computation of Flows Around Space Configurations.* High Performance Scientific and Engineering Computing, **LNCSE 21**:131–138, 2001. Springer-Verlag, Editors: Breuer, M., Durst, F., and Zenger, C.

[8] Hirschel, E. H. *Heat Loads as a Key Problem of Hypersonic Flight.* Z. Flugwiss. Weltraumforsch., **16(6)**:349–356, 1992.

[9] Hänel, D., Henze, A., and Krause, E. *Supersonic and Hypersonic Flow Computations for the Research Configuration ELAC I and Comparison to Experimental Data.* Z. Flugwiss. Weltraumforsch., **17(2)**:90–98, 1993.

[10] Jacobs, M. *Wärmeübergangsmessung am Modell eines Raumtransportsystems in supersonischer Strömung.* 1997. Dissertation, Aerodynamisches Institut, Rheinisch-Westfälische Technische Hochschule Aachen.

[11] Janovsky, R. *Analyse und Bewertung horizontal startender Raumtransportsysteme.* 12. VDI-Verlag GmbH, Düsseldorf, 1995. No. 242, Dissertation, Rheinisch-Westfälische Technische Hochschule Aachen.

[12] Kittel, C. *Introduction to Solid State Physics.* John Wiley & Sons, Inc., sixth edition, 1986.

[13] Miller, D. S and Wood, R. M. *Leeside Flows over Delta Wings at Supersonic Speeds.* Journal of Aircraft, **21(9)**:680–686, 1984.

[14] Peake, D. J. and Tobak, M. *Three-Dimensional Interactions and Vortical Flows with Emphasis on High Speeds.* AGARD-AG-252. 1980. Technical Report, Advisory Group for Aerospace Research & Development.

[15] Ting, C.-C. *Strömungs- und Wärmeübergangsmessung für das zweistufige Raumtransportsystem ELAC.* 2003. Dissertation, Aerodynamisches Institut, Rheinisch-Westfälische Technische Hochschule Aachen.
[16] Ting, C.-C., Henze, A., and Schröder, W. *Heat Transfer Measurements in Supersonic Flow using the Liquid Crystal Display Technique.* (**AIAA 2002-5200**), October 2002. 11th AIAA/AAAF International Conference, Orleans France.

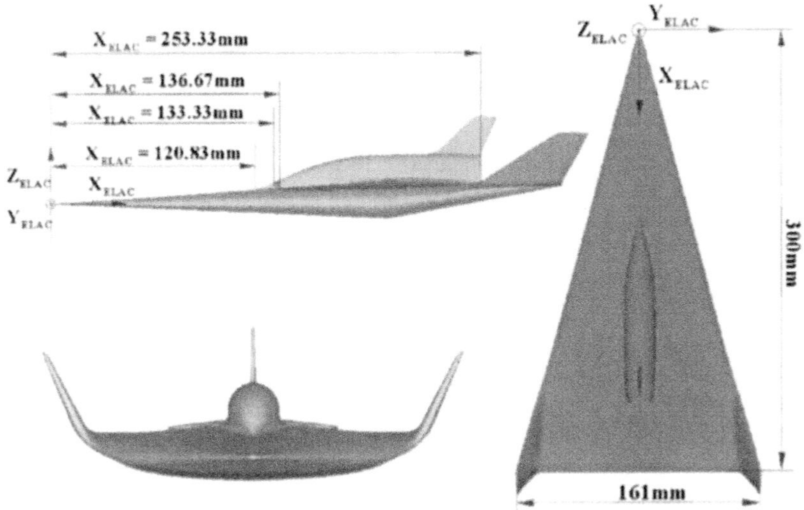

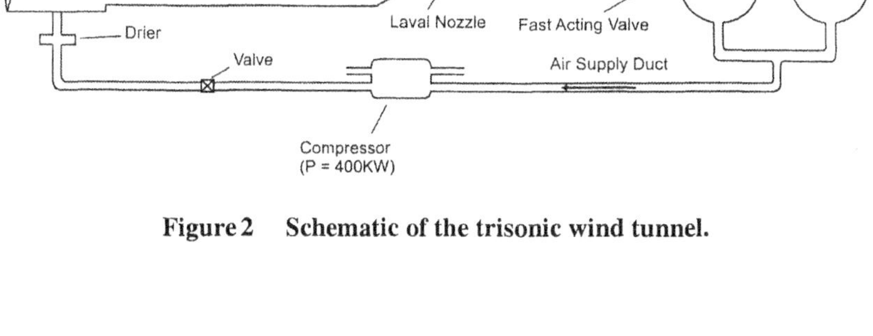

Figure 2 Schematic of the trisonic wind tunnel.

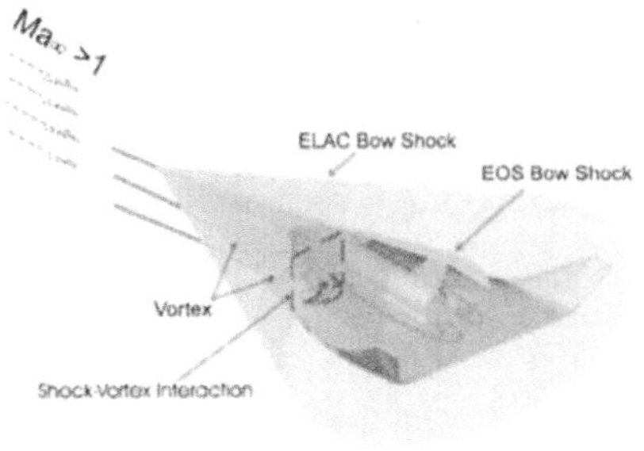

Figure 3 Schematic of ELAC1c flow field.

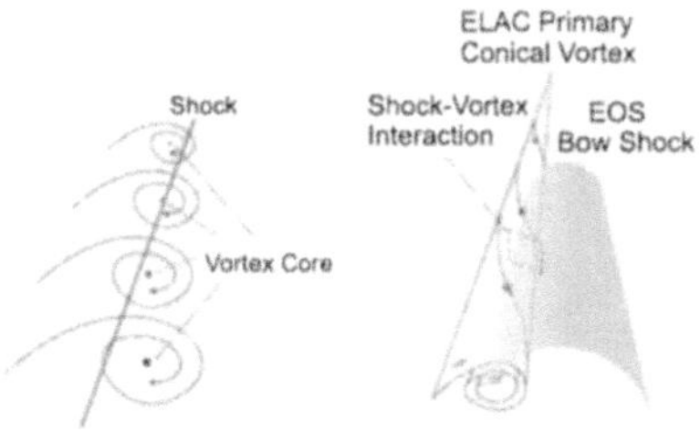

Figure 4 Schematic of shock-vortex interaction.

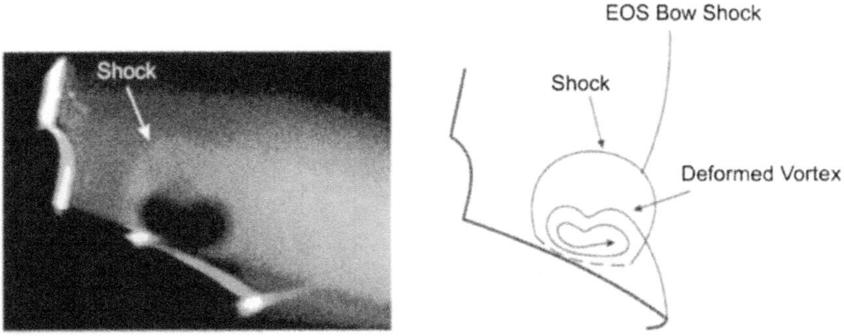

(a) Vapor screen image at $Ma = 2$, $x/L = 0.7$, and $\alpha = 10°$.

(b) Schematic of shock-vortex interaction from Figure 5(a).

Figure 5 Distortion of the vortex due to shock-vortex interaction.

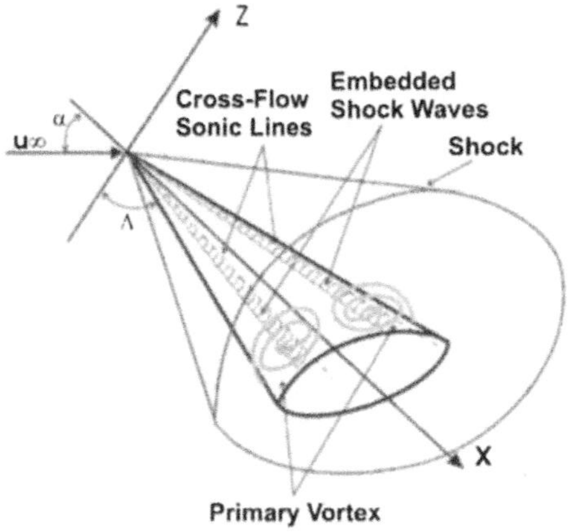

Figure 6 Schematic illustration of embedded shocks.

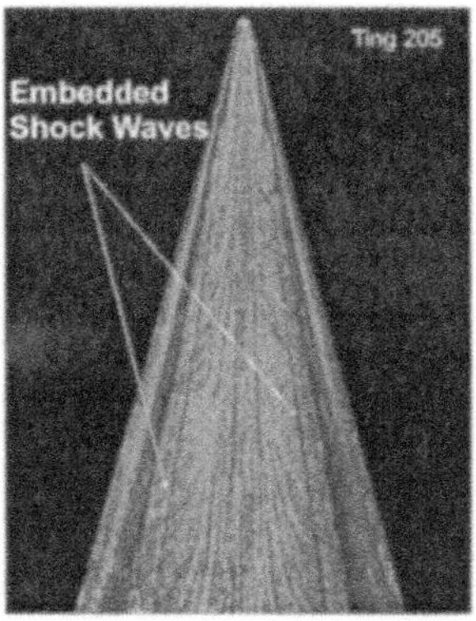

Figure 7 Oil film pattern at Ma=1.5 and $\alpha = 15°$.

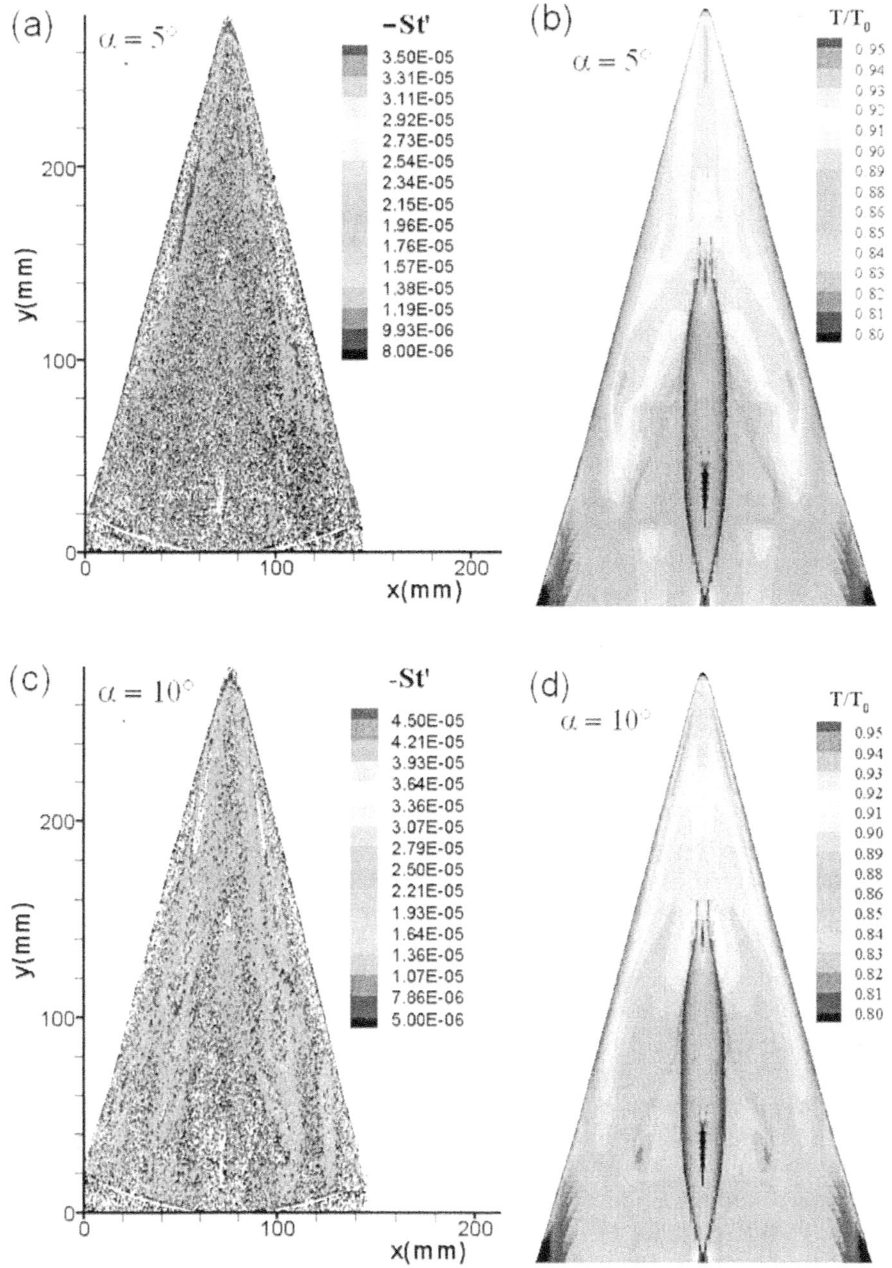

Figure 8 Experimental heat transfer (a, c) and numerical wall temperature (b, d) results for $Ma_\infty = 2$.

Numerical Simulation of Turbulent Flows Inside a Hypersonic Inlet

Birgit U. Reinartz and Josef Ballmann

Mechanics Laboratory, Aachen University,
RWTH Aachen, Templergraben 64, 52064 Aachen, Germany

Summary

Results of 2D and 3D numerical simulations of a mixed–compression inlet for a scramjet propulsion system under Mach 2.41 and 3.0 test conditions are presented. The computations are performed with the DLR FLOWer code solving the Reynolds averaged Navier–Stokes equations with different turbulence models. The details of corresponding inlet experiments, e.g., leading edge bluntness and tripping wires, are included in the simulation to assess their influence on the test data.

1 Introduction

One of the key interests of the Collaborative Research Center SFB 253 is the technological challenge of reusable hypersonic airbreathing propulsion systems. In hypersonic flows, shock/boundary layer interactions may cause severe separations and, when taking place inside the inlet, those separations will lead to a system unstart. Therefore, the compression shock system of the inlet is of great interest and is studied numerically as well as experimentally within the frame of the Research Center [4]. A schematic view of the complex flow field with shock/shock and shock/boundary layer interactions is shown in Fig. 1.

The test model is specified by the lower stage of the two–stage to orbit reference configuration of the SFB ('ELAC'). The design is based on a mixed–compression inlet with two external compression ramps and a consecutive isolator section, which is a constant area duct, responsible for internal compression and flow homogenization. To achieve the target test conditions, the first outer ramp had to be omitted during the experimental investigations [2]. Instead, an angle of attack of the freestream of $\alpha = 10°$ was specified (see Fig. 2). Due to the relatively large width–to–height ratio of the inlet–isolator model of 3.5 (see Table 1), a planar flow was assumed and all measurements were taken in the center plane. The test model is assembled of different parts allowing for easy variations of the geometry. To smooth the transitions between the different modules, the model was coated by an aluminium foil, which caused a slight bluntness of the leading edge of the compression ramp.

The objective of the present CFD analysis is to complement the experimental investigations and to enhance the understanding of the experimental results. Therefore, the model geometry and the details of the test set–up are simulated with great care. Furthermore, the influence of various parameters, not completely determined by the experiments, e.g., the turbulent state of the boundary layer and the nose radius on the leading edge, are investigated. In correspondence to the experiments, two different inflow conditions are analyzed (see Table 2) and the computed results are evaluated by the test data.

2 Numerical Method

2.1 Navier–Stokes Solver FLOWer

The computations are performed with the DLR FLOWer code [3] including some extensions for hypersonic flows [5]. The code uses a finite volume technique with upwind discretization on block-structured grids to solve the unsteady 3D RANS equations for compressible fluid flow. Due to the moderate test conditions, the air is considered to be a perfect gas with constant ratio of specific heats ($\gamma = 1.4$). An AUSM Flux Vector splitting with MUSCL extrapolation and either a modified van Albada or a van Leer limiter function is used for the inviscid fluxes. Time integration is performed by an explicit five-stage Runge-Kutta time-stepping scheme. To enhance convergence, multigrid method, implicit residual smoothing, and local time stepping are applied.

2.2 Turbulence Modeling

Due to the high Reynolds number achieved in the hypersonic test bed, turbulent flow is assumed for the numerical simulations and a freestream turbulence intensity of 0.5% is specified. To investigate the influence of turbulence modeling, various models implemented in the FLOWer code are applied to the test case. Detailed information on the models is given by Coratekin [1]; here, only the basic features are presented. The following turbulence models are used: the original Spalart-Allmaras and its modification, the SALSA model (both are one-equation models), Wilcox k-ω model with some extensions for hypersonic as well as separated flows, the SST (Shear Stress Transport) k-ω model, and an explicit algebraic Reynolds–Stress model (EARSM). Except for EARSM, the discretization of the turbulence equations is performed using an AUSM upwind scheme for the convective fluxes together with an implicit time integration scheme (DDADI). For EARSM, which is implemented in a former version of FLOWer, only central spatial discretizations and Runge–Kutta time integration are available.

2.3 Boundary Conditions

At the inflow boundary, the supersonic test conditions given in Table 2 are prescribed and the turbulent values are determined by the specified freestream turbulence intensity. For the supersonic outflow, the variables are extrapolated from the interior onto the boundary. At solid walls, the no-slip condition is enforced. Additionally, depending on the turbulence model, either the turbulent kinetic energy or the eddy viscosity is set to zero. If applicable, the specific dissipation rate is set proportional to the wall shear stress and the surface roughness [1]. Since the total temperature of the flow is in the range of ambient temperature and due to the long measuring times of thirty minutes or more, an adiabatic energy boundary condition is used at the wall.

3 Computations

Figure 3 shows the computational domain for 2D simulations. In addition, the displayed grid detail indicates the refinement of the grid in certain areas, e.g., towards the walls and near the onset of geometry induced shock waves. The flow resolution was tested by performing a grid sensitivity analysis with three different grid refinement levels: coarse (497×65), medium (993×129), and fine (1985×257). In Fig. 4, the determined deviations in surface pressure distributions are negligible for the fine and medium grids, whereas some details of the distribution are missed by the coarse grid. The turbulent resolution of the medium grid is $y^+ \leq 2$, which seems sufficient for the investigation of surface pressure distributions.

The density residual and the turbulence residual (L2-norm) are monitored in Fig. 5. After 15000 iterations, the residuals have dropped to their minimum value and an additional 5000 iterations are performed to ensure that all other variables are converged, too. Additionally, the deviation between computed inflow and outflow mass flux is monitored and an upper limit of 1% is enforced. This additional criterion is necessary to ensure convergence of the quasi-steady computations even for widely separated (i.e. unsteady) flows. Validation of the CFD results is achieved by the available test data.

The computations are performed on the new SunFire SMP–cluster of the RWTH Aachen using up to 26 processors as well as on a NEC SX–5 cluster. The parallelization is jointly based on FLOWer's block–based MPI (message passing interface) formulation as well as on Sun's shared memory auto–parallelization. A CFL number of 4 is realized and the CPU times required for a 2D simulation and a 3D simulation are approximately 8 h and 42 h, respectively.

4 Results

4.1 Influence of Turbulence Model and Nose Radius

To assess the influence of turbulence modeling, different models have been applied to the $M_\infty = 2.41$ flow. In Fig. 7, the various obtained distributions of pressure coefficient along the compression ramp surface are presented. The steep pressure drop at $x = 0.046$ m correlates to the expansion around the ramp shoulder. Slightly downstream of the shoulder, the impingement of the lip shock induces separation (see Fig. 1). Only the Spalart-Allmaras and the SALSA model seem to predict the onset of separation correctly, which is correlated to the first pressure increase after the expansion. The second increase denotes reattachment and the subsequent pressure peak is caused by the impingement of the cowl deflection shock. The following peaks are induced as the various shocks are reflected downstream. The pressure peaks are best met by the extended k-ω and the SST k-ω model. However, all computed pressure increases seem more or less to be shifted downstream compared to the test data, indicating that the boundary layer thickness might not be determined correctly. Therefore, the slight bluntness of the leading ramp edge is simulated by a finite nose radius, causing the separation of the ramp shock and increasing the boundary layer thickness. The aluminium foil, which was used to coat the surface, measured 0.1 mm. Considering also the original finite thickness of the model leading edge, two different values for the nose radius were analyzed: $r = 0.45$ mm and 0.25 mm. With increasing thickness, the boundary layer flow looses energy and, thus, is less able to resist incipient separation. Consequently, the predicted onset of separation moves upstream with increasing nose radius, as is shown in Fig. 6. However, the thicker boundary layer also damps the shock impingement and reduces the pressure peaks. The overall best agreement with the test data is achieved by a medium nose radius of $r = 0.25$ mm, therefore, this model geometry and the extended k-ω turbulence model were used for all further simulations.

4.2 Tripping Wire

For the $M_\infty = 3.0$ test case, the expansion around the ramp shoulder is approx. 30% stronger than for the $M_\infty = 2.41$ flow. To prevent relaminarization of the boundary layer, a tripping wire was mounted on the compression ramp during the experiments. Because the flow pattern is altered not only in the vicinity of the wire but also downstream of its position, the wire has to be incorporated in the numerical simulations, too. Figure 8 shows the grid in the vicinity of the wire where 32 nodes are distributed along the wire surface. Furthermore, the altered flow field with two additional shock waves is shown. The qualitative agreement of the computed Mach contour lines with the experimentally obtained schlieren picture is very good, as shown in Fig. 10. The positions of the various shock waves and the thicknesses of the boundary layers are well predicted. A more quantitative comparison is presented in Fig. 9, where the computed and measured pressure distributions along the compression ramp and the cowl surface are shown. Especially, the size of the separation is determined correctly. The agreement of the cowl surface distribution is excellent and the only discrepancy from the test data is located in the diffuser section ($x \approx 0.16$ m), outside the mixed compression inlet, where the resolution of the grid is reduced.

4.3 Effect of Side Walls

The design of the experiment assumed a predominantly 2D flow behavior and, therefore, measurements were only taken in the center plane. To assess the possible influence of the swept side walls on the flow in the center plane, a 3D numerical simulation of one half of the inlet model (Fig. 11) was performed. Comparison of surface pressure distributions (Fig. 12) and Mach number contour lines (Fig. 13) confirm the $M_\infty = 2.41$ flow in the symmetry plane to be almost identical to corresponding 2D simulations. However, when the overall aerodynamic performance of the inlet, characterized by total pressure recovery and captured mass flow, is investigated, side wall effects have to be taken into account. In Fig. 14, a separation zone on the side wall caused by the passing lip shock (LS) is shown that leads to additional mass flow spillage. Due to the thick ramp boundary layer (RBL), the development of the corner vortex (CV) becomes visible in Fig. 15. The vortex intensifies the overall heat load and causes the shock system to be displaced, as can be seen in the vertical extension of the separation shock (SS) and the reattachment shock (RS). Whereas, the boundary layer, initiated on the tip of the engine cowl, is still small and, therefore, the passing cowl deflection shock wave (CDS) remains undisturbed.

5 Conclusion

2D and 3D simulations of a mixed–compression inlet are presented. Various turbulence models are tested and Wilcox k-ω model, with some extensions for hypersonic and for separated flows, is selected, yielding the best agreement with the available test data. The numerical analysis stresses the importance to observe even small details of the experimental set–up in the simulation. In this respect, the influence of a slight bluntness of the leading edge of the compression ramp and of a tripping wire is investigated. Furthermore, the CFD analysis confirms that the flow in the center plane is indeed unaffected by side wall effects. However, it also shows that for a performance analysis of the engine inlet, the effect of the swept side walls have to be taken into account since they influence the mass capture ratio and lead to a displacement the shock system.

References

[1] Turan A. Coratekin. Numerical simulation of turbulent shock/boundary-layer interactions in hypersonic flows. *Fortschritt-Berichte VDI*, 7(381), 2000.

[2] Carsten D. Herrmann and Wolfgang W. Koschel. Experimental investigation of the internal compression of a hypersonic intake. AIAA Paper 02–4130, 2002.

[3] Norbert Kroll, Cord-Christian Rossow, Klaus Becker, and Frank Thiele. The megaflow project. *Aerospace Science and Technology*, 4:223–237, 2000.

[4] Birgit U. Reinartz, Josef Ballmann, Carsten Herrmann, and Wolfgang Koschel. Analysis of hypersonic inlet flows with internal compression. AIAA Paper 02-5230, 2002.

[5] Birgit U. Reinartz, Joern van Keuk, Turan Coratekin, and Josef Ballmann. Computation of wall heat fluxes in hypersonic inlet flows. AIAA Paper 02-0506, 2002.

Table 1 Main dimensions of inlet model.

Quantity	Value
Inner width:	52 mm
Isolator height h:	15 mm
Isolator length l:	79.3 mm
Second ramp angle δ_2:	21.5°
Cowl angle δ_3:	9.5°
Divergence angle δ_4:	5°

Table 2 Test conditions for inlet.

Quantity	Conditions 1	Conditions 2
M_∞ [-]	2.41	3.0
Re_∞ [1/m]	$5.07\cdot10^7$	$4.87\cdot10^7$
$p_{t,0}$ [kPa]	540	620
$T_{t,0}$ [K]	305	290

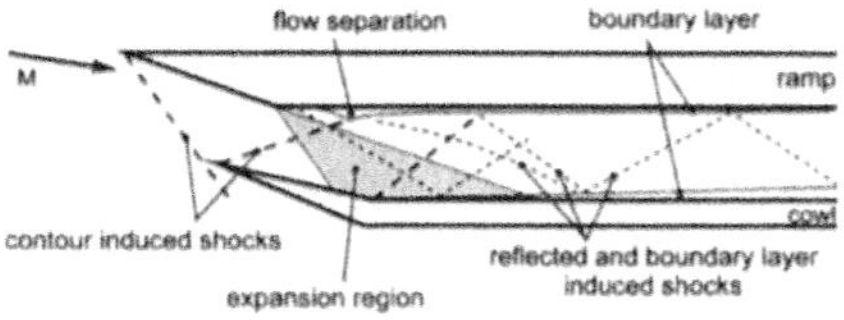

Figure 1 Schematic of flow field.

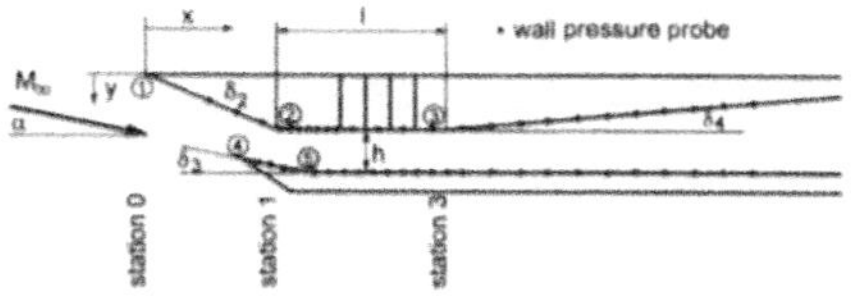

Figure 2 Planar inlet-isolator model.

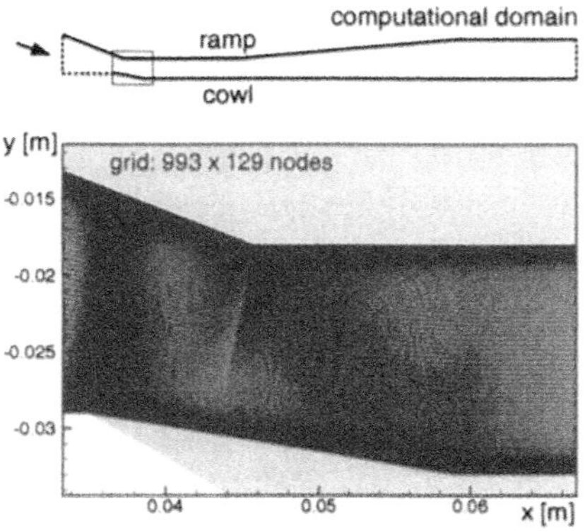

Figure 3 Computational domain and grid detail for supersonic inlet.

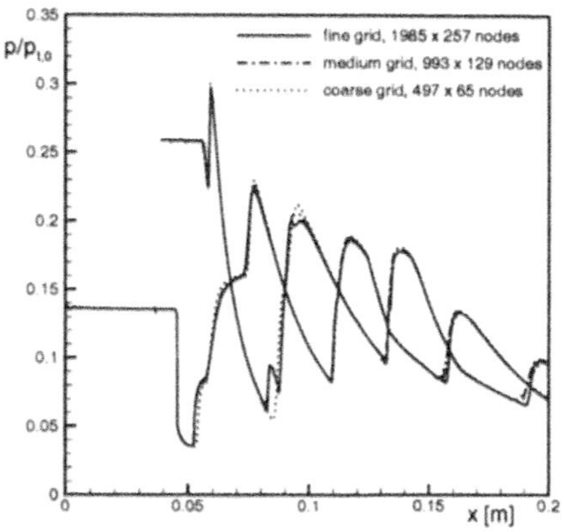

Figure 4 Surface pressure distributions for refined grids.

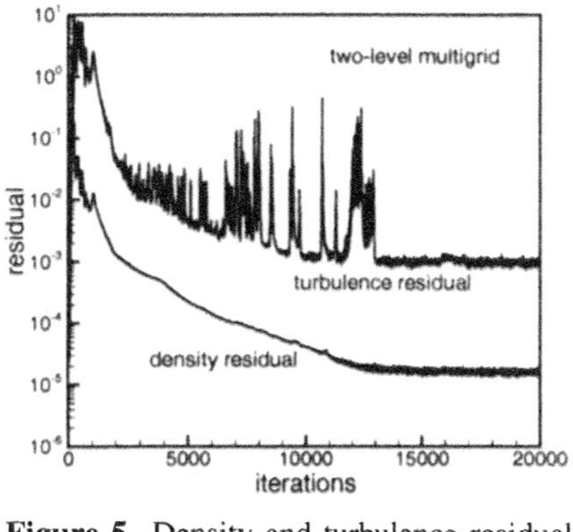

Figure 5 Density and turbulence residual for supersonic inlet computations.

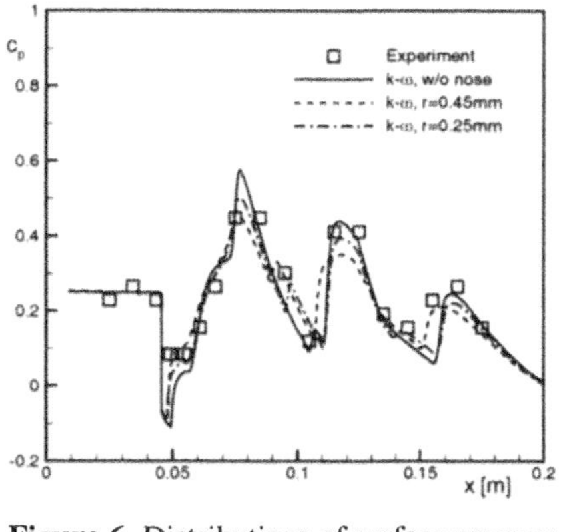

Figure 6 Distributions of surface pressure coefficient for different nose radii.

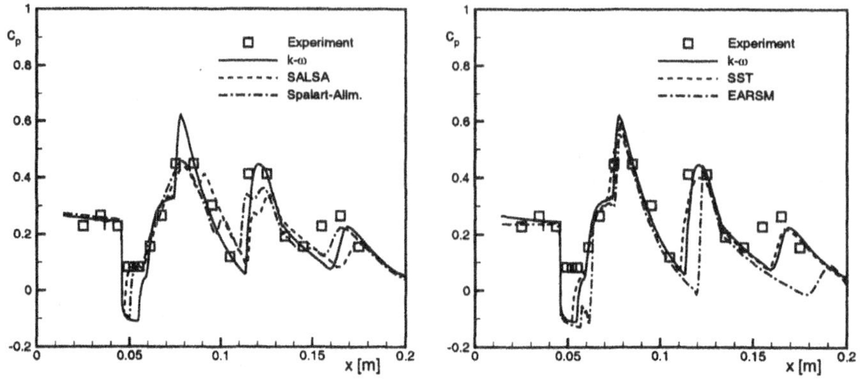

Figure 7 Distributions of surface pressure coefficient for different turbulence models.

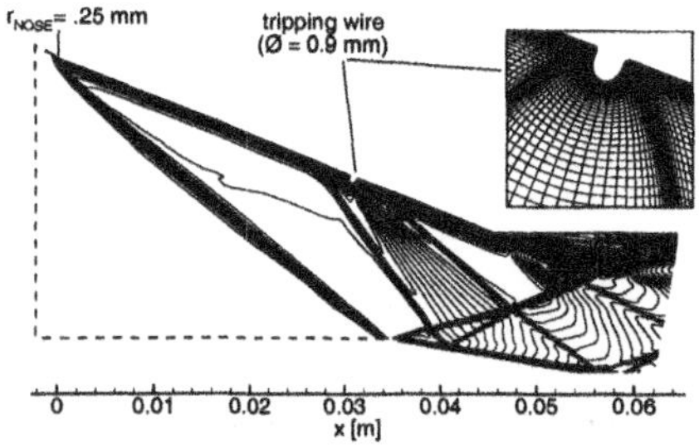

Figure 8 Flow field and grid detail in the vicinity of mounted wire.

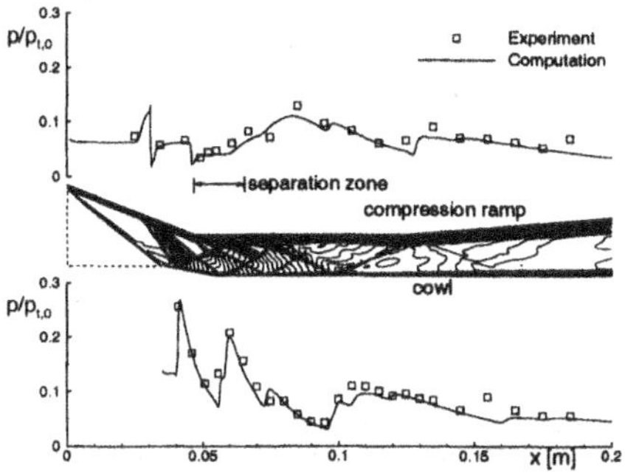

Figure 9 Mach number contour lines and normalized surface pressure distributions.

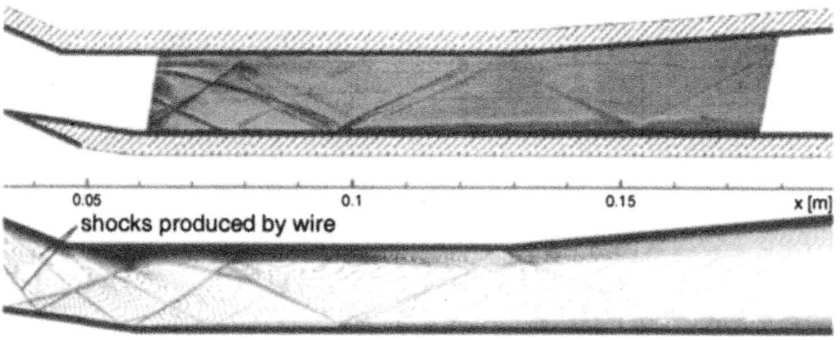

Figure 10 Schlieren picture (top) and corresponding Mach number contours (bottom).

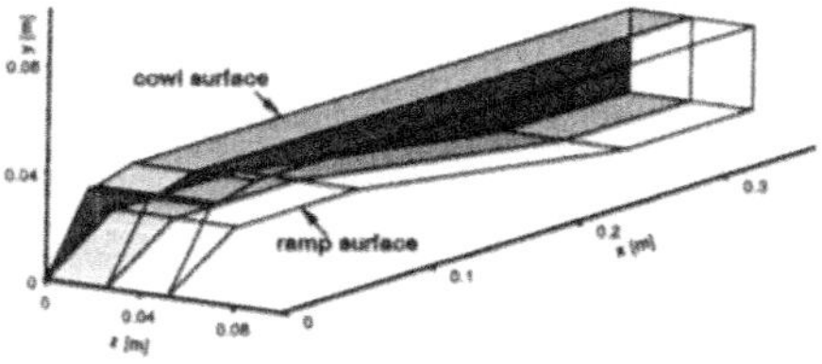

Figure 11 Numerical representation of one half of the 3D test model incl. side walls.

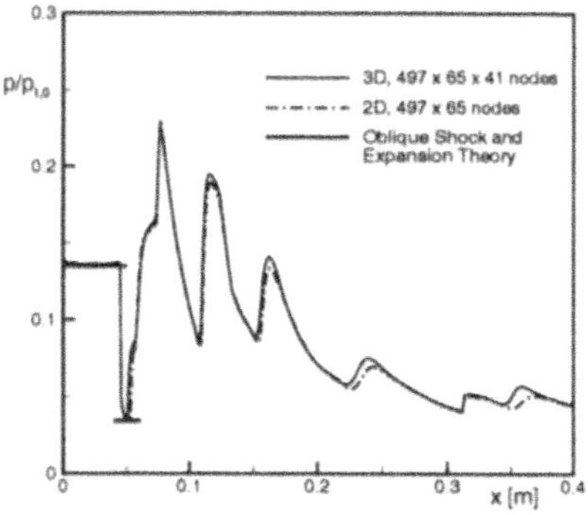

Figure 12 Wall pressure distribution along the body surface and in the symmetry plane of the 3D computation.

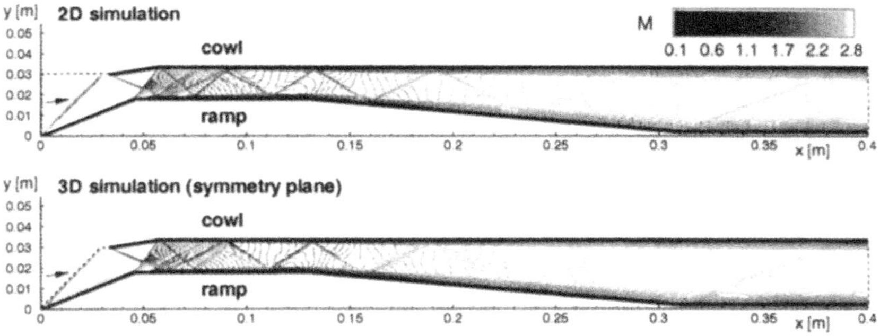

Figure 13 Mach number distribution in symmetry plane for 2D and 3D simulation.

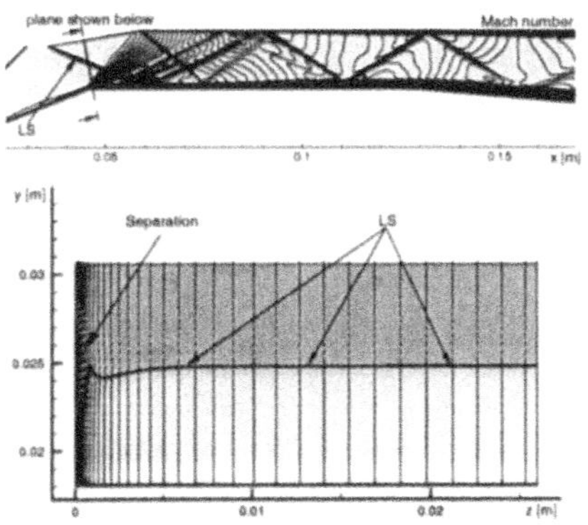

Figure 14 Mach number distribution in symmetry plane (top) and in one vertical plane (bottom) showing a separation bubble occurring on the side wall due to lip shock.

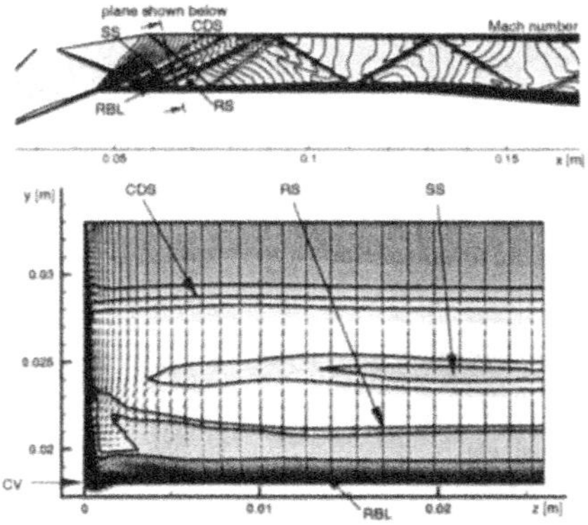

Figure 15 Mach number distribution in symmetry plane (top) and in one vertical plane (bottom) showing flow pattern with several shock waves and a corner vortex.

PSP Measurements on a Linear Plug Nozzle: Aerodynamic Investigations on Linear Plug Nozzle Configurations

CHR. KLEIN[1], S. NIEBERGALL[2], W.E. SACHS[1], U. HENNE[1], R.H. ENGLER[1]

[1] DLR Göttingen, Institut für Aerodynamik und Strömungstechnik, Bunsenstr. 10, D-37073 Göttingen, Germany

[2] TU-Dresden, Hochgeschwindigkeitskanal-Merkers, Gartenstr. 1, D-36460 Merkers, Germany

Summary

Pressure field measurements on a Linear Plug Nozzle model were conducted to evaluate the feasibility and accuracy of measurements using DLR's Pressure Sensitive Paint (PSP) system in the high-speed wind tunnel at Merkers (HWK) of TU-Dresden. A part of the nozzle surface was coated with binary PSP paint. Results show that the optical pressure measurement system with binary paint can be used in the HWK facility. In addition, the paper gives a summary of the thrust vector calculations based on measured static pressure distributions on the plug surface using thrust vector control.

1 Introduction

The physical basis of all Pressure Sensitive Paint (PSP) formulations is the oxygen-quenching process in which excited luminophore molecules are deactivated by oxygen. Luminophore molecules are placed in a binder layer on the model surface and are excited by an appropriate light source. Oxygen from an airflow diffuses into the binder layer. Oxygen concentration in the layer is a function of local static pressure on the upper surface of the binder. The resulting luminescent intensity distribution reveals the static pressure distribution on the model [2]. The PSP method gives quantitative as well as qualitative pressure distribution images of the observed model surfaces [3] without significant disturbances. This method can also be used for flow visualization and to provide detailed static pressure information about the model's aerodynamics. PSP techniques have been widely used in wind tunnels for investigations in transonic and supersonic flow [9].

The objectives of the aerodynamic testing of the plug nozzle model are to acquire new knowledge about:

- Expansion in the axial direction for different nozzle configurations,
- Expansion with and without side fences in the spanwise direction,
- Qualitative flow phenomena using Schlieren optical method,
- Thrust vector analysis using Scanivalve pressure measurements,

2 PSP Basis

PSP techniques are based on the deactivation of photochemical-excited organic molecules (luminophores), by oxygen molecules. This oxygen-quenching process of luminescence was first discovered and described by Kautsky and Hirsch [6] in 1935. The behavior of photoluminescence of a luminophore quenched by oxygen molecules can be described by the Stern-Volmer relation [2]:

$$I_0/I = 1 + K_q\, PO_2\,, \qquad (1)$$

where I_0 is the photoluminescence in the absence of oxygen (vacuum), I the detected photoluminescence, K_q the quenching constant, and PO_2 is the partial pressure of oxygen. K_q is a function of temperature.
In the most widely-used pressure sensitive paint formulations, the luminophores (e.g. ruthenium, pyrene) are located in a polymeric binder material (e.g. silicon rubber). The binder compound is permeable to the oxygen molecules. To calculate the static pressure values from a measured intensity distribution for such paint formulations the second order approximation is more useful:

$$p = A(T) + B(T)(I_{ref}/I) + C(T)(I_{ref}/I)^2, \qquad (2)$$

where p is the local static pressure, A,B and C are temperature dependent calibration coefficients of the pressure sensitive paint formulation, which can be determined in laboratory or pressurized wind tunnel and I_{ref} is the corresponding intensity value for a constant reference pressure. Thus, using a ratio of images taken at two pressure conditions ("wind-on" and "wind-off") allows the determination of static pressures over the surface of interest. Problems occurring when PSP is applied using the intensity method are: model displacement, uncertainties in the temperature field on the model surface, excitation light instability, the spread of exposure time, and the luminescence light scattering on the adjacent model parts or test section walls (self illumination). Alignment of "wind off" and "wind on" images does not totally eliminate the influence of model displacement and deformation [3] and [7]. The influence of the temperature field uncertainties can be minimized by direct measurements of the temperature distributions on the model surface, by using PSP formulations with small temperature sensitivity, by estimation of the temperature field at the base of measured pressure fields and some calculations of heat flux on the model surface [4].

3 Experimental Setup

The measurements were performed in the high-speed wind tunnel at Merkers (HWK) of TU-Dresden in the salt mine environment, see Figure 1. The chamber for nozzle tests is integrated between the high pressure and the low pressure reservoir of the HWK, see cipher 4 in Figure 1. Dry cold gas was used. In contrast to the tests in the FESTIP program [1], two subsonic diffusers were newly installed at the entry to the low pressure reservoir, see cipher 8 in Figure 1. This increases the useable dynamic pressure ratio greatly. The maximum static pressure ratio between the low and high pressure sides is 1: 50. Because of ejector effects on the nozzle flow in the test chamber and the new subsonic diffusers, dynamic pressure ratios of 1: 250 are possible. The practicable pressure ratio depends on a constant mass flow rate controlled by the pipe diameter to the low pressure reservoir.

For the tests the plug nozzle model "LION" (**LI**near **O**ptimized Internal/External Expansion **N**ozzle) of length L3 was used. The LION contour is a linear plug nozzle contour with pre-expansion based on a modified contour calculation [5]. In Figure 2, a spatial drawing of the investigated plug nozzle including the system of coordinates and the definition of the direction of torque is shown. The values u and v are special variables for calculating the vector describing the plug surface. Thrust vector control can be simulated by partial blocking of the pre-expansion module which is divided into three parts, see Figure 3. Blocked module parts are presented black. All configurations have been investigated for nozzle pressure ratios p_c/p_a of 30, 60 and 110. The special case is configuration C2. This configuration was investigated without blocked modules but with side fences. The height of the module exhaust is 22.7 mm. The height of the side fences is $h_e/2$ = 11.35 mm normal to the wall contour.

The accuracy of the thrust vector analysis using scanivalve data is determined by the number and the location of pressure taps. Because the majority of flow fields are symmetrical to the symmetry plane, a symmetrical instrumentation of pressure taps was used.

For PSP tests, a wall plate element was built for the fast implementation of the excitation illuminators and CCD camera observation. The B1 binary paint formulation from Optrod Ltd., based on pyrene as luminophore, was used for the measurements. This paint consists of a screen layer, adhesive layer and an active layer with a total thickness about 50 μm and a response time of 0.5s. The paint was applied directly in the test section of HWK with a spray-gun. The emission of the PSP depends on the excitation intensity. That means that non-uniformity of excitation light and model movement or model deformations relative to the light source will lead directly to an error in the pressure measurements. To minimize this error a binary paint composition is necessary. Binary composition provides the possibility to simultaneously acquire a reference image, which contains information about the local excitation intensity. Using a Xenon flash-lamp light source for excitation and scientific-grade CCD cameras, exposure times of approximately 5s can be realized for each pair of "pressure/reference" images. These images were acquired simultaneously by two cameras. The model surface temperature during each run was estimated to be 21°C. Pyrene-based PSP formulations have a temperature coefficient in the range of 0.3 to 0.5 %/°C. Uncertainties of 1°C in the temperature distribution on the model surface will create errors of about 10 mbar. Therefore one of the most important remaining systematic errors for the PSP measurements comes from temperature influences.

4 Method of solution – thrust and calculation of torque

The thrust S can be calculated by using the impulse and pressure force in the exhaust area of the nozzle, or the impulse and the pressure force at the throat plus the pressure force at the nozzle wall surface. The first method is somewhat complicated, and because of the requirement to measure the velocity vector at the nozzle exhaust area, generally not as exact as the second method. For this special problem it is difficult to define a nozzle exhaust area, therefore, we have to use the second method. The entry to the area of control is the throat, where the flow data and the geometry are both well known. The thrust producing surface components are presented and defined in Figure 4.

Thrust force calculations depend on the total pressure. For comparison and discussion of results between different configurations a dimensionless number is needed. The thrust coefficient $c_f = S/(p_0 * A_t)$ is independent of the total pressure and will be used for comparison of results between different configurations.

The following method of solution was used [11]:

- Calculation of thrust component in the throat
- Description of the surface through a vector, determination of tangential and normal vectors
- Calculation of thrust component of plug and plug side module surface forces by:
  - Interpolation of pressure values at grid points
  - Description of the function p(u,v) through linear interpolation
  - Calculation of force vector
- Calculation of thrust component on module 2 surface
- Calculation of thrust component of the bottom
- Assumptions for calculating the thrust of configurations C5, C6:
  - On the underside of the plug, for equal pressure ratios (p_c/p_a), the Mach numbers are equal to those for configuration C2
  - For configuration C6 the wall pressure values on the plug side surface of the module 1 (see Figure 4) are calculated by using the average pressure ratios for the configurations C3, C4 and C5

The configurations C3 ... C6 produce torque which can be used for control. In aeronautics and aerospace engineering the torque is indicated by the names of axes. For calculation of the torque, the components of each force vector are multiplied by the distance to the corresponding axis. The force vectors are located at the centre of gravity of each small integration area.

5 Results

For the data reduction of the PSP images, the newly developed DLR-software package "ToPas" has been used. ToPas has the following features:

1. 2D and 3D analysis software based on UNIX and OpenGL.
2. Automatic marker recognition using improved cross correlation method.
3. All 2D and 3D functions are available interactively and in batch mode.

4. Very high degree of automation.
5. Stable numerical 3D alignment, including perspective and lens correction.
6. Fast processing. 2D and 3D results are presented immediately after measurement.

A typical PSP result for the investigated nozzle model for Configuration 2 is given in Figure 5. Figure 5 shows quantitatively the pressure distribution on the PSP coated area of the nozzle model. The comparison of the measured static pressure values using the PSP technique and the conventional pressure tap technique (Scanivalve) shows a good agreement.

Figure 6 presents the thrust coefficient contribution of each thrust producing nozzle component. The main thrust component is the impulse and pressure force at the throat. The second highest portion is the thrust component from the pressure forces on the module 1 and module 2 surfaces (see Figure 4). The component of the module 2 surface is negative, because the contour in this module 2 is negative (see Figure 4). The comparison between configuration C1 and C2 are of special interest. The side fences distinguish configuration C2. The coefficients at the throat and the module 1,2 surfaces are identical between C1 and C2 as expected. The thrust coefficient at the plug surface is increased by using side fences [10]. For this reason the total thrust coefficient for configuration C2 is higher than for C1. The force on the bottom is very small in all cases.

Configurations C3 and C4 cause a torque: a rolling torque in C3 and a yawing torque in C4. The rolling torques around the x-axis for the different pressure ratios are depicted in Figure 7. In C3 the rolling torque contributions of impulse and pressure force in the throat as well as the contribution from the module 2 surface act against the rolling torque contributions of the plug side module surface and the plug (see Figure 7) [10]. For decreasing pressure ratios p_c/p_a the contribution by the plug is also decreasing. For this reason we observed for configuration C3 that decreasing pressure ratios result in an increasing resultant torque.

6 Conclusion

The PSP technique has been successfully implemented in the high-speed wind tunnel at Merkers (HWK) of TU-Dresden. The newly developed DLR-software package "ToPas" has been used. The plug nozzle with the "LION" contour of length L3 was investigated. The paper gives a summary of the thrust vector calculations based on Scanivalve pressure results. The qualitative improvement in axial thrust by using side fences (between C1 and C2) is shown now quantitatively. The flow fields of configurations C5 and C6 are asymmetric with respect to the x,z-plane. Based on consistent assumptions it was possible to calculate thrust vector control, in spite of missing knowledge of the pressure distribution at the uninstrumented plug side (lower side).

It is now very interesting to compare these Scanivalve based results to results based on PSP measurements. The evaluation of PSP data will be done by SNECMA/Vernon. Initial analysis shows good correspondence between thrust vector results based on Scanivalve and PSP data.

Acknowledgements

We would like to thank ESA (ESTEC), Noordwijk for their support. Prime contractor of these ESTEC financed investigations was Astrium, Ottobrunn. The authors would like to thank in particular Dr. Gerald Hagemann from Astrium and Dr. Richard Schwane from the European Space Agency.

References

[1] Detsch, F., Niebergall, S. (2001). *Status of HWK Plug Nozzle Investigations*, in Heller, W., Klingenberg, J. (eds.) Advances in Fluid Mechanics. ISBN 3-86005-276-4.

[2] Engler, R.H., Hartmann, K., Troyanovski, I. & Vollan, A. (1992). *Description and assessment of a new optical pressure measurement system (OPMS) demonstrated in the high speed wind tunnel of DLR in Göttingen*, DLR-FB 92-24.

[3] Engler, R.H. & Klein, Chr. (1997). *First Results Using the New DLR PSP System –Intensity and Lifetime Measurements*, Conference "Wind Tunnels and Wind Tunnel Test Techniques", The Royal Aeronautical Society, Cambridge UK, ISBN 185768 048 0, 43.1-43.9.

[4] Fonov, S.D., Radchenko, V.N. & Mosharov, V.E. (1997). *Luminescent Pressure Sensors in Aerodynamic Experiments*, TsAGI's books, No. 8005.

[5] Hagemann, G., Immich, H., Terhardt, M. (1998). *Flow Phenomena in Advanced Rocket Nozzels – The Plug Nozzle*, AIAA Paper 98-3522.

[6] Kautsky, H. & Hirsch, A. (1935). *Nachweis geringster Sauerstoffmengen durch Phosphoreszenztilgung*, Z. f. anorg. und allg. Chemie, Band 222, 126-134.

[7] Le Sant, Y., Delegise, B., Mebarki, Y. & Merienne, M-C. (1997). *An Automatic Image Alignment Method Applied to Pressure Sensitive Paint Measurements*, 17th ICIASF Congress, Monterey, CA., USA, IEEE publication 97CH36121, 57-65.

[8] McLachlan, B.G., Kavandi, J.L., Callis, J.B., Gouterman, M., Green, E. & Khalil, G. (1993). *Surface Pressure Field Mapping Using Luminescent Coatings*, Experiments in Fluids 14, 33-41.

[9] Morris, M.J., Donovan, J.K. Kegelmann, J.T., Schwab, S.D., Levy, R.L. & Crites, R.C. (1993). *Aerodynamic Applications of Pressure Sensitive Paint*, AIAA Journal, Vol. 31, No.3, 419-425.

[10] Niebergall, S. (2002). *Investigation of Linear Plug Nozzle Flows – Thrust Vector Control – PSP Tests*, ESA ESTEC Contract No. 15412/01/NL/PA, Technical Report No. 5.

[11] Wohlleben, M. (2002). *Thrust Vector Calculation on Linear Plug Nozzle Configurations*, Master Thesis. TU Dresden, High Speed Wind Tunnel Laboratory.

Figures

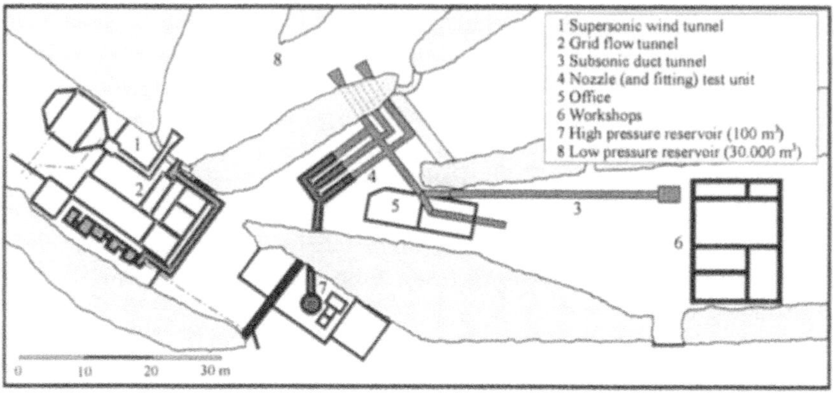

Figure 1 Ground plan of the HWK laboratory of TU Dresden in the salt mine in Merkers/Thuringia.

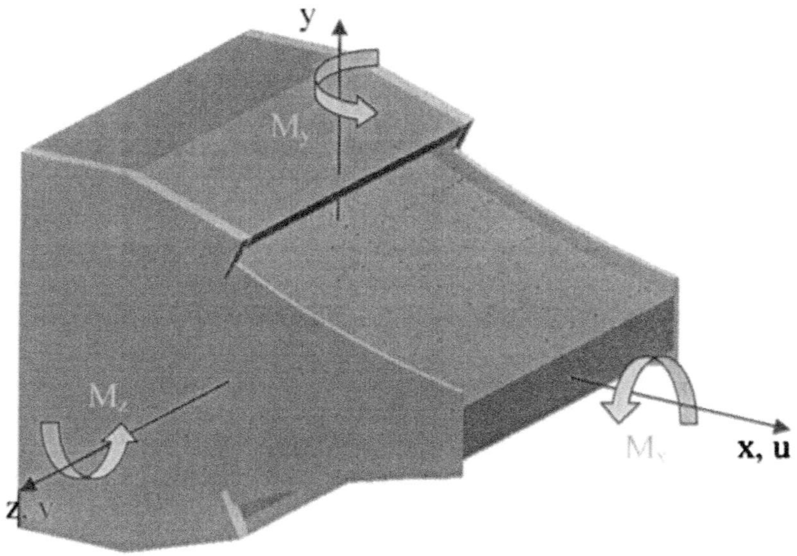

Figure 2 Plug nozzle configuration C2

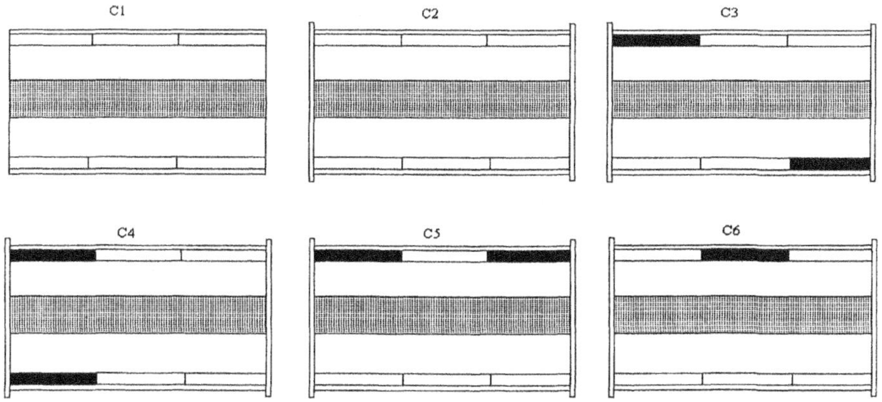

Figure 3 Investigated configurations

$$S = S_{At} + S_{module1} + S_{module2} + S_{plug} + S_{bottom} - S_u$$

S_{At} : Thrust component: impulse and pressure force at the throat (vena contracta)
$S_{module1}$: Thrust component: Force on plugside module surface
$S_{module2}$: Thrust component: Force on outer module surface
S_{plug} : Thrust component: $\sum F_{plug}$
S_{bottom} : Thrust component: $\sum F_{bottom}$
S_u : Thrust component of ambient pressure:
- acts against the other thrust components
- consideration in calculation through subtraction of ambient pressure

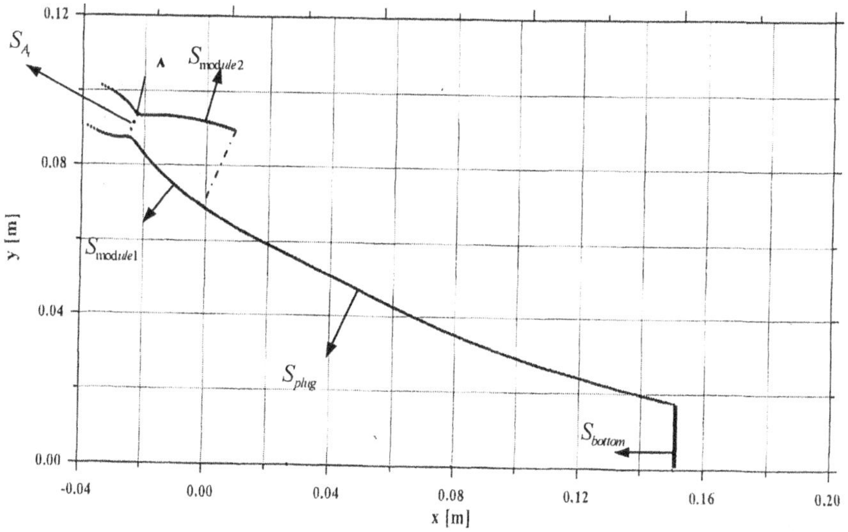

Figure 4 Thrust producing surface components

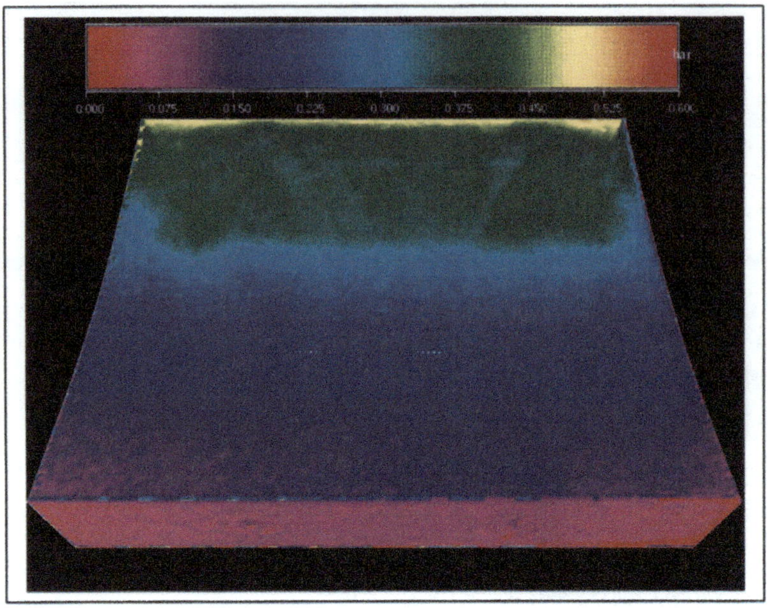

Figure 5 Visualization of the pressure distribution on the nozzle

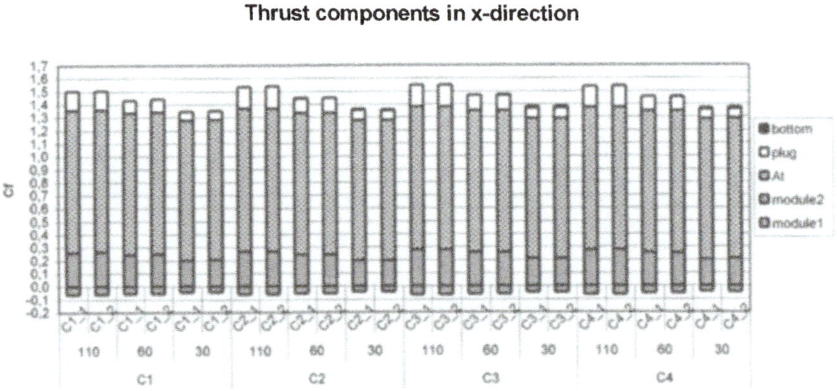

Figure 6 Thrust coefficient of Configuration C1 … C4

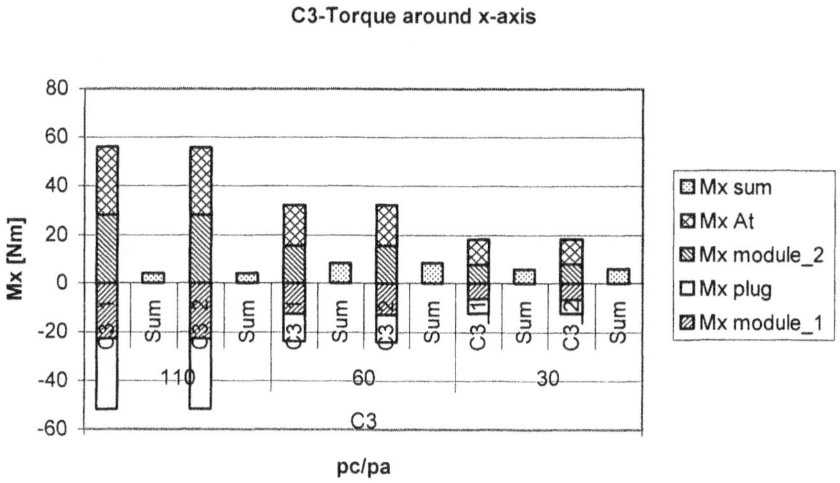

Figure 7 Torque components around the x-axis of C3

Application of an Air Collection and Enrichment System aboard a TSTO Space Transport

S. LENTZ, W. STAUDACHER
University of the German Armed Forces Munich
Institute for Aeronautics
Werner-Heisenberg-Weg 39
D-85577 Neubiberg, Germany

Summary

Main Focus is the investigation of in-flight LOX-collection aboard a (**T**wo-**S**tage-**T**o-**O**rbit) TSTO reusable space transport very similar to the SÄNGER concept. In previous research activities a suitable design point for vehicle and LOX-collection system has been determined. Starting from this, variations of essential design parameters are discussed to identify and evaluate influences and risks on the complete system.

1. Introduction

Collection and enrichment of aerial oxygen during atmospherical ascent of reusable winged space transportation has its origins in the sixties of the last century. Concerning a reorientation towards fully reusable TSTO launch vehicles LOX-collection may have the potential to reduce development and operational costs and hence can give rise of attractiveness for commercial space transportation. Application of an ACES (**A**ir-**C**ollection-and-**E**nrichment-**S**ystem) provides significant reductions in take-off gross weight due to empty LOX-tanks of the orbital stage. However, the collection device represents an additional system causing an increase in complexity and affects the vehicle during operation. Identification of risks and cost drivers for the design of space transports using in-flight LOX-collection are essential. Therefore, sensitivity analysis needs to be performed. At the Institute for Aeronautics at the University of the German Armed Forces the design tool CADEHYP can be used for such tasks. It resembles a numerical code (upgrade of the industrial version developed by the second author and used for the numerical design process of the SÄNGER-TSTO) which is scaling a so called basic-configuration along a dedicated mission-profile. By means of parameter-variations (e.g. overall-weight and drag as well as fuel consumption and thrust) effects on the vehicle can be examined. This procedure is most notably reasonable especially in the early design phase, when necessary system-components like engines or now the ACES are not well defined [1, 2, 3, 4].

2. Benefits and Risks using in flight LOX-collection

Operation of an ACES – as primary effect – provides significant reduction in take-off gross weight (TOGW) of the space transport due to empty LOX-tanks of the orbital stage. This may result in structural and empty weight savings particularly for the first stage. The required oxygen for the upper stage does not need to be accelerated at the beginning of the mission. It can be collected at comparatively high velocities [3, 4].

Another benefit is the possibility to confine the use of airbreathing propulsion to Mach numbers of 4 or even less with pure turbojets instead of combined cycle engines. Using LH_2 for cooling and liquefaction of oxygen prior to generating thrust could yield to additional synergy-effects.

However, the aforementioned potentials cause an increase in complexity of the first stage due to application of an additional system besides the engines. Further, ACES is critical to overall success – failure means mission abort.

Concerning operation, such a device directly influences the so-called food chain of drag, thrust, fuel consumption and vehicle volume (see figure 1). Enlarging one of these attributes will cause the others to rise and hence will increase vehicle weight. As worst case this can lead to a divergence of the entire design.

3. Basic-Configuration and Initial Conditions

The basic-configuration is a SÄNGER-like TSTO with a turbo-ramjet as propulsion system. Scaling of the vehicle is performed along the original distance mission (2280 km) starting in south or central Europe [1]. LOX-collection takes place during supersonic cruise at M = 2.5 and a dynamic pressure of q = 50 kPa (optimal ACES operating conditions). Throughout acceleration to a separation-Mach number of 6.8 engine-mode is switched to ramjet operation above M = 3.5. Cruise back is carried out at M = 4.5 and descent occurs along a q = 30 kPa trajectory.

For a SÄNGER-like TSTO staging at a Mach number of 6.8 the amount of LOX stored in the orbiter is 80 tons (see table 1). Concerning collection during a supersonic cruise at M = 2.5 the ACES was scaled for an airflow of 256 kg/s_{Air}. With respect to a specific weight range of 10-15 $kg/kg/s_{Air}$ and specific volume range of 0.12-0.16 $m^3/kg/s_{Air}$ the data given in table 2 are considered for the device [5].

In previous investigations it has been proven that significant weight reductions are achievable. Compared to a vehicle without LOX-collection device, 75 tons of TOGW and 10 tons of empty weight savings were obtained. Data for the design point of both the vehicle and the ACES are given in table 1 and table 2 [5].

4. ACES Influences on the Vehicle

To perform sensitivity analysis vehicle influencing variables of the ACES have to be defined. For this the mode of operation will be outlined at first.

4.1 ACES Mode of Operation

The ACES is based on the chemical engineering process of oxygen extraction using liquid hydrogen as coolant [3, 4]. It consists of an arrangement of cryogenic heat exchangers, condensers, turbo components and the separation device [5]. According to figure 2 during operation air enters the vehicle via an intake (I) and is decelerated. An impulse is generated acting contrary to flight-direction thus resulting in additional drag. The air flow is cooled down to saturation conditions where it partly condenses. Therefore a certain amount of hydrogen is necessary, which is – depending on design of the ACES – above or below consumption of the engines. Oxygen is separated and stored in the LOX-tanks of the upper stage. During this time vehicle weight is increasing about the amount of LOX-mass minus the spent LH_2 mass. Finally nitrogen enriched air is expelled through a nozzle (N) for drag compensation.

4.2 Influencing Variables

Concerning the aforementioned mode of operation following areas of influence can be defined. First of all these are mass and volume of the device and related components which are assembled into the vehicle. During operation these are drag and fuel consumption. Additionally the purity grade of the extracted oxygen is related to the specific impulse of the upper stage rocket engines. A 1% dilution of oxygen-purity diminishes specific impulse of LH_2/LOX rocket engines by 1.5 sec. Compared to ground-based oxygen extraction facilities, ACES can reach a purity of 90% - 98%. If specific impulse of rocket engines decreases due to impure oxygen, more LOX and LH_2 are required to meet the Δv-demand of the upper stage. Orbiter size will increase which results in additional weight and drag for the complete system [3].

For sensitivity studies the following design parameters of the ACES come into consideration:

- weight
- volume
- drag
- LH_2-consumption
- LOX purity grade (variation of orbiter size)

5. Sensitivity Analysis

As mentioned above sensitivity analysis can be performed to examine vehicle behaviour on possible changes of design parameters of both the complete system and critical components. The ACES parameters will be varied around the given design point (see table 2). The resulting relative TOGW changes will be described over relative design parameter variations of the ACES.

In figure 3 the parameters weight, volume and LOX-purity are displayed. Increasing volume or weight of the collection device results in a TOGW rise of the complete system. The slope of the weight parameter is steeper than that of volume. This means that mass changes have a more detrimental influence on the vehicle than volume changes. Due to complex coherences regarding variations in the LOX-purity (orbiter size is related to drag, weight and other parameters) different aspects of the vehicle are affected which result in the most significant influence.

Sensitivities of drag and LH_2-consumption are shown in figure 4. Compared to the other three parameters both indicate a discontinuity in the design point. Increasing values result in a rise of TOGW of the complete system, but reductions display no influence. Both ACES and engines are connected via hydrogen as coolant and fuel respectively. In the design point they consume equal amounts of fuel and hence are operating at optimum conditions from the vehicle point of view. If ACES-consumption exceeds that of the turbojets, LH_2 will be expelled unused (in regard to thrust generation) and consequently will be wasted. For the opposite condition (engine consumption prevails) all the hydrogen is used for thrust generation and no influence on the vehicle can be observed. Considering ACES drag (already described in figure 1) a rise provokes a need for more thrust and thus for more fuel. As long as the engines consume more LH_2 than the ACES it will affect TOGW of the vehicle. Since hydrogen demand of the LOX-collection device remains constant, further drag reductions again lead to a waste of fuel because not all LH_2 is required for thrust generation.

The influences on the vehicle can now be determined by use of sensitivity factors.

$$X_s\,[-] = \left(\frac{\Delta X\,[\%]}{\Delta s\,[\%]} \right)_{DP}$$

X_s resembles the approximate derivative of a parameter X of the complete vehicle (e.g. TOGW or empty vehicle) with respect to an increment Δs of a specific parameter s of the considered system (ACES, propulsion system or vehicle) in the design point (DP). The magnitude of X_s provides an indicator for the level of influence on the complete system (the higher, the worse).

In table 3 the sensitivity factors are given. LOX-purity is the most influential parameter due to its complex relation towards the size of the orbital stage. LH_2-consumption and drag are additional influential variables. Both provoke a rise in overall fuel demand which has to be stored, carried and accelerated by the vehicle and hence enlarge tank volume, weight, drag and so forth. Thus they are called critical sensitivities in contrast to dead weight variations which are non critical.

6. Conclusion

In-flight LOX-collection aboard a TSTO reusable space transport can provide TOGW and empty weight savings. However these are strongly dependent on performance parameters of the ACES. The most sensitive parameter is the purity of the extracted LOX because it directly affects specific impulse and hence size of the upper stage. A scaled up orbital stage does not only stand for a simple rise in weight and drag of the complete system it also means a higher design risk (primarily for the orbiter but also for the complete system) and can yield to conceptual problems concerning stage integration.

The next influential parameter is the LH_2-consumption or (expressed as performance parameter of the ACES) the collection ratio Γ. Since ACES and engines are connected via hydrogen in the manner described above, only deviations from the design point in terms of increasing consumption affect TOGW. The same applies for the drag of the LOX-collection device; only an increase influences the vehicle. Noncritical in contrast are deviations in weight and volume of the ACES.

Regarding the design process of a space transportation system performance parameters of the ACES in particular LOX-purity, collection ratio and drag should be well known at an early stage. However when ACES is frozen, deviations from the design point will have severe influences on the complete system. This will yield to higher risks and costs.

7. References

[1] N.N.: Leitkonzept Sänger, Referenz-Daten-Buch, interner Bericht, MBB, 1993

[2] M. Hornung, S. Lentz, W. Staudacher: Arbeits- und Ergebnisbericht des Teilbereichs C8, SFB 255, Transatmosphärische Flugsysteme, Arbeits- und Ergebnisbericht Juli 1998 – Juni 2002, Technische Universität München, Universität der Bundeswehr München, 2001, pp. 629-668

[3] J.L. Leingang, J.Q. Maurice, L.R. Carreiro: In Flight Oxidizer Collection System for Airbreathing Space Boosters, in S.N.B. Murthy, E.T. Curran (eds.) Developements in High-Speed-Vehicle Propulsion Systems, Progress in Astronautics and Aeronautics, Volume 165, AIAA, 1996, pp. 333 – 384

[4] V.V. Balepin: Air Collection Systems, in S.N.B. Murthy, E.T. Curran (eds.) Developments in High-Speed-Vehicle Propulsion Systems, Progress in Astronautics and Aeronautics, Volume 165, AIAA, 1996, pp. 385 – 420

[5] S. Lentz, W. Staudacher: Entwurf eines wiederverwendbaren zweistufigen Raumtransportsystems mit einer Anlage zur Sauerstoffgewinnung während des Fluges, DGLR Jahrestagung 2002 Stuttgart 23.09–26.09, DGLR-JT2002-180, 2002

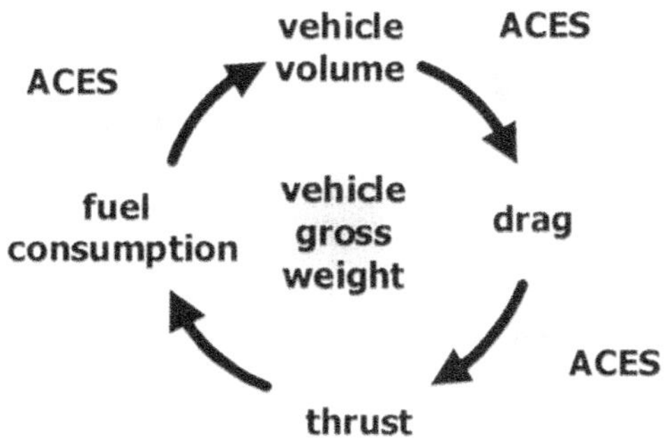

Figure 1 Food chain and influence of ACES.

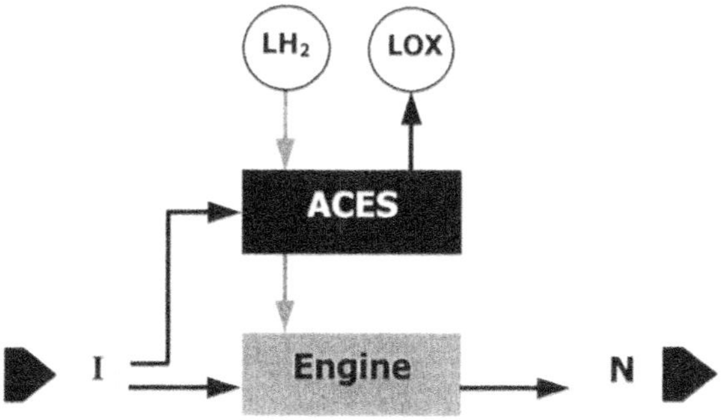

Figure 2 Schematic of ACES and involved mass flows.

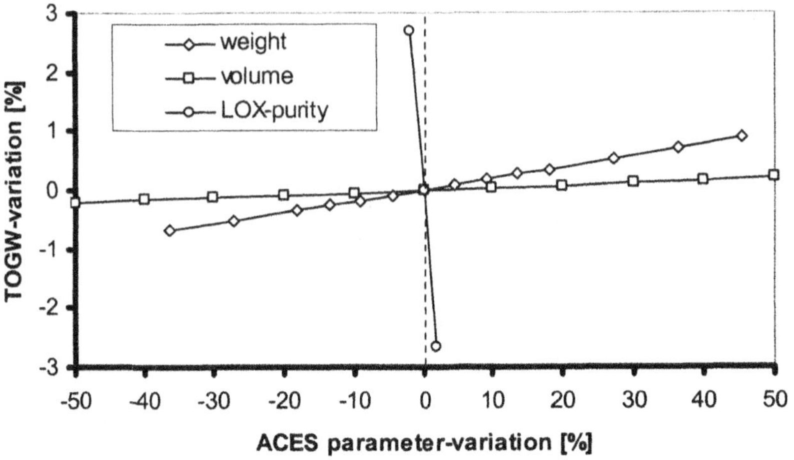

Figure 3 Parameter variation of weight, volume and LOX-purity.

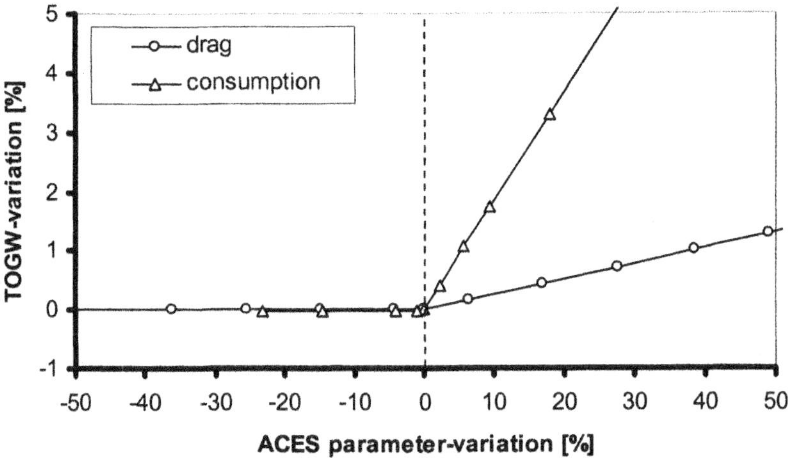

Figure 4 Parameter variation of drag and LH_2 consumption.

Table 1 Design point weight data of the TSTO space transport without ACES (left) and with ACES (right) [5]
(TO: Orbiter weight at take off; SEP: Orbiter weight at stage separation).

Parameter	w/o ACES	with ACES
Empty weight	164 to	149 to
Fuel weight	95 to	107 to
Op. reserves	13 to	14 to
Orbiter (TO)	115 to	40 to
Orbiter (SEP)	115 to	120 to
Collected LOX weight	-	80 to
TOGW	**387 to**	**310 to**

Table 2 Design point data of ACES and bandwidth of parameter variation [5].

Parameter	Value	Variation
Weight	3185 kg	- 38 % to 45 %
Volume	40 m^3	- 50 % to 50 %
Collection Ratio $\Gamma = \dot{m}_{LOX}/\dot{m}_{LH_2}$	3.0	
LH_2-consumption	18.2 kg/s	- 25 % to 50 %
LOX-concentration	98 %	- 2 % to 2 %
ACES-Drag	96 kN	- 38 % to 50 %

Table 3 Sensitivity factors based on ACES parameter variations with respect to TOGW of the vehicle (*only for increasing values).

Parameter	Sensitivity [-]
LOX-purity	-1.31
LH_2 consumption	0.183*
Drag	0.026*
Weight	0.019
Volume	0.004

The Small Disturbance Navier-Stokes Equations: Development of an Efficient Method for Calculating Unsteady Air Loads

A. Pechloff, M. Iatrou, C. Weishäupl and B. Laschka

Lehrstuhl für Fluidmechanik - Technische Universität München
Boltzmannstraße 15 - 85747 Garching - Germany
alexander@flm.mw.tum.de

Summary

A concise derivation of the laminar small disturbance Navier-Stokes equations is presented and their numerical implementation described. The method's validity is shown for a harmonically pitching NACA 0012 airfoil in subsonic flow as an initial test case.

1 Introduction

With the time accurate numerical solution of the Reynolds-averaged Navier-Stokes equations an adequate means for computing unsteady viscous flow around aircraft wings exists today. However, even when employing the resources of modern super-computers this method remains time consuming, making it an expensive computational approach. Especially the aeroelastic analysis process relies on the evaluation of unsteady air loads for a multitude of parameter variations, necessitating more efficient solvers. So far the problem of computational cost has been circumvented by using linear potential or, as of late, small disturbance Euler methods. Yet, if the flow field is dominated by viscous phenomena such as flow separation or shock boundary layer interaction, then correct predictions fail with these otherwise efficient solvers. Therefore, it is reasonable to extend an existing small disturbance Euler method, which allows the high quality resolution of non-linearities in the transonic regime, to viscous flows. The small disturbance Euler solver FLM-SDEu [3] developed at the Lehrstuhl für Fluidmechanik of the Technische Universität München will serve as the basis for incorporating the additional viscous terms.

2 Fundamentals

2.1 The Triple Decomposition

The small disturbance Navier-Stokes equations are derived through the so called triple decomposition [5]. The unsteady flow is separated into a steady mean flow, a harmonic time dependent part and a turbulent one. Let us consider an arbitrary instantaneous flow quantity Φ and its decomposition, with $\bar{}$ denoting the mean, $\tilde{}$ and $'$ distinguishing the harmonic and turbulent quantity, respectively:

$$\Phi(x,y,t) = <\Phi(x,y,t)> + \Phi'(x,y,t) = \bar{\Phi}(x,y) + \tilde{\Phi}(x,y,t) + \Phi'(x,y,t) \quad (1)$$

By inserting the decomposed quantity into the instantaneous Navier-Stokes equations and separately applying the phase average

$$<\Phi(x,y,t)> = \lim_{N\to\infty} \frac{1}{N} \sum_{n=1}^{N} \Phi(x,y,t+n\cdot T) \quad (2)$$

and the time average

$$\bar{\Phi}(x,y) = \frac{1}{T} \int_{t}^{t+T} \Phi(x,y,\tau) d\tau \,, \quad (3)$$

under consideration of certain mathematical identities [6], we obtain two sets of equations. One governs the unsteady flow, with turbulence having been averaged out, while the other describes the steady mean flow. Subtracting the phase average from the time average yields the governing equation for the time dependent harmonic fluctuation, or as symbolized with the arbitrary flow quantity:

$$\tilde{\Phi}(x,y,t) = <\Phi(x,y,t)> - \bar{\Phi}(x,y) \quad (4)$$

Both phase and time averaging allow the isolation of Reynolds stress terms within the respective convective flux. By subtracting the time averaged term from the phase averaged term, an oscillating component of the Reynolds stress can be obtained. The system of differential equations can be closed by approximating this new correlation term under Boussinesq's assumption. Introducing an eddy viscosity the Reynolds stress is expressed through the mean and harmonic fluctuation of the velocity gradients [6]. Consequently, the perturbed shear stress tensor can be separated into a laminar and turbulent part, for example component $\tilde{\tau}_{xy}$:

$$\tilde{\tau}_{xy} = (\bar{\mu}_{\mathrm{l}} + \bar{\mu}_{\mathrm{t}})\left(\frac{\partial \tilde{v}}{\partial x} + \frac{\partial \tilde{u}}{\partial y}\right) + (\tilde{\mu}_{\mathrm{l}} + \tilde{\mu}_{\mathrm{t}})\left(\frac{\partial \bar{v}}{\partial x} + \frac{\partial \bar{u}}{\partial y}\right) + (\tilde{\mu}_{\mathrm{l}} + \tilde{\mu}_{\mathrm{t}})\left(\frac{\partial \tilde{v}}{\partial x} + \frac{\partial \tilde{u}}{\partial y}\right) \quad (5)$$

In the following only the set of laminar equations will be considered. The triple decomposition becomes a valid method of describing fluid flow, when a known harmonic motion of a wing or body imposes an oscillatory component on the flow field development and higher harmonics of the response are deemed negligible. As the frequency of the motion is known, the fluctuating quantity can be modeled through the harmonic function.

2.2 The Governing Equations of the Perturbed Flow

Applying the triple decomposition to the Navier-Stokes equations yields the strong conservation form of the perturbed flow governing equations. As it is intended to use these equations in a computational method treating harmonic rigid body oscillations and eigenmodes, a formulation in a curvilinear body fitted coordinate system is adopted:

$$\frac{\partial \tilde{\mathbf{Q}}}{\partial \tau} + \frac{\partial \tilde{\mathbf{F}}}{\partial \xi} + \frac{\partial \tilde{\mathbf{G}}}{\partial \eta} = \frac{\partial \tilde{\mathbf{F}}_{\mathbf{v}}}{\partial \xi} + \frac{\partial \tilde{\mathbf{G}}_{\mathbf{v}}}{\partial \eta} \tag{6}$$

$\tilde{\mathbf{Q}}$ denotes the unsteady vector of the harmonically perturbed conservative variables within the transformed system, while $\tilde{\mathbf{F}}$, $\tilde{\mathbf{G}}$ represent the convective and $\tilde{\mathbf{F}}_{\mathbf{v}}$, $\tilde{\mathbf{G}}_{\mathbf{v}}$ the viscous fluxes of the perturbed flow in ξ- and η-direction, respectively. As the transformation rule between the cartesian and curvilinear coordinate system is supplied by the metrics [3], an unsteady component is introduced through the body's harmonic motion and the resulting deformation of the computational grid. Therefore, the motion of the spatial coordinate for an arbitrary grid point can be decomposed in a mean and harmonically perturbed component:

$$x(\xi,\eta,\tau) = \bar{x}(\xi,\eta) + \tilde{x}(\xi,\eta,\tau)\,, \qquad y(\xi,\eta,\tau) = \bar{y}(\xi,\eta) + \tilde{y}(\xi,\eta,\tau) \tag{7}$$

By formulating the determinant J of the coordinate transformation's Jacobian, the volume V_{xy} of a computational cell is obtained and decomposed for the unsteady motion:

$$J = \det\left(\frac{\partial(x,y,t)}{\partial(\xi,\eta,\tau)}\right) = V_{\mathrm{xy}}\,, \qquad J = \bar{J} + \tilde{J} \tag{8}$$

Consequently, the contravariant metric terms, which represent the spatial and time derivative components of the cell face normal vectors can be written as:

$$J\xi_{x,y,t} = \overline{J\xi_{x,y,t}} + \widetilde{J\xi_{x,y,t}}\,, \qquad J\eta_{x,y,t} = \overline{J\eta_{x,y,t}} + \widetilde{J\eta_{x,y,t}} \tag{9}$$

This allows a further decomposition of the transformed state and flux vectors of the perturbed flow, establishing connectivity between the cartesian and curvilinear formulation of the governing equations:

$$\begin{aligned}
\tilde{\mathbf{Q}} &= \bar{J}\tilde{\mathbf{q}} + \tilde{J}\bar{\mathbf{q}} + \tilde{J}\tilde{\mathbf{q}} \\
\tilde{\mathbf{F}} &= \overline{J\xi_t}\tilde{\mathbf{q}} + \overline{J\xi_x}\tilde{\mathbf{f}} + \overline{J\xi_y}\tilde{\mathbf{g}} + \widetilde{J\xi_t}\bar{\mathbf{q}} + \widetilde{J\xi_x}\bar{\mathbf{f}} + \widetilde{J\xi_y}\bar{\mathbf{g}} + \widetilde{J\xi_t}\tilde{\mathbf{q}} + \widetilde{J\xi_x}\tilde{\mathbf{f}} + \widetilde{J\xi_y}\tilde{\mathbf{g}} \\
\tilde{\mathbf{G}} &= \overline{J\eta_t}\tilde{\mathbf{q}} + \overline{J\eta_x}\tilde{\mathbf{f}} + \overline{J\eta_y}\tilde{\mathbf{g}} + \widetilde{J\eta_t}\bar{\mathbf{q}} + \widetilde{J\eta_x}\bar{\mathbf{f}} + \widetilde{J\eta_y}\bar{\mathbf{g}} + \widetilde{J\eta_t}\tilde{\mathbf{q}} + \widetilde{J\eta_x}\tilde{\mathbf{f}} + \widetilde{J\eta_y}\tilde{\mathbf{g}} \\
\tilde{\mathbf{F}}_{\mathbf{v}} &= \overline{J\xi_x}\tilde{\mathbf{f}}_{\mathbf{v}} + \overline{J\xi_y}\tilde{\mathbf{g}}_{\mathbf{v}} + \widetilde{J\xi_x}\bar{\mathbf{f}}_{\mathbf{v}} + \widetilde{J\xi_y}\bar{\mathbf{g}}_{\mathbf{v}} + \widetilde{J\xi_x}\tilde{\mathbf{f}}_{\mathbf{v}} + \widetilde{J\xi_y}\tilde{\mathbf{g}}_{\mathbf{v}} \\
\tilde{\mathbf{G}}_{\mathbf{v}} &= \overline{J\eta_x}\tilde{\mathbf{f}}_{\mathbf{v}} + \overline{J\eta_y}\tilde{\mathbf{g}}_{\mathbf{v}} + \widetilde{J\eta_x}\bar{\mathbf{f}}_{\mathbf{v}} + \widetilde{J\eta_y}\bar{\mathbf{g}}_{\mathbf{v}} + \widetilde{J\eta_x}\tilde{\mathbf{f}}_{\mathbf{v}} + \widetilde{J\eta_y}\tilde{\mathbf{g}}_{\mathbf{v}}
\end{aligned} \tag{10}$$

It becomes apparent, that not only the cartesian state and flux vectors for the perturbed flow ($\tilde{\mathbf{q}}$, $\tilde{\mathbf{f}}$, $\tilde{\mathbf{g}}$, $\tilde{\mathbf{f}}_{\mathbf{v}}$, $\tilde{\mathbf{g}}_{\mathbf{v}}$), but also their mean flow counterparts ($\bar{\mathbf{q}}$, $\bar{\mathbf{f}}$, $\bar{\mathbf{g}}$, $\bar{\mathbf{f}}_{\mathbf{v}}$, $\bar{\mathbf{g}}_{\mathbf{v}}$) contribute to the terms of Equation (6) through the unsteady part of the decomposed metric.

2.3 Linearization and Insertion of the Time Law

In the following the governing equations for the perturbed flow (6) are linearized for small disturbances (SD). As can be seen in (10) terms of second order appear within the transformed state and flux vector, where products of the perturbed metrics and cartesian vectors are formed. With regard to small perturbations introduced into the flow field by the body's motion, these terms are neglected and only linear perturbation terms remain. Under consideration that the temporal metrics $\overline{J\xi_t}$, $\overline{J\eta_t}$ of the mean, i.e. reference state, are non-existent, the state vector and fluxes in the curvilinear coordinate system are collected into two groups. The first group, denoted by $^{(1)}$, summarizes terms containing only the perturbed flow quantities, while group $^{(2)}$ retains the terms formed with the perturbed metrics:

$$
\begin{aligned}
\tilde{\mathbf{Q}}^{(1)} &= \bar{J}\tilde{\mathbf{q}}^*\,, & \tilde{\mathbf{Q}}^{(2)} &= \tilde{J}\bar{\mathbf{q}}^* \\
\tilde{\mathbf{F}}^{(1)} &= \overline{J\xi_x}\tilde{\mathbf{f}}^* + \overline{J\xi_y}\tilde{\mathbf{g}}^*\,, & \tilde{\mathbf{F}}^{(2)} &= \widetilde{J\xi_t}\bar{\mathbf{q}}^* + \widetilde{J\xi_x}\bar{\mathbf{f}}^* + \widetilde{J\xi_y}\bar{\mathbf{g}}^* \\
\tilde{\mathbf{G}}^{(1)} &= \overline{J\eta_x}\tilde{\mathbf{f}}^* + \overline{J\eta_y}\tilde{\mathbf{g}}^*\,, & \tilde{\mathbf{G}}^{(2)} &= \widetilde{J\eta_t}\bar{\mathbf{q}}^* + \widetilde{J\eta_x}\bar{\mathbf{f}}^* + \widetilde{J\eta_y}\bar{\mathbf{g}}^* \\
\tilde{\mathbf{F}}_{\mathbf{v}}^{(1)} &= \overline{J\xi_x}\tilde{\mathbf{f}}_{\mathbf{v}}^* + \overline{J\xi_y}\tilde{\mathbf{g}}_{\mathbf{v}}^*\,, & \tilde{\mathbf{F}}_{\mathbf{v}}^{(2)} &= \widetilde{J\xi_x}\bar{\mathbf{f}}_{\mathbf{v}}^* + \widetilde{J\xi_y}\bar{\mathbf{g}}_{\mathbf{v}}^* \\
\tilde{\mathbf{G}}_{\mathbf{v}}^{(1)} &= \overline{J\eta_x}\tilde{\mathbf{f}}_{\mathbf{v}}^* + \overline{J\eta_y}\tilde{\mathbf{g}}_{\mathbf{v}}^*\,, & \tilde{\mathbf{G}}_{\mathbf{v}}^{(2)} &= \widetilde{J\eta_x}\bar{\mathbf{f}}_{\mathbf{v}}^* + \widetilde{J\eta_y}\bar{\mathbf{g}}_{\mathbf{v}}^*
\end{aligned}
\tag{11}
$$

Furthermore the cartesian vectors are linearized [6], as denoted by *. Reformulating the governing equations yields

$$
\begin{aligned}
&\frac{\partial\tilde{\mathbf{Q}}^{(1)}}{\partial\tau} + \frac{\partial(\tilde{\mathbf{F}}^{(1)} - \tilde{\mathbf{F}}_{\mathbf{v}}^{(1)})}{\partial\xi}) + \frac{\partial(\tilde{\mathbf{G}}^{(1)} - \tilde{\mathbf{G}}_{\mathbf{v}}^{(1)})}{\partial\eta} = \\
&-\left[\frac{\partial\tilde{\mathbf{Q}}^{(2)}}{\partial\tau} + \frac{\partial(\tilde{\mathbf{F}}^{(2)} - \tilde{\mathbf{F}}_{\mathbf{v}}^{(2)})}{\partial\xi} + \frac{\partial(\tilde{\mathbf{G}}^{(2)} - \tilde{\mathbf{G}}_{\mathbf{v}}^{(2)})}{\partial\eta}\right]\,,
\end{aligned}
\tag{12}
$$

with the $^{(2)}$ terms being isolated on the right hand side of the equation as an additional source term, only being dependent on the mean, i.e. reference, solution. Symbolically, the harmonic time law is defined as

$$
\tilde{\Phi}(x,y,\tau) = \hat{\Phi}(x,y)\cdot e^{ik\tau}\,,
\tag{13}
$$

$\hat{\Phi}$ representing the complex amplitude and k the non-dimensionalized frequency of the oscillation. On this basis the vectors and metrics can be decomposed, e.g.

$$
\tilde{\mathbf{Q}}^{(1)} = \hat{\mathbf{Q}}^{(1)}\cdot e^{ik\tau}\,,\ \ \tilde{\mathbf{F}}_{\mathbf{v}}^{(1)} = \hat{\mathbf{F}}_{\mathbf{v}}^{(1)}\cdot e^{ik\tau}\,,\ \ \widetilde{J\xi}_{x,y,t} = \widehat{J\xi}_{x,y,t}\cdot e^{ik\tau}\,,
\tag{14}
$$

and inserted into (12). Consequently, the unsteady part of the perturbation $e^{ik\tau}$ can be extracted and eliminated completely from the equation. Recasting the governing equations for the complex amplitude of the state vector $\hat{\mathbf{Q}}^{(1)}$

finally yields

$$\frac{\partial \hat{\mathbf{Q}}^{(1)}}{\partial \tau} + \frac{\partial (\hat{\mathbf{F}}^{(1)} - \tilde{\mathbf{F}}_{\mathbf{v}}^{(1)})}{\partial \xi}) + \frac{\partial (\hat{\mathbf{G}}^{(1)} - \tilde{\mathbf{G}}_{\mathbf{v}}^{(1)})}{\partial \eta} = \tag{15}$$
$$- \left[ik\hat{\mathbf{Q}}^{(1)} + ik\hat{\mathbf{Q}}^{(2)} + \frac{\partial (\hat{\mathbf{F}}^{(2)} - \hat{\mathbf{F}}_{\mathbf{v}}^{(2)})}{\partial \xi} + \frac{\partial (\hat{\mathbf{G}}^{(2)} - \hat{\mathbf{G}}_{\mathbf{v}}^{(2)})}{\partial \eta} \right] .$$

As $\hat{\mathbf{Q}}^{(1)}$ is not time dependent the derivative $\frac{\partial \tilde{\mathbf{Q}}^{(1)}}{\partial \tau}$ actually would disappear. However, it is kept in the formulation in order to retain the equations time hyperbolic character. With the derivation of the SD-Navier-Stokes equations an unsteady problem has been reduced to a steady one for the amplitude quantities. Coupling of the inphase (real) and out of phase (imaginary) part of the complex equation system occurs solely through the term $ik\hat{\mathbf{Q}}^{(1)}$.

3 Numerics

In constructing the algorithm for solving (15) special care has to be taken in order to retain the numerical properties of the reference Navier-Stokes code FLM-NS [1]. On one hand FLM-NS will supply the mean steady state solution of the perturbation problem to the small disturbance method FLM-SDNS, while on the other hand being employed to solve the non-linear Navier-Stokes equations in the time domain for the unsteady problem. As the finite volume approach, TVD character, Roe's convective flux difference splitting and Chakravarthy's method of calculating the viscous fluxes are the main numeric features of the reference solver FLM-NS, they are modified according to the small disturbance theory for use in FLM-SDNS. Especially Chakravarthy's application of the Gaussian integral theorem to shear stress evaluation is expanded under the assumption of small disturbances [2], making it a vital part of Equation (15) discrete formulation. Integration of the small disturbance Navier-Stokes equations in the frequency domain is accomplished with a new highly stable implicit scheme [7], which has been derived from the LU-SSOR method implemented in the reference solver. As a result of the equivalent integration and discretization properties comparability between the non-linear method and the small disturbance method is maintained. As the first step in the numerical solution process two computational grids are generated, one for the reference position of the wing and the other for the perturbed extremum position. In the reference grid the mean steady state flow is produced with FLM-NS and then used as initial solution for the non-linear computation of the unsteady problem. The required mesh deformation is realized through interpolation between the reference and the extremum position according to Equation (7). Through Fourier analysis the first harmonics are extracted from the complete unsteady solution during post-processing.

In contrast the FLM-SDNS solution procedure subtracts reference from extremum position in order to obtain the amplitudes of the coordinates and consequently the perturbed metrics. Together with the mean solution provided beforehand by FLM-NS the invariant part of the source term is calculated only once at the beginning of the process. Employing FLM-SDNS the amplitude of the unsteady flow solution is obtained directly in form of the perturbed state vector's real- and imaginary part.

4 Computational Results

For the initial phase of code validation the harmonic pitching motion of a NACA 0012 airfoil about an axis at 25% chord length ($x_{\mathrm{m,ref}}$), with $\alpha_0 = 0°$ as mean angle of attack and $\Delta\alpha = 1°$ as amplitude, is selected. The sinusoidal oscillation occurs with a reduced frequency of $k_{red} = 0.5$ at $Ma = 0.3$ and $Re = 2000$ freestream conditions. The computational domain is discretized using an elliptically smoothed structured mesh in C-topology composed of 384 cells in circumferential and 96 cells in normal direction. The off-body distance of the first grid line is $5 \cdot 10^{-4}$ chord lengths, guaranteeing a $y^+ < 0.5$ for the chosen Reynolds number. The deformed grid for the airfoil's upper dead center position ($\alpha_0 + \Delta\alpha$) retains all characteristics of the reference grid (α_0). The simulation of the unsteady problem with FLM-NS is performed by calculating three oscillations, each discretized with 100 physical time intervalls. This insures that transient phenomena have been eliminated and the flow field is periodic. Each physical time step is converged through dual-time-stepping until a normalized density residual of $1 \cdot 10^{-6}$ is reached. In accordance the linearized method FLM-SDNS will also terminate its calculation when having reached this level. A comparison between FLM-NS and FLM-SDNS is performed for the first harmonic of the pressure and skin friction distribution on the airfoil contour in Fig. 1 and Fig. 2, respectively. Real and imaginary parts of the pressure c_{p}^1 and friction coefficient c_{f}^1 obtained through FLM-SDNS generally show good conformity to the Fourier analyzed FLM-NS solution. However, slight deviations between the two methods can be found. For one $Re\, c_{\mathrm{p}}^1$ of FLM-SDNS exhibits a zero-crossing farther away from the trailing edge than FLM-NS. Second, $Re\, c_{\mathrm{f}}^1$ of FLM-SDNS falls short of the peak skin friction calculated by FLM-NS. In and out of phase components of the lift coefficient ($Re\, c_{\mathrm{l}}^1$, $Im\, c_{\mathrm{l}}^1$), as well as of the pitching moment ($Re\, c_{\mathrm{m}}^1$, $Im\, c_{\mathrm{m}}^1$) are calculated with both solvers and compared in Table 1. The small disturbance values deviate from the non-linear unsteady ones in a range of 1% to 4.5%, estabilishing good agreement. FLM-SDNS is faster than FLM-NS by a factor of 5.3. However, this advantage in computational time is bought by memory requirements that are 3.1 times higher than for the non-linear method. FLM-SDNS achieves a 75% degree of vectorization with the implicit integration on the Fujitsu VPP of the Leibniz Rechenzentrum. Efficiency of FLM-SDNS is judge by performing a parameter variation in regard to the test case's reduced frequency. As can be seen in Fig. 3, with increasing k_{red} the CPU time ratio between FLM-NS and FLM-SDNS drops significantly, being mainly attributed to the non-linear solver's increased performance at higher frequencies.

5 Conclusion

An efficient method (FLM-SDNS) for calculating unsteady air loads on the basis of the small disturbance Navier-Stokes equations was presented and an initial test case was calculated to show its validity. However, further extensive validation of the small disturbance code has to be undertaken. It is intended to employ the Navier-Stokes solver FLOWer [4] as an additional validation tool. For the near future the implementation of a turbulence model accessible to the small disturbance approach and the application to wings of various aspect ratio is planned.

References

[1] Cvrlje T., Breitsamter C., Weishäupl C., Laschka B.: Euler and Navier-Stokes Simulation of Two Stage Hypersonic Vehicle Longitudinal Motions, Journal of Spacecraft and Rockets, Vol. 37. No. 2, March - April 2000.

[2] Iatrou. M., Weishäupl C., Laschka B.,: Entwicklung eines instationären Navier-Stokes-Verfahrens bei kleinen Störungen für aeroelastische Problemstellungen. DFG-Bericht, Technische Universität München, TUM-FLM-2002/9, 2002.

[3] Kreiselmaier E., Laschka B.: Small Disturbance Euler Equations: Efficient and Accurate Tool for Unsteady Load Predictions. Journal of Aircraft, Vol. 37, No. 5, Sept.-Oct. 2000.

[4] Kroll N.: FLOWer Installation and User Handbook, Institut für Entwurfsaerodynamik, DLR, 1998.

[5] Laschka B: Unsteady Flows - Fundamentals and Applications, AGARD Conference Proceedings No. 386, Unsteady Aerodynamics - Fundamentals and Applications to Aircraft Dynamics, Göttingen, FRG 1985.

[6] Pechloff A.: Triple Decomposition of the Two-dimensional Navier-Stokes Equations in Cartesian Coordinates and Linearization for Small Disturbances, technical report, Technische Universität München, TUM-FLM-2001/4, 2001.

[7] Pechloff A.: An Implicit Method for Solving the Small Disturbance Navier-Stokes Equations, technical report, Technische Universität München, TUM-FLM-2002/21, 2002.

Table 1 Lift and moment coefficients (c_l^1, c_m^1) at $k_{red} = 0.5$

method	Rec_l^1	Imc_l^1	Rec_m^1	Imc_m^1	CPU time [h]
FLM-NS	0.573E-01	0.195E-01	-0.181E-02	0.437E-02	12.75
FLM-SDNS	0.598E-01	0.197E-01	-0.187E-02	0.453E-02	2.40

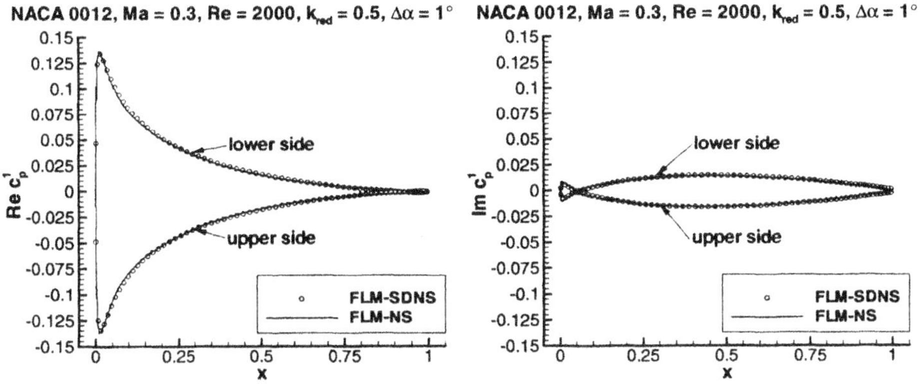

Figure 1 Comparision between FLM-SDNS and FLM-NS for c_{p}^{1}

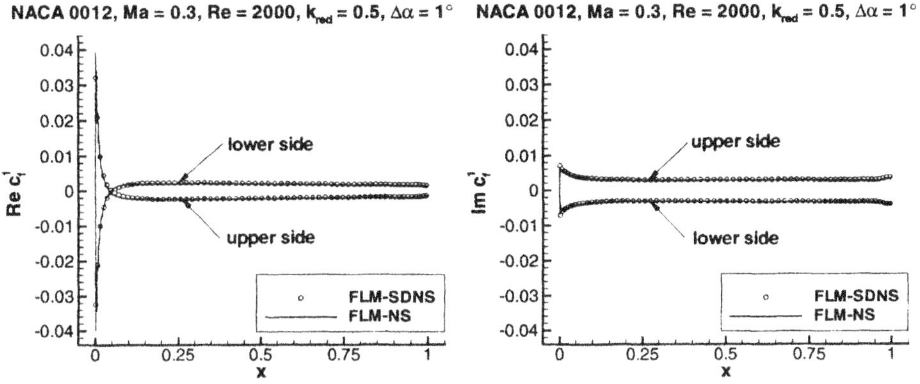

Figure 2 Comparision between FLM-SDNS and FLM-NS for c_{f}^{1}

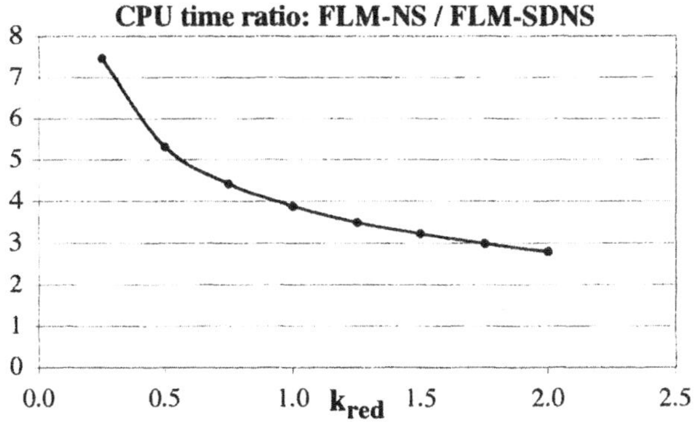

Figure 3 Ratio of computational time between FLM-NS and FLM-SDNS

Development of an Efficient Linearised Navier-Stokes Flow Solver for Turbomachinery Applications

Paul Petrie-Repar

DLR-AE, Bunsenstraße 10, 37073 Göttingen, Germany,
Paul.Petrie-Repar@dlr.de

Summary

Flutter analysis for turbomachinery requires the determination of the unsteady flow due to blade motion. One method of achieving this is to transform the unsteady flow equations to the frequency domain. The resulting so-called time-linearised Navier-Stokes equations require significantly less computational effort than time-domain methods (one to two orders of magnitude). Here, an efficient and robust method for solving the time-linearised Navier-Stokes equations using GMRES with preconditioning is presented. The Spalart & Allmaras turbulence model was included in the flow equations. The accuracy of the method is validated by examining standard test cases.

1 Introduction

Blade flutter in turbomachinery is highly undesirable as it leads to high cycle fatigue and shorter blade life. In order to design blading that will be free of flutter, the design engineer requires a tool that is capable of predicting blade flutter. Unsteady flow simulation is an essential component of flutter analysis. For transonic flow or separated flow, it is necessary to consider viscous effects. Time-marching of the Navier-Stokes equations [3] is one method of simulating the flow, however this is computationally expensive and hence has a limited use in real design. An alternative to the time-marching method is the so called time-linearised method [2]. Here the unsteady flow is assumed to be the steady flow plus small harmonic linear perturbations. This assumption allows the equations to be transformed to the frequency domain. The resulting equations require significantly less computational effort to solve than time-marching methods, typically 1 to 2 orders of magnitude faster. The aim of the current work is to develop a fast and accurate linearised Navier-Stokes flow solver that is capable of being used in the design process. The formulation of the linearised equations is similar to the methods presented by Clark & Hall [2] and Sbardella & Imregun [9].

2 Problem Description

The problem is to determine whether a blisk (bladed disk, Fig 1) will suffer from blade flutter at a given flow condition. Flutter is a sustained oscillation due to the interaction between aerodynamic forces, elastic forces and inertia forces [1]. Typically, in turbomachinery applications, the structural forces dominate and the aerodynamic forces are generally not large enough to significantly alter the natural modes of the blading. In this case, the determination of the structural modes and the unsteady aerodynamics can be decoupled. If the following assumptions are made: the unsteady flow is linear, the blade spacing around the blisk is equal, and each blade is physically identical (identical natural modes) then the flutter analysis for a blisk with N blades can be performed by examining the unsteady flow response to N Travelling Wave Modes [6]. For each Travelling Wave Mode the blades are oscillating at constant and equal amplitude with a phase difference (interblade phase angle) between neighbouring blades equal to $\sigma = 2\pi n/N$, $(0 < n < N - 1)$. It has been shown [6] that a blisk will suffer from blade flutter if the aerodynamic damping is negative for any of the possible inter-blade phase angles. The aerodynamic damping quantifies the work done by the structural forces on the fluid per cycle.

The current method assumes that the flow through the blisk is two-dimensional. The flow domain (Fig. 1) is created by making a circumferential slice through the blisk. Only one blade channel needs to be considered due to the periodic Travelling Wave Modes.

3 Flow Model

The two-dimensional Reynolds-averaged Navier-Stokes equations with the Spalart & Allmaras one-equation turbulence model are used to model the unsteady flow. The unsteady flow equation for a moving control volume can be written,

$$\frac{\partial}{\partial t}\int_{\vartheta(t)} \mathbf{U}\, d\vartheta + \oint_{S(t)} \mathbf{H}\, dS = \int_{\vartheta(t)} \mathbf{S}\, d\vartheta \tag{1}$$

where $\mathbf{U}$ is the vector of conserved variables, ϑ is the volume of the control volume, S is the control surface, $\mathbf{H}$ is the flux of conserved variables at the control surface, t is time, and $\mathbf{S}$ is the vector of source terms due to the turbulence model.

3.1 Boundary Conditions

Each vertex along a periodic boundary has a corresponding vertex on the other periodic boundary such that $\mathbf{x}_{upper} = \mathbf{x}_{lower} + \mathbf{g}$ where $\mathbf{g}$ is a constant vector and $|\mathbf{g}|$ is equal to the pitch of the blading. For steady-state flow, the periodic boundaries are treated as if they were joined. For unsteady flow the amplitudes of the flow perturbations at corresponding points on the periodic boundaries are assumed to be equal but the phases of perturbations are assumed to be shifted by the interblade phase angle.

For steady-state flow, total pressure, total temperature and flow angle are prescribed at the upstream far-field boundary and static pressure is prescribed at the downstream far-field boundary. For unsteady flow, the boundary conditions at the far-field (inlet and outlet) are described in terms of wave motion. Waves that originate from outside the flow domain are not allowed to enter the flow domain at the far-field boundaries and waves that originate from inside the flow domain (due to blade oscillation) are allowed to pass through the far-field boundary free of reflection. The method of Giles [5] is used to apply this boundary condition.

On the blade surface standard slip (inviscid) and non-slip (viscous) boundary conditions are applied.

3.2 Discretisation

The flow domain is discretised by a multi-block structured grid. The flow equation is discretised using a cell-centred finite-volume scheme. The semi-discretised flow equation can be written,

$$\frac{d}{dt}[\mathbf{U}\,\vartheta] + \sum_{i=1}^{4} \mathbf{H}_i\, S_i = \mathbf{S}\,\vartheta. \tag{2}$$

where $\mathbf{U}$ is the flow state at the cell centre, ϑ is the volume of the cell, S_i is the area of a cell interface between two cells, and $\mathbf{H}_\mathbf{i}$ is the flux at the mid-point of a cell interface. The flux is determined by reconstructing the flow at the mid-point of a cell interface using MUSCL interpolation with a modified van Albada limiter [12] and Roe's approximate Riemann solver [11].

4 Linearisation and Transformation to Frequency Domain

The transformation of the flow equations to the frequency domain is achieved using the method of Hall [2]. The grid points are assumed to oscillate harmonically about their steady-state positions

$$\mathbf{x} = \overline{\mathbf{x}} + \Re\{\tilde{\mathbf{x}} e^{\jmath\omega t}\}. \tag{3}$$

The motion of the grid points on the blade surface is prescribed and the grid points on the outer boundaries are assumed to be at rest. The values of $\tilde{\mathbf{x}}$ for the internal points can be found by solving Laplace's equation for each ordinate, $\nabla^2(\tilde{\mathbf{x}} \cdot \mathbf{e}_\mathbf{i}) = 0$. The unknown unsteady flow perturbations are also assumed to be harmonic

$$\mathbf{U} = \overline{\mathbf{U}} + \Re\{\tilde{\mathbf{U}} e^{\jmath\omega t}\}. \tag{4}$$

The unsteady flow perturbation $\tilde{\mathbf{U}}$ does not refer to a fixed point in space but to the flow state at the cell centre which does move with time. Note that $\tilde{\mathbf{U}}$ and $\tilde{\mathbf{x}}$ are complex. The estimated flux at the mid-point of a cell interface is split into two components:

$$\mathbf{H} = \mathbf{H_s} - \mathbf{w} \cdot \hat{\mathbf{n}}\,\mathbf{U} \tag{5}$$

where $\mathbf{H_s}$ is the flux through the cell interface when the mesh is stationary, and $\mathbf{w} \cdot \hat{\mathbf{n}}\,\mathbf{U}$ is the additional flux due to grid motion. The stationary flux is assumed to be a linear function of the unknown surrounding cell-centred flow values and the positions of the surrounding grid points.

$$\mathbf{H_s} = \overline{\mathbf{H_s}} + \Re\{\sum \frac{\partial \mathbf{H_s}}{\partial \mathbf{U}_i}\tilde{\mathbf{U}}_i e^{j\omega t} + \sum \nabla_i \mathbf{H_s} \cdot \tilde{\mathbf{x}}_i e^{j\omega t}\} \quad (6)$$

The linearisation of the turbulent source term is performed in a similar manner

$$\mathbf{S} = \overline{\mathbf{S}} + \Re\{\sum \frac{\partial \mathbf{S}}{\partial \mathbf{U}_i}\tilde{\mathbf{U}}_i e^{j\omega t} + \sum \nabla_i \mathbf{S} \cdot \tilde{\mathbf{x}}_i e^{j\omega t}\}. \quad (7)$$

The perturbations of the cell volumes and surfaces due to the prescribed motion of the grid points can be described as

$$\vartheta \simeq \overline{\vartheta} + \Re\{\tilde{\vartheta} e^{j\omega t}\} \quad (8)$$

$$S \simeq \overline{S} + \Re\{\tilde{S} e^{j\omega t}\}. \quad (9)$$

The resulting first order terms after the substitution of the prescribed grid motion, the assumed harmonic unsteady flow perturbations and the linearisations into the semi-discretised unsteady Navier-Stokes (Eqn. 2) is

$$j\omega\, \tilde{\mathbf{U}}\, \overline{\vartheta} + \sum\sum \frac{\partial \mathbf{H_s}}{\partial \mathbf{U}_i}\, \tilde{\mathbf{U}}_i\, \overline{S_j} - \sum \frac{\partial \mathbf{S}}{\partial \mathbf{U}_i}\, \tilde{\mathbf{U}}_i\, \overline{\vartheta} \;=\; (\overline{\mathbf{S}} - j\omega\, \overline{\mathbf{U}})\, \tilde{\vartheta}$$
$$+j \sum \omega\, \tilde{\mathbf{x}} \cdot \overline{\mathbf{n}}\, \overline{\mathbf{U}}\, \overline{S} - \sum \left(\sum \nabla_i \mathbf{H_s} \cdot \tilde{\mathbf{x}}_i + \overline{\mathbf{H_s}} \tilde{S}\right) + \left(\sum \nabla_i \mathbf{S} \cdot \tilde{\mathbf{x}}_i\right) \overline{\vartheta}. \quad (10)$$

Note that the zero-order terms cancel because they result in the steady-state equation and it is assumed that the residual of the steady-state equation is zero. All other higher-order terms are neglected.

5 Solution Method

The steady-state solution is found by time-stepping (explicitly or implicitly) the unsteady flow equation (Eqn. 1) until the steady-state residual has been reduced several orders of magnitude. Local time-stepping is used to increase the rate of convergence. The traditional method [2, 9] of solving the time-linearised flow equation (Eqn. 10) is to add a pseudo-time derivative term to the left side of the equation. The equation now has a similar form to the unsteady flow equation and hence the same method used to find the steady-state solution can be employed. Instead of the traditional pseudo-time-stepping, the current scheme uses GMRES with preconditioning to solve the linearised flow equations.

5.1 GMRES with Preconditioning

The Generalized Minimum RESidual (GMRES) method was proposed by Saad [7] in order to solve large, sparse and nonsymmetric linear systems. GMRES belongs to the class of Krylov based iterative methods. Consider the linear system

$$\mathbf{A}\mathbf{x} = \mathbf{b} \quad (11)$$

where $\mathbf{A}$ is a square nonsingular $n \times n$ complex matrix, and $\mathbf{b}$ is a complex vector of length n. Let $\mathbf{x}_0 \in \mathcal{C}^n$ be an initial guess for this linear system and $\mathbf{r}_0 = \mathbf{b} - \mathbf{A}\mathbf{x}_0$ be its corresponding residual. The GMRES algorithm builds an approximation of the solution of Equation 11 under the form

$$\mathbf{x}_l = \mathbf{x}_0 + \mathbf{V}_l \mathbf{y} \tag{12}$$

where $\mathbf{V}_l$ is an orthonormal basis for the Krylov space of dimension l

$$\mathcal{K}_l = \text{span}\left\{\mathbf{r}_0, \mathbf{A}\mathbf{r}_0, \ldots, \mathbf{A}^{l-1}\mathbf{r}_0\right\}, \tag{13}$$

where $\mathbf{y}$ belongs to $\mathcal{C}^l$. The vector $\mathbf{y}$ is determined so that 2-norm of the residual $||\mathbf{r}_l||_2$ is minimal over $\mathcal{K}_l$. The 2-norm is defined as

$$||\mathbf{x}||_2 = \sqrt{\sum_{i=1}^{n} |x_i|^2}. \tag{14}$$

For more information see Saad [7].

The convergence rate of Krylov subspace methods is highly dependent on the condition number of the matrix $\mathbf{A}$. The smaller the condition number, the better the convergence. Preconditioning can be used to reduce the condition number of the linear system.

The current scheme uses ILU(p) [7] to right precondition the linear system. The preconditioner is calculated by performing an incomplete Gaussian elimination. The method is incomplete because the non-zero pattern on the preconditioner is determined by the non-zero pattern of the original matrix and the level of fill p. In order to construct the preconditioner, it is necessary to explicitly determine and store the governing matrix

$$\mathbf{A} = j\omega\,\mathbf{I} + \sum\sum \frac{\partial \mathbf{H_s}}{\partial \mathbf{U}_i}\overline{S_j} - \sum \frac{\partial \mathbf{S}}{\partial \mathbf{U}_i}\overline{\vartheta}. \tag{15}$$

The governing matrix is constructed from the Jacobians of the flux calculation and the source term. Both Jacobians are calculated numerically and the Jacobian of the flux calculation includes the action of MUSCL and the limiter.

6 Test Cases

6.1 Standard Configuration 10

Standard configuration 10 [10] is a two-dimensional compressor cascade of a modified NACA0006 profile at a stagger angle of 45 degrees with a gap to chord ratio of 1.0. The blades are harmonically pitching about the mid-chord axis at a reduced frequency $\omega^* = 0.5$ based on chord. The inlet conditions are Mach number $M_1 = 0.7$ and flow angle $\beta_1 = 55.0^o$. The flow is assumed to be inviscid. Figure 2 compares the aerodynamic damping calculated by the current method with two previous numerical results. The agreement is good, in particular the peaks due to resonance are well captured. This is a good indication that the non-reflecting boundary conditions have been implemented correctly.

6.2 Standard Configuration 11

Standard Configuration 11 [8] is a two-dimensional turbine cascade. An off-design transonic condition is examined which is characterised by a separated flow region on the suction side with detachment immediately downstream of the leading edge and re-attachment at approximately 30 per cent chord. A shock is also present on the suction surface at approximately 80 percent chord. The inflow conditions are: flow angle $\beta_1 = -34.0$ and inlet Reynolds number is 1.1×10^6. The isentropic Mach number at the exit is 0.99. The unsteady pressure response due to harmonic bending normal to chord with a reduced frequency based on chord $\omega^* = 0.309$, and interblade phase angle $\sigma = 180^o$ is shown in Figure 3. The agreement between the current RANS solution and the previous numerical work [9] is good. It can be seen that the viscous effects are significant in the separated flow region and that the RANS solution is closer to the experimental data than the Euler solution.

7 Computational Performance

All calculations were performed on a personal computer with two Intel Pentium IV processors (jobs run in serial mode) with a clock speed of 1.7GHz (Linpack [13] performance 230 Mflops). An inviscid calculation (3029 cells) for Standard Configuration 10 (Section 6.1) requires less then four seconds. The viscous calculation (17518 cells) for Standard Configuration 11 (Section 6.2) required 86 seconds and 495 Mb RAM. The breakdown of the computational resources for this case are: calculation of Jacobians required 25 seconds and 106 Mb RAM, calculation of preconditioner ILU(3) required 20 seconds and 250 Mb RAM, and GMRES required 36 seconds and 106 Mb RAM. Note that only 57 explicit time-steps (two stage) can be performed in the same computational time. Even with the introduction of implicit time-stepping, it would be difficult to beat the computational performance of GMRES.

8 Conclusions

A linearised Navier-Stokes flow solver which includes the Spalart & Allmaras turbulence model that is suitable for flutter investigations has been presented. GMRES with preconditioning has been used to solve the unsteady linearised flow equations in a robust and efficient manner. A three-dimensional version of this method is planned.

References

[1] AGARD: "Multilingual Aeronautical Dictionary". AGARD. ISBN 92-835-1666-7. 1980.

[2] W.S. Clack, K.C. Hall: "A Time-Linearized Navier-Stokes Analysis of Stall Flutter". Journal of Turbomachinery. **122**, 2000, pp. 467-476.

[3] S. Weber, M.F. Platzer: "A Navier-Stokes Analysis of the Stall Flutter Characteristics of the Buffum Cascade". Journal of Turbomachinery. **122**, 2000, pp. 769-776.

[4] P.R. Spalart, S.R. Allmaras: "A One-Equation Turbulence Model for Aerodynamic Flows". In Proceedings of 30th Aerospace Sciences Meeting & Exhibit. 1992, AIAA-92-0439.

[5] M. Giles: "Non-Reflecting Boundary Conditions for the Euler Calculations". AIAA Journal. **28**, 1990, pp. 2050-2058.

[6] F. Lane: "System Mode Shapes in the Flutter of Compressor Blade Rows". Journal of the Aeronautical Sciences. **23**, 1956. pp. 54–66.

[7] Y. Saad: "Iterative Methods for Sparse Linear Systems". 2000. URL: http://www-users.cs.umn.edu/~saad/books.html.

[8] T.H. Fransson, M. Jöcker, A. Bölcs, P. Ott. "Viscous and Inviscid Linear/Nonlinear Calculations Versus Quasi 3D Experimental Cascade Data for a New Aeroelastic Turbine Standard Configuration". Journal of Turbomachinery. **121**, 1999, pp. 717-725.

[9] L. Sbardella, M. Imregun: "Linearised Unsteady Viscous Turbomachinery Flows Using Hybrid Grids". Journal of Turbomachinery. **123**, 2001, pp. 568-582.

[10] T.H. Fransson, J.M. Verdon: "Updated Report on Standard Configurations for Unsteady Flow Through Vibrating Axial-Flow Turbomachine Cascades". Report. Royal Institute of Technology, Stockholm, Sweden. URL: http://www.egi.kth.se/ekv/stck.

[11] P.L. Roe: "Approximate Riemann Solvers, Parameter Vectors, and Difference Schemes". Journal of Computational Physics. **43**, 1981, pp. 357-372.

[12] I.A. Johnston: "Simulation of Flow Around Hypersonic Blunt-Nosed Vehicles for the Calibration of Air Data Systems". Ph.D. Thesis. Department of Mechanical Engineering. University of Queensland. Australia. 1999.

[13] H. Meuer, E. Strohmaier, J. Dongarra, H.D. Simon: "TOP500 Computer Sites". URL: http://www.top500.org/.

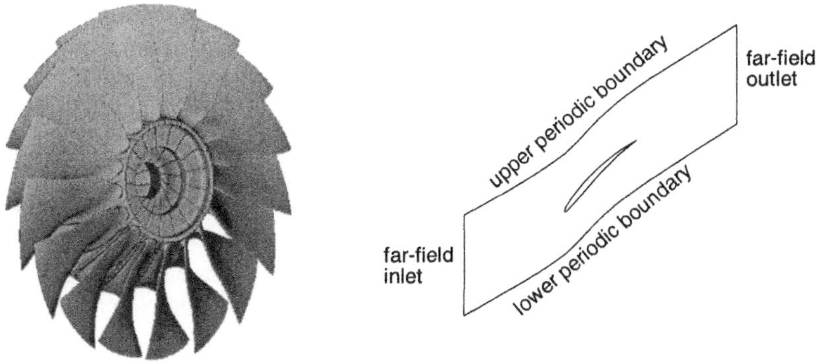

Figure 1 A blisk (left) and typical flow domain (right) with boundary description.

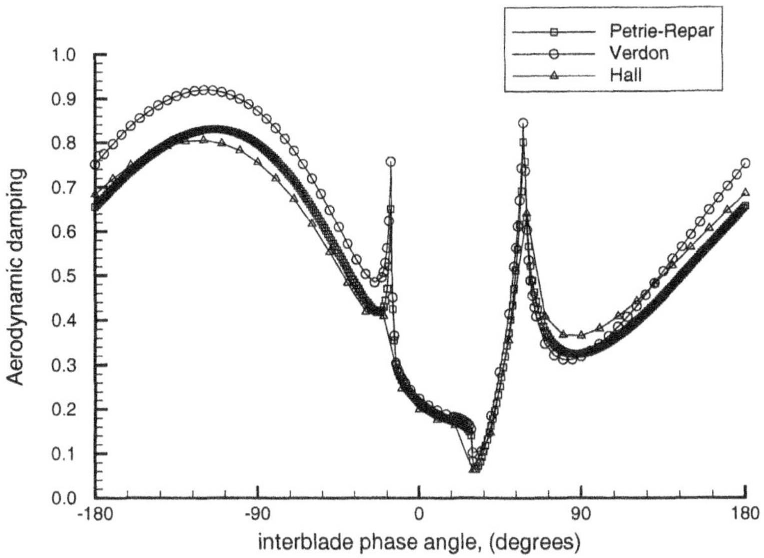

Figure 2 Aerodynamic damping for Standard Configuration 10 $M_1 = 0.7$, $\beta_1 = 55.0^o$ and $\omega^* = 0.5$.

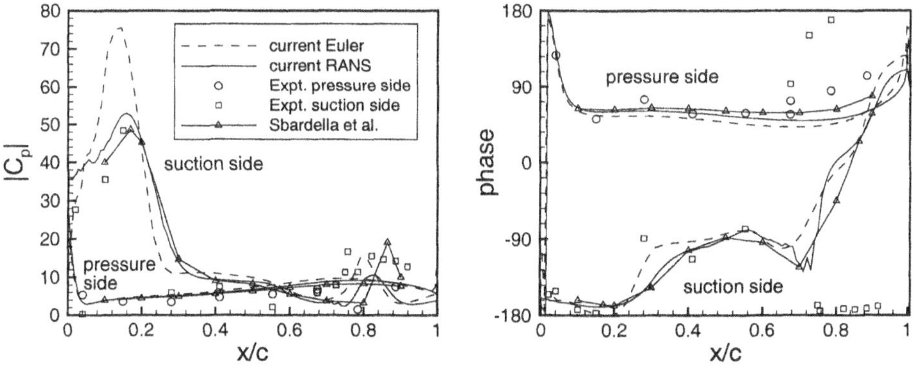

Figure 3 Unsteady pressure response for Standard Configuration 11.

DNS of Transonic Flow about an Oscillating Airfoil

Igor Klutchnikov and Josef Ballmann

Mechanics Department of RWTH Aachen, University of Technology,
Templergraben 64, D-52056 Aachen, Germany

Summary

The objective of the work presented in this paper is direct numerical simulation (DNS) of three-dimensional turbulent compressible fluid flow about oscillating boundaries with application to problems of aeroelasticity. In a first step a high order in space and second order in time difference method which was originally designed for DNS of compressible fluid flow with stationary boundary conditions has been extended for simulating turbulent flow on dynamic grids fitted to a body which oscillates about an axis fixed in space. The first tests of the method concern transonic flow about the supercritical airfoil NLR7301.

1 Introduction

Flutter experiments with supercritical airfoils in the transonic regime are dominated by periodic movements and changes of strength of the occuring transonic shocks, together with shock-boundary layer interaction and possibly trailing edge flow separation. Numerical simulation of some of the extensive wind tunnel experiments performed by the group of Schewe [1] have been tried by many authors who employed Reynolds averaged Navier-Stokes solvers using various formulations for turbulence modelling [2]. But none of them has been able to reproduce quantitatively the results of the experiments. This little encouraging situation may have many reasons, one of them can be that the turbulence models used might not be able to reproduce correctly shock-boundary layer interaction or flow separation at a boundary surface being in unsteady motion. This situation was the motivation to extend the method developed in [4] for DNS of transonic flow about an oscillating airfoil.

2 Governing Equations

The Navier-Stokes equations for unsteady three-dimensional compressible fluid flow about an airfoil oscillating in the x,y-plane about the z-axis fixed in space are transformed from Cartesian coordinates x, y, z ant time t to time-dependent body fitted curvilinear coordinates $\xi(x, y, t), \eta(x, y, t), \zeta = z$ and time $\tau = t$. All physical tensor quantities like momentum density and fluxes remain represented in their Cartesian components in spite of their dependence on the variables ξ, η, ζ, τ. This involves scalar products of the cartesian base vectors and the contravariant base vectors belonging to the curvilinear coordinates ξ, η. Furthermore, due to the transformation of the partial derivative with respect to time, the grid velocity with cartesian components $x_{,\tau}$ and $y_{,\tau}$ occurs in combination with the gradient operator which is expressed in the variables ξ, η, ζ. After some algebra, this leads to the following formulation of the Navier-Stokes equations in conservative differential but not strong conservation form for the solution vector of conservative variables $\mathbf{U}$

$$\mathbf{U}_{,\tau} + \frac{1}{J}\vartheta \mathbf{U}_{,\xi} + \frac{1}{J}\omega \mathbf{U}_{,\eta} + \frac{1}{J}\mathbf{F}^c_{,\xi} + \frac{1}{J}\mathbf{G}^c_{,\eta} + \mathbf{H}^c_{,z} = \frac{1}{J}\mathbf{F}^d_{,\xi} + \frac{1}{J}\mathbf{G}^d_{,\eta} + \mathbf{H}^d_{,z}. \quad (1)$$

Variable behind comma denotes partial differentiation with respect to the variable. $\mathbf{F}^c$, $\mathbf{G}^c$, $\mathbf{H}^c$ and $\mathbf{F}^d$, $\mathbf{G}^d$, $\mathbf{H}^d$ represent the convective and dissipative fluxes

$$\mathbf{U} = \begin{bmatrix} \rho \\ \rho u \\ \rho v \\ \rho w \\ \rho E \end{bmatrix}, \quad \mathbf{F}^c = \begin{bmatrix} \rho\alpha \\ \rho u\alpha + p y_{,\eta} \\ \rho v\alpha - p x_{,\eta} \\ \rho w\alpha \\ (\rho E + p)\alpha \end{bmatrix}, \quad \mathbf{G}^c = \begin{bmatrix} \rho\beta \\ \rho u\beta - p y_{,\xi} \\ \rho v\beta + p x_{,\xi} \\ \rho w\beta \\ (\rho E + p)\beta \end{bmatrix}, \quad \mathbf{H}^c = \begin{bmatrix} \rho w \\ \rho u w \\ \rho v w \\ \rho w^2 + p \\ (\rho E + p) w \end{bmatrix},$$

$$\mathbf{F}^d = \begin{bmatrix} 0 \\ \tau_{xx} y_{,\eta} - \tau_{xy} x_{,\eta} \\ \tau_{xy} y_{,\eta} - \tau_{yy} x_{,\eta} \\ \tau_{xz} y_{,\eta} - \tau_{yz} x_{,\eta} \\ (u_j \tau_{xj} - q_x) y_{,\eta} - \\ -(u_j \tau_{yj} - q_y) x_{,\eta} \end{bmatrix}, \quad \mathbf{G}^d = \begin{bmatrix} 0 \\ \tau_{xy} x_{,\xi} - \tau_{xx} y_{,\xi} \\ \tau_{yy} x_{,\xi} - \tau_{xy} y_{,\xi} \\ \tau_{yz} x_{,\xi} - \tau_{xz} y_{,\xi} \\ (u_j \tau_{yj} - q_y) x_{,\xi} - \\ -(u_j \tau_{xj} - q_x) y_{,\xi} \end{bmatrix}, \quad \mathbf{H}^d = \begin{bmatrix} 0 \\ \tau_{xz} \\ \tau_{yz} \\ \tau_{zz} \\ u_j \tau_{zj} - q_z \end{bmatrix},$$

$$J = x_{,\xi} y_{,\eta} - x_{,\eta} y_{,\xi}, \qquad \alpha = u y_{,\eta} - v x_{,\eta}, \qquad \beta = v x_{,\xi} - u y_{,\xi},$$
$$\vartheta = x_{,\eta} y_{,\tau} - x_{,\tau} y_{,\eta}, \qquad \omega = x_{,\tau} y_{,\xi} - x_{,\xi} y_{,\tau},$$

u, v and w are the Cartesian velocity components, ρ, p and E are density, pressure and total specific energy, $\boldsymbol{\tau}$ and $\mathbf{q}$ are the stress tensor and heat flux vector, respectively.

3 Numerical Method

As computational domain we take a C-H shaped grid space around the rectangular wing segment with infinite width in z direction which is oscillating in time t in the (x,y) about the 40 percent line of chord. The numerical method used is of order N=4,6,8,10... in space and employs central differences for spatial discretization together with an explicit time integration [4], [5]. This explicit time integration is either done with a two-step Richtmyer scheme of second order or with a Runge-Kutta method of higher order. The results in this work have been calculated using the Richtmyer scheme solving the following difference equations

$$\mathbf{U}^{n+1/2}_{i_r+1/2} = L_r \mathbf{U}^n_{i_r+1/2} - 0.5\lambda_r \mathbf{A}^n_{ri_r+1/2}(R_r \mathbf{U}^n_{i_r+1/2})$$

$$\mathbf{U}^{n+1}_{\bullet} = \mathbf{U}^n_{\bullet} - \sum_{r=1}^{3} \lambda_r \{(\mathbf{F}^{cn+1/2}_{ri_r+1/2} - \mathbf{F}^{cn+1/2}_{ri_r-1/2})$$
$$- (\mathbf{F}^{dn}_{ri_r+1/2} - \mathbf{F}^{dn}_{ri_r-1/2}) - (S_r \mathbf{U}^n_{i_r+1/2} - S_r \mathbf{U}^n_{i_r-1/2})$$
$$+ \sigma_r (\mathbf{U}^n_{i_r+1/2} - \mathbf{U}^n_{i_r-1/2})\}, \quad (2)$$

$$(\mathbf{F}_1, \mathbf{F}_2, \mathbf{F}_3) = (\mathbf{F}, \mathbf{G}, \mathbf{H}), \quad (\sigma_1, \sigma_2, \sigma_3) = (\vartheta, \omega, 0) \quad (3)$$

$$(\lambda_1, \lambda_2, \lambda_3) = (\frac{\Delta t}{J^n_{\bullet} \Delta\xi}, \frac{\Delta t}{J^n_{\bullet} \Delta\eta}, \frac{\Delta t}{\Delta z}). \quad (4)$$

wherein $\mathbf{A}_r$ is the flux Jacobian, and L_r, R_r and S_r are special discrete operators in space, which have been introduced first by Zalesak [6]. Upper indices n and $n+1/2$ correspond to time levels and lower indices i_r, $r = 1, 2, 3$ are running in the space directions $r = 1, 2, 3$ from $i_r = 1$ to $i_r = i_{rmax}$. A detailed description of the space operators in the method can be found in [4], [5]. Compared with the method for problems with stationary boundary conditions only the last term in eq.(2) is additional. Furthermore all metric dependent quantities like e.g. $J^n_{\bullet}$ are to be updated in every time step.

Boundary conditions are the no slip condition at solid walls, and on the artificial boundaries, which are introduced to have a finite computational domain, extrapolation or the concept of local simple waves are used optionally. Outside of the computational domain fictitious points are introduced to maintain the order N in the complete computational domain.

The solution procedure for initiating 3D simulations performs as follows: first, a two-dimensional flow is calculated in the main flow plane (x,y-plane). Then, the z component of the velocity near the airfoil boundary is disturbed in a non-stationary fashion for a short period of time. If these disturbances are damped out later on, then the solution falls back to two-dimensional flow and does not become turbulent. For a Reynolds number as in the case discussed in this paper we expect the flow to be transient to three-dimensional unsteady flow in the airfoil boundary layer, and not to return to stationary two-dimensional flow.

Validation tests of the method have been conducted for the high order and FCT algorithm, on the one hand by comparison with analytical solutions of the linear Burgers' equation which are not presented here and on the other hand by comparison with experimental and numerical results for channel flow at M=0.2. The results of the latter correspond to experiments and DNS in straight channels. Reynolds number is based on the channel width h and the bulk mean velocity (Re_h). Fig.2 a) and b) display results of the method compared with those from references [8], [9], [7], [10].

4 Results

4.1 Grid Resolution

Before application, extensive dynamic grid tests have been performed to prove the fulfilment of the geometric conservation law. For this purpose dynamic grid deformations according to large amplitude oscillations of the NLR7301 profile were chosen, but with free boundary inflow/outflow conditions along the oscillating contour. The tests performed perfectly without any disturbance by the dynamic grid.
An important point in application of DNS is that all important vortical scales must be resolved. Because of its lesser numerical effort, it is worthwile to study the grid resolution quality in 2D flow simulation first, without claiming to do true DNS which has to be conducted in 3D, of course. It is assumed that 3D simulation introduces no relevant smaller length scales which was confirmed by previous simulations. The 2D flow simulations about the supercritical airfoil NLR7301 having been conducted first were performed under stationary boundary conditions at zero angle of attack on five different grids g_1, g_2, g_3, g_4 and g_5, with emphasis on the asymptotics of developing separation spots, their location and size. Mach number and macroscale Reynolds number (based on chord length) were chosen $M = 0.7$ and $Re_c = 2.14 \cdot 10^6$, according to experiments [3]. Based on these results the macro-scale Reynolds number Re_c can be replaced by another Reynolds number more appropriate to characterize the evolution of turbulence in the boundary layer, e.g. Re_τ. The basic grid g_1 is a C-type grid with 512x80 grid points where the 80 grid lines $\eta = const$ are distributed according to an 1.5-exponential law. For grid g_2 an g_3 these 80 grid lines are redistributed such that the point density is threefold for g_2 and sixfold for g_3 in the domain of the first ten $\eta = const$ lines of grid g_1. In grid g_4 the number of points was doubled in η-direction with twelvefold the density of the grid lines of g_1 near the airfoil, whilest in grid g_5 the same number of grid points as for grid g_1 was chosen but distributed according to a 2.5-exponential law.

4.2 Flow Simulation

In Fig.1 a) to c) instantaneous flow vector fields in the x, y-plane are displayed for the three grids g_1, g_4 and g_5 in this order, where in Fig.1 a) and c) every grid line $\eta = const$, in Fig.1 b) every eighth line $\eta = const$ is presented. All pictures exhibit small vortical structures on the upper and lower side of the profile. Their size decreases from g_1 to g_4 and shows then an asymptotic behaviour (figures b) and c)). This observation is confirmed by the statistical mean of the velocity component u in main flow direction. Its contour lines are presented in Fig.1 d) to f) for the grids g_1, g_4, g_5 using the same grey scale level gradation. The values found for boundary layer related quantities along the first grid line for the grids g_1 to g_4 range Re_τ from 700 to 400, U_τ from 0.005 to 0.0025, y^+ from 12 to 0.9, while $x^+_{max} < 130$ was found. For grid g_5 the respective values are $Re_\tau < 400$, $U_\tau < 0.0028$, $y^+ < 1$ and $x^+ < 60$.
To reduce the value of x^+ the AGARD test case [3] with the NLR7301 airfoil at

stationary conditions with $M = 0.7$, $Re_c = 2.14 \cdot 10^6$, angle of attack $\alpha = 2.5^o$ and free transition we chose 1280x130 grid points and found almost $x^+ < 20$ and $y^+ < 1$. Results on this grid are displayed in Fig.3. A 2D simulation of this flow has been started from scratch which has not yet completely reached its final state in Fig.3 but one can already observe all essential features of the flow. Fig.3 a) and b) present the instantaneous contours of Mach number and pressure which exhibit a local supersonic region closed by a fluctuating shock structure and along the profile followed by Mach number and pressure fluctuations corresponding with the vortical structures extending downstream from the foot region of the shock. Fig.3 c) displays the instantaneous vorticity contours. The instantaneous pressure distribution along the surface has already reached a state not to far from the mean measured in the experiment as can be seen by comparison in Fig.3 d).
3D flow simulation about the oscillating rectangular wing segment with profile NLR7301 was started on a C-H-type dynamic grid built up by staggering 40 planes of grid g_5 in the z-direction with initial data according to a well established 2D solution for the flow parameters $M = 0.7$, $Re_c = 2.14 \cdot 10^6$, $\alpha = 0^o$ and the airfoil at rest. Then the angle of attack is abruptly changed to a mean value of $\alpha = 2.5^o$ about which immediately a harmonic oscillation with frequency 80 Hz starts. Periodic boundary conditions are imposed on the boundary planes $z = const$. For a short period of time a forced velocity disturbance w in the z-direction is superposed to the 2D flow field in the leading edge region and thereafter removed. Fig.4 a) and b) show instantaneous distributions of Mach number (a) and pressure (b) on the third grid surface and the vorticity distribution in close neighbourhood of the airfoil (c). The forced disturbance in z-direction is already removed since several hundred thousands of time iteration steps and, as can be observed, the flow is still three-dimensional, and it will remain three-dimensional. Concerning the abrupt change of the flow conditions the flow is still in its adaptation phase. A transonic region is developing on the upper side near the leading edge which is closed by a shock. We emphasize the streaky structure of the Mach isolines behind the shock in Fig.4 a) indicating the presence of longitudinal vortices which can be recognized in Fig.4 c) and lead to transition in the rear part of the airfoil.

5 Conclusions

In this work we presented a high order in space finite difference method for three-dimensional transonic flow about an oscillating airfoil which operates on a dynamic grid. The method has been tested successfuely and is now being applied to a basic problem of aeroelasticity according to [1].

Acknowledgements

The authors thank the computing centers of RWTH Aachen and Forschungszentrum Jülich for the kind permission to use their powerful equipment. We also thank Dr. Schewe and his group for many fruitful discussions motivating this work.

References

[1] G. Schewe, A. Knipfer, H. Mai, G. Dietz. "Experimentelle und numerische Untersuchung nichtlinearer Effekte beim transsonischen Flattern". DLR Göttingen, DLR IB 232-2001 J03, 2001.

[2] G. Britten. "Numerische Aerostrukturdynamik von Tragflügeln großer Spannweite". Dissertation, RWTH Aachen, 2002.

[3] AGARD-R-702, Compendium of Unsteady Aerodynamic Measurements, 1982.

[4] I. Klutchnikov. "Direct Numerical Simulation of Turbulent Compressible Fluid Flow". Habilitation Thesis, Russian Academy of Sciences, Moscow, 1998.

[5] I. Klutchnikov, J. Ballmann. "DNS of Turbulent Compressible Fluid Flow with a High Order Difference Method". AIAA paper 2001-2542, 2001.

[6] S.T. Zalesak. "A Physical Interpretation of the Richtmyer Two-Step Lax-Wendroff Scheme, and its Generalization to Higher Spatial Order". In: Advances in Computer Methods for Partial Differential Equations, Vichnevetsky R., Stepleman R.S. (eds.), IMACS, Rutgers University, New Brunswick, NJ, Vol. V, 1984, pp. 491-496.

[7] J. Kim, P. Moin, R.D. Moser. "Turbulence statistics in fully developed channel flow at low Reynolds number". J. Fluid Mech., 177, 1987, pp. 133-166.

[8] G.A. Whan, R.R. Rothfus. Amer. Inst. Chem. Eng. J., Vol.5, N2, 1959, pp. 205-208.

[9] V.C. Patel, M.R. Head. "Some observations on skin friction and velocity profiles in fully developed pipe and channel flows". J. Fluid Mech., Vol. 38, N1, 1969, pp. 181-201.

[10] P.R. Spalart. "Direct simulation of a turbulent boundary layer up to R_Θ=1410". J. Fluid Mech., 187, 1988, pp. 61-98.

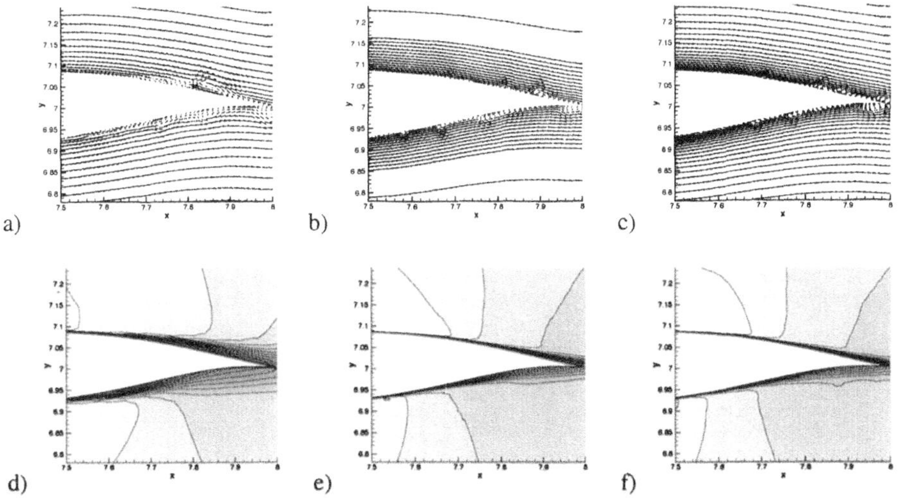

Figure 1 2D simulation. $M = 0.7$, $Re_c = 2.14 \cdot 10^6$, $\alpha = 0^o$. Instantaneous flow vector fields (pictures a) to c)) and mean distributions of velocity component u (pictures d) to f)) in the (x, y) plane of the airfoil for three grid cases: (a,d)-g1, (b,e)-g4, (c,f)-g5

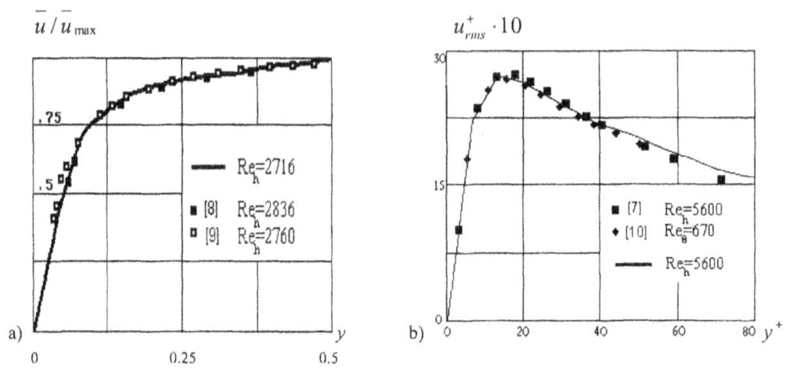

Figure 2 Numerical results for channel flow at $M = 0.2$ and Re as indicated in the figures, compared with experiments from [8], [9] and DNS from [7] and [10]

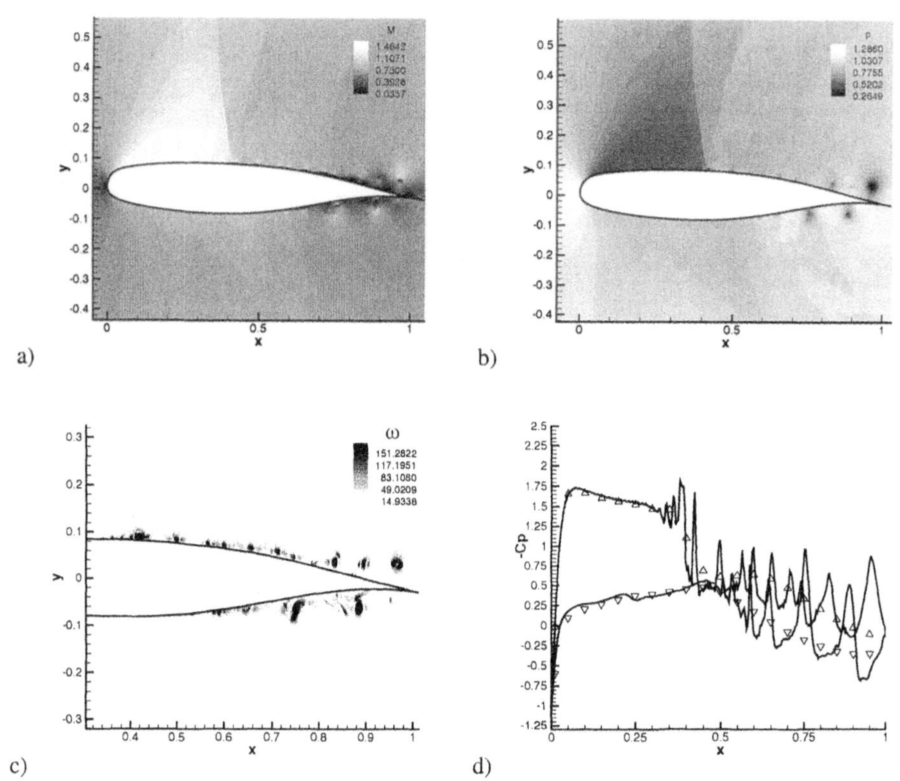

Figure 3 2D simulation. $M = 0.7$, $Re_c = 2.14 \cdot 10^6$, $\alpha = 2.5^o$. Instantaneous distribution of Mach number (a), pressure (b) and vorticity (c) in the (x, y) mid-plane of the 3D computational domain. Instantaneous distributions of pressure coefficient (d) on the airfoil surface. Symbols represent the experimental mean distribution [3].

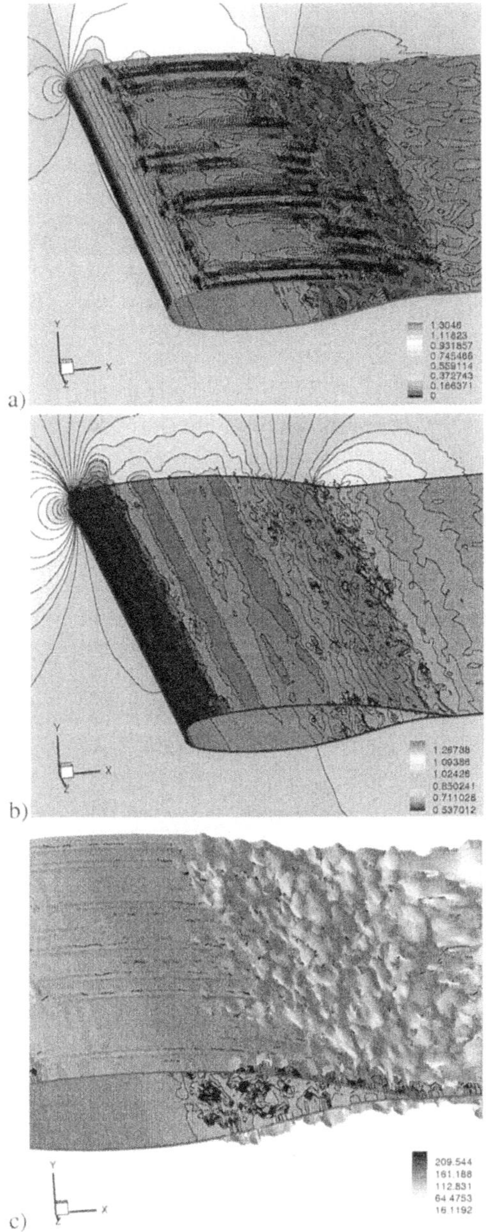

Figure 4 3D simulation. $M = 0.7$, $Re_c = 2.14 \cdot 10^6$, $\alpha = 2.5^o$, $\Delta\alpha = 0.5^o$. Instantaneous contours of the distributions of Mach number (a), pressure (b) and isosurface of vorticity (c)

On the Structure and Stability of Three-Dimensional Laminar Separation Bubbles on a Swept Plate

Tilman Hetsch and Ulrich Rist

Institut für Aerodynamik und Gasdynamik, Universität Stuttgart,
Pfaffenwaldring 21, 70550 Stuttgart, Germany.
hetsch@iag.uni-stuttgart.de

Summary

A family of short leading-edge separation bubbles on a swept flat plate are investigated by means of direct numerical simulation (DNS) and spatial linear stability theory. An infinite swept plate with appropriate pressure distribution is a simple but suitable model for the mid-chord section of a swept wing or high lift device. A similar setup had therefore been chosen by Horton for his extensive experiments in this field in 1968. The main goal of the present study is to identify the influence of the sweep angle on the structure and stability of the flow through its systematic variation from 0° up to 60°. Due to the independence principle this influence on the base flow characteristics is only moderate. However, it does have a noticeable impact on stability. In contrast to the unswept two-dimensional (2D) case, *oblique* disturbance waves experience the highest amplification in swept cases.

1 Introduction

Due to strong adverse pressure gradients and low Reynolds numbers, separation bubbles are often encountered with high lift devices or on laminar airfoils of gliders. For instance, a transitional separation bubble was observed at the slat of an Airbus A310 [1]. Although the sweep angle of a modern passenger plane is of the order of 30°, so far numerical and experimental studies of separation bubbles have almost exclusively addressed the two-dimensional case.

The largest body of experimental data devoted to three-dimensional separation bubbles known to the authors was published by Horton in the late 1960s [2, 3]. Apart from Horton's benchmark data, literature on three-dimensional transitional separation bubbles is still rare. Davis and his coworkers successfully validated their boundary layer code with an inviscid-viscid interaction technique against Horton's data and thus confirmed his experiments [4]. More recently, Kaltenbach and Janke published a DNS of a transitional separation bubble in the flow behind a swept, rearward-facing step, with emphasis on its turbulent properties and sweep effects [5]. Up to an angle of 40° the mean flow followed the independence principle. The transition was reported to be triggered by a Kelvin-Helmholtz type instability of the shear-layer. Generally spatial growth rates increased with increasing sweep angle, though Kaltenbach and Janke found only a weak influence in their case.

Little is known about the transition mechanisms inside a 3D separation bubble, which are the key for understanding the global phenomena like control strategies or even the location of transition. There is the open question to which degree our knowledge of 2D separation bubbles can be extended to 3D configurations. The present study represents a first step in this direction. We investigate the effect of sweep on the structure and stability of a steady, laminar separation bubble. The resulting "stream pipe" can serve as a *base flow* for unsteady disturbance calculations in the future.

2 Numerical Method and Case Description

For the calculation of leading-edge separation bubbles on an infinite swept plate a body fitted coordinate system is most suitable. The z- and x-direction are the parallel and normal to the leading-edge, respectively. All quantities in the paper are nondimensionalised by the reference length $\overline{L} = 0.05\,\mathrm{m}$ and the chordwise free-stream velocity $\overline{U}_\infty = 30\,\frac{\mathrm{m}}{\mathrm{s}}$, which was held constant for all cases. Different sweep angles Ψ were realized by varying the spanwise free-stream velocity $\overline{W}_\infty = \overline{U}_\infty \tan(\Psi)$. The calculations were performed with a global Reynolds number of $Re = \overline{U}_\infty \overline{L}/\nu = 100000$, although for physical interpretation the flow is better characterised by $Re_{\delta_1} = \overline{U}_\infty\, \delta_1(x_o)/\nu = 331$, based on the displacement thickness at the inflow. The rectangular integration domain is shown in Figure 2(a). As the leading-edge is itself not included in the calculation domain, a Falkner-Scan-Cooke (FSC) profile without pressure gradient is prescribed at the inflow boundary $x_o = 0.37$. The wall-normal coordinate y ranges from 0 to $y = 0.059 = 18 \cdot \delta_1(x_o)$.

The calculation of the base flow is based on the three-dimensional, incompressible Navier-Stokes equations in a velocity-vorticity formulation for an infinite swept plate ($\boldsymbol{\omega} := -\mathrm{rot}(u, v, w,)^T$):

$$
\begin{aligned}
\frac{\partial \omega_x}{\partial t} + \frac{\partial (v\,\omega_x - u\,\omega_y)}{\partial y} &= \frac{1}{Re}\frac{\partial^2 \omega_x}{\partial x^2} + \frac{\partial^2 \omega_x}{\partial y^2}, & \frac{\partial^2 u}{\partial x^2} &= -\frac{\partial^2 v}{\partial x \partial y}\\
\frac{\partial \omega_y}{\partial t} - \frac{\partial (v\,\omega_x - u\,\omega_y)}{\partial x} &= \frac{1}{Re}\frac{\partial^2 \omega_y}{\partial x^2} + \frac{\partial^2 \omega_y}{\partial y^2}, & \frac{1}{Re}\frac{\partial^2 v}{\partial x^2} + \frac{\partial^2 v}{\partial y^2} &= -\frac{\partial \omega_z}{\partial x}\\
\frac{\partial \omega_z}{\partial t} + \frac{\partial (u\,\omega_z)}{\partial x} + \frac{\partial (v\,\omega_z)}{\partial y} &= \frac{1}{Re}\frac{\partial^2 \omega_z}{\partial x^2} + \frac{\partial^2 \omega_z}{\partial y^2}, & \frac{\partial^2 w}{\partial x^2} &= -\frac{\partial \omega_y}{\partial x}.
\end{aligned}
$$

The key assumption for an infinite swept wing is uniformity ($\frac{\partial}{\partial z} \equiv 0$) in spanwise direction for laminar calculations. This leads to a decoupling of the 2D-quantities u, v and ω_z, appearing in unswept calculations, from the new 3D-quantities ω_x, ω_y and w in the equation system above, which is known as the *independence principle*.

As a consequence, u, v, ω_z and any quantity derived directly from them do not depend on the sweep angle. The equations are discretised with fourth order central finite differences and integrated in a semi-implicit manner until steady state is reached. The steady calculation of the separation bubble is justified by its small size (e.g. the maximum backflow is $0.3\,\%\,\overline{U}_\infty$) and computational experience from [6].

Note that due to the total absence of disturbances the base flow is purely laminar. The flow separates and re-attaches following the adverse pressure gradient which is implicitly introduced by the deceleration of the chordwise potential velocity $u_e(x)$ shown in Figure 3(c). There are two constant u_e plateaus at the beginning and at the end of the domain, respectively, which are connected smoothly by an analytical function. The whole setup represents the 3D extension of the two-dimensional leading-edge bubble investigated in [6], only with a slightly lower u_e deceleration of 8.95 % $\overline{U}_\infty$. More details concerning the general numerical method can be found in [7].

3 Flow Topology and Validation

The flow topology of the three-dimensional separation bubble shown in Figure 2(c) is exactly the same as the one sketched by Horton (Figure 2(b)). Just like in the experiments, the independence principle enforces leading-edge-parallel separation and re-attachment lines, here at the 2D-positions $x_S = 1.75$ and $x_R = 2.13$ for all sweep angles. The internal flow follows helical paths, resulting from the superposition of a w-component and the closed streamline loops known from 2D calculations. The symmetrical cross-section in the x-y plane of the separation bubble is simply a result of the laminar scenario. As demonstrated in [6] for the 2D case, the typical triangle-shaped outline observed in experiments develops under the presence of unsteady disturbances which trigger transition.

For the special case of vanishing z derivatives Moore has proven, that the general form of a "displacement surface" in 3D flows simplifies to the 2D definition of the displacement thickness $\delta_1(x)$ at any given z position [8]. Therefore it is correct to calculate δ_1, the momentum thickness δ_2 and the boundary layer thickness δ_{99} from u only, as in the unswept case. Figure 3(a) demonstrates that our code respects the independence principle, for the results are independent of Ψ. At the same time one can see that δ_{99} reaches only 65 % of the height of the integration domain, so that it is well justified to assume potential flow at the upper boundary.

The step size independence of the numerical solution was investigated by grid refinement studies for the case $\Psi = 45°$. The standard resolution of 1394 and 97 grid points in x- and y-direction was independently halved and doubled. The maximum relative error with respect to the fine solution was then determined for all flow quantities $q \in \{u, v, w, \omega_x, \omega_y, \omega_z\}$ separately. The variation of the x-resolution yielded that the standard resolution was finer than necessary. The maximum deviation of the coarse solution to the fine one was 0.0017 %, so 698 grid points would have been sufficient. The maximum error in y occurring in v is shown in Figure 3(b). Clearly, the coarse resolution of only 49 grid points in y is insufficient. But the maximum error of the standard resolution compared to the fine one is still 0.07 % and only visible in an enlarged cutout of Figure 3(b). It follows that the resolution of the calculation is quite satisfactory.

4 Base Flow: 3D Structure and Influence of Sweep

From the independence principle it is clear that u, v, ω_z, separation and reattachment positions as well as boundary layer thickness are identical to the 2D case for any sweep angle. This 2D flow is merely *superimposed* with a w-velocity component which grows in strength with increasing sweep angle. The separation bubble has only an indirect influence on w through the displacement thickness. This is indicated by the w-isolines displayed in Figure 3(c) which generally follow the trend of δ_1. As seen in Figure 3(e) the w profiles increase linearly from the wall, and in close-ups immediate to the wall no influence of the bubble is visible in sharp contrast to the behaviour or u and v in this region. This was to be expected from the continuity equation which simplifies to its 2D form because $\partial / \partial z \equiv 0$. For the same reason the general definitions of ω_x and ω_y simplify to $\omega_x = -\frac{\partial w}{\partial y}$, $\omega_y = \frac{\partial w}{\partial x}$, so that these quantities depend directly on w. In a short region around the separation bubble the w profiles exhibit an inflexion point (e.g. Figure 3(e)) which is more easy to spot in the crossflow profiles w_s. As an overview, Figure 1 shows the downstream development of the streamline-oriented velocity components u_s and w_s for $\Psi = 30°$: due to constant u_e at the upper boundary there is no crossflow present at the inflow. Within the bubble, a strong crossflow-component w_s is induced by the local u_e deceleration. Finally, near the outflow the flow is asymptotically approaching FSC-characteristics, but a weak crossflow still exists.

The effect of sweep on the base flow can exemplarily be seen in the w profiles at the separation line in Figure 3(d). Increasing sweep increases the absolute value of a 3D quantity without changing its form. It simply scales a sweep-independent prototype profile, as demonstrated in Figure 3(e). This also holds for ω_x and ω_y as derivatives of w. Because of $w_e \equiv \overline{W}_\infty / \overline{U}_\infty$ the nondimensionalised profiles of w, ω_x, ω_y scale with the global factor $\tan(\Psi)$ like the values of w at the upper boundary. After normalisation all swept profiles coincide with the one at $\Psi = 45°$, which can be seen as the generic solution. This is a general consequence of the independence principle for infinite swept plates, not restricted to separation bubbles: As u and v are known from the 2D solution, the w-momentum equation is *linear* in w:

$$u \frac{\partial w}{\partial x} + v \frac{\partial w}{\partial y} = \frac{1}{Re} \left(\frac{\partial^2 w}{\partial x^2} + \frac{\partial^2 w}{\partial y^2} \right).$$

5 The Influence of Sweep on the Flow Stability

Despite of the only moderate influence of sweep on the laminar flow it has a noticeable influence on its stability. The inflection point in the swept crossflow profiles gives rise to an inviscid instability which is not present in the unswept case. For example, steady crossflow vortices with a frequency of 0 Hz become weakly amplified. Figure 4 summarises the results from three-dimensional spatial linear stability theory. Figure 4(e) gives an overview of the whole integration domain. The separation and re-attachment points are marked with *S* and *R*, respectively. The graph shows the spatial amplification rate $-\alpha_i$ as a function of x and the frequency β of the disturbance. It is valid for all sweep angles as a vanishing spanwise wavenumber $\gamma = 0$ cancels the influence of w in the Orr-Sommerfeld and the Squire equation. At the separation point $x_{\text{sep}} = 1.75$ the spatial amplification is displayed in cross sections for $\Psi = 0°, 15°, 30°$ and $45°$ in Figures 4(a) through 4(d). The initially symmetrical distribution in the planar case is more and more distorted and the amplification maximum moves to oblique waves with positive γ. In accordance with the findings in [5] increasing growth rates are observed for increasing sweep angles. In Figure 4(f) the position of the amplification maxima are marked relative to the 2D maximum. All relevant data of these maxima are summarised in table 1. It shows that the propagation direction of the most amplified wave is approximately 10° smaller than the local free-stream direction. Therefore Tollmien-Schlichting waves, not crossflow disturbances are most important in this case. The maximal amplification for $\Psi = 45°$ is 7 % higher than in the 2D case.

6 Conclusions

A family of three-dimensional laminar separation bubbles on an infinite swept plate has been successfully calculated and validated. The quantities w, ω_x, ω_y which determine the three-dimensional character of the flow have been analysed in detail. An systematic investigation of the effect of sweep up to $\Psi = 60°$ on structure and up to $\Psi = 45°$ on stability of the flow yielded the following results:

- Due to the independence principle the separation bubble is only moderately influenced by the sweep angle: The 3D quantities w, ω_x and ω_y are superimposed on the 2D quantities u, v and ω_z without changing the general flow topology. The sweep angle merely scales the "generic" solution $\Psi = 45°$ in w, ω_x and ω_y with a global factor of $tan(\Psi)$. These features make the calculated flow an ideal candidate for disturbance investigations: All properties of the flow are now known. As no new features of the flow appear for higher sweep angles, the effects of sweep are isolated.
- Despite of the simple structure and although two-dimensional disturbances are amplified *earlier* according to Squire's theorem, the amplification rates of oblique waves are up to 7 % *stronger* in the case $\Psi = 45°$. For 3D flows it is therefore generally not sufficient to restrict oneself to the investigation of 2D waves when determining basic properties like N-factors.

Acknowledgements

The authors would like to thank Professor Gaster and Dr. Horton from the Queen Mary College, University of London, for providing us with a copy of Horton's thesis and allowing us to reprint a figure from it. The financial support by the Deutsche Forschungsgemeinschaft (DFG) under contract number Ri 680/12-1 is gratefully acknowledged.

References

[1] E. Greff: "In-flight Measurements of Static Pressures and Boundary-Layer State with Integrated Sensors", J. Aircraft, *28*, No. 5, (1991), pp. 289-299.

[2] H. Horton: "Some Results of Investigations of Separation Bubbles". AGARD **CP-4**, (1966), pp. 779–811.

[3] H. Horton: "Laminar Separation Bubbles in Two and Three Dimensional Incompressible Flow". PhD thesis, Department of Aeronautical Engineering, Queen Mary College, University of London, (1968).

[4] R. Davis, J. Carter, E. Reshotko: "Analysis of Transitional Separation Bubbles on Infinite Swept Wings". AIAA Vol. **25**, No. 3, (1987), pp. 421–428.

[5] H. Kaltenbach, G. Janke: "Direct numerical simulation of flow separation behind a swept, rearward-facing step at $Re_{\mathrm{H}} = 3000$". Phys. Fluids, **12**, (2000), pp. 1220–2337.

[6] U. Rist: "Zur Instabilität und Transition laminarer Ablöseblasen". Habilitation, Universität Stuttgart, Shaker Verlag, (1999).

[7] P. Wassermann: "Mechanisms and passive control of crossflow-vortex-induced transition in a three-dimensional boundary layer". J. Fluid Mech., **456**, (2002), pp. 49–84.

[8] F. Moore: "Displacement Effect of a Three-Dimensional Boundary Layer". NACA Report 1124, (1953).

Table 1 Parameters of the most amplified disturbance at x_{sep} for various sweep angles Ψ: spatial amplification rate $-\alpha_i$, chordwise wavenumber α_r, spanwise wavenumber γ, frequency β, local propagation direction $\varphi = \arctan(\gamma/\alpha_r)$, local free-stream direction $\varphi_e = \arctan(w_e/u_e)$, difference $\Delta\varphi = \varphi - \varphi_e$.

Ψ	$-\alpha_{i_{\mathrm{Max}}}$	α_r	γ	β	φ	φ_e	$\Delta\varphi$
0°	10.89	47.32	0.00	17.17	0.0°	0.0°	0.0°
15°	10.95	46.54	5.76	17.53	7.0°	15.9°	−8.9°
30°	11.17	42.73	14.43	18.99	18.6°	31.6°	−13.0°
45°	11.65	36.20	25.01	23.90	34.6°	46.8°	−11.2°

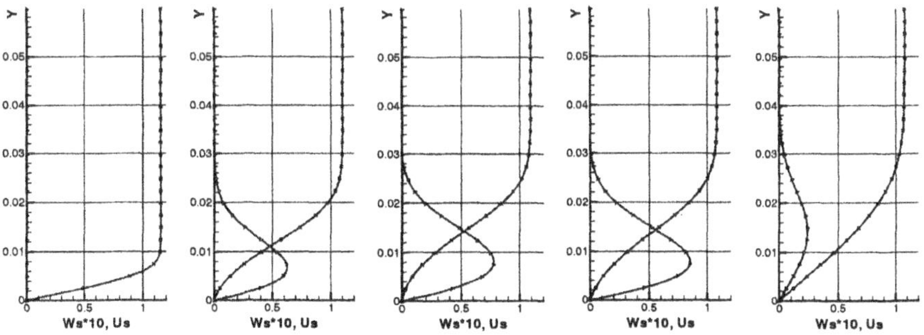

Figure 1 Streamline-oriented velocity profiles u_s (square), w_s (triangle) at the inflow, at separation, at the middle of the bubble, at reattachment, far downstream at $x = 4.4$.

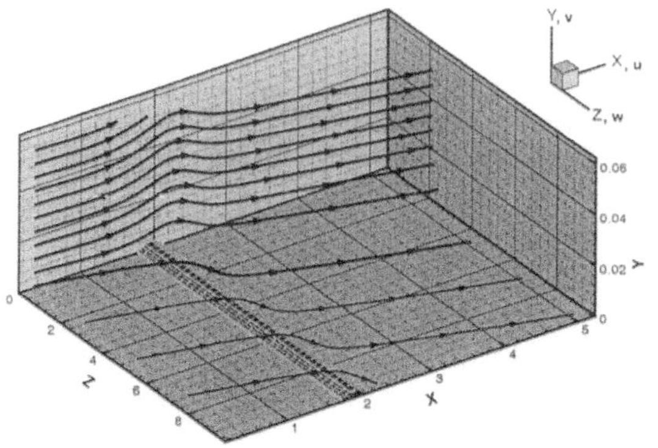

(a) $\Psi = 30°$: the entire integration domain.

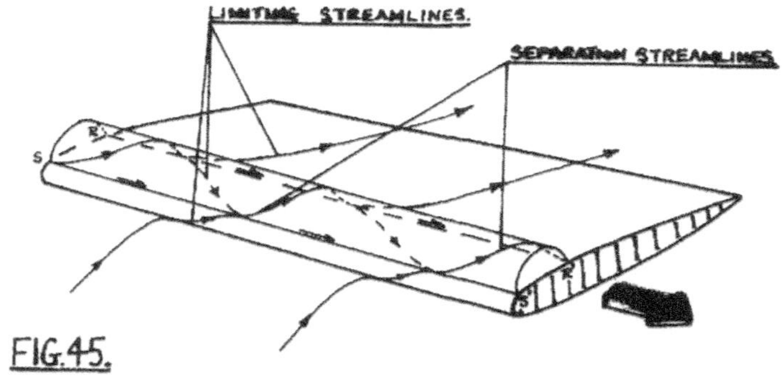

(b) Horton's model of a leading-edge bubble taken from [3].

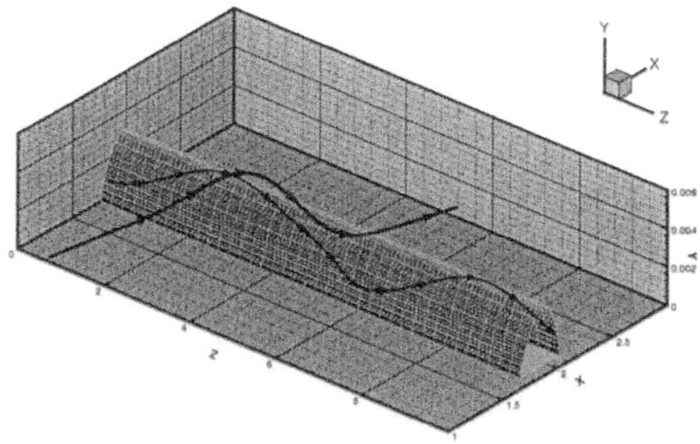

(c) Internal & external streamlines in simulation.

Figure 2 Global flow topology of a separation bubble on a infinite swept plate.

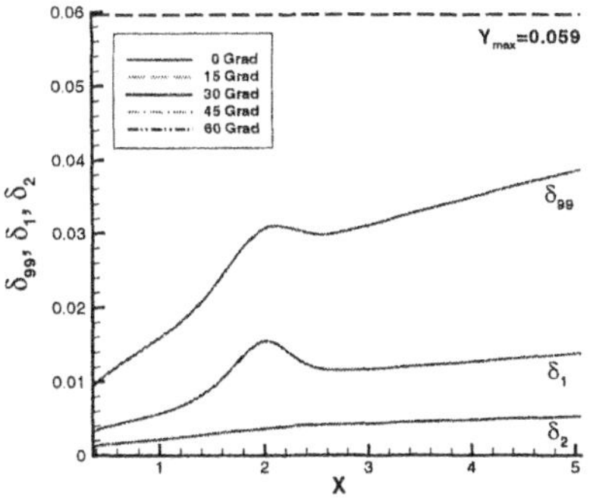

(a) The sweep independent boundary layer thicknesses.

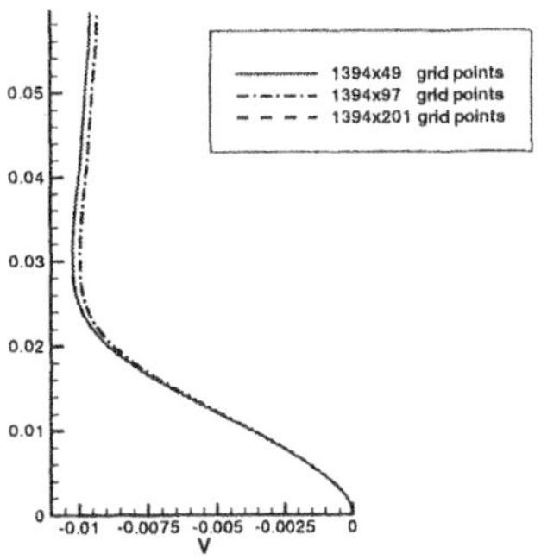

(b) Step size test with 49, 97 and 201 grid points in y .

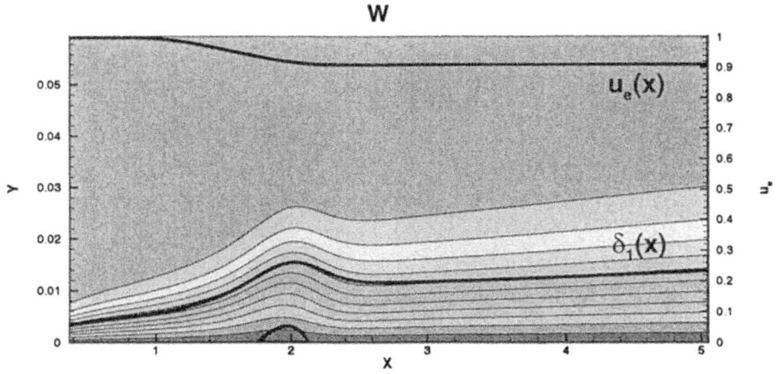

(c) Overview over the w-component and u at the upper boundary.

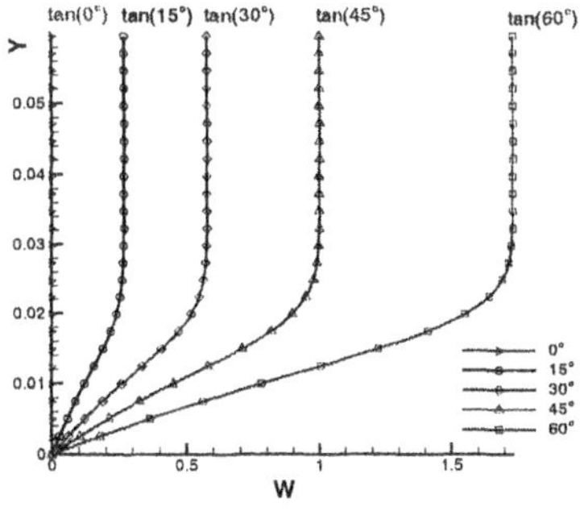

(d) w profiles at $x_{\text{sep}} = 1.75$. Upper BC: $w_e = \tan(\Psi)$.

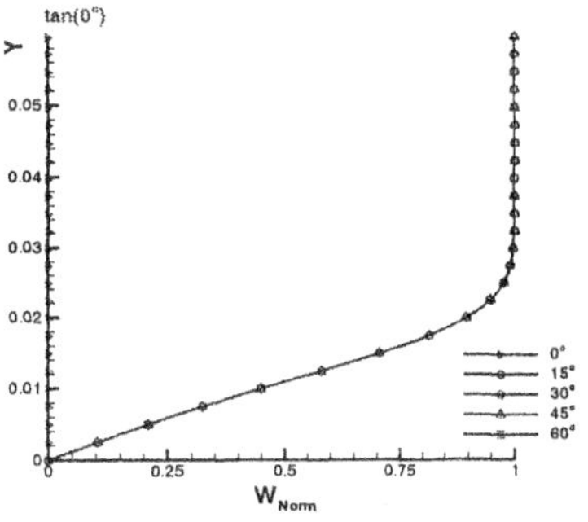

(e) Normalised w profiles at $x_{\text{sep}} = 1.75$, note inflection point.

Figure 3 Laminar flow properties and influence of sweep

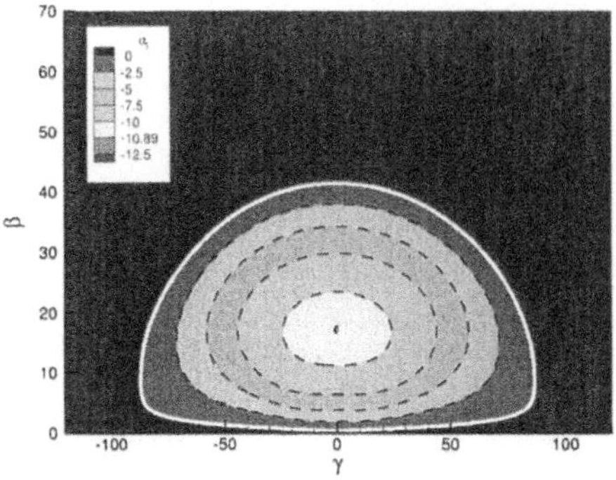

(a) Stability diagram for $\Psi = 0°$.

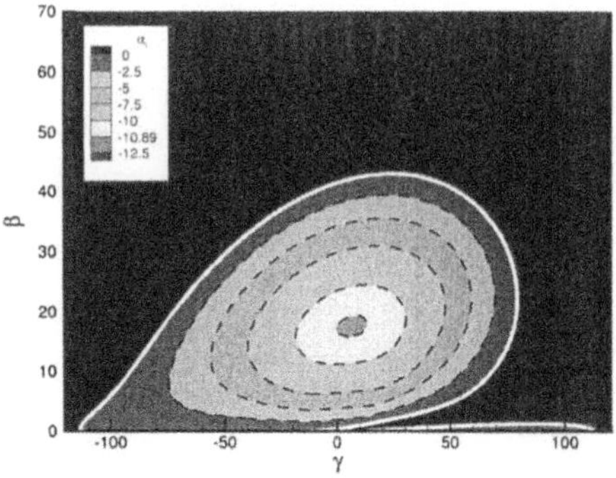

(b) Stability diagram for $\Psi = 15°$.

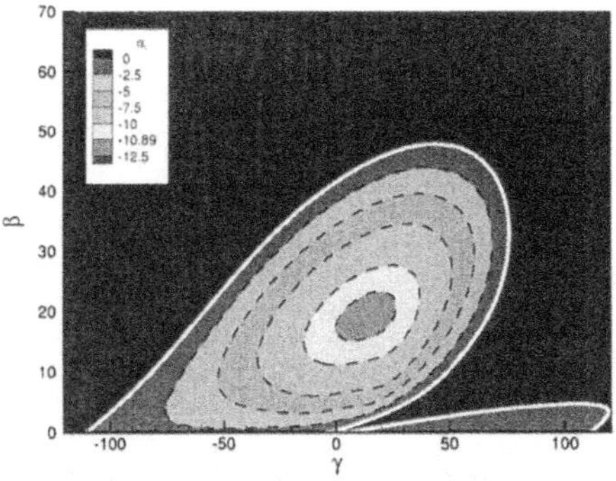

(c) Stability diagram for $\Psi = 30°$.

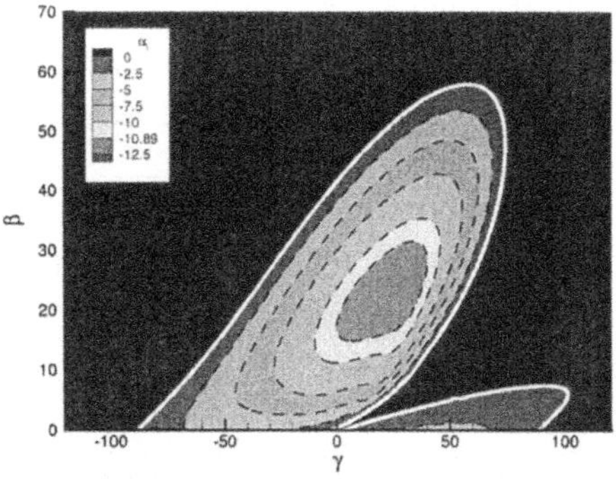

(d) Stability diagram for $\Psi = 45°$.

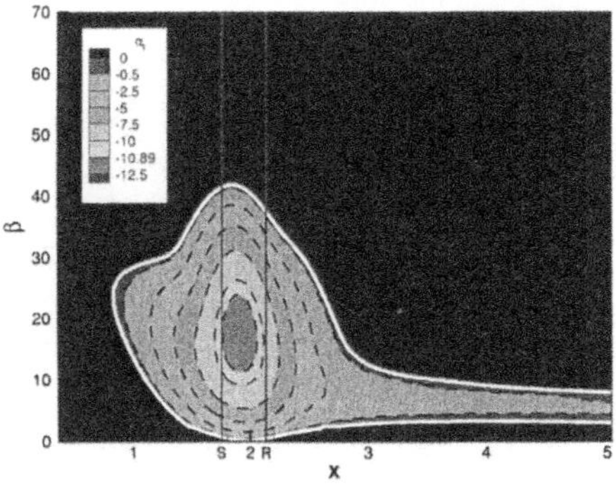

(e) Stability diagram $\alpha_i(x, \beta)$ with fixed $\gamma = 0$ (for all Ψ).

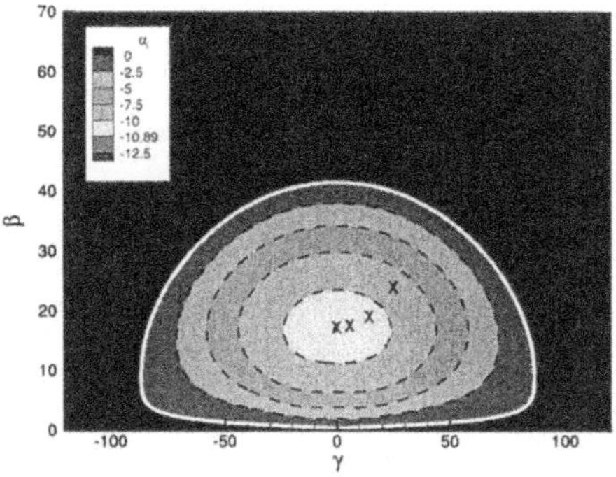

(f) Position of $-\alpha_i$-Maxima for all Ψ's: plotted into the $\Psi = 0°$ case.

Figure 4 Linear stability theory results at $x = 1.75$ (separation line) versus frequency β and spanwise wavenumber γ. The white curve corresponds to neutral stability. The innermost contour level is kept constant at the maximal spatial growth rate $-\alpha_i = 10.89$ of the case with sweep angle $\Psi = 0°$.

Receptivity Study of the Attachment–Line Instability on a Blunt Body in Supersonic Flow

Jörn Sesterhenn and Rainer Friedrich

Fachgebiet Strömungsmechanik, TU–München, D–85757 Garching,
joern.sesterhenn@alumni.ethz.ch, r.friedrich@lrz.tum.de

Summary

The receptivity on a yawed parabolic leading edge in supersonic flow is studied by direct numerical simulation. We investigate the response of the boundary layer due to entropy disturbances ahead of the shock. These disturbances interact with the shock and generate entropy, vorticity and acoustic disturbances of the non–homogeneous base flow. In turn they impinge on the leading edge and induce a boundary–layer response akin to the first eigenmodes of the flat plate.

We use a characteristic–type numerical scheme which was developed for direct numerical simulations of compressible transitional and turbulent flows [14]. In these flows numerical dissipation and dispersion rather than the capturing of shocks is the primary issue and a primitive variable formulation is employed in order to achieve high accuracy and performance. The bow shock is treated explicitly with a shock–fitting procedure [1].

1 Introduction

By *receptivity*, we understand the entrainment process of free–stream perturbations into a boundary layer and the subsequent response of this layer which eventually leads to transition. While linear stability analysis computes eigenfunctions of an idealised problem and asks whether they are going to grow or decay, receptivity studies ask how finite arbitrary external disturbances are converted into suitable candidates for growing instabilities. They may even *by–pass* the classical linear growth and trigger more rapid transition mechanisms. Possible conversion mechanisms are connected to the leading edge, curvature changes or surface roughnesses of the body in question [5].

In this paper we discuss the stagnation–line flow along a curved leading edge, as depicted in figure 1(left). Empirically, the attachment–line instability was found to be important for laminar/turbulent transition on swept wings since the early fifties and was investigated experimentally and theoretically by several authors. See [11] for a review.

The results of their findings might be summarised as follows:
For a Reynolds number

$$Re = \frac{w_\infty}{\sqrt{\frac{2u_\infty}{R}\nu}} \tag{1}$$

the incompressible swept flow over a flat plate is linearly stable up to $Re \approx 580$ [3]. Here w_∞ is the sweep velocity and $\frac{2u_\infty}{R}$ is taken to be the velocity gradient $\frac{\partial u_\infty}{\partial x} \equiv -\frac{\partial v_\infty}{\partial y}$ of the potential solution for stagnation point flow. Using a trip–wire to introduce large disturbances into the boundary layer one can observe a subcritical instability for Reynolds numbers as low as $Re \approx 250$ [11]. For a curved leading edge, theoretical investigations [9] indicate a stabilisation due to curvature. For a finite curvature radius of the order of 10^3 boundary–layer thicknesses the critical Reynolds number increases to $Re \approx 637$. From experiments up to $\overline{M} = 3.3$ [10], the sweep–Machnumber $\overline{M} = w_\infty / c_\infty$ is believed to be slightly stabilising. Finally the wall temperature has major influence on the growth rate of the disturbances [7], and wall cooling is found to stabilise an otherwise unstable flow configuration [13]. To facilitate a comparison, the viscosity is evaluated at recovery temperature.

The first DNS studying this subject was performed by Spalart [15] for the case of an incompressible fluid impinging on a flat plate. We want to extend these studies in several directions: We consider a **compressible fluid** over a **curved leading edge** under **supersonic flow conditions**.

In the paper we present our numerical method (Sec.2.1) along with a description of boundary conditions (Sec. 2.2) and the validation of the scheme (Sec. 2.3). Further, we show the base flow solution (Sec.3), the kind of disturbance we want to investigate (Sec.4) as well as interaction studies (Sec. 5).

2 Description of the numerical scheme

2.1 Interior Scheme

The Navier–Stokes equations in Cartesian coordinates may be written in the following form:

$$
\begin{aligned}
\frac{\partial p}{\partial t} &= -\frac{\rho c}{2}\left((X^+ + X^-) + (Y^+ + Y^-) + (Z^+ + Z^-)\right) + \\
&\quad + \frac{p}{C_v}\left(\frac{\partial s}{\partial t} + X^s + Y^s + Z^s\right), \\
\frac{\partial u}{\partial t} &= -\left(\frac{1}{2}(X^+ - X^-) + Y^u + Z^u\right) + \frac{1}{\rho}\frac{\partial \tau_{1j}}{\partial x_j}, \\
\frac{\partial v}{\partial t} &= -\left(X^v + \frac{1}{2}(Y^+ - Y^-) + Z^v\right) + \frac{1}{\rho}\frac{\partial \tau_{2j}}{\partial x_j}, \\
\frac{\partial w}{\partial t} &= -\left(X^w + Y^w + \frac{1}{2}(Z^+ - Z^-)\right) + \frac{1}{\rho}\frac{\partial \tau_{3j}}{\partial x_j}, \\
\frac{\partial s}{\partial t} &= -(X^s + Y^s + Z^s) + \frac{R}{p}\left(-\frac{\partial q_i}{\partial x_i} + \Phi\right).
\end{aligned}
\tag{2}
$$

with the following abbreviations:

$$
X^+ := (u + c)\left(\frac{1}{\rho c}\frac{\partial p}{\partial x} + \frac{\partial u}{\partial x}\right); \quad X^- := (u - c)\left(\frac{1}{\rho c}\frac{\partial p}{\partial x} - \frac{\partial u}{\partial x}\right);
$$

$$X^v := u\frac{\partial v}{\partial x}; \quad X^w := u\frac{\partial w}{\partial x}; \quad X^s := u\frac{\partial s}{\partial x};$$

$$Y^+ := (v+c)\left(\frac{1}{\rho c}\frac{\partial p}{\partial y} + \frac{\partial v}{\partial y}\right); \quad Y^- := (v-c)\left(\frac{1}{\rho c}\frac{\partial p}{\partial y} - \frac{\partial v}{\partial y}\right);$$

$$Y^u := v\frac{\partial u}{\partial y}; \quad Y^w := v\frac{\partial w}{\partial y}; \quad Y^s := v\frac{\partial s}{\partial y};$$

$$Z^+ := (w+c)\left(\frac{1}{\rho c}\frac{\partial p}{\partial z} + \frac{\partial w}{\partial z}\right); \quad Z^- := (w-c)\left(\frac{1}{\rho c}\frac{\partial p}{\partial z} - \frac{\partial w}{\partial z}\right);$$

$$Z^u := w\frac{\partial u}{\partial z}; \quad Z^v := w\frac{\partial v}{\partial z}; \quad Z^s := w\frac{\partial s}{\partial z}.$$

By insertion, the reader may verify that these equations are equivalent to the Navier–Stokes equations in their usual Cartesian form. (We use Cartesian coordinates here only for a more concise description. The equations for arbitrary coordinates on a timedependent grid are developed in [14].) ρ,c,R,C_v are the density, the speed of sound, the specific gas constant and the specific heat at constant volume. Each quantity $X^\pm$, $Y^\pm$ etc. describes the transport of information on a selected direction along the Monge cone constructed for the Euler equations. Accordingly, each of these terms is discretised with an upwind scheme, in our case a compact scheme of fifth order accuracy. The terms τ_{ij}, q_i and Φ are the viscous stress tensor, the conductive heat flux vector and the rate of energy dissipation. These terms are of dissipative nature and discretised centrally by a sixth order compact scheme [8]. Time advancement of the equations is done with a Runge–Kutta scheme of 3rd or 4th order.

This special form of writing the Navier–Stokes equations has two major advantages. Firstly, it serves as the basis for the proper upwinding and thus the stabilisation of the numerical scheme with a minimum of numerical dissipation. Secondly, it provides a very convenient way of prescribing proper boundary conditions both at solid boundaries of the flow domain as well as at block boundaries in a parallel computation.

2.2 Boundary conditions

Consider **a subsonic outlet** in a plane at $x = x_1$. If the flow is from left to right, four characteristics, corresponding to X^+, X^s, X^v and X^w will leave the computational domain. They must not be prescribed and their values are properly evaluated from the interior of the domain. X^- is unknown and needs to be prescribed in such a way that the pressure assumes the desired value $p(t)$. We may differentiate $p(t)$ and use $(2)_1$. This yields an equation for the unknown X^-:

$$\begin{aligned} X^- = & -\left(X^+ + (Y^+ + Y^-) + (Z^+ + Z^-)\right) \\ & + \frac{2p}{\rho c C_v}\left(\frac{\partial s}{\partial t} + X^s + Y^s + Z^s\right) - \frac{2}{\rho c}\dot{p}(t) \end{aligned} \tag{3}$$

All quantities on the right hand side (including $\frac{\partial s}{\partial t}$ from $(2)_5$) are known. Furthermore no assumption or simplification of the boundary conditions was necessary. The only difference to the interior scheme is the fact that both, convective and dissipative terms, are discretised one–sided. For the convective terms this is in full accordance with their upwind direction. For the dissipative term this is all we can do.

The **bow shock** is treated in essentially the same way. The "only" difference lies in the fact that four characteristics, corresponding to X^+, X^s, X^v and X^w will *enter* the computational domain which is identified as the high pressure side of the shock. Additionally one has to solve for the shock acceleration and distort the grid accordingly. In short, the Rankine–Hugoniot equations in a coordinate system fixed to the shock are differentiated with respect to time and the resulting system of equations is solved for the four unknowns as functions of the upstream values $\dot{p}_1(t), \dot{\mathbf{u}}_1(t)$, and $\dot{s}_1(t)$ — index 1 denotes the low pressure side of the shock — as well as the X^- wave which enters the shock from the interior of the computational domain. The procedure can be found for example in [2] up to the point where one has determined $\dot{p}_2(t), \dot{\mathbf{u}}_2(t), \dot{s}_2(t)$ and $\dot{\mathbf{u}}_s(t)$, where the index 2 denotes the high pressure side of the shock and $\mathbf{u}_s(t)$ is the local shock velocity. The values for X^+, X^s, X^v and X^w can be computed from the non–Cartesian form of equation(2):

$$X^+ := \frac{2}{h_1}\left(\frac{\tau_x}{\rho_2}+\frac{h_1}{2}X^- - \dot{u}_2(t) - \frac{g_{13}}{h_3 2}(Z^+ - Z^-) - Z^u - \frac{g_{12}}{2h_2}(Y^+ - Y^-) - Y^u\right),$$

$$X^v := \frac{\tau_y}{\rho_2} - \dot{v}_2(t) - \frac{g_{23}}{2h_3}(Z^+ - Z^-) - Z^v - \frac{h_2}{2}(Y^+ - Y^-) - \frac{g_{12}}{2h_1}(X^+ - X^-),$$

$$X^w := \frac{\tau_z}{\rho_2} - \dot{w}_2(t) - \frac{h_3}{2}(Z^+ - Z^-) - \frac{g_{23}}{2h_2}(Y^+ - Y^-) - Y^w - \frac{g_{13}}{2h_1}(X^+ - X^-),$$

$$X^s := \sigma\frac{R}{p_2} - \dot{s}_2(t) - Z^s - Y^s.$$

For the definition of the metric terms g_{ij} and h_i and the definition of the waves $X^{\pm}$, X^s, X^v and X^w for the curvilinear case, see [14].

The location of the shock is determined by integration of $\dot{\mathbf{u}}_s(t)$ with the same time–advancement scheme as the flow variables.

2.3 Validation

For validation of our numerical method, we present two computations. The stationary flow about an unswept circular cylinder at $M \approx 6$ and $Re \approx 2000$, as well as the interaction of an entropy spot with a shock. The latter is deferred to Sec. 4, because we use the same disturbance for our receptivity studies.

In figure 1 (right) the temperature distribution in the zone between the shock and an isothermal cylinder at $M = 5.73$ and $Re_D = 2050$ is shown. The shock forms the inflow boundary in the present shock–fitting algorithm and is not plotted. This test case was shown by Kopriva [6]. He used a spectral code together with a similar shock–fitting technique. We have used a comparable grid of 30x60 points. The quantitative agreement between his and our solution is very good.

3 Base flow solutions

We compare three compressible flow situations. Firstly (α_1) a flat plate at $\overline{M} = 0.3$, $Re = 644$ with an adiabatic wall and isothermal wall at recovery temperature[7]. The second case (α_2) is essentially the same except for a curvature $\delta/R = 1/212$ and the fact that the wall is slightly cooled. [13]. An adiabatic computation is under way. The third flow–case (β) is a supersonic flow with $\delta/R = 1/247$, $\overline{M} = 1.25$ and a heated wall.

Figure 3(a) shows the base flow along the stagnation line. The markers are the similarity solution which very closely correspond to the two weakly compressible cases (not shown). The other lines correspond to the supersonic flow, both for the adiabatic wall and the heated wall presented herein. The normal flow velocities do not collapse with the incompressible solution if the viscosity is evaluated at recovery temperature. A suitable temperature can be found empirically but it does not correspond to the empirical reference temperature suggested by Poll [11].

4 Entropy disturbance

The three–dimensional base flow is disturbed by an entropy spot. I.e., we prescribe the free–stream entropy as

$$s = s_\infty + \epsilon\, C_p\, e^{-\left(\frac{\|\mathbf{x}-\mathbf{x}_0\|}{C}\right)^2}. \tag{4}$$

After interaction with the shock, the spot is compressed in streamwise direction. The emanating vorticity field consists of a vortex ring. In the 2D case it is represented by two counter–rotating vortices (Figure 2, right graph). Their sense of rotation depends on the sign of ϵ in equation (4). A secondary pair of weaker vortices follows. The acoustical field consists of a rarefaction or compression wave, also depending on the sign of ϵ (Figure 2, left graph).

This disturbance was investigated first for the two–dimensional planar case and also serves as a validation of the shock–fitting algorithm [1]. The plots in figure 2 show a superposition of the numerical results and a linear theory as developed by Fabre [1] on the basis of Ribner's results [12].

5 Results

Case α_1 is unstable with both boundary conditions. However, the isothermal wall exhibits a considerably larger growth rate [7]. The other cases (α_2) and (β) are stable, the second one due to the cooled wall, because the Reynolds numbers lie within the unstable region. The third case is situated outside of the instability region of the *incompressible* flow. The heating was obviously not able to enforce the instability. However, all three cases exhibit the same eigenmodes. The most significant "Görtler–Hämmerlin" mode [4] with linear growth in chordwise direction can be detected in all cases. This is shown in figure 3 (b)—(d).

6 Conclusions

Supersonic attachment–line flow exhibits qualitatively the same features as its flat plate relatives. Also in this case, the "Görtler–Hämmerlin" modes, which were analytically introduced for feasibility reasons only, are found.

References

[1] David Fabre, Laurent Jacquin, and Jörn Sesterhenn. Linear interaction of a cylindrical entropy spot with a shock. *Physics of Fluids*, August 2001.

[2] G.Erlebacher, M.Y.Hussaini, and T.L.Jackson. Nonlinear strong shock interactions: A shock–fitted approach. *Theoret.Comput.Fluid Dynamics*, 11:1–29, 1998.

[3] P. Hall, M. Malik, and D.I. Poll. On the stability of an infinite swept attachment–line boundary layer. *Proc. R. Soc. Lond.*, A(395):229–245, 1984.

[4] G. Hämmerlin. *50 Jahre Grenzschichtforschung*, chapter Zur Instabilitätstheorie der ebenen Staupunktströmung, pages 315–327. Vieweg, Braunschweig, 1955.

[5] E. Kerschen. Boundary layer receptivity. In *AIAA Paper No. 89-1109*, 1989.

[6] David A. Kopriva. Spectral solution of the viscous blunt–body problem. *AIAA Journal*, 31(7), July 1993.

[7] Anne Le Duc. *Simulation numerique Directe des Instabilites dans l'Ecoulement Tridimensionel Compressible Hiemenz*. PhD thesis, Laboratoire de Mecanique des Fluides, EC Lyon & Fachgebiet Strömungsmechanik, TU München, 2001.

[8] Sanjiva K. Lele. Compact finite difference schemes with spectral-like resolution. *Journal of Computational Physics*, 103:16–42, 1992.

[9] Ray-Sing Lin and Mujeeb R. Malik. On the stability of attachment–line boundary layers. part 2. the effect of leading edge curvature. *J. Fluid Mech.*, 333:125–137, 1997.

[10] Akira Murakami, Egon Stanewsky, and Paul Krogmann. Boundary–layer transition on swept cylinders at hypersonic speeds. *AIAA Journal*, 34(4):649–654, 1996.

[11] D.I.A. Poll. Transition description and prediction in three-dimensional flows. In *Special Course on Stability and Transition of laminar Flow*, pages 5/1–5/23. AGARD, 1984.

[12] H.S. Ribner. Convection of a pattern of vorticity through a shock wave. Technical Report TR 1164, NACA, 1953.

[13] Florian Schwertfirm. DNS von Instabilitäten in der dreidimensionalen kompressiblen Strömung entlang der Vorderkante eines gepfeilten Tragflügels. Master's thesis, TU-München, Fachgebiet Strömungsmechanik, August 2002.

[14] Jörn Sesterhenn. A characteristic–type formulation of the Navier–Stokes equations for high order upwind schemes. *Computers & Fluids*, 30(1):37–67, 2001.

[15] P.R. Spalart. Direct numerical study of leading edge contamination. Technical Report CP-438, Fluid Dyn. of 3D Turb. Shear Flows and Transition, AGARD, 1988.

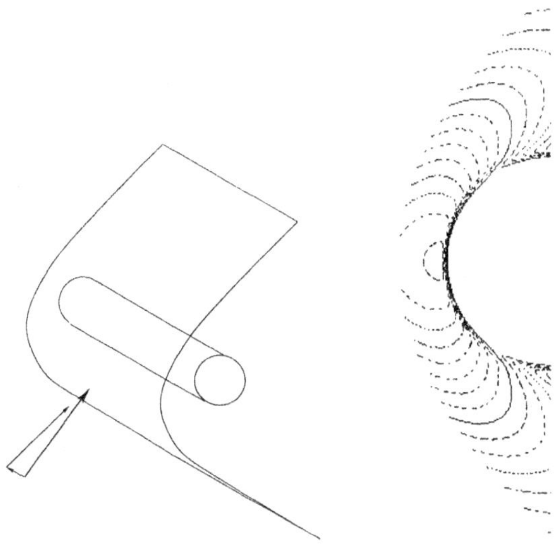

Figure 1 Left: Base flow geometry. Right: The temperature distribution T/T_{wall} between shock and cylinder at $M = 5.73$. Lines are drawn at intervals of 0.25 and increase towards the stagnation point. The solid line corresponds to a value $T/T_{\mathrm{wall}} = 5.5$

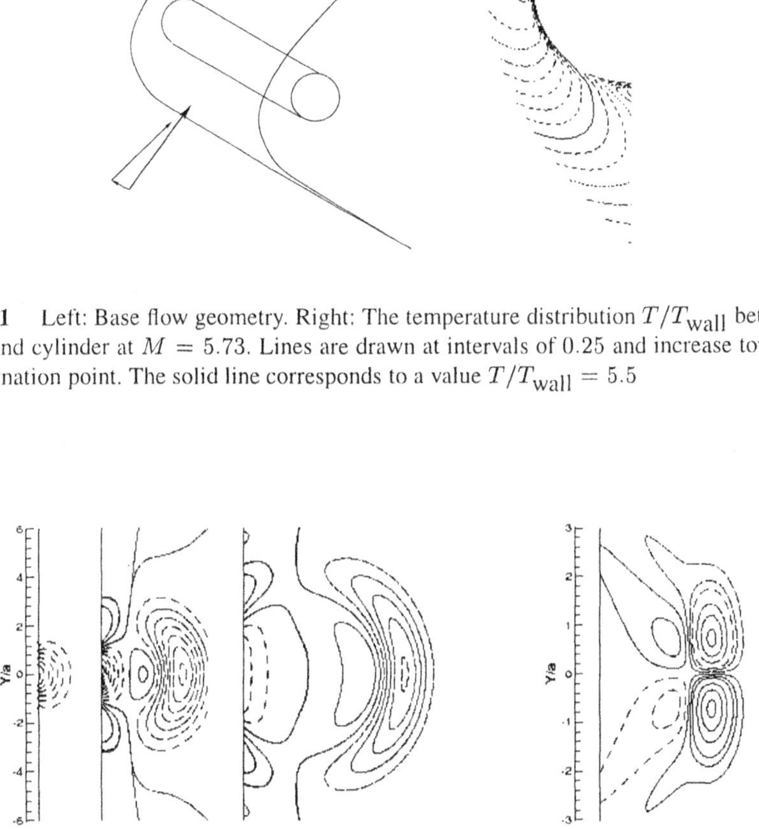

Figure 2 Interaction of an entropy spot with a planar shock. We show three instances of the pressure and one of the vorticity fields for $M_1 = 2$. The plots display a superposition of results achieved by linear theory and numerical simulation. The pressure plots correspond to three different non–dimensional times $t = 0, 2, 4$ and the vorticity field to $t = 4$. Reproduced from [1]. See this article for further details.

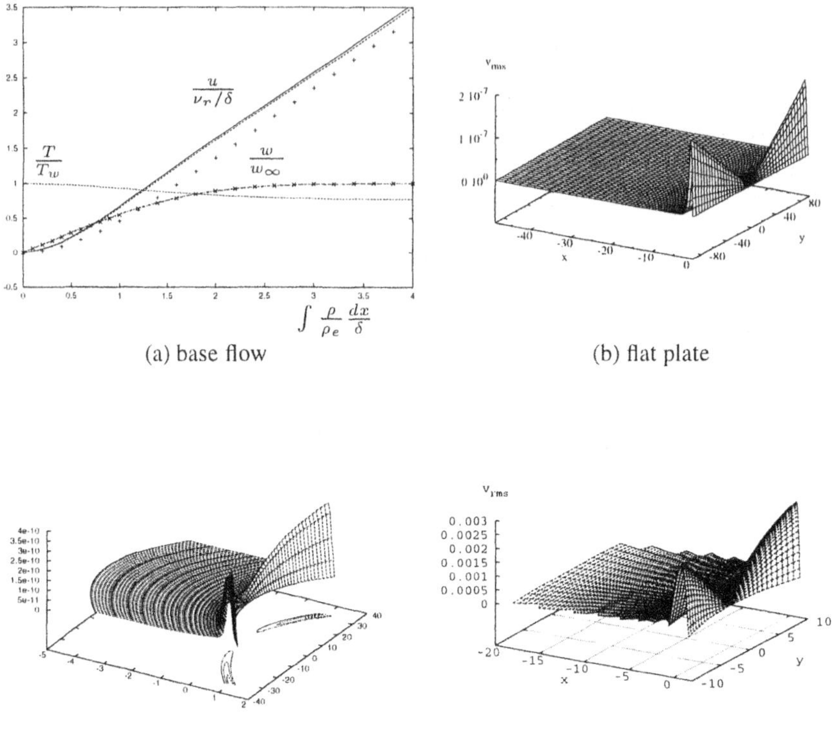

(a) base flow

(b) flat plate

(c) curved plate

(d) supersonic case

Figure 3 Base flow and disturbances

Numerical Investigation of the Flow in the Vicinity of an Isolated 3D Surface Roughness

Anke Wörner, Ulrich Rist, and Siegfried Wagner

Universität Stuttgart, Institut für Aerodynamik und Gasdynamik
Pfaffenwaldring 21, D-70550 Stuttgart, Germany

Summary

In the present paper, a detailed numerical investigation of the flow field in the vicinity of an isolated 3D surface roughness is presented. The Direct Numerical Simulations (DNS) shown are based on the velocity-vorticity formulation of the complete Navier–Stokes equations for incompressible fluids. The small surface roughness is modeled within the cartesian grid using an immersed boundary technique.

In the wake of the surface irregularity, a pair of counter-rotating streamwise vortices is visible at each side of the roughness. The pair of vortices results in a high-speed streak downstream of the spanwise edge of the roughness, which is flanked by two low-speed streaks. The whole vortex system looks very similar to horseshoe vortex systems that were observed in experiments around surface mounted cylinders or cubes.

1 Introduction

The specific fuel consumption of any aircraft is directly related to its drag. A major portion of aircraft drag is due to friction which is confined to the wall boundary layer of the flow. Since a turbulent boundary layer produces much higher skin friction than a laminar one, the overall skin friction is highly influenced by the location of laminar-turbulent transition. By means of control (i.e. delay) of laminar-turbulent transition a reduction of the wall friction is possible. A better understanding of the mechanisms of laminar-turbulent transition is the key for being able to actively or passively delay transition on an airfoil.

As long as the external perturbations in the freestream are small, the process of laminar-turbulent transition can be subdivided into four main stages. The first stage, the so called receptivity, is the penetration of external perturbations into the boundary layer where they are tuned to boundary layer disturbances. The second stage is the linear amplification of these initially created disturbances, the third one is the nonlinear development and the last one the breakdown to turbulence. This process of transition from the laminar to the turbulent state may be influenced by external perturbations in the freestream such as vortices, sound, entropy or particles as well as by internal perturbations in the boundary layer itself such as vibrations or surface non-uniformities.

In the manufacturing process of an airfoil, surface discontinuities such as steps at junctions or small humps are unavoidable. Additionally, surface roughness may be caused by contamination of the airfoil surface by insects, ice or mud. All these surface discontinuities can influence the location of transition on the airfoil via two dominating mechanisms. First, they are possible sources of receptivity, which means that they provide the small length scale which is necessary for the conversion process of large scale external perturbations into small scale boundary layer disturbances (see [1], [2] or [4]). A second aspect is however, that they are also able to either locally stabilize or destabilize the boundary layer.

The study presented here shall provide a first insight into the influence of an isolated 3D surface non-uniformity on the transition process. As a first step, the perturbations that are created by the presence of the surface non-uniformity itself are studied in detail using direct numerical simulations.

2 Numerical Method

The DNS code used for the present investigation is based on a code first developed by Fasel and further improved by Rist, Kloker et al., which is used for the investigation of transition phenomena on a flat plate.

The code is based on the velocity-vorticity formulation of the complete Navier–Stokes equations for incompressible fluids. The flow field is discretised using sixth order accurate compact finite differences in streamwise (x-) and wall-normal (y-) direction. An equidistant grid is used in x-direction, whereas grid stretching is applied in y-direction, leading to a very fine grid in the near wall region. In spanwise (z-) direction a spectral representation of the flow field is used. The time integration is done using a fourth order, four step Runge–Kutta scheme. A sketch of the integration domain including a buffer domain at the outflow boundary to avoid reflections at this boundary is shown in figure 1.

The localised surface irregularity within the cartesian grid is modeled using a technique which is related to Peskin's immersed boundary approach [8], and which has already successfully been applied for example by Goldstein et al. [3], Linnick [7] or von Terzi et al. [10]. In this approach, the effect of the surface irregularity on the surrounding flow field is modeled with an external force field which enforces no-slip and no-through-flow at selected grid points or at selected points between the grid points at every time step.

The vorticity transport equations with an external force field $\boldsymbol{F} = (F_x, F_y, F_z)$ can be written as

$$\begin{aligned}
\frac{\partial \omega_x}{\partial t} + \frac{\partial}{\partial y}(v\omega_x - u\omega_y) - \frac{\partial}{\partial z}(u\omega_z - w\omega_x) &= \frac{1}{Re}\Delta\omega_x + \frac{\partial F_y}{\partial_z} - \frac{\partial F_z}{\partial y} \\
\frac{\partial \omega_y}{\partial t} - \frac{\partial}{\partial x}(v\omega_x - u\omega_y) + \frac{\partial}{\partial z}(w\omega_y - v\omega_z) &= \frac{1}{Re}\Delta\omega_y + \frac{\partial F_z}{\partial_x} - \frac{\partial F_x}{\partial_z} \qquad (1)\\
\frac{\partial \omega_z}{\partial t} + \frac{\partial}{\partial x}(u\omega_z - w\omega_x) - \frac{\partial}{\partial y}(w\omega_y - v\omega_z) &= \frac{1}{Re}\Delta\omega_z + \frac{\partial F_x}{\partial y} - \frac{\partial F_y}{\partial x}
\end{aligned}$$

Here u,v and w denote the streamwise, wall-normal and spanwise velocity components and ω_x, ω_y and ω_z the streamwise, wall-normal and spanwise components of vorticity. t denotes the time and Re is a Reynolds number which is determined by the choice of the reference length and velocity used for non-dimensionalisation.

The force term on the right hand side is the force exerted by the surface irregularity on the fluid and can be written as

$$\boldsymbol{F} = \oint \boldsymbol{f}(\boldsymbol{x}_S, t) \cdot \delta(\boldsymbol{x} - \boldsymbol{x}_S)\, dS. \tag{2}$$

For numerical reasons, the delta function is approximated by a Gaussian function as follows:

$$\delta \approx e^{-\left(\frac{x-x_S}{\sigma_x}\right)^2 - \left(\frac{y-y_s}{\sigma_y}\right)^2 - \left(\frac{z-z_s}{\sigma_z}\right)^2} \tag{3}$$

The surface body force $\boldsymbol{f}$ is determined from the relation

$$\boldsymbol{f}(\boldsymbol{x}_S, t) = \alpha_S \cdot \int_0^t \boldsymbol{v}(\boldsymbol{x}_S, t')\, dt' + \beta_S \cdot \boldsymbol{v}(\boldsymbol{x}_S, t) \tag{4}$$

for surface points $\boldsymbol{x}_S$, velocity $\boldsymbol{v}$, time t and negative constants α_S and β_S which represents a feedback scheme in which the velocity is used to iteratively approach the desired value.

If the surface of the irregularity is located between the grid points, the velocity at this location is interpolated from the values at the neighbouring grid points using a 4th order Lagrangian interpolation procedure.

Considering the fact that a spectral representation of the flow field in spanwise direction is used for the present investigation, it is clear, that, introducing point forces into such a scheme will cause some numerical problems. These result from the fact that singularities tend to produce significant spatial oscillations even in a temporally steady flow and have already been encountered by Goldstein et al. [3]. To eliminate these oscillations, a mild spectral smoothing in spanwise direction is applied. The spectral coefficients of the vorticity components are multiplied by

$$e^{-(k/K)^{20}}, \tag{5}$$

where k is the grid point index in spanwise direction and K the number of Fourier modes used in this direction. This is a low-pass filter. The decay constant of 20 is chosen so that there will be a sharp cut-off of the highest modes and so that the highest wavenumber will be reduced by $1/e$. Of course, with higher resolution this smoothing will have less influence on the flow and will probably no longer be necessary if a really huge resolution could be used. The maximum resolution used in the present investigation was restricted by the available amount of memory and computation time. The resolution used can be described as follows: The height of the integration domain is about 4 times the boundary layer thickness at the inflow boundary, its length about 45 times and its width about 8 times the radius of the roughness element. The number of grid points is 1802 in streamwise and 193 in wall-normal direction. In spanwise direction 80 Fourier modes are used.

3 DNS Results

As a first step towards receptivity calculations, the flow field in the vicinity of a localised, three-dimensional roughness element on a flat plate was calculated and investigated in detail. The surface irregularity is located at $x = 2.471$, which corresponds to $Re_{\delta_1} = 855$ for the undisturbed flow (Blasius: $Re_{\delta_1} = \sqrt{Re \cdot x}$). It's height, non-dimensionalised with the local displacement thickness of the undisturbed boundary layer at the location of the irregularity, is $h/\delta_1 = 0.5$. The surface roughness is rotationally symmetric with a radius of $b/\delta_1 = 8.42$ and the following shape:

$$h(x, z) = h_0 \cdot cos^3 \left(\frac{\pi r}{2b} \right) \tag{6}$$

with

$$r = \sqrt{(x - x_1)^2 + z^2}. \tag{7}$$

Here x denotes the streamwise and z the spanwise coordinate, h_0 is the maximum height in the center of the roughness, b the radius of the roughness and x_1 denotes the distance of the center of the roughness from the leading edge of the flat plate. In spanwise direction, the roughness is located at $z_1 = 0$. The shape of the surface irregularity under investigation is illustrated in figure 2. First calculations showed that a very fine spatial resolution, especially in the vicinity of the surface non-uniformity was absolutely necessary to gain reasonable results. This led to the use of grid stretching in wall-normal direction to keep the computational effort within a reasonable range.

What we observe in our DNS up to now is that the flow field in the vicinity of the surface irregularity is governed by the following features: In the wake of the surface irregularity a pair of counter-rotating vortices is visible on each side of the roughness. The outer vortex seems to originate directly upstream of the roughness and seems to be wrapped around in a similar way as a horseshoe vortex. The inner vortex seems to originate downstream of the outer one at the spanwise edge of the roughness element. The whole flow pattern looks very similar to horseshoe vortex systems that were observed in experiments around surface mounted cylinders or cubes (see [9]). A visualisation of the vortical structures in the flow field using the λ_2-method [5] is shown in figure 3. To provide a better insight into the details of the flow field in the near wall region, the wall-normal direction is stretched by a factor of 5. Figure 4 shows selected streamlines together with contours of the streamwise velocity component u in zy-planes. The y-direction is again stretched, in contrast to the previous plot, this time by a factor of 10. The illustration shows, that the pair of vortices results in a high-speed streak downstream of the spanwise edge of the roughness, which is flanked by two low-speed streaks. In the middle of the two counter-rotating vortices, high speed fluid from the outer region of the boundary layer is transported towards the wall, whereas on the outer edge of the two vortices, low speed fluid from the near wall region is transported away from the wall. This can be seen more clearly in a cutting plane shown in figure 5 located at $x = 2.57$, downstream of the roughness, where the location of the vortices (visualised by contours of the streamwise vorticity ω_x) is compared to the location of the streaks (visualised by contours of u', i.e. the deviation of u from the value for a flat plate without roughness). Here, solid lines denote positive values of u' (acceleration) and dashed lines negative values (deceleration). One can clearly see, that the high-speed streak is located in the middle of the vortices, whereas the low-speed streaks are located on the outer edges. An additional observation is, that all disturbances created by the presence of the surface irregularity are limited to the region downstream of the roughness, which means that they do not spread in spanwise direction. This observation is illustrated in figure 6, where ω_z at the wall is plotted versus x and z. It is in agreement to experimental findings reported by Kendall in [6].

4 Conclusion

As a first step towards calculations on non-linear 3D roughness receptivity, the flow field in the vicinity of a localised 3D surface non-uniformity was studied in detail using direct numerical simulations. For the modeling of the surface non-uniformity, an immersed boundary technique was applied.

By the presence of the surface non-uniformity, a pair of counter-rotating streamwise vortices was formed on each spanwise edge of the roughness element. Each pair of vortices resulted in a high-speed streak in the middle of the two vortices flanked by two low-speed streaks on the outer edges. The vortex system, that was generated by the presence of the localised surface non-uniformity, looked very similar to horseshoe vortex systems observed in experiments in the vicinity of surface mounted cylinders or cubes.

In a next step, investigations of the interaction of a planar Stokes wave with the present roughness element will be performed. This will yield a deeper insight into the transition process behind a 3D roughness element.

Acknowledgements

The financial support by the German Research Council (DFG) under contract number Ri 680/7-2 is gratefully acknowledged. The computations were carried out at the HLRS Stuttgart on a NEC SX-5.

References

[1] M. Choudhari and C.L. Streett: "A finite Reynolds-number approach for the prediction of boundary-layer receptivity in localized regions". Phys. Fluids, A **4**, 1992, pp. 2495-2514.

[2] J.D. Crouch: "Localized receptivity of boundary layers". Phys. Fluids, A **4**, 1992, pp. 1408-1414.

[3] D. Goldstein, R. Handler and L. Sirovich: "Modeling a no-slip flow boundary with an external force field". J. Comp. Phys., **105**, 1993, pp. 354-366.

[4] M.E. Goldstein: "Scattering of acoustic waves into Tollmien-Schlichting waves by small streamwise variations in surface geometry". J. Fluid Mech., **154**, 1985, pp. 509-529.
[5] J. Jeong and F. Hussain: "On the identification of a vortex". J. Fluid Mech., **285**, 1995, pp. 69-94.
[6] J.M. Kendall: "Laminar boundary layer velocity distortion by surface roughness: effect upon stability". AIAA-Paper 81-0195.
[7] M.N. Linnick: "Investigation of actuators for use in active flow control". M. Sc. Report, University of Arizona, 1999.
[8] C.S. Peskin: "Numerical analysis of blood flow in the heart". J. Comp. Phys., **25**, 1977, pp. 220-252.
[9] R. Sedney: "A survey of the effects of small protuberances on boundary-layer flows". AIAA Journal, vol. **11**, no. 6, 1973, pp. 782-792.
[10] D.A. von Terzi, M.N. Linnick, J. Seidel and H. Fasel: "Immersed Boundary Techniques for High-Order Finite-Difference Methods". AIAA-Paper 2001-2918.

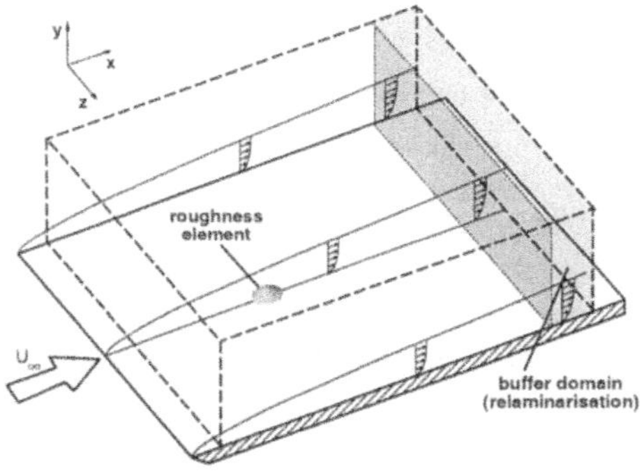

Figure 1 Sketch of the integration domain

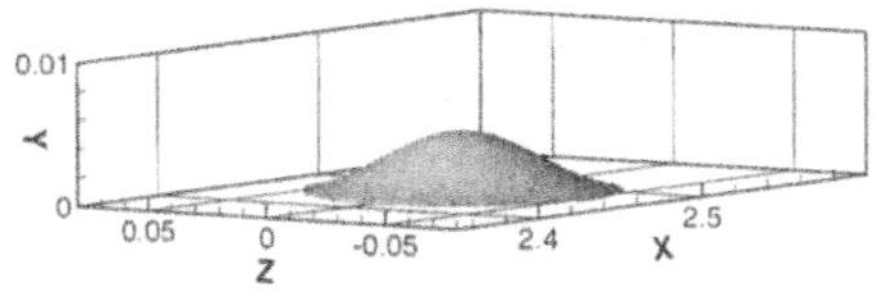

Figure 2 Shape of the localised roughness element; wall-normal direction stretched by a factor of 5 in relation to streamwise and spanwise direction

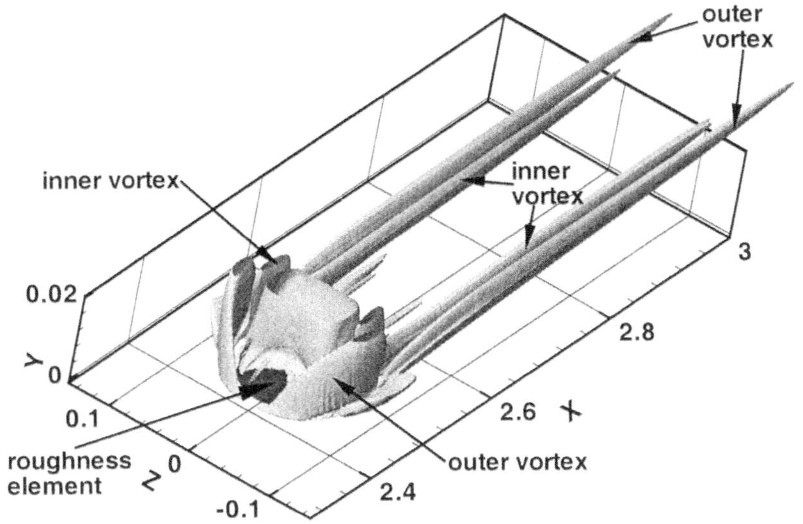

Figure 3 Flow field in the vicinity of a localised hump, located at $x = 2.471$, with $h = 0.004275$, $b = 0.072$; visualisation of the vortex system generated by the localised hump by a λ_2-Isosurface ($\lambda_2 = -1$); y-direction stretched by a factor of 5 in relation to x- and z-direction

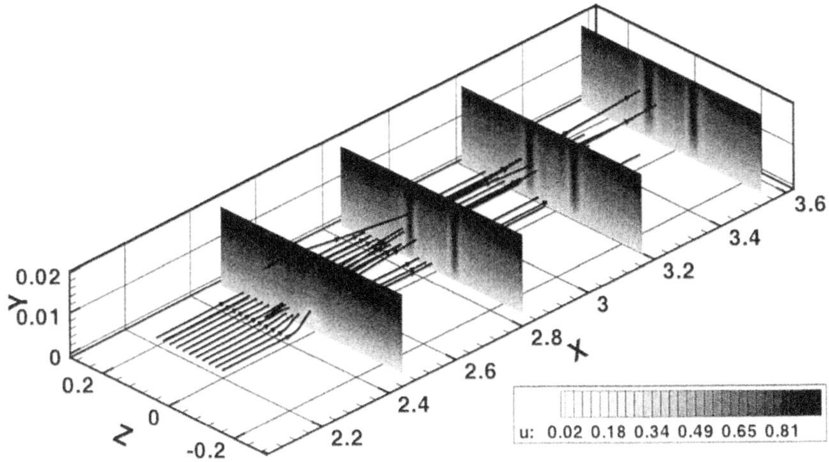

Figure 4 Flow field in the vicinity of a localised hump, located at $x = 2.471$, with $h = 0.004275$, $b = 0.072$; selected streamlines and contours of the streamwise velocity u in zy-planes; y-direction stretched by a factor of 10 in relation to x- and z-direction

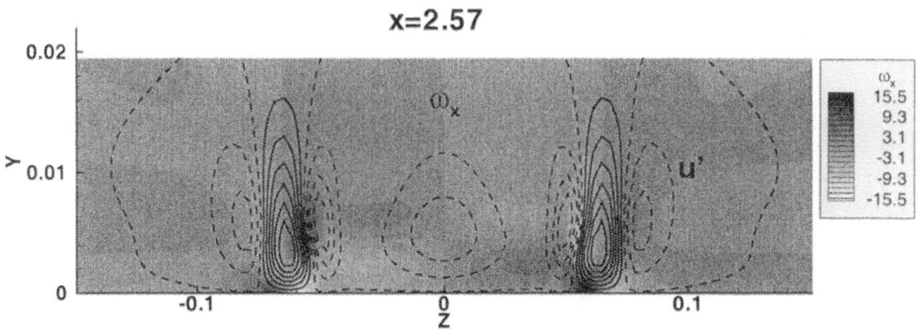

Figure 5 Comparison of ω_x-contours with u'-contours downstream of the surface non-uniformity at $x = 2.57$; solid lines: positive values of u', dashed lines: negative values of u'

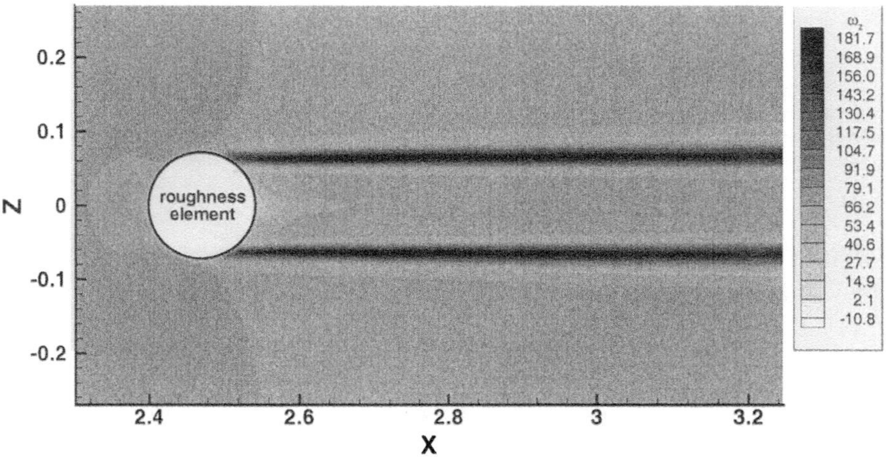

Figure 6 Contours of ω_z at the wall ($y = 0$); surface non-uniformity located at $x = 2.471$, $z = 0$

Separation Control in Diffuser Flow Using Periodic Excitation

ERIK WASSEN and FRANK THIELE

Technical University Berlin, Hermann–Föttinger–Institute of Fluid Mechanics, Sekr. HF1, Müller–Breslau–Str. 8, 10623 Berlin, Germany

Summary

Large–Eddy Simulations were performed for the turbulent separated flow in an axisymmetric diffuser. In order to decrease the size of the separation region an active control technique was applied using blowing/suction through slits on the diffuser's circumference. The slit position and slit geometry as well as the frequency and amplitude of the control flow were varied. Using this technique, the length of the separation zone could be reduced by up to 37%. At the same time, the pressure loss in the diffuser could be reduced by 20%.

1 Introduction

Axisymmetric expansions are devices that are used in a large variety of engineering applications. Their main purpose is to decelerate the flow and recover pressure, or in other words, to convert part of the kinetic energy in the flow into potential energy. Due to design constraints in many practical situations the diffuser angle and length is chosen such that separation of the flow in the expansion cannot be avoided. Separation of the flow and the development of a recirculation zone cause additional losses and are thus counter–productive with respect to the diffuser's purpose of recovering potential energy.

For small diffuser angles the flow separation is pressure–induced whereas for larger angles the separation is geometry–induced. The transition from pressure–induced to geometry–induced separation has been found to take place at diffuser half–angles of $\alpha \approx 14^o - 22^o$ [3].

From a variety of experimental works it has been known that the separated flow in expansions is receptive to periodic perturbations. Obi et al. [4] investigated the periodic perturbation of a shear layer in a plane asymmetric diffuser. They showed that, while the location of the separation point remained unaffected by the perturbations, the size of the recirculation zone could be reduced significantly. This can be explained by an increased momentum exchange perpendicular to the main flow direction. Using sinusoidal perturbations applied through a slit along the edge of a backward–facing step, Chun & Sung [2] were able to decrease the length of the separation zone by up to 70%.

It is the goal of the present work to investigate separation control in axisymmetric expansions in order to reduce the size of the separation region and the amount of pressure loss in the expansion. The technique being used is sinusoidal blowing/suction through a slit or several slits, respectively, on the circumference of the expansion. Large–Eddy Simulations (LES) have been carried out to calculate the turbulent flow in the axisymmetric diffuser. It is the advantage of LES that it can resolve the dominant, energy–containing three–dimensional and transient structures in the turbulent flow, while avoiding the excessive numerical effort required by Direct Numerical Simulations (DNS).

In Section 2 an overview of the simulation procedure and the parameters used in this work are given. Simulation results for different control parameters are presented in Section 3.

2 Numerical Procedure and Parameters

The simulation algorithm is based on the RANS code ELAN3D developed at Technical University Berlin [7] that has been extended for the use of LES [6,5]. The numerical procedure uses a finite–volume discretization of the Navier-Stokes equations for incompressible flow. The flow domain is discretized by block–structured grids which allows for a high flexibility to calculate flows in complex three–dimensional geometries. The code has been parallelized in order to be used on todays most powerful massively–parallel computer architectures.

In the present work investigations focus on an 18°–diffuser (half–angle) which, as stated above, is in the range between pressure–induced and geometry–induced separation. The flow in diffusers within this range is believed to be most receptive to excitation. In Figure 1 details of the geometry of the diffuser are given. The diffuser's length is $L = 8\,d$, where d is the inflow diameter. The diameter at the outflow cross section is $D = 1.6\,d$. At the first glance, the length of the inflow pipe, $l = d$, seems rather short. It will be shown in Section 3, however, that the excitation has very little effect upstream of the expansion edge. The Reynolds number for all simulations was $Re_b = u_b\,d/\nu = 10^4$, where u_b is the bulk velocity at the inflow cross section, and ν is the kinematic viscosity of the fluid. Figure 1 also shows the grid and block structure of the flow domain. The geometry is discretized using a total of $2 \cdot 10^6$ grid cells, divided into 72 equally sized blocks. Note that in Figure 1 only every other grid point is shown. In the axial (x–) direction grid points are spaced equally. In wall units, the size of a grid cell in x–direction is $\Delta x^+ = \Delta x\, u_\tau/\nu = 12.7$, where $u_\tau = \sqrt{\tau_w/\rho}$ is the shear velocity, τ_w is the time averaged wall shear stress at the inflow cross section, and ρ is the fluid density. In the radial direction, the maximum cell size in the center of the diffuser is $\Delta y^+_{max} = \Delta z^+_{max} = 17.6$, and the minimum cell size at the wall is $\Delta y^+_{min} = \Delta z^+_{min} = 0.76$. The maximum cell size in circumferential direction is $\Delta l^+_{circ} = 19.7$. There are five grid points in the near–wall region within a distance of $h^+ = 10$ from the wall.

The inflow into the diffuser is a fully developed turbulent pipe flow. The inflow data for each time step have been obtained from a previously calculated LES for a pipe flow at the same Reynolds number.

3 Results

The numerical simulations in the present work were carried out in close collaboration with experiments conducted at the Institute of Aeronautics and Astronautics at Technical University Berlin. In Figure 2 a comparison is shown between numerical and experimental results for profiles of the time averaged streamwise velocity $\overline{u}/u_b$ at different streamwise positions. The results agree very well in the expansion region of the diffuser where the recirculation zone is located and which is of particular interest in the present work. There is some disagreement further downstream where the numerical results underestimate the experimental results in the region near the diffuser center.

In order to control the separation in the axisymmetric expansion an active technique was applied, using sinusoidal blowing/suction through slits on the diffuser's circumference. In preliminary simulations and experiments the slit position and the perturbation amplitude, $A = v_p/u_b$, where v_p is the peak velocity of the secondary flow, were varied. It was found that the most effective slit position was on the edge of the expansion. Also, a noticeable effect of the perturbation on the separation zone was only found for relatively large amplitudes $A \approx O(1)$.

In the present work two different slit geometries were used: a ring slit and a set of four slit segments, see Figure 3. The latter were spaced equally along the circumference of the expansion, and each segment had a length of one eighth of the circumference. Hence, for the same perturbation amplitude the slit segments have half the throughput of secondary fluid. In all cases the width of the slits was $b = 0.02\,d$, and the direction of blowing/suction was inclined by 45° with respect to the axial direction with no circumferential component, as shown in the close up view of the slit in Figure 3. In the simulations, the Strouhal number of the perturbation was kept constant at $St = f\,d/u_b = 0.27$, where f is the excitation frequency. This Strouhal number is close to the optimal Strouhal number observed experimentally in [1].

In figure 4 the friction coefficient $\overline{c}_f$ in axial direction is shown. Besides time averaging, $\overline{c}_f$ was also averaged in circumferential direction. Negative values of $\overline{c}_f$ indicate recirculating flow. Four different cases are shown in Figure 4: the unperturbed base flow, one ring slit case for an amplitude of $A = 1.0$, and two slit segment cases for amplitudes of $A = 1.0$ and $A = 2.0$, respectively. In all control cases the length of the recirculation region is significantly reduced by 26–37%. The peak that can be observed in all cases at $x/d \approx 2$ marks the end of the diffuser's expansion region. The backward flowing fluid in the vicinity of the wall is decelerated as it approaches the corner, resulting in a decrease of the absolute value of $\overline{c}_f$ which is negative in this region. The relative reduction of recirculation length l_r for the different cases is given Table 1. Also, this table shows the relative reduction of pressure loss Δp_{tot}. For two of the test cases the pressure loss could be reduced by as much as 20%. For the slit segment case of $A = 1.0$, however, only a reduction of 6.6% was achieved, even though this case still shows a significant reduction of recirculation length of 26%. This implies that decreasing the size of the recirculation region does not necessarily lead to a similar decrease of pressure loss. Further research and a larger statistical basis is needed to clarify the relation between l_r, Δp_{tot} , and the design of the control mechanism.

In Figure 5 contour lines of the time averaged streamwise velocity $\overline{u}/u_b$ are shown for the base flow and one control case. Because of the symmetry of the diffuser only the lower hålf is shown. It can be seen that for the control case the deceleration of the flow in the expansion happens further upstream. Also, the reduced size of the separation region for the active control case is clearly visible. In addition, the location of the reattachment point, i.e. the end of the separation region, has been marked.

Profiles of $\overline{u}/u_b$ at six different positions in streamwise direction are shown in Figures 6 and 7 for three different cases, the base flow and two control cases. Again only one half of the diffuser is shown, with $y/d = 0$ marking the centerline and $-0.8 \leq y/d \leq -0.5$ marking the diffuser wall. For the first location, $x/d = 0.5$ (Fig. 6), only one line is visible because all profiles coincide. This shows that the perturbations have very little effect upstream of the expansion region and that the use of a relatively short inflow tube is justified. Further downstream it can be seen that for the control cases the flow velocity is lower in the center and higher towards the wall as compared to the base flow. Obviously, the excitation increases the momentum exchange in radial direction and causes the profiles to be more homogeneous. In particular, it can be observed in Figure 6 at $x/d = 1.3$ and $x/d = 1.7$, and in Figure 7 at $x/d = 2.4$, that the radial extension of the recirculation zone is decreased drastically, in addition to the decrease of its axial extension as described earlier.

In Figure 8 profiles of the Reynolds stress $\overline{u'u'}/u_b^2$ are shown for three different locations. Like in Figure 6, the profiles coincide upstream of the expansion region at $x/d = 0.5$, indicating again that there is hardly any effect of the perturbations on the upstream flow. At the two other locations the perturbed flow shows increased Reynolds stress levels, and the peak of $\overline{u'u'}/u_b^2$ becomes broader for the perturbed case. This indicates that the blowing/suction enhances the mixing and the momentum exchange between the detached flow and the fluid in the recirculation zone, thus increasing the width of the shear layer.

4 Conclusion

Large–Eddy Simulations were performed for the separated flow in an axisymmetric expansion. An active control technique was applied using periodic blowing/suction through slits in order to reduce the size of the recirculation zone. Different slit designs and perturbation amplitudes were investigated. The size of the separation region could be reduced significantly. Its streamwise length was decreased by up to 37%, going along with a decrease in pressure loss of 20%. However, it was shown that reducing the size of the separation region not necessarily leads to a comparable reduction of pressure loss. Further research is needed to improve the efficiency of the control technique.

Acknowledgements

The work presented in this paper is supported by DFG within the scope of the Collaborative Research Center (Sonderforschungsbereich) 557, project B3. All simulations were carried out under a grant from Konrad–Zuse–Zentrum für Informationstechnik Berlin (ZIB) on the Cray T3E at ZIB.

References

[1] A. Brunn and W. Nitsche. "Separation control in an axisymmetric diffuser flow by periodic excitation". In Engineering Turbulence Modelling and Experiments 5, W. Rodi and N. Fueyo, Eds., Elsevier Sci. Publ., 2002, pp. 587–596.

[2] K. Chun and H. Sung. "Control of turbulent separated flow over a backward–facing step by local forcing". Experiments in Fluids, 21, 1996, pp. 417–426.

[3] W. Nitsche and M. Haberland. "On turbulent separated flows in axisymmetric diffusers". In Notes on Numerical Fluid Mechanics, Numerical Flow Simulation I, Volume 40, Vieweg Verlag, 1993, pp. 116–124.

[4] S. Obi, H. Ohimuzi, K. Aoki and S. Masuda. "Turbulent separation control in a plane asymmetric diffuser by periodic perturbation". In Engineering Turbulence Modelling and Experiments, Volume 2, Elsevier Science Publ., 1993, pp.464–471.

[5] S. Schmidt. "Large–Eddy Simulation of turbulent flows in complex geometries and at high Reynolds numbers". Dissertation, Technical University Berlin, 2000 (in German).

[6] S. Schmidt, L. Xue and F. Thiele. "Large–Eddy simulation of turbulent pipe flow on semi–structured grids". In 13th Australian Fluid Mechanics Conference (Monash University, Melbourne, Australia), Volume 2, M. Thompson and K. Hourigan, Eds., 1998, pp. 651–654.

[7] L. Xue. "Development of an efficient parallel solution algorithm for the three–dimensional simulation of complex turbulent flows". Dissertation, Technical University Berlin, 1998 (in German).

Table 1 Relative reduction of recirculation length and relative reduction of pressure loss.

Case	$(l_r - l_{r,base}) / l_{r,base}$	$(\Delta p_{tot} - \Delta p_{tot,base}) / \Delta p_{tot,base}$
base flow	0	0
ring slit, $A = 1.0$	-0.31	-0.195
slit segments, $A = 1.0$	-0.26	-0.066
slit segments, $A = 2.0$	-0.37	-0.201

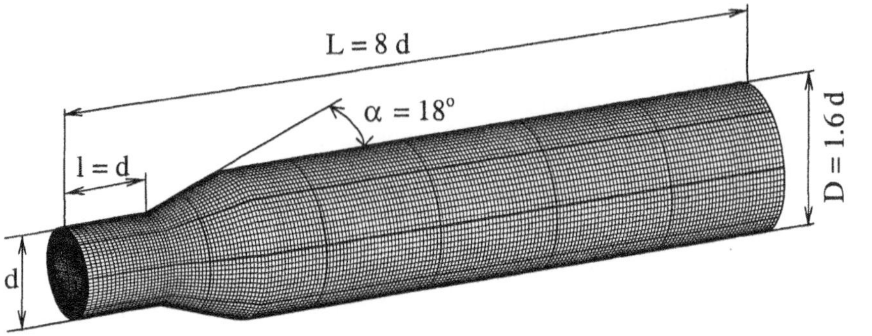

Figure 1 Grid geometry and dimensions for axisymmetric diffuser.

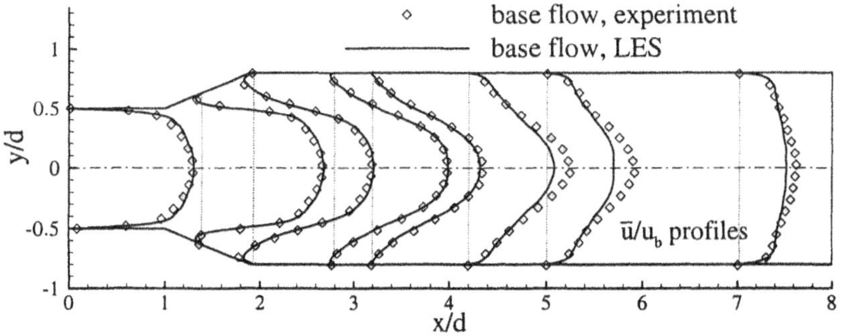

Figure 2 Profiles of streamwise velocity $\overline{u}/u_b$ for the base flow, comparison between simulation and experiment.

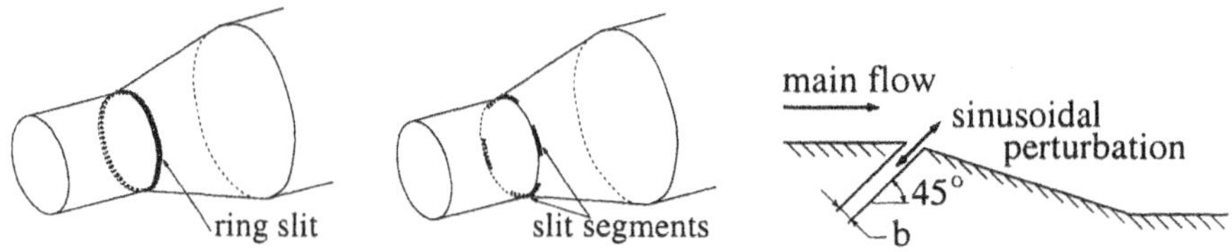

Figure 3 Position and geometry of the ring slit (left) and the slit segments (middle); close up view of the slit (right).

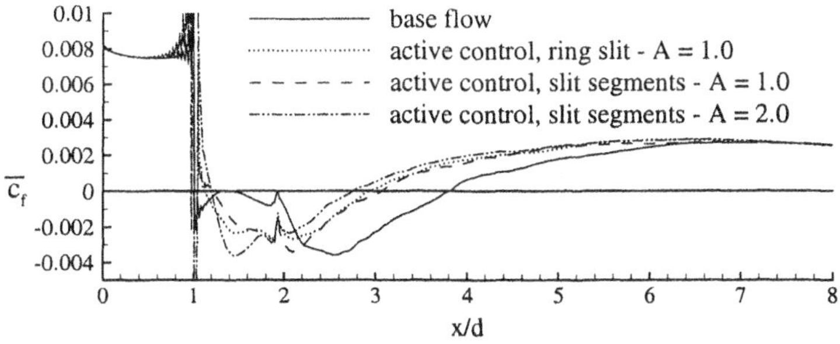

Figure 4 Friction coefficient in streamwise direction.

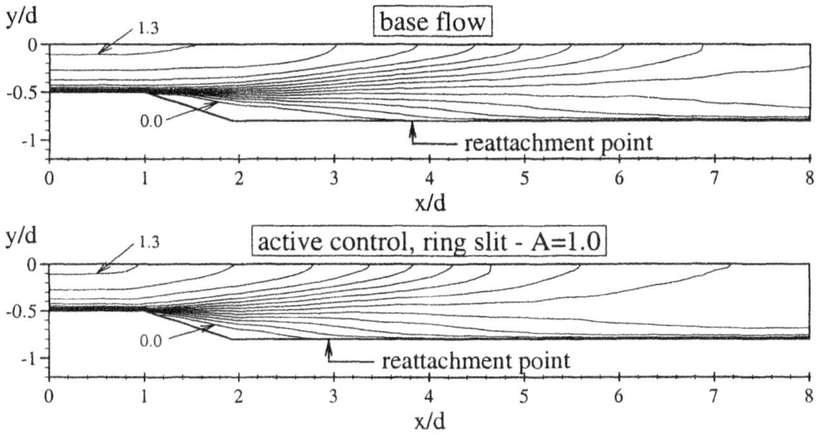

Figure 5 Contour lines of streamwise velocity $\overline{u}/u_b$; minimum level is 0.0, maximum level is 1.3, increment is 0.1.

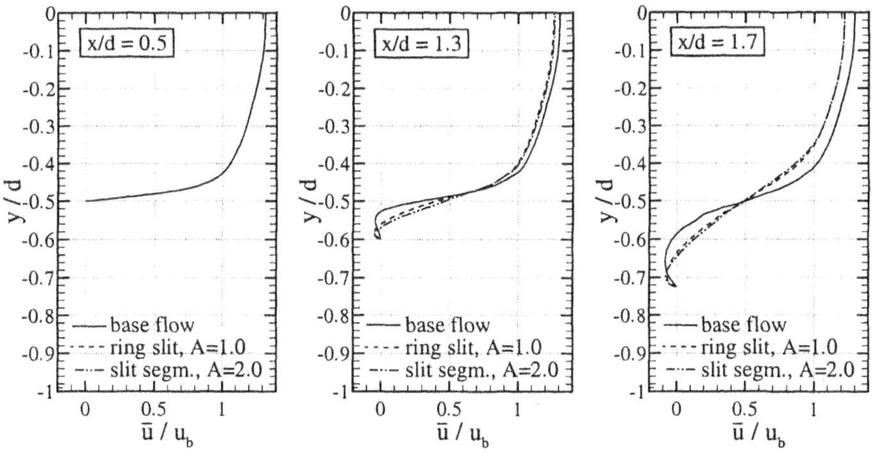

Figure 6 Profiles of streamwise velocity $\overline{u}/u_b$ for three different cases at streamwise positions $x/d = 0.5$, $x/d = 1.3$, and $x/d = 1.7$.

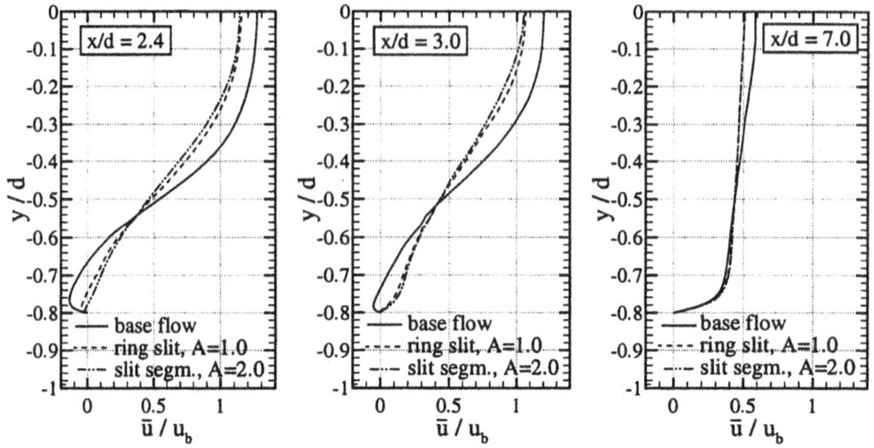

Figure 7 Profiles of streamwise velocity $\overline{u}/u_b$ for three different cases at streamwise positions $x/d = 2.4$, $x/d = 3.0$, and $x/d = 7.0$.

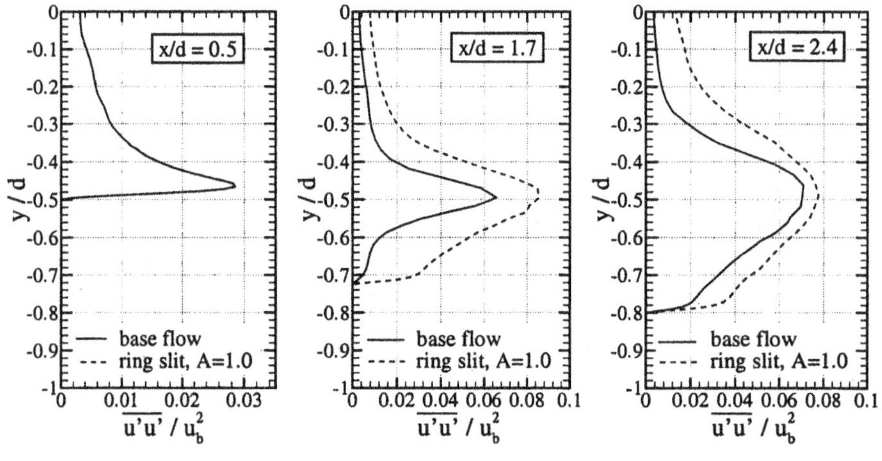

Figure 8 Profiles of Reynolds stress $\overline{u'u'}/u_b^2$ at streamwise positions $x/d = 0.5$, $x/d = 1.7$, and $x/d = 2.4$.

Active Control of the Separated Shear Layer in an Axisymmetric Diffuser Flow by Means of Periodic Forcing

A. BRUNN and W. NITSCHE

Department of Aerodynamics, Institute of Aeronautics and Astronautics
Technical University Berlin, Marchstr. 12-14, 10587 Berlin, Germany
andre.brunn@tu-berlin.de

Summary

This paper describes experimental investigations on active control of a separated shear layer in a turbulent axisymmetric diffuser flow. The research was aimed at influencing the separated flow in the diffuser in order to reduce the separation length by means of periodic pressure perturbations at the diffuser entrance edge. The measurements were conducted in a closed pipe flow channel with an optically fully accessible test section to use laser optical velocity measurement methods. The overall separation length could be significantly reduced to about 40% of the unforced base flow, while an enlarged intensity of the turbulent impulse transport inside the shear layer could be observed.

1 Introduction

The phenomenon of flow separation occurs in several kinds of flow. A very frequently investigated flow device with an included separation region is a diffuser. Flow separations result in pressure losses and a reduced performance of diffusers and similar flow devices.

In engineering there is often accepted a compromise between short diffuser length and small pressure losses because the separation process in diffusers primarily depends on the diffusor half angle. So the separation point is geometrically fixed at the diffuser entrance for $\alpha > 22°$ and the separation disappears at $\alpha < 6°$ at turbulent inflow conditions [1]. Between these values of 6° and 22° the separation is pressure induced and located in the diffuser.

The size of the separation bubble is characterized by the recirculation length x_S (see Figure 1). In addition to the diffuser half-angle the recirculation length depends on the Reynolds number of the entrance flow and the diffuser expansion ratio.

For preventing separation or reducing the length of a separation bubble, passive methods such as vortex generators as well as active methods were proposed and investigated in several ways. Here the momentum transport from the outer region of the flow into the separation bubble was intensified by means of induced longitudinal vortices [6].

The experimental investigations on separated shear flows showed that large vortex structures in the separated shear layer can also be induced by periodical excitation at the wall. The development of these vortex structures depends on forcing amplitude and frequency [5].

The effect of local forcing on flow structures in plane asymmetric diffuser flows has been studied by Coller et al. [4] or Obi et al. [8] and over a backward facing step by Chun & Sung [3]. The most effective frequencies were observed to be close to the vortex shedding frequency of the unforced flow. The re-attachment length could be significantly reduced and the amount of pressure recovery was increased. It was demonstrated by experiment and reduced-order models [4] that separation control can be achieved by forcing vortical shear layer structures. The studies of Sigurdson [10] showed effective forcing frequencies with associated vortex wavelength comparable to the hight of the separation bubble.

In the present study the active separation control in an overcritical axisymmetric diffuser flow by periodical jet-excitation will be investigated.

2 Experimental Setup

The experiments were performed in the turbulent flow through an axisymmetric diffuser with a half-angle of 18° an inlet diameter of D_1 = 50 mm and an outlet diameter of D_2 = 80 mm (Figure 1). The Reynolds number based on the inlet diameter was set to Re_D=10^4. To provide a fully developed turbulent inlet flow a small tripping wire was placed $60{\cdot}D_1$ upstream of the diffuser entrance to fix the laminar-turbulent transition far upstream.

The measurements were conducted in a closed pipe flow channel with an optically fully accessible test section to apply laseroptical velocity measurement methods and flow visualization techniques.

White oil was used as the working fluid with particular optical specifications to adjust the fluid refractive index equal to that of the Plexiglas channel walls by variation of the fluid temperature [2]. A standard Laser Doppler Velocimeter (LDV) was used for the time-dependent velocity measurements and frequency spectra of the velocity fluctuations, while Particle Image Velocimetry (PIV) was applied for measurements of instantaneous and phase-averaged velocity fields. The digital PIV system consists of a frequency-doubled Nd:YAG-laser ($\lambda = 532nm$, 25 mJ energy per pulse), a CCD-Cross Correlation-Camera (12 bit greyscale, SVGA, 200 ns full-frame repetition) and a synchronization unit [2]. For the phase-averaged velocity measurements the PIV system was triggered by the excitation mechanism.

The forcing mechanism consists of an external pump and a rotating valve to generate the periodic pressure perturbations which are transfered into a ring slit at the diffuser entrance edge (Figure 2). The oscillating wall jets are without net mass flow.

3 Forcing Parameters

During the investigations the forcing parameters Strouhal number St_D (based on the inlet diameter D_1 and the streamwise mean flow velocity u_0) and forcing amplitude $A = v_S/u_0$ (v_s is the forcing velocity at the slit exit) were varied. The perturbations during the forcing should amplify preferably vortex structures in the separated shear layer to increase the growth of vortices and intensify the entrainment process.

These vortices appear due to the mechanism of instability in the free shear layer after the flow has separated from the wall. This initial instability (Kelvin-Helmholtz-Instability) leads to a roll-up of the shear layer into spanwise vortices. The fundamental wavelength of these vortex structures depends on the boundary layer conditions upstream of the separation point. The vortical structures grow and are shed from the separation region (vortex shedding).

In Figure 3 the instantaneous flow field of the unforced base flow measured by PIV at the lower diffuser wall is shown, where the appearance of these growing vortices is obvious.

In order to detect characteristic frequencies in the shear layer, measurements of streamwise velocity fluctuations were made by means of LDV. While $St_D = 0.4$ represents the initial Kelvin-Helmholtz-Instability, the value of $St_D = 0.23$ corresponds approximately to the shedding frequency of the separated shear layer in axisymmetrical flows [5], but none of these single-frequencies could be detected clearly. However, a significant low frequency band measured near the re-attachment point appeared in the range of $0.05 \leq St_D \leq 0.5$ [2]. Therfore all control experiments with excited diffuser flow were performed in a corresponding Strouhal number range. The forcing amplitude was set to $0.1 \leq A \leq 0.66$. At lower amplitudes no effect of flow excitation could be detected, the upper limit was set by the pump power.

4 Results

The time-averaged velocity field with streamlines in the x-y-plane of the diffuser is shown in Figure 4. The overall separation length was measured to be two times the diffuser inlet diameter. This value corresponds very well with LES-calculations done by Schmidt & Thiele [9] for the same axisymmetric diffuser configuration and other results of investigations on plane diffuser configurations e.g. by Makiola [7].

In the time-averaged flow field the primary vortices appear nearly symmetrically at the upper and lower diffuser wall. However, an asymmetric and alternating expansion of the separation bubble in the diffuser could be observed by means of instantaneous PIV-measurements in earlier investigations [2].

During the local forcing the overall separation length was significantly reduced to approximately 45% of the base flow, as can be seen in Figure 5, and the separation point was distinctly shifted downstream towards the diffuser end. The isolines of $u = 0$ in the Figures 4 and 5 mark the recirculation region. Compared with Figure 4 (unforced base flow), the streamlines of the forced flow show that the separation bubble is strongly reduced and nearly disappears.

A significant reduction in the re-attachment (x_R) length was achieved at almost all forcing amplitudes tested, i.e. $0.1 \leq A \leq 0.66$, Figure 6. However, the most effective reduction was observed at reduced frequencies close to the vortex shedding (St_D = 0.24). In this case the averaged re-attachment length was reduced to about 65% compared to the unforced base flow while the overall separation length was reduced to 40%. This result is confirmed by numerical calculations (LES) of Wassen & Thiele [11] investigating the same diffuser configuration. In this numerical approach a reduction of the re-attachment length to 63% of the base flow was observed for St_D = 0.27 and $A = 2$, while a pressure recovery gain of about 20 % was achieved [11].

An indicator for the vortex growth in the shear layer is the cross flow intensity. Due to an enlargement of the cross flow the momentum transport between the separation region and the mean flow is intensified.

Chun & Sung [3] observed a vortex merging in a separated shear layer due to local forcing. This vortex merging caused an increase in turbulence and cross flow intensity and enlarged significantly the entrainment. In Figure 7 the normalized cross flow velocity distribution v/u_0 is shown. Compared to the base flow, the momentum in cross flow direction is significantly enlarged by the forcing. At frequencies close to the vortex shedding (Figure 7, left-below) the cross flow intensity is increased just inside the diffuser, while at lower frequencies the intensity increases further downstream. So the significant reduction of the separation length is the result of intensified momentum transport in the shear layer due to the forcing at frequencies close to the vortex shedding.

5 Conclusions

In this study experimental investigations on separation control of a separated shear layer in an axisymmetric turbulent 18° -diffuser were performed. The aim was to reduce the separation length by means of local forcing at the diffuser edge. The most effective reduction of the separation length was observed at St_D = 0.24 . This Strouhal number corresponds to the vortex shedding frequency of the unforced base flow. Under these conditions the time-averaged overall separation length could be significantly reduced to about 40% of the separation length of the unforced base flow.

The influence of the periodic excitation on the increased vortex spreading became obvious in terms of increased cross flow intensity in the separated shear layer downstream from the diffuser entrance edge.

Acknowledgements

This research was funded by the German Science Foundation (DFG) within the scope of the Sonderforschungsbereich Sfb 557 "Control of Complex Turbulent Shear Flows". This support is thankfully acknowledged by the authors.

References

[1] P. Bartsch, M. Nasseri and W. Nitsche. "Investigation on Turbulent Transport Processes in Relaxing Diffuser Flows". In Engineering Turbulence Modelling and Experiments, Vol. 2, 1993, W. Rodi and F. Martelli (Eds.), Elsevier Science Ltd., pp. 479-488.

[2] A. Brunn and W. Nitsche. "Separation control in an axisymmetric diffuser flow by periodic excitation". In Engineering Turbulence Modelling and Experiments, Vol. 5, 2002, W. Rodi and N. Fueyo (Eds.), Elsevier Science Ltd., pp. 587-596.

[3] K.B. Chun and H.J. Sung. "Control of Turbulent Separated Flow over a Backward Facing Step by Local Forcing". Exp. in Fluids, Vol. 21, 1996, pp. 417-426.

[4] B.D. Coller, B.R. Noack, S. Narayanan, A. Banaszuk and A.I. Khibnik. "Reduced-Basis Model for Active Separation Control in a Planar Diffuser Flow". AIAA-paper, No. 2000-2562, 2000.

[5] M. Kiya, M. Shimizu and O. Mochizuki. "Sinusoidal forcing of a turbulent separation bubble". J. Fluid Mech., Vol. 342, 1997, pp. 119-139.

[6] J.C. Lin, F.G. Howard, D.M. Bushnell and B.V. Selby. "Comparative Study of Control Techniques for Two-Dimensional Low-Speed Turbulent Flow Separation". In Separated Flows and Jets, 1990, V.V. Kozlov and A.V. Dovgal (Eds.), Springer-Verlag Berlin, pp. 429-474.

[7] B. Makiola. "Experimentelle Untersuchungen zur Strömung über die schräge Stufe". Fortschr.-Ber. Nr. 204, 1992, VDI-Verlag Düsseldorf.

[8] S. Obi, H. Ohizumi, K. Aohi and S. Masuda. "Turbulent Separation Control in a Plane Asymmetric Diffuser by Periodic Perturbation". In Engineering Turbulence Modelling and Experiments, Vol. 2, 1993, W. Rodi and F. Martelli (Eds.), Elsevier Science. Ltd., pp. 633-642.

[9] S. Schmidt and F. Thiele. "Flow control in an axial-diffuser by blowing and suction". Euromech Colloqium Shear-Flow Control, No. 415, 24-26 July 2000, Book of Abstracts, TU Berlin.

[10] L.W. Sigurdson. "The structure and control of a turbulent reattaching flow". J. Fluid Mech., Vol. 298, 1995, pp. 139-165

[11] E. Wassen and F. Thiele. "Separation Control in Diffuser Flow Using Periodic Excitation". To be published in Notes on Numerical Fluid Mechanics, Proceedings of the 13th STAB/DGLR Symposium Munich, 12-14 Nov. 2002, Springer-Verlag

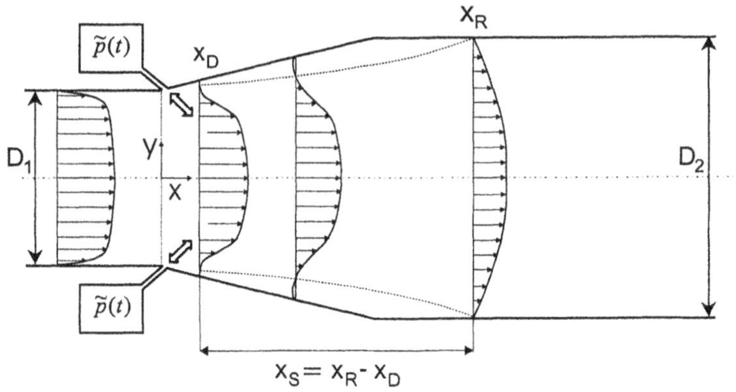

Figure 1: The geometrical boundary conditions of the axisymmetric diffuser configuration with single slit actuator situated at the diffuser edge

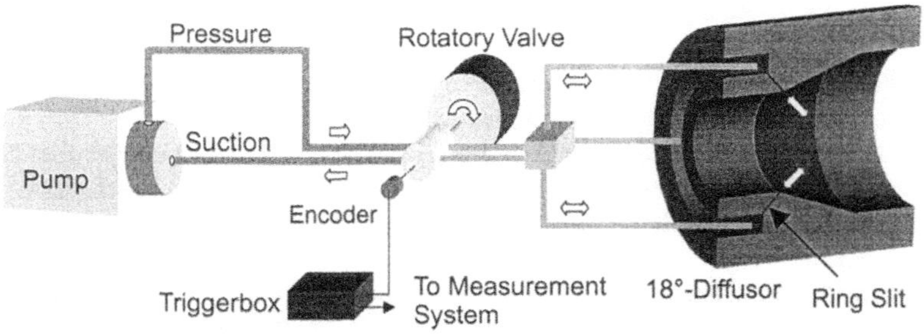

Figure 2: Excitation mechanism with pump and rotatory valve to generate an oscillating fluid flow in the ring slit at the diffuser entrance edge

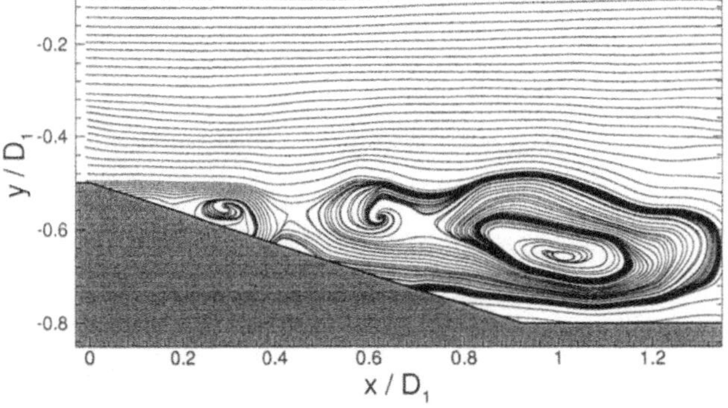

Figure 3: Instantaneous flow field with streamlines shows the roll-up of the separated shear layer at the lower diffuser wall (PIV)

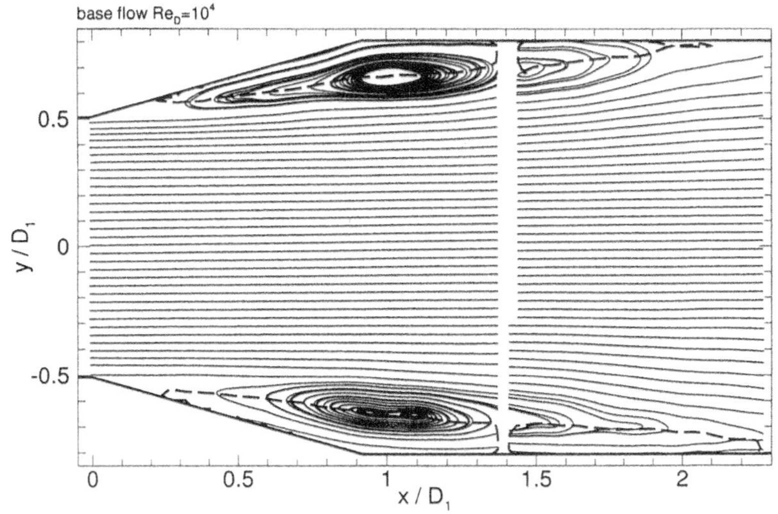

Figure 4: Time-averaged velocity field (PIV) of the unforced base flow. The dashed lines mark the isolines $u = 0$.

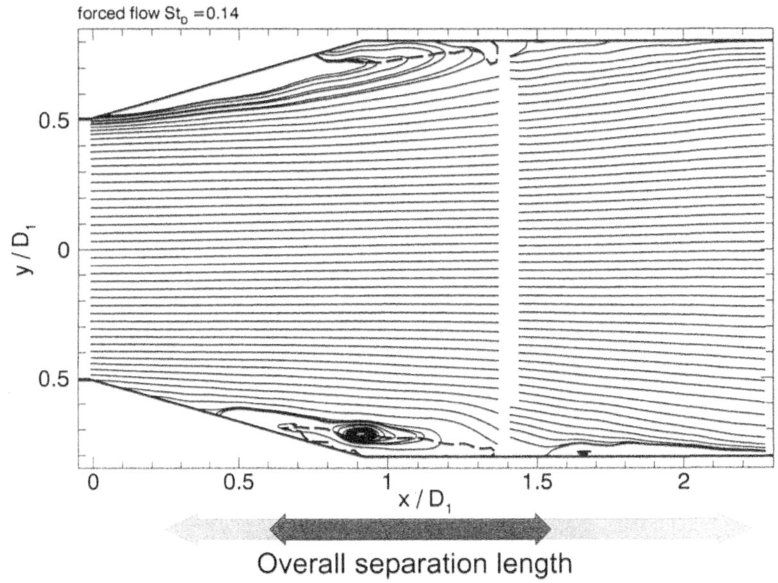

Figure 5: Time-averaged velocity field (PIV) with local forcing ($A = 0.38$; $St_D = 0.14$), the arrows show the reduced separation length in comparison to the unforced base flow

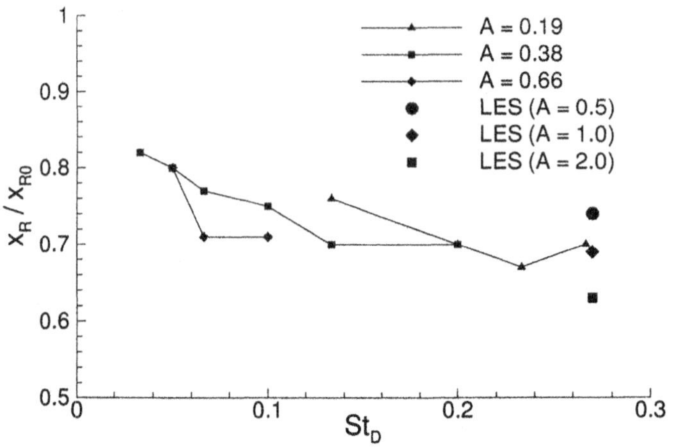

Figure 6: Reduction in re-attachment length (x_R, see Fig. 1) compared with the unforced base flow in relation to forcing parameters (A, St_D) with added numerical data [11]

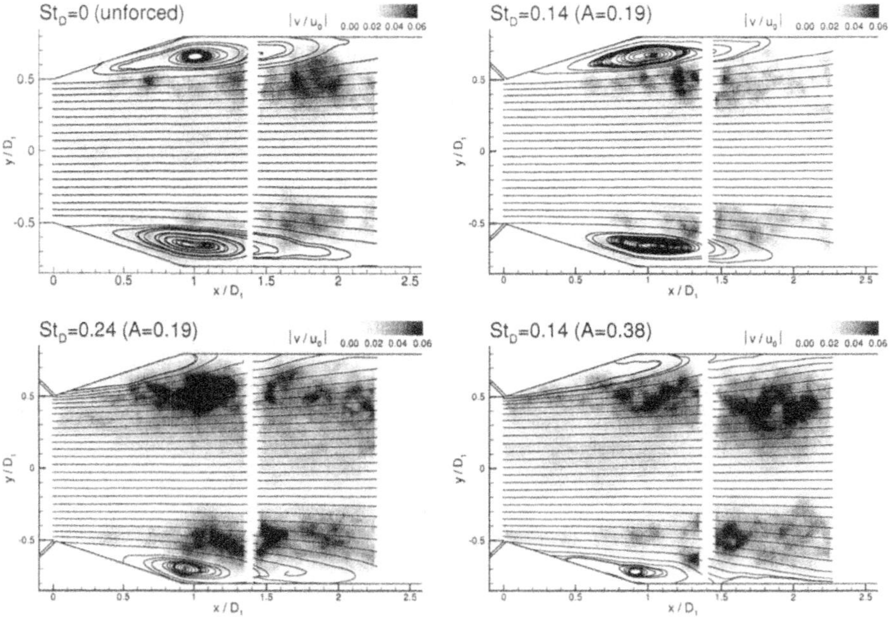

Figure 7: Mean cross flow velocities in the separated shear layer: unforced base flow (above, left) and at different forcing cases

DNS and Scaling Laws from New Symmetry Groups of ZPG Turbulent Boundary Layer Flow

G. KHUJADZE, M. OBERLACK
Hydromechanics and Hydraulics Group, Darmstadt University of Technology, Petersenstr. 13, 64287 Darmstadt, Germany

Introduction

The Lie group approach developed by Oberlack (see [1] and references therein) is used to derive *new scaling laws* for zero pressure gradient turbulent boundary layer flow. A direct numerical simulation (DNS) is performed for its verification. The approach unifies a large set of scaling laws (invariant solutions) for the mean velocity of stationary parallel turbulent shear flows and extends the work done by Oberlack. From the two-point correlation equations the knowledge of the symmetries allows us to rederive a broad variety of invariant solutions for turbulent flows: the logarithmic law of the wall, an algebraic law, the viscous sublayer, the linear region in the centre of a turbulent Couette flow and in the centre of a rotating channel flow, and a new exponential mean velocity profile that is found in the mid-wake region of flat-plate boundary layers. The present focus is on the exponential law and corresponding new scaling laws for one- and two-point correlations.

Theoretical analysis

The present analysis is based on the two-point correlation equation with mean velocity profile $\bar{u}_1 \equiv \bar{u}_1(x_2)$, where x_2 is the wall normal coordinate [2].

$$\frac{\bar{D}R_{ij}}{\bar{D}t} + R_{kj}\frac{\partial \bar{u}_i(\mathbf{x},t)}{\partial x_k} + R_{ik}\frac{\partial \bar{u}_j(\mathbf{x},t)}{\partial x_k}\bigg|_{\mathbf{x}+\mathbf{r}} + [\bar{u}_k(\mathbf{x}+\mathbf{r},t) - \bar{u}_k(\mathbf{x},t)]\frac{\partial R_{ij}}{\partial r_k}$$
$$+\frac{\partial \overline{pu'}_j}{\partial x_i} - \frac{\partial \overline{pu'}_j}{\partial r_i} + \frac{\partial \overline{u'_i p}}{\partial r_j} + \frac{\partial R_{(ik)j}}{\partial x_k} - \frac{\partial}{\partial r_k}[R_{(ik)j} - R_{i(jk)}] = 0, \tag{1}$$

where $R_{ij} = \overline{u'_i(\mathbf{x},t)u'_j(\mathbf{x}+\mathbf{r},t)}$ is a two-point correlation tensor and $\frac{\bar{D}}{\bar{D}t} = \left(\frac{\partial}{\partial t} + \bar{u}_k\frac{\partial}{\partial x_k}\right)$ is the mean substantial derivative.

In the present study Lie's procedure is used to find symmetry transformations and self-similar solutions of the equation. Lie's procedure to find symmetry transformations and the derivation of self-similar solutions may be divided into three parts. The first one, the computation of the determining system, is completely algorithmic and has been aided by the Lie group software package [3]. The second part, the solution of the system, which is a linear over-determined system of partial differential equations, has been done by hand. The last part is the computation of the similarity variables.

The result of the first step is an over-determined set of linear partial differential equations (not presented here). The solution of the system is the infinitesimal form of the symmetry transformation:

$$\xi_{x_2} = -c_1 x_2 - c_2, \tag{2}$$

$$\xi_{r_1} = -c_1 r_1 - c_3, \tag{3}$$

$$\xi_{r_2} = -c_1 r_2 - c_4, \tag{4}$$

$$\xi_{r_3} = -c_1 x_3 - c_5, \tag{5}$$

$$\eta_{\bar{u}_1} = (c_6 - c_1 + c_7)\bar{u}_1 + f(x_2, \mathbf{r}), \tag{6}$$

$$\eta_{R_{ij}} = (c_1 - c_6) R_{ij}, \tag{7}$$

$$\eta_{\overline{pu}_i} = c_7 \overline{pu}_i, \tag{8}$$

$$\eta_{Z_{ij}} = (c_1 + c_7) Z_{ij}. \tag{9}$$

As a result all symmetries are found in the infinitesimal form.

For the present problem we focus on scaling symmetries, Galilean invariance and the translation group:

$$X_1 = -x_2 \frac{\partial}{\partial x_2} - r_i \frac{\partial}{\partial r_i} - \bar{u}_1 \frac{\partial}{\partial \bar{u}_1} + R_{ij} \frac{\partial}{\partial R_{ij}} + \cdots, \tag{10}$$

$$X_2 = \bar{u}_1 \frac{\partial}{\partial \bar{u}_1} - R_{ij} \frac{\partial}{\partial R_{ij}}, \tag{11}$$

$$X_3 = \bar{u}_1 \frac{\partial}{\partial \bar{u}_1} + \overline{u_i p} \frac{\partial}{\partial \overline{u_i p}} + \overline{pu}_i \frac{\partial}{\partial \overline{pu}_i} + \cdots, \tag{12}$$

$$X_4 = \frac{\partial}{\partial x_2}, \tag{13}$$

$$X_5 = f(x_2, \mathbf{r}) \frac{\partial}{\partial \bar{u}_1}, \tag{14}$$

where the dots denote that also higher order correlations are involved in the corresponding symmetry.

Once the infinitesimals of a transformation are known, their global form can be determined by integrating the first order system (using Lie's first theorem [4]):

$$\frac{d\mathbf{x}^*(\varepsilon)}{d\varepsilon} = \xi(\mathbf{x}^*(\varepsilon), \mathbf{y}^*(\varepsilon)); \qquad \frac{d\mathbf{y}^*(\varepsilon)}{d\varepsilon} = \eta(\mathbf{x}^*(\varepsilon), \mathbf{y}^*(\varepsilon)); \tag{15}$$

with the initial conditions:

$$\varepsilon = 0 : \mathbf{x}^*(\varepsilon) = \mathbf{x} \text{ and } \mathbf{y}^*(\varepsilon) = \mathbf{y},$$

where $\mathbf{x}^*$ and $\mathbf{y}^*$ correspond to the new independent and dependent variables.

Thus, using Lie's first theorem we obtain the global transformations for all variables involved in (1)

$$\tilde{x}_2 = x_2 e^{-\varepsilon_1}, \quad \tilde{r}_i = r_i e^{-\varepsilon_1}, \quad \tilde{\bar{u}}_1 = \bar{u}_1 e^{-\varepsilon_1}, \quad \tilde{R}_{ij} = R_{ij} e^{\varepsilon_1}, \quad \tilde{\bar{z}}_i = \bar{z}_i, \cdots \tag{16}$$

$$\tilde{x}_2 = x_2, \quad \tilde{r}_i = r_i, \quad \tilde{\bar{u}}_1 = \bar{u}_1 e^{\varepsilon_2}, \quad \tilde{R}_{ij} = R_{ij} e^{-\varepsilon_2}, \quad \tilde{\bar{z}}_i = \bar{z}_i, \cdots \tag{17}$$

$$\tilde{x}_2 = x_2, \quad \tilde{r}_i = r_i, \quad \tilde{\bar{u}}_1 = \bar{u}_1 e^{\varepsilon_3}, \quad \tilde{R}_{ij} = R_{ij}, \quad \tilde{\bar{z}}_i = \bar{z}_i e^{-\varepsilon_3}, \cdots \tag{18}$$

$$\tilde{x}_2 = x_2 + \varepsilon_4, \quad \tilde{r}_i = r_i, \quad \tilde{\bar{u}}_1 = \bar{u}_1, \quad \tilde{R}_{ij} = R_{ij}, \quad \tilde{\bar{z}}_i = \bar{z}_i, \cdots \tag{19}$$

$$\tilde{x}_2 = x_2, \quad \tilde{r}_i = r_i, \quad \tilde{\bar{u}}_1 = \bar{u}_1 + \varepsilon_5 f(x_2, \mathbf{r}), \quad \tilde{R}_{ij} = R_{ij}, \quad \tilde{\bar{z}}_i = \bar{z}_i, \cdots \tag{20}$$

where $z_i \equiv \{\overline{pu}_i, \overline{u_i p}\}$. The variables $\varepsilon_1 - \varepsilon_5$ are the group parameters of the corresponding transformations.

The most interesting fact with respect to the latter groups is that *three independent scaling groups* (16)-(18) have been computed. First two symmetry groups correspond to the scaling symmetries of the Euler equations. The last one (equation (18)) is a new scaling symmetry. This is in striking contrast to the Euler and Navier-Stokes equations, which only admit two and one scaling groups, respectively. Equations (19) and (20) represent the translation symmetry in space and the generalized Galilean transformation, respectively.

The corresponding characteristic equations for the invariant solutions (see [1]) read

$$\frac{dx_2}{\xi_{x_2}} = \frac{dr_i}{\xi_{r_i}} = \frac{d\bar{u}_1}{\eta_{\bar{u}_1}} = \frac{dR_{ij}}{\eta_{R_{ij}}} = \frac{d\overline{pu}_i}{\eta_{\overline{pu}_i}} = \frac{d\overline{u_i p}}{\eta_{\overline{u_i p}}} \cdots \tag{21}$$

In the paper [1] it has been shown that the symmetry breaking of the scaling of space leads to a new exponential scaling law which physically corresponds to the outer part of a boundary layer flow, the so-called wake region.

Imposing the assumption of symmetry breaking of the scaling of space, i.e. eliminating (10) from the infinitesimals and integrating the corresponding equations we obtain an extended set of scaling laws for the mean velocity and two-point correlation function as follow:

$$\bar{u}_1 = a_1 + a_2 e^{-a_3 x_2}, \tag{22}$$

$$R_{ij} = e^{-a_4 x_2} B_{ij}(\mathbf{r}), \cdots \tag{23}$$

where $a_1 - a_4$ are constants comprising several group parameters and constants of integration and B_{ij} is a function of $\mathbf{r}$ only.

For positive a_3 the velocity law (22) converges to a constant velocity for $x_2 \to \infty$. For plane shear flows this may only be applicable to a boundary layer type of flow. In this flow the symmetry breaking length scale is boundary layer thickness.

In normalized and nondimensional variables the exponential scaling law (22) may be re-written in the following form

$$\frac{\bar{u}_\infty - \bar{u}}{u_\tau} = \alpha \exp\left(-\beta \frac{x_2}{\Delta}\right), \tag{24}$$

One can derive the scaling laws for the Reynolds stresses from the equation (23) when $\mathbf{r} = 0$:

$$\overline{u_i' u_j'} = b_{ij} \exp\left(-a\frac{x_2}{\Delta}\right), \tag{25}$$

where α, β, b_{ij} and a are universal constants and Δ is the Rotta-Clauser length scale ($\Delta \equiv \frac{\bar{u}_\infty}{u_\tau}\delta^*$).

DNS and scaling law validation

The code for DNS was developed at KTH, Stockholm (for details see [5]) using a spectral method with Fourier decomposition in the horizontal direction and Chebyshev discretization in the wall normal direction. Time integration is performed using a third order Runge-Kutta method for the advective and forcing terms (necessary for the fringe region) and Crank-Nicolson for the viscous terms.

Since the boundary layer is spatially developing in the downstream direction, it is necessary to use non-periodic boundary conditions in the streamwise direction. This is done by adding a fringe region downstream of the physical domain. Usage of this method allows to rescale the boundary layer growth normal to the wall. The fringe region is implemented by the extra term in the Navier-Stokes equations which is non-zero only in the fringe region. In this way the physical domain and fringe region together satisfy periodic boundary conditions (about fringe region technique see [6]).

The simulation starts with a laminar boundary layer at inflow which is triggered to transition by a random volume force near the wall. All quantities are non-dimensionalised by the free-stream velocity u_∞ and the displacement thickness δ^* at $x = 0$ where the flow is laminar and at this point $Re_{\delta^*}|_{x=0} = 600$. The size of the computational box (in these units) is $450 \times 30 \times 34$ (including fringe region). The useful region was confined to $150 - 300$ which corresponds to Re_{δ^*} from 800 to 1100 or Re_θ from 540 to 750.

Two simulations were performed with 7.9 and 31.2 million modes, respectively. The simulations were run for a total of 10000 time units (δ^*/u_∞), and the sampling for the statistics was performed during the last 5000 time units. DNS was performed on the IBM computer with 32 parallel processors (Regatta-H) at TU Darmstadt.

The comparison between two resolutions (for $u_{i,rms}$) is shown in Figure 1. The difference between results is small even though the number of points in each direction for coarser grid is reduced by a factor of almost 1.59. Thus the coarser grid is sufficient to resolve the flow presented in the paper.

The comparison of the Reynolds stresses with Spalart's DNS (1988) data [7] ($Re_{\delta^*} \approx 1000$) is presented in Figure 2. The results agree more closely for $\overline{u_2' u_2'}, \overline{u_3' u_3'}$. In a case of $\overline{u_1' u_1'}$ one can see more difference between the results due to the following factors. Resolution in plus units in our and Spalart's DNS were $\Delta X^+ \approx 17$, $\Delta Z^+ \approx 9$ and $\Delta X^+ \approx 20$, $\Delta Z^+ \approx 6.7$, respectively. Hence in the critical wall normal direction resolution is slightly better. Statistics accumulation time in Spalart's DNS was $200\delta^*/u_\infty$, which is considerebly smaller than in our DNS.

The mean velocity of the turbulent boundary layer data is plotted in Figure 3. It shows good agreement between DNS data and the scaling law (24). As one can see the velocity defect law in the outer part of the boundary layer decreases more rapidly than what was derived from the theoretical result. There may be two reasons for this. First it might be the result of the low Reynolds number phenomenon in DNS while for the theoretical results we assumed the large Reynolds number limit. Second it is argued in [8] that non-parallel effects become dominant in the outer part of the wake region. The theoretical results were derived assuming a fully parallel flow. For these reasons, we have a relatively small "coincidence region" of the theoretical and DNS results.

In Figures (4-6) Reynolds normal stresses as obtained from DNS results and the scaling law (25) are compared. Constants in the exponential scaling law for Reynolds stresses (equation 25) are different for the different Reynolds stress tensor components. The last means that the "coincidence region" for each component of Reynolds stress tensor is located in the different area of the outer part of the boundary layer flow that is caused by the low Reynolds number phenomenon.

Conclusion

The most interesting theoretical result is that new symmetries of the two-point correlation equations are found which are not intrinsic to the Navier-Stokes or Euler equations. These symmetries have in turn be used to derive new scaling laws for the two-point and Reynolds stress quantities. A low Reynolds number DNS shows the validity of the scaling laws.

References

[1] Oberlack M. "A unified approach for symmetries in plane parallel turbulent shear flows". J.Fluid Mech., **427**, pp. 299-328, (2001).

[2] Oberlack M. "Asymptotic expansion, symmerty groups and invariant solutions of laminar and turbulent wall bounded flows" Z.Angew.Math.Mech, **80** 11-12, pp. 791-800, (2000).

[3] Carminati J., Vu K. "Symbolic Computation and differential equations: Lie symmetries". J.Sym. Comp., 29, pp. 95-116 (2000).

[4] Bluman G.W., Kumei S. "Symmetries and differential equations". Applied mathematical sciences 81, Springer-Verlag (1989).

[5] Lundbladh A. et al. "An efficient spectral method for simulation of incompressible flow over a flat plate". *Tech. Rep.* 1999:11. KTH, Stokholm.

[6] Nordstrom J. et al. "The fringe region technique and the fourier method used in the direct numerical simulation of spatially evolving viscous flows". SIAM J. Sci. Comput., 20, pp. 1365-1393 (1999).

[7] Spalart P.R. "Direct simulation of a turbulent boundary layer up to $R_\theta = 1410$". J. Fluid Mech. 187, pp. 61-98 (1988).

[8] Lindgren B., Österlund J. M., Johansson A. V. "Evaluation of scaling laws derived from Lie group symmetry methods in zero pressure-gradient turbulent boundary layers". Submitted to J. of Fluid Mech. (2002).

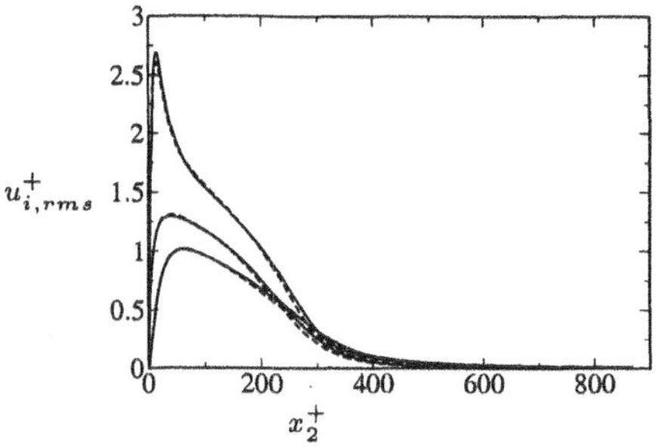

Figure 1: $u^+_{1,rms}, u^+_{2,rms}, u^+_{3,rms}$. Dashed lines correspond to 7.9 (480 × 129 × 128), solid lines to 31.2 (800 × 217 × 180) million grid points.

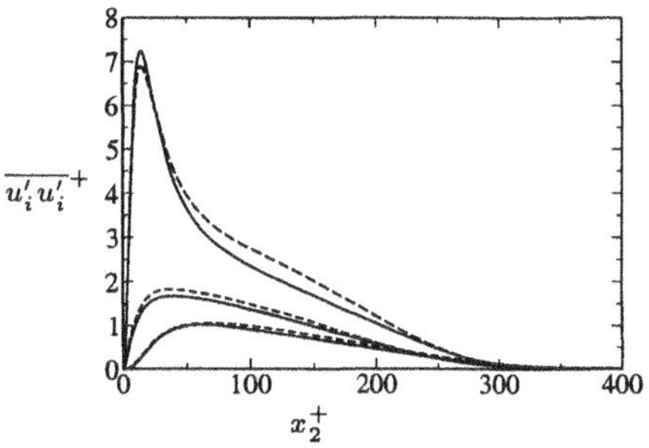

Figure 2: Reynolds stresses $\overline{u'_i u'_i}$. Dashed lines correspond to Spalart's data (1988), solid ones to the present DNS ($Re_{\delta^*} \approx 1100$).

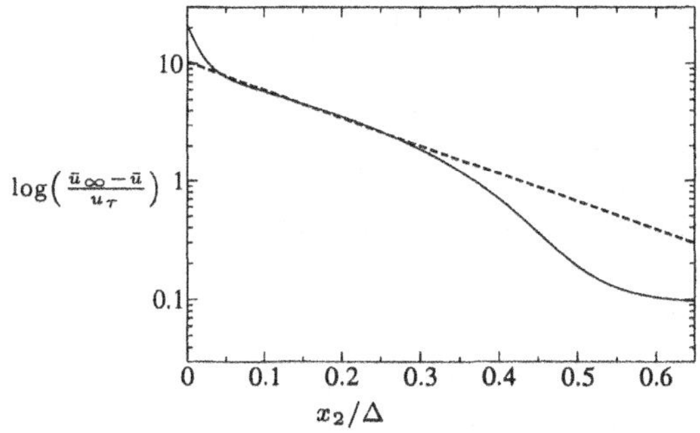

Figure 3: Mean velocity profile in linear-log scaling. Dashed line corresponds to DNS, solid one to the equation 24.

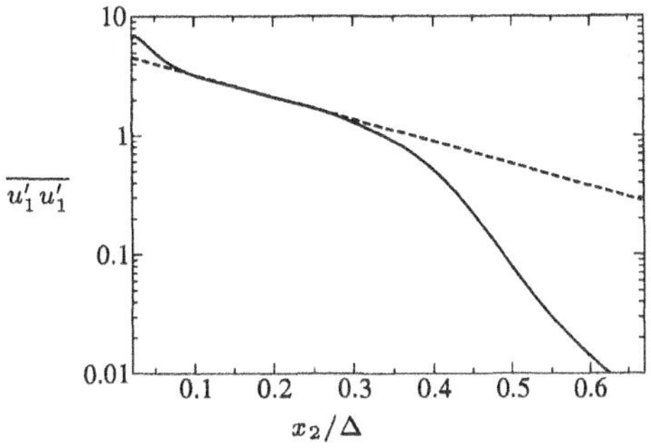

Figure 4: $\overline{u'_1 u'_1}$ Reynolds stress. Dashed line corresponds to DNS, solid one to the equation 25.

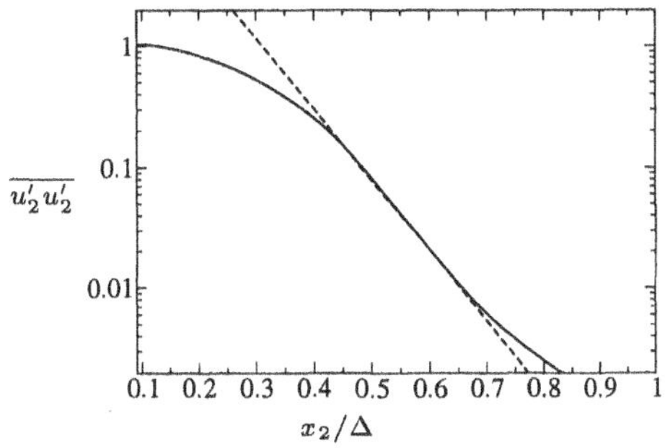

Figure 5: Reynolds stress $\overline{u'_2 u'_2}$. Dashed line corresponds to DNS, solid one to the equation 25.

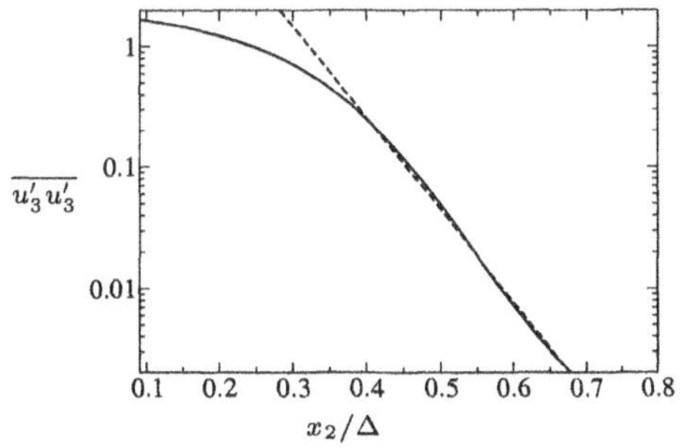

Figure 6: Reynolds stress $\overline{u'_3 u'_3}$. Dashed line corresponds to DNS, solid one to the equation 25.

Supersonic Turbulent Channel Flow with Passive Scalar Transport

Holger Foysi[1] and R. Friedrich[1]

Fachgebiet Strömungsmechanik, TU München, Boltzmannstr. 15, 85748 Garching, Germany

Summary

Direct numerical simulations of compressible supersonic turbulent channel flow with passive scalar transport are performed at the Reynolds numbers Re_τ of 455 and 544, at a Mach number of 1.5 and Prandtl and Schmidt numbers of 0.7 and 1.0, respectively. A mean scalar gradient is imposed across the mean flow. Budgets of the scalar variance and the scalar flux components are presented for both Reynolds numbers. A scaling of the Van-Driest transformed scalar similar to u^+_{VD} in the viscous is observed. Furthermore scalings of the Reynolds stress components and the pressure-rate-of-strain tensor that collapse both compressible and incompressible data in the outer layer are derived and investigated in more detail.

1 Introduction

Incompressible fully developed turbulent channel flow has been directly simulated for the first time by Kim *et al.* [7] at a Reynolds number Re_τ, based on the friction velocity, u_τ, and the channel half width, h of 180. Kim *et al.* [8] were also the first authors to study the transport of a passive scalar via DNS in a channel flow configuration at the same Reynolds number. Two types of boundary conditions were used, in the first case the scalar was internally generated with isothermal walls at the same temperature, in the second case a different scalar concentration on each of the isothermal walls was imposed. The Prandtl numbers were varied between 0.1 and 2. Recently Johannsson *et al.* [6] performed a similar DNS, imposing a mean scalar gradient, at a Reynolds number of $Re_\tau = 295$. Not much is known so far about the influence of compressibility on the transport of a passive scalar although its thorough understanding is prerequisite to the understanding of active scalar transport and combustion processes. Investigations of compressibility effects in supersonic channel flow have been presented first by Coleman *et al.* [2] in 1995, by Huang *et al.* [4] and later by Lechner *et al.* [5]. They all used isothermal walls, a domain size of (L_x,L_y,L_z) = (4π,$4\pi/3$,2)h in streamwise, spanwise and wall normal direction and Mach numbers of 1.5 ([2],[4],[5]) and 3 ([2],[4]). Huang *et al.* demonstrated, that terms responsible for compressibility effects, like the pressure-dilatation correlation or the compressible dissipation rate are negligibly small. They showed that the strong Reynolds analogy does not apply and that the dominant compressibility effect is due to mean property variation.

In the first part, this paper addresses the response of the Reynolds stress components and the pressure-rate-of-strain tensor to Mach and Reanolds number variations in fully developed compressible supersonic channel flow. In the second part budgets of the scalar variance and streamwise scalar flux are shown for supersonic flow only, as well as scalings of the Favre averaged scalar.

1.1 Numerical method and computational details

The compressible Navier-Stokes equations are solved in a pressure-, velocity-, entropy-form [14]. This form has been choosen because in a compressible flow the vorticity fluctuations interact with entropy and pressure fluctuations as shown by Kovasznay [9]. Following Sesterhenn [14], the Navier-Stokes equations are written in the form

$$
\begin{aligned}
\frac{\partial p}{\partial t} &= -\frac{\rho c}{2}\left((X^+ + X^-) + (Y^+ + Y^-) + (Z^+ + Z^-)\right) + \frac{R}{C_v}\left(-\frac{\partial q_i}{\partial x_i} + \Phi\right),\\
\frac{\partial u}{\partial t} &= -\left(\frac{1}{2}(X^+ - X^-) + Y^u + Z^u\right) + \frac{1}{\rho}\frac{\partial \tau_{1j}}{\partial x_j},\\
\frac{\partial v}{\partial t} &= -\left(X^v + \frac{1}{2}(Y^+ - Y^-) + Z^v\right) + \frac{1}{\rho}\frac{\partial \tau_{2j}}{\partial x_j},\\
\frac{\partial w}{\partial t} &= -\left(X^w + Y^w + \frac{1}{2}(Z^+ - Z^-)\right) + \frac{1}{\rho}\frac{\partial \tau_{3j}}{\partial x_j},\\
\frac{\partial s}{\partial t} &= -\left(X^s + Y^s + Z^s\right) + \frac{R}{p}\left(-\frac{\partial q_i}{\partial x_i} + \Phi\right).
\end{aligned}
\tag{1}
$$

with the following abbreviations:

$$
\begin{aligned}
&X^+ := (u+c)\left(\frac{1}{\rho c}\frac{\partial p}{\partial x} + \frac{\partial u}{\partial x}\right); \quad X^- := (u-c)\left(\frac{1}{\rho c}\frac{\partial p}{\partial x} - \frac{\partial u}{\partial x}\right);\\
&X^v := u\frac{\partial v}{\partial x}; \quad X^w := u\frac{\partial w}{\partial x}; \quad X^s := u\frac{\partial s}{\partial x};\\
&Y^+ := (v+c)\left(\frac{1}{\rho c}\frac{\partial p}{\partial y} + \frac{\partial v}{\partial y}\right); \quad Y^- := (v-c)\left(\frac{1}{\rho c}\frac{\partial p}{\partial y} - \frac{\partial v}{\partial y}\right);\\
&Y^u := v\frac{\partial u}{\partial y}; \quad Y^w := v\frac{\partial w}{\partial y}; \quad Y^s := v\frac{\partial s}{\partial y};\\
&Z^+ := (w+c)\left(\frac{1}{\rho c}\frac{\partial p}{\partial z} + \frac{\partial w}{\partial z}\right); \quad Z^- := (w-c)\left(\frac{1}{\rho c}\frac{\partial p}{\partial z} - \frac{\partial w}{\partial z}\right);\\
&Z^u := w\frac{\partial u}{\partial z}; \quad Z^v := w\frac{\partial v}{\partial z}; \quad Z^s := w\frac{\partial s}{\partial z}.
\end{aligned}
$$

This formulation resembles the characteristic formulation of the Euler equations. The terms appearing in the Eulerian part are discretized according to the sign of the eigenvalues of the Euler equations with fith-order upwind or downwind schemes [1]. These schemes are constructed in such a way, that numerical dissipation plays a role only at high wave numbers, therefore damping spurious modes. For the viscous terms sixth-order central compact schemes are used [10], the time integration is performed by using a third-order low storage Runge-Kutta method [15].

The code has been efficiently vectorized and parallelized. For the parallel solution of the tri- and pentadiagonal systems the algorithm of Erlebacher et al [3] has been used and extended. In this way a performance of 2.5 GFlops per node on the Hitachi SR80000-F1 was achieved. By using a homogenous body force $\bar{f}_1 = -\partial\bar{p}/\partial x_1 = \tau_w/h$ instead of the mean pressure gradient, the flow is made periodic in streamwise direction. The viscosity varies as the 0.7th power of the temperature,the Mach numbers, Reynolds numbers, grid resolution and domain size are given in table 1.

2 Results

2.1 Scaling of the Reynolds-stress and pressure-rate-of-strain tensor

It is well known ([2]) that the outer scaling for the Reynolds shear stress $\bar{\rho}R_{13} = \overline{\rho u_1'' u_3''}$ is τ_w (the subscript w refers to wall values), but this can be shown to apply to all Reynolds stress components. Integrating the averaged total enthalpy equation

$$\frac{\partial}{\partial x_j}(\overline{\rho u_j H}) = \frac{\partial}{\partial x_j}\left(-\bar{q}_j + \tilde{u}_i\bar{\sigma}_{ij} + \overline{u_i''\sigma_{ij}'}\right) \tag{2}$$

from the wall $x_3 = 0$ to the centerline $x_3 = h$ and replacing the streamwise pressure gradient by the homogeneous body force one gets an estimate for the wall heat flux

$$q_w \simeq \tau_w u_{av} \tag{3}$$

where u_{av} is the averaged velocity defined by $u_{av} = \int_0^1 \tilde{u}_1 d(x_3/h)$. Integrating the averaged internal energy equation

$$-\overline{p\frac{\partial u_j}{\partial x_j}} - \frac{\partial \bar{q}_j}{\partial x_j} - \frac{\partial}{\partial x_j}\left(\bar{\rho}\widetilde{u_j'' e''}\right) - \bar{\sigma}_{ik}\frac{\partial \tilde{u}_i}{\partial x_j} - \overline{\sigma_{ik}''\frac{\partial u_i''}{\partial x_j}} = 0 \tag{4}$$

from $x_3 = 0$ to $x_3 = h$ yields a second estimate after some algebra

$$q_w \simeq \int_0^h \bar{\rho}\epsilon dx_3 \simeq \tau_w u_{av}. \tag{5}$$

Using the balance equation for the Reynolds stresses $\bar{\rho}R_{ij} = \overline{\rho u_i'' u_j''}$:

$$\frac{\partial \bar{\rho}\tilde{u}_k R_{ij}}{\partial x_k} = P_{ij} + \Pi_{ij} - \bar{\rho}\epsilon_{ij} + \frac{\partial T_{ijk}}{\partial x_k}, \tag{6}$$

noticing that Π_{ij} and $\bar{\rho}\epsilon_{ij}$ are of the same order of magnitude and that local isotropy, $\bar{\rho}\epsilon_{ij} \approx 2/3\bar{\rho}\epsilon\delta_{ij}$, holds in the outer layer, it follows that the pressure-rate-of-strain tensor $\Pi_{ij} := \overline{p'(u_{i,j}' + u_{j,i}')}$ scales like $\tau_w u_{av}/h$. Figures 1 and 2 show R_{11} and Π_{11} normalized by τ_w and $\tau_w u_{av}/h$, respectively, for the two compressible flow cases A,C and the incompressible channel flow data of Moser *et al.* [13]. One can see a perfect collapse of the data in the outer layer.

2.2 Passive scalar transport

A mean scalar gradient has been imposed on the flow, using an initial profile of the form

$$\xi(x_1, x_2, x_3, 0) = \log_{10}\left\{\frac{z_0 + x_3}{z_0 - x_3}\right\}/\log_{10}\left\{\frac{z_0 + 1}{z_0 - 1}\right\}, \quad z_0 = 1.007 \tag{7}$$

and the bondary conditions $\xi(x_1, x_2, -1, t) = 1, \quad \xi(x_1, x_2, 1, t) = -1$. This corresponds to an injection of the scalar from one wall and a removal from the other. The following transport equation is integrated at each time step

$$\frac{\partial \xi}{\partial t} + \rho u_i \frac{\partial \xi}{\partial x_i} = \frac{1}{\rho}\frac{\partial}{\partial x_k}\left(\rho D \frac{\partial \xi}{\partial x_k}\right). \tag{8}$$

with $\rho D = \mu/Sc$. Figure 3 provides an impression of instantaneous large scale motions in supersonic turbulent channel flow at $Re_\tau = 544$, via contours of the scalar normalized with its wall value. Figure 4 shows the normalized Favre averaged scalar profile $\tilde{\xi}/\xi_w$ for the Reynolds numbers $Re_\tau = 455$ and $Re_\tau = 544$. Integrating the mean scalar transport equation from $x_3 = 0$ to $x_3 = h$ and neglecting correlations of the form $\overline{\mu'\xi'}$ gives in '+' coordinates

$$\overline{\rho(u_3^+)''(\xi^+)''}/\rho_w - \frac{1}{Sc}\frac{\bar{\mu}}{\mu_w}\frac{d\xi^+}{dx_3^+} \simeq 1 \tag{9}$$

where the wall normal scalar flux $\overline{\rho u_3''\xi''}$ has been normalized by the wall value of the molecular flux

$$\chi_w = \overline{\frac{\mu}{Sc}\frac{\partial \xi}{\partial x_3}}\bigg|_w = \rho_w u_\tau \xi_\tau .$$

This definition introduces the friction scalar ξ_τ in analogy to the friction velocity u_τ and gives $\xi^+ = \xi/\xi_\tau$. As in incompressible flow [6] an upper limit for the production term of the scalar variance

$$P_\xi = \overline{\rho(u_3^+)''(\xi^+)''}/\rho_w \frac{\partial \xi^+}{\partial \bar{x}_3^+} \simeq \left(1 - \frac{1}{Sc}\frac{\bar{\mu}}{\mu_w}\frac{\partial \xi^+}{\partial \bar{x}_3^+}\right)\frac{\partial \xi^+}{\partial \bar{x}_3^+} \tag{10}$$

can be derived. After multiplication of (10) by $\frac{\bar{\mu}}{\mu_w}\frac{\partial \xi^+}{\partial x_3^+}$ the maximum of the resulting equation is found to be $\frac{\partial \xi^+}{\partial x_3^+} = \frac{0.5 Sc}{\bar{\mu}/\mu_w}$. Inserting this into (10) gives

$$Max\left(\frac{\bar{\mu}}{\mu_w}P_\xi\right) = 0.25 Sc, \tag{11}$$

which is the same upper limit as in the incompressible case [6]. The upper limit of P_ξ, however, varies with Ma, via mean property variation.

For the viscous sublayer, after neglecting the scalar flux in (9), one gets

$$\tilde{\xi}_w^+ - \tilde{\xi}_{VD}^+ = Sc\, \tilde{u}_{VD}^+ \tag{12}$$

and for $x_3^+ > 30$, after using Prandtl's mixing length theory

$$\tilde{\xi}_w^+ - \tilde{\xi}_{VD}^+ = \frac{1}{K_\xi} \ln x_3^+ + B_\xi, \tag{13}$$

with the Van Driest transformed velocity and scalar

$$\tilde{\xi}_{vd}^+ = \int_0^{\tilde{\xi}^+} \sqrt{\frac{\bar{\rho}}{\rho_w}} d\tilde{\xi}^+ \qquad \tilde{u}_{VD}^+ = \int_0^{\tilde{u}^+} \sqrt{\frac{\bar{\rho}}{\rho_w}} d\tilde{u}^+. \tag{14}$$

Figure 5 shows the Van Driest transformed scalar for a Reynolds number of $Re_\tau = 544$, as well as a line fitting the profile in the log-region. The value for constant K_ξ of 0.26 is slightly lower than that found by Johannsson *et al.* [6], namely 0.33. Figures 6 and 7 show different terms in the budgets of the scalar variance $\overline{\rho(\xi'')^2/2}$, normalized by $\rho u_\tau^2 \xi_{av}^2/\nu$ and the streamwise scalar flux $\overline{\rho u_1 \xi''}$, normalized by $\rho u_\tau u_{av} \xi_\tau/\nu$. The normalization has been chosen such that the data for the different Reynolds numbers collapse in the outer layer. In the near wall region small Reynolds number effects can be observed. In the variance budget a nearly perfect balance between production and dissipation can be seen.

Conclusion

Using the scalings derived in section 2.1, both the compressible and incompressible data has been shown to collapse in the outer layer. Introducing a Van Driest transformed scalar, the Favre averaged scalar obleys similar scalings in the viscous sublayer and in the log-region as the velocity. Simulations at a higher Mach number of 4 are presently performed and will help to quantify compressibility effects in further detail.

Acknowledgements

The authors are grateful to Prof. S. Sarkar for his contribution to the first part of our work and to Dr. Brehm of the Munich Computer Centre for helpful advice related to the Hitachi SR8000.

References

[1] N.A. Adams, K. Shariff: "A high-resolution hybrid compact-ENO scheme for shock-turbulence interaction problems". JCP **127**, 1996, pp. 27

[2] G.N. Coleman, J. Kim, R.D. Moser: "Turbulent supersonic isothermal-wall channel flow". J. Fluid Mech. **305**, 1995, pp. 159-183

[3] T.M. Eidson, G. Erlebacher: "Implementation of a Fully-balanced Periodic Tridiagonal Solver on a Parallel Distributed Memory Architecture", ICASE Rept. 94-37. Concurrency Practices & Experience **7**, No 4, 1995, pp. 273-302

[4] P.G. Huang, G.N. Coleman, P. Bradshaw: "Compressible turbulent channel flows: DNS results and modelling". J. Fluid Mech. **305**, 1995, pp. 185-218
[5] R. Lechner, J. Sesterhenn, R. Friedrich: "Turbulent supersonic channel flow". J. Turbulence **2**, 2001, pp. 1-25
[6] A.V. Johansson, P.M. Wikström: "DNS and Modelling of Passive Scalar Transport in Turbulent Channel Flow with Focus on Scalar Dissipation Rate Modelling". Flow, Turb. and Combustion **63**, 1999, pp. 223-245
[7] J. Kim, P. Moin, R.D. Moser:"Turbulence statistics in fully developed channel flow at low Reynolds number". J. Fluid Mech. **177**, 1987, pp. 133
[8] J. Kim, P. Moin: "Transport of passive scalars in turbulent channel flow". *Turbulent Shear Flows 6* Springer Verlag, Berlin, 1989, pp. 85-96
[9] L.S. Kovasznay: "Turbulence in supersonic flow". J. Aeronaut. Sci. **20**, 1953, pp. 657
[10] S. Lele: "Compact Finite Difference Schemes with Spectral-like Resolution". J. Comput. Physics **103**, 1992, pp. 16-42
[11] Q. Li, N.D. Sandham, G.N. Coleman:"DNS of high-speed compressible channel flows". *Advances in Turbulence IX*, 2002, pp. 359
[12] N.N. Mansour, J. Kim, P. Moin: "Reynolds-stress and dissipation-rate budgets in a turbulent channel flow". J. Fluid Mech. **194**, 1988, pp. 15-44
[13] R.D. Moser, J. Kim, N.N. Mansour: "Direct numerical simulation of turbulent channel flow up to $Re_\tau = 590$". Phys. Fluids **11**, 1999, pp. 943-945
[14] J. Sesterhenn: "A characteristic-type formulation of the Navier–Stokes equations for high order upwind schemes". Computers & Fluids **30**, 2001, pp. 37
[15] J.H. Williamson: "Low-storage Runge-Kutta schemes", J. Comp. Physics **35**, 1980, pp. 48-56

Table 1 Parameters for the different DNS simulations

	M	Re	Re_τ	L_{x1}/h	L_{x2}/h	L_{x3}/h	N_{x1}	N_{x2}	N_{x3}
A	1.5	3000	221	4π	$4\pi/3$	2	192	128	151
B	1.5	5000	455	2π	$2\pi/3$	2	256	128	201
C	1.5	6000	544	2π	$2\pi/3$	2	256	128	201

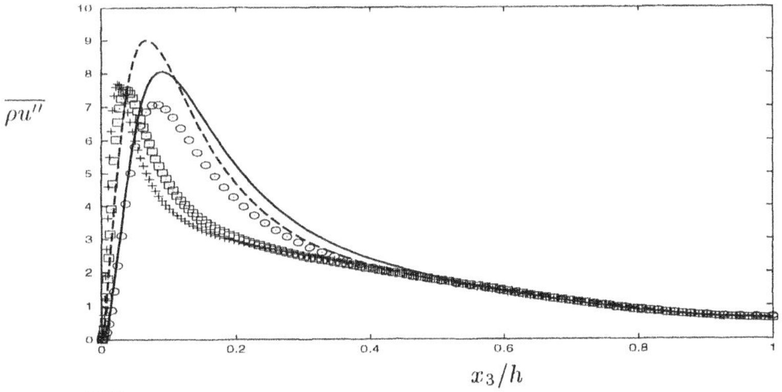

Figure 1 $\overline{\rho u''}$ normalized by τ_w. Compressible cases are ——, case A; – – –, case C. Incompressible cases are: ⊙ $Re_\tau = 180$, □ $Re_\tau = 395$, + $Re_\tau = 590$ (Moser *et al.* (1999))

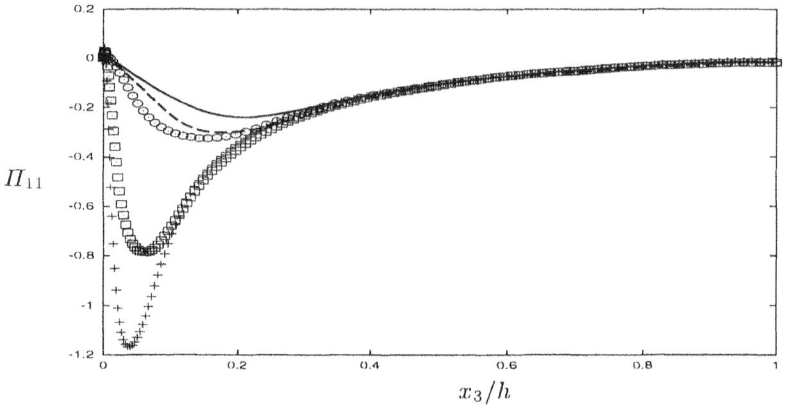

Figure 2 Π_{11} normalized by $u_{av}\tau_w/H$. Compressible cases are ——, case A; - - -, case C. Incompressible cases are: ⊙ $Re_\tau = 180$, □ $Re_\tau = 395$, + $Re_\tau = 590$ (Moser *et al.* (1999))

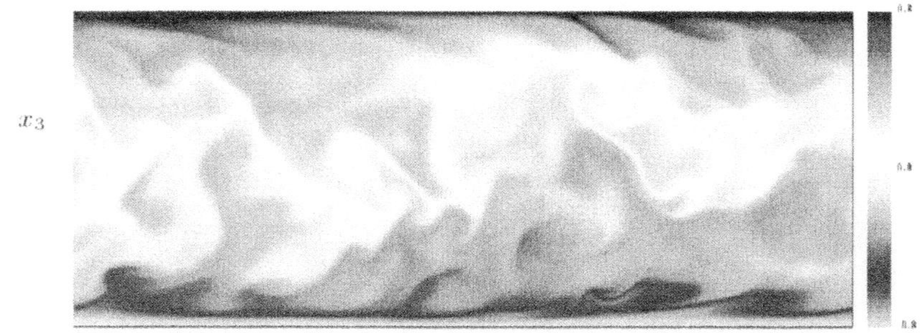

Figure 3 Scalar concentration in the x_1-x_3 - plane, case C

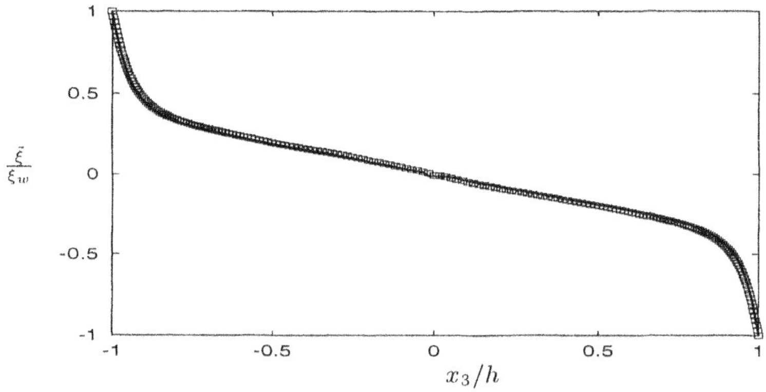

Figure 4 Favre averaged scalar, normalized by ξ_w. ——, Case C; - - -, Case B

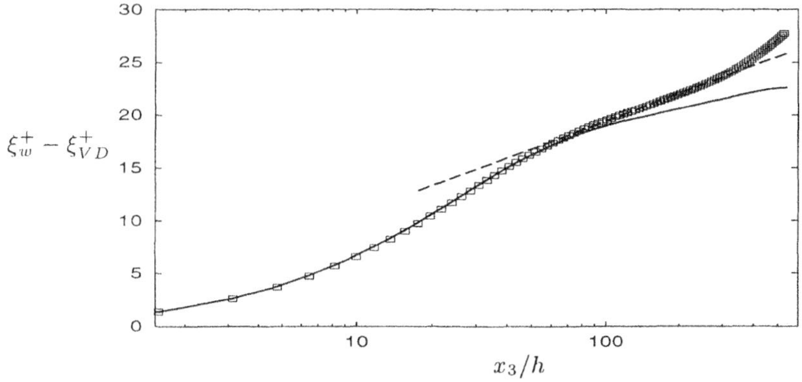

Figure 5 Van Driest transformed scalar. □ case B; ——, Van Driest transformed velocity; - - -, $3.8 \log x_3^+ + 1.95$

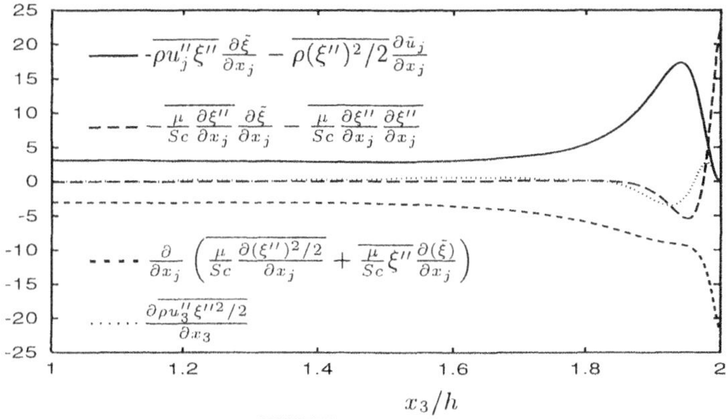

Figure 6 Terms in the $\overline{\rho \xi''^2/2}$-budget, normalized by $\rho_w u_\tau \xi_{av}^2/\nu$

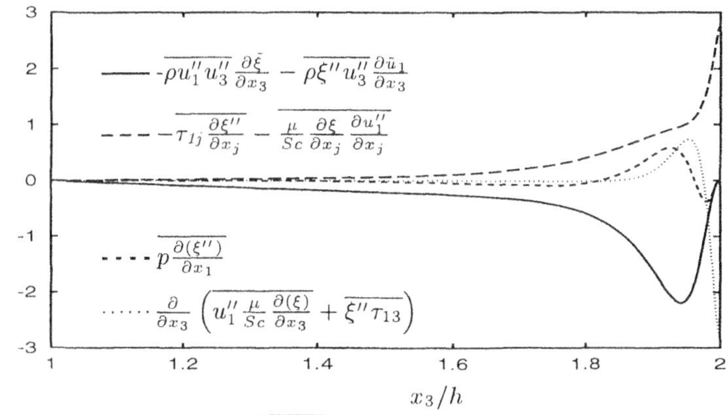

Figure 7 Terms in the $\overline{\rho u_1'' \xi''}$-budget, normalized by $\rho_w u_\tau u_{av} \xi_\tau/\nu$

Flow Computation at All Speeds

C.-C. Rossow
DLR, Institut für Aerodynamik und Strömungstechnik
Lilienthalplatz 7, D-38108 Braunschweig, Germany

Summary

An approach to compute flows over the whole Mach number range is proposed. Pressure is used as a dependent variable, and the derivation exploits the fact that the divergence-free constraint on the velocity field for incompressible flow derives from the energy equation. In the compressible regime, the implicit elliptic equation transforms into an explicit hyperbolic equation, leading to the common formulation used in compressible codes. The method strictly respects the strong conservation form, and reliable shock capturing is established by the use of a proven approximate Riemann solver for flux computation.

1. Introduction

Numerical methods suited to cover the whole Mach number range are still under development. To extend a compressible code towards incompressible flow, the system of unsteady, compressible equations is usually preconditioned with a suitable matrix such, that the disparity in the eigenvalues is reduced, and the integration with a time marching process is still applicable. Preconditioning has received widespread attention, and a variety of suitable preconditioning matrices were derived, e.g. [2,9,10]. However, in regions of very low local Mach number, e.g. near stagnation points and in recirculation regions, the preconditioning matrix may become singular [3], and the robustness of the computation may be impaired.
As an alternative, incompressible codes are extended towards compressible flows [1,4,5]. Here, usually the continuity equation is manipulated to yield a Poisson equation for a pressure correction. These so-called pressure-based methods naturally respect the constraint of a divergence-free velocity field, a condition which is not explicitly accounted for in the preconditioning approach [9]. The extension to compressible flows is then achieved by incorporating convective terms into the pressure correction equation and to use Laplace's equation to obtain a time-derivative of pressure. However, usually the strong conservation of the governing equations is not strictly respected, and the computation of cell interface fluxes is not achieved with established Riemann solvers.
To avoid the shortcomings of preconditioning and pressure-based methods, in the present work a method will be developed which employs a unique formulation throughout the Mach number regime. The governing equations are solved in strong conservation form, the cell interface flux computation is based on the approximate solution of the Riemann problem, the divergence-free constraint is accounted for in the incompressible limit, and the solution method adapts to the change in character of the governing equations from hyperbolic to elliptic.

2. Governing Equations

We consider the two-dimensional Euler equations for compressible flow. The Euler equations read in divergence form:

$$\begin{aligned}
\frac{\partial \rho}{\partial t} &+ div(\rho\vec{q}) & &= 0 \\
\frac{\partial(\rho u)}{\partial t} &+ div(\rho u\vec{q}) &+ \frac{\partial p}{\partial x} &= 0 \\
\frac{\partial(\rho v)}{\partial t} &+ div(\rho v\vec{q}) &+ \frac{\partial p}{\partial y} &= 0 \\
\frac{\partial(\rho E)}{\partial t} &+ div(\rho H\vec{q}) & &= 0
\end{aligned} \quad , \quad \text{where } \vec{q} = \begin{pmatrix} u \\ v \end{pmatrix}. \tag{1}$$

In compressible codes, the equation of state is used to compute pressure:

$$\frac{p}{\rho} = R \cdot T \tag{2}$$

In this work however, pressure is directly solved for. To respect conservation, pressure changes $\delta(p)$ have to be computed from changes of conservative variables by employing the chain rule:

$$\delta(p) = (\gamma - 1)\left[\frac{q^2}{2}\delta(\rho) - u\delta(\rho u) - v\delta(\rho v) + \delta(\rho E)\right]. \tag{3}$$

Using definition (3), an equation for pressure can be derived to replace the energy equation in the system (1) to directly solve for pressure

$$\frac{\partial p}{\partial t} + (\gamma - 1)\cdot\left[\frac{q^2}{2}div(\rho\vec{q}) - u\left(div(\rho u\vec{q}) + \frac{\partial p}{\partial x}\right) - v\left(div(\rho v\vec{q}) + \frac{\partial p}{\partial y}\right) + div(\rho H\vec{q})\right] = 0 \tag{4}$$

or in abbreviated form introducing $\vec{Q}$ as the vector of flux balances with

$$\frac{\partial p}{\partial t} + (\gamma - 1)\left[\frac{q^2}{2}Q(\rho) - uQ(\rho u) - vQ(\rho v) + Q(\rho E)\right] = 0. \tag{5}$$

The equation of state (2) is then used to compute specific total energy or enthalpy. Note that approaching the limit of incompressible flow, eq. (4) reduces to

$$\frac{\partial p}{\partial t} + (\gamma - 1)[\, div(\rho H\vec{q})\,] = 0 \tag{6}$$

3. Derivation of the Present Approach

The present method is derived by considering the discrete flux through a computational cell. Using the basic Flux Difference Splitting (FDS) scheme of [7], the inviscid flux density vector $\vec{F}$ normal to a cell interface S may be written as:

$$\vec{F} = \frac{1}{2}\left(\vec{F}^L + \vec{F}^R\right) - \frac{1}{2}\left|\overline{\overline{A}}\right| \Delta\vec{W} \quad , \tag{7}$$

where $\vec{F}^L$ and $\vec{F}^R$ are the left and right states of the inviscid flux density vector normal to the cell interface, $\overline{\overline{A}}$ is the corresponding flux Jacobian, and $\Delta\vec{W}$ denotes the differences in conservative variables between left and right states of a cell interface. In [8], the expression $\left|\overline{\overline{A}}\right| \cdot \Delta\vec{W}$ was expanded in terms of the interface Mach number M_0, which is bounded by −1 and 1. The discrete enthalpy flux density normal to a cell face *S* can then be expressed following [8] as

$$F^{\rho H} = \frac{1}{2} \cdot \left[\left(\rho H q_n^L + \rho H q_n^R \right) - \frac{H}{c'} \left(1 - \left| M_o \right| \right) \Delta p + ODT \right], \tag{8}$$

where q_n denotes the velocity normal to *S*, *ODT* stands for other dissipative terms, and *c'* is the preconditioned speed of sound for appropriate scaling of the dissipative terms [8,9]. The pressure difference Δp denotes the difference between pressure on the left and right side of *S*.

Using the FDS discretization of [7,8] to obtain the intermediate state at cell interfaces, the discrete divergence div_δ of $\rho H\vec{q}$ in a cell can be expressed as:

$$\mathrm{div}_\delta(\rho H\vec{q}) = \frac{1}{vol} \cdot \sum_{i=1}^{ifaces} F_i^{\rho H} \cdot S_i = \frac{1}{vol} \cdot \sum_{i=1}^{ifaces} \frac{1}{2} \cdot \left[\left(\rho H q_n^L + \rho H q_n^R \right) - \frac{H}{c'} \left(1 - \left| M_o \right| \right) \Delta p + ODT \right] \cdot S_i \tag{9}$$

where *ifaces* denotes the number of faces of the computational cell with *i* as the running index, and *vol* is the corresponding cell volume.

Using definition (9), the discrete analogue of eq. (4) or eq. (5) is expressed as

$$\frac{\delta p}{\delta t} + (\gamma - 1) \left[\frac{q^2}{2} Q_\delta(\rho) - u Q_\delta(\rho u) - v Q_\delta(\rho v) + div_\delta(\rho H\vec{q}) \right] = 0 \tag{10}$$

where $\vec{Q}_\delta(\rho, \rho u, \rho v)$ denote the discrete flux balances of the continuity and momentum equations, respectively. Rearranging with use of eq. (9) leads to:

$$\frac{\delta p}{\delta t} = -(\gamma - 1) \left[\frac{1}{vol} \cdot \sum_{i=1}^{ifaces} \frac{1}{2} \cdot \left[\left(\rho H q_n^L + \rho H q_n^R \right) - \frac{H}{c'} \left(1 - \left| M_o \right| \right) \Delta p + ODT \right]_i S_i + \frac{q^2}{2} Q_\delta(\rho) - u Q_\delta(\rho u) - v Q_\delta(\rho v) \right] \tag{11}$$

Introducing an implicit formulation for pressure by using $p^{n+1} = p^n + \delta p$ yields

$$\frac{\delta p}{\delta t} - (\gamma - 1) \frac{1}{vol} \cdot \sum_{i=1}^{ifaces} \frac{1}{2} \cdot \left[\frac{H}{c'} \left(1 - \left| M_o \right| \right) \Delta(\delta p) \right]_i S_i = RHS \tag{12}$$

where

$$RHS = -(\gamma - 1) \left[\frac{q^2}{2} Q_\delta(\rho) - u Q_\delta(\rho u) - v Q_\delta(\rho v) + div_\delta(\rho H\vec{q}) \right] \tag{13}$$

Equation (12) reflects the role of pressure for compressible and incompressible flow in the discrete formulation: For compressible flow at Mach numbers above one, the second term on the left hand side of eq. (12) vanishes, and eq. (12) becomes purely explicit. The system of equations (1) is solved simultaneously with an explicit time stepping scheme, and the simultaneous solution of the system (1) reflects the tight coupling of all variables in compressible flow. However, when the Mach number tends to zero, eq. (12) requires an implicit solution for p. This reflects the fact that for $M \to 0$ pressure fluctuations propagate infinitely fast compared to the advection speed $\vec{q}$.

Note that for incompressible flow, by virtue of eq. (6), eq. (12) becomes a Poisson equation for pressure similar to the pressure correction equation derived in [9] to respect the divergence-free constraint. However, in eq. (12) the enthalpy flux $\rho H\vec{q}$ is regarded and not the mass flux $\rho\vec{q}$, since in [9] the Poisson equation was derived from the continuity equation. In [6] it was shown for inviscid flow by asymptotic analysis, that the divergence-free constraint for the incompressible velocity field, $div(\vec{q}) = 0$, derives from the energy equation, but not from the continuity equation, which is now reflected by the discrete method.

4. Implementation of the Current Approach

In conventional compressible codes with Runge-Kutta time stepping schemes, the set of governing equations (1) is solved simultaneously to update the conservative variables $\vec{W} = (\rho, \rho u, \rho v, \rho E)$:

$$\vec{W}_j^{(m)} = \vec{W}_j^{(0)} + \alpha^{(m)} \cdot \vec{R}_j\left(\vec{W}^{(m-1)}\right), \tag{14}$$

where the superscripts (m) denote the stage count with m running from 1 to the maximum number of stages, the subscripts j correspond to the location at mesh points in the flow field, $\alpha^{(m)}$ is the stage coefficient of the (m)-stage, and $\vec{R}_j\left(\vec{W}^{(m-1)}\right)$ represent the conservative residuals of continuity, momentum and energy equations evaluated using the flow variables from the previous $(m$-1)-stage.

To solve for another set of variables $\vec{U} = (\rho, \rho u, \rho v, p)^T$, eq. (14) becomes

$$\vec{U}_j^{(m)} = \vec{U}_j^{(0)} + \alpha^{(m)} \cdot \vec{R}_j\left(\vec{U}^{(m-1)}\right), \tag{15}$$

where the explicit residual for the pressure equation is obtained after evaluation of the conservative flux balances $\vec{Q}$ and application of the chain rule eq. (3).

In order to implement the formulation of the pressure equation (12) into the Runge-Kutta scheme (15), in each stage m a preliminary pressure update δp_j is computed from eq. (12). Having computed δp_j, pressure is then updated in the Runge-Kutta framework as

$$p_j^{(m)} = p_j^0 + \alpha^{(m)} \cdot \delta p_j . \tag{16}$$

Note that for Mach numbers greater than unity, the second term on the left hand side of eq. (12) vanishes, and the pressure update from eq. (16) becomes purely explicit and identical to the update by the basic Runge-Kutta scheme of eq. (15).

Throughout this work, the MAPS+ approximate Riemann solver developed in [8] is used for spatial discretization. The preconditioned speed of sound c' in eq. (8) is evaluated using the freestream velocity as reference, see [8,9].

5. Application of the Method

First, the present approach is used to compute one dimensional nozzle flow. In order to clearly identify capabilities and limits, only a first order implementation was investigated, and results of the basic compressible code used in [8,9] serve as reference. Note, that the preconditioned speed of sound c' in eq. (8) is computed identically to the preconditioned reference code of [9]. The first test case is the nearly incompressible flow in a converging/diverging nozzle with an inflow Mach number of $M_{in} = 0.00845$. Fig. 1 shows the Mach number distribution along the nozzle axis. Since the pressure equation is based on the conservative flux balances according to eq. (4), converged results of the reference code and the present approach are identical. In Fig. 2 the convergence histories are displayed. For the reference solution the preconditioning matrix of [9] was employed. In the second test case, compressible flow in a Laval nozzle was computed. Fig.3 shows the Mach number distribution in the Laval nozzle for both the reference code and the present approach. The identical convergence behavior shown in Fig. 4 confirms the statement that at high speed the present approach reverts to an explicit compressible formulation. In order to give further evidence that the present approach consistently blends to a compressible formulation suitable for reliable shock capturing, nozzle flow with a pre-shock Mach number of $M = 32$ was computed, similarly to the investigations of [8]. The axial Mach number distribution and convergence behavior are shown in Figs. 5 and 6.
From the computation of the 1D test cases it can be concluded that the present approach is well suited for computations over the whole Mach number range, without any artificial adaptation of the formulation to a specific flow regime.

Next, the method is applied to inviscid flows around the NACA0012 airfoil using the code of [9] as reference and numerical framework for the implementation of the current approach. For the computations, a structured O-mesh with 160 cells around the airfoil and 32 cells in normal direction was chosen. To check the behavior of the present scheme with varying Mach number, for an angle of attack of $\alpha = 0°$, free-stream Mach number was varied from $M_\infty = 0.001$ to $M_\infty = 0.85$. Fig.7 displays surface pressure distributions obtained for selected Mach numbers, which illustrate the change in the flow characteristics. The convergence behavior of the present method and the reference code with preconditioning can be assessed from Fig.8, where the time steps required to reduce the density residual by 8 orders of magnitude on the finest mesh are displayed as a function of free-stream Mach number. At lower Mach numbers, the present approach converged significantly faster than the reference code with preconditioning. This is a result of the implicit treatment of pressure through eq. (12), which accounts at low Mach number for the divergence-free constraint. For higher Mach numbers, the convergence rates of the different approaches become similar, as may be expected since the present approach becomes similar to the original compressible formulation.

In the next case the robustness of the method was assessed. Following the route of investigations in [9], at $M_\infty = 0.001$ the angle of attack was increased from α=12° to α=14° and α=16°. In all cases, computations were started from free-stream conditions. The corresponding convergence behavior of density residual and lift are shown in Figs.9 and 10. For α=12° and α=14°, convergence rates are similar. At α=16° the residual stalled, but total lift converged to a steady value. In [9], the same cases were computed using the reference code with preconditioning, and using a pressure correction method. The pressure correction method achieved converged solutions up to α=14°, and a lift-converged solution for α=16°. The reference code with preconditioning did not yield a stable solution even for α=14°. This indicates that the robustness of the present approach for incompressible flows is comparable to that of classical pressure correction schemes, and superior to the preconditioned approach.
In further studies, it was found that the reference value for the preconditioned speed of sound could be lowered substantially below the freestream velocity, in contrast to preconditioned codes [8,9]. However, to establish a clear comparison of the methods, c' was evaluated identically in the present study. The influence of bounds for c' will be subject of future work.

6. Conclusion

In the present work, a computational approach was derived which consistently covers incompressible and compressible flows. For compressible flows, the method is similar to standard explicit codes solving the compressible equations. For incompressible flows, an implicit solution for pressure is established. The implicit solution is derived from pressure terms arising in the enthalpy flux balance when the Mach number approaches zero, with the pressure terms identified from the approximate solution of the Riemann problem. The resulting equation is similar to a pressure correction equation which in the present approach is however derived from the energy equation. The method was found to be applicable over the whole Mach number regime. Compared to preconditioning methods, it yielded improved robustness and efficiency for incompressible flow. At high Mach numbers, the robustness and accuracy of compressible codes was retained

References

1. Bijl, H. and Wesseling, P.: A Unified Method for Computing Incompressible and Compressible Flows in Boundary Fitted Coordinates, *J.Comp.Phys.,* Vol.141, pp.153-177, 1998.
2. Choi, Y.-H. and Merkle, C.L.: The Application of Preconditioning to Viscous Flows, *J.Comp.Phys.*, Vol.105, pp 207-223, 1993.
3. Darmofal, D.L. and Van Leer, B.: Local preconditioning: manipulating mother nature to fool father time, *In Computing the Future 2: Computational Fluid Dynamics and Transonic Flow,* ed. D. Caughey, M. Hafez, 1998.
4. Demirdzic, I., Lilek, Z. and Peric, M.: A Collocated Finite Volume Method for Predicting Flows at All Speeds, *Int.J.Numer.Meth.Fluids,* Vol.16, pp.1029-1050, 1993.

5. Karki, K.C. and Patankar, S.V.: Pressure based calculation procedure for viscous flows at all speed in arbitrary configurations, *AIAA Journal,* Vol.27, pp.1167-1174, 1989.
6. Klein, R.: Semi-Implicit Extension of a Godunov-Type Scheme Based on Low Mach Number Asymptotics I: One-Dimensional Flow, *J.Comp.Phys.,*Vol.121, pp.213-237, 1995.
7. Roe, P.L.: Approximate Riemann Solvers, Parameter Vectors and Difference Schemes, *J. Comp. Phys.,* Vol. 43, pp.357-372, 1981.
8. Rossow, C.-C.: A Flux Splitting Scheme for Compressible and Incompressible Flows, *J. Comp. Physics,* Vol. 164, pp.104-122, 2000.
9. Rossow, C.-C.: A Blended Pressure/Density Based Method for the Computation of Incompressible and Compressible Flows, *J.Comp.Phys.,*Vol.185, pp.375-398, 2003.
10. Weiss, J.M. and Smith, W.A.: Preconditioning Applied to Variable and Constant Density Flows, *AIAA Journal,* Vol. 33, pp.2050-2057, 1995.

Figures

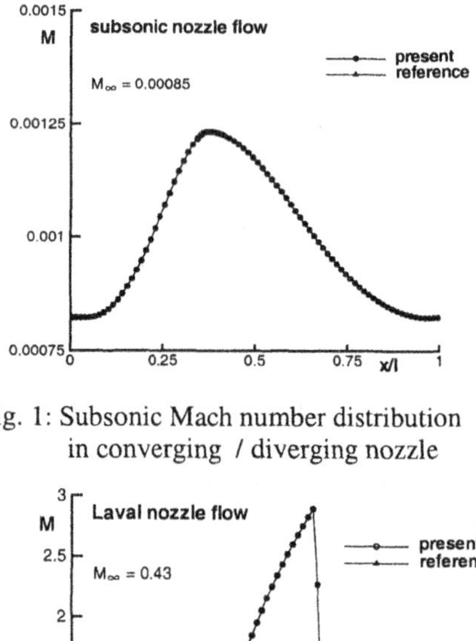

Fig. 1: Subsonic Mach number distribution in converging / diverging nozzle

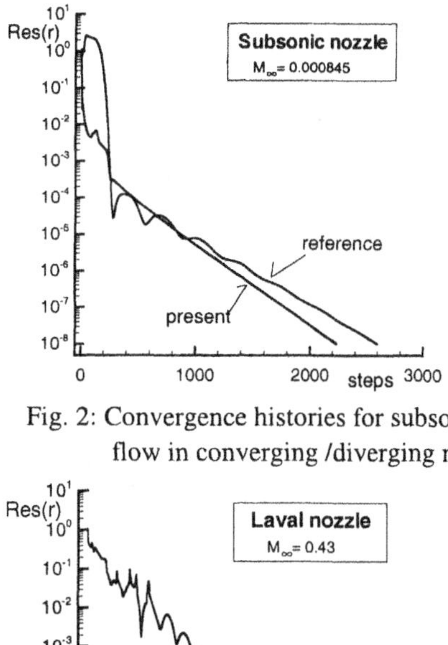

Fig. 2: Convergence histories for subsonic flow in converging /diverging nozzle

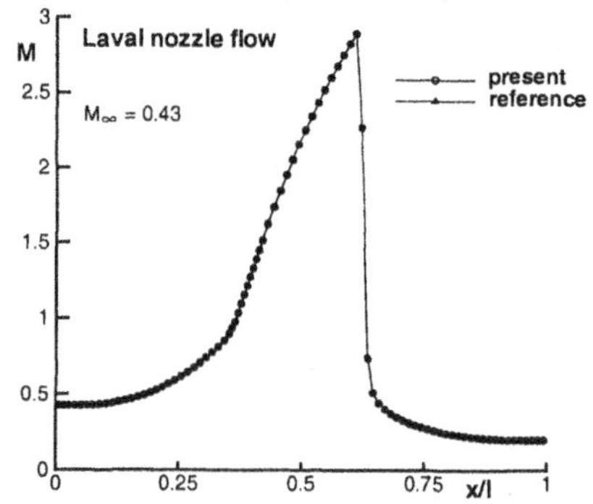

Fig. 3: Mach number distribution for 1D Laval nozzle

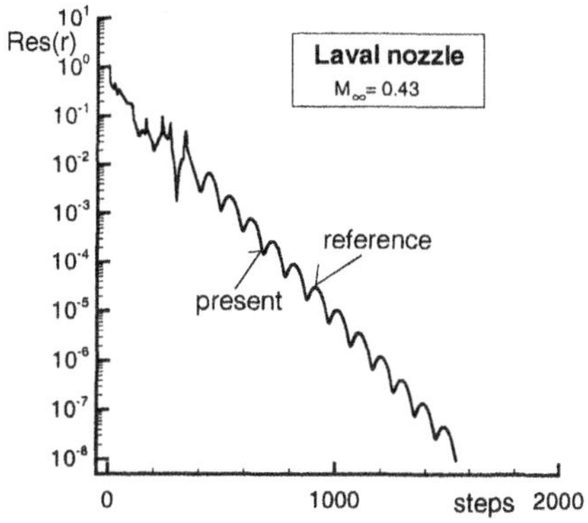

Fig. 4: Convergence histories for 1D Laval nozzle

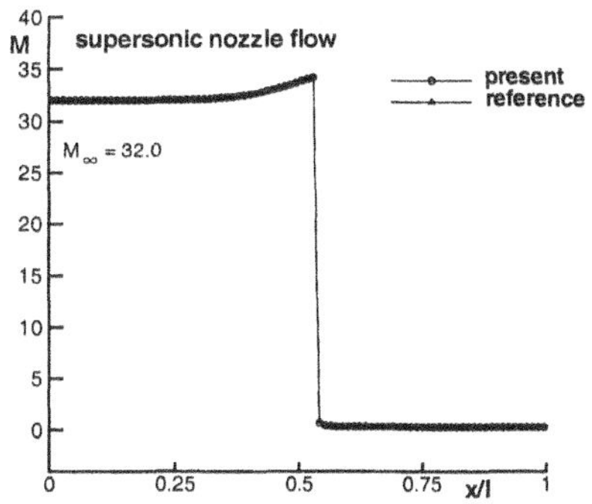

Fig. 5: Mach number distribution for supersonic nozzle flow

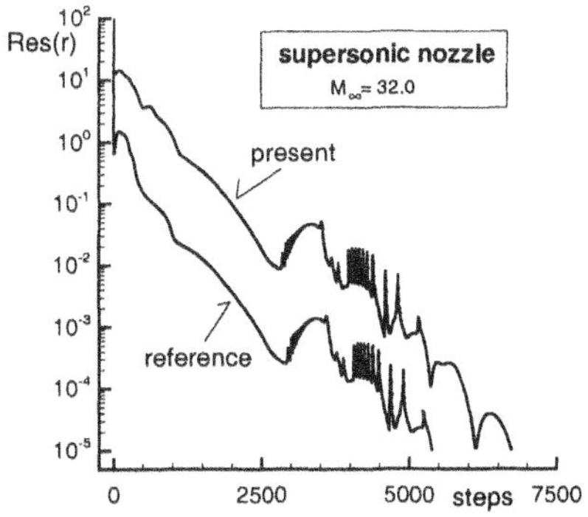

Fig. 6: Convergence histories for supersonic nozzle flow

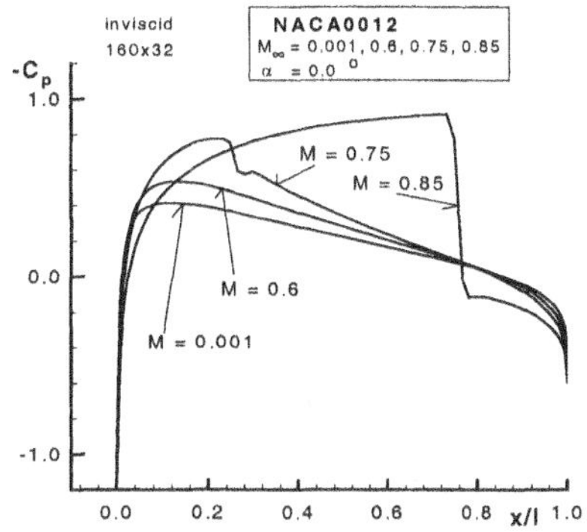

Fig. 7: Surface pressure distribution for Mach number variation of inviscid flow around NACA0012 airfoil at $\alpha = 0°$

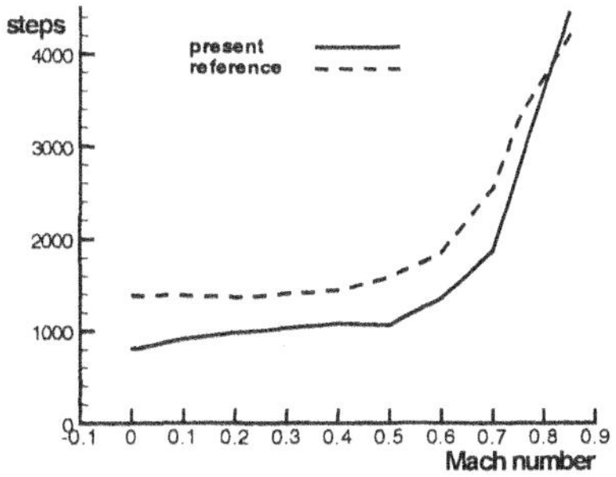

Fig. 8: Number of time steps as function of free-stream Mach number for inviscid flow around NACA0012 airfoil

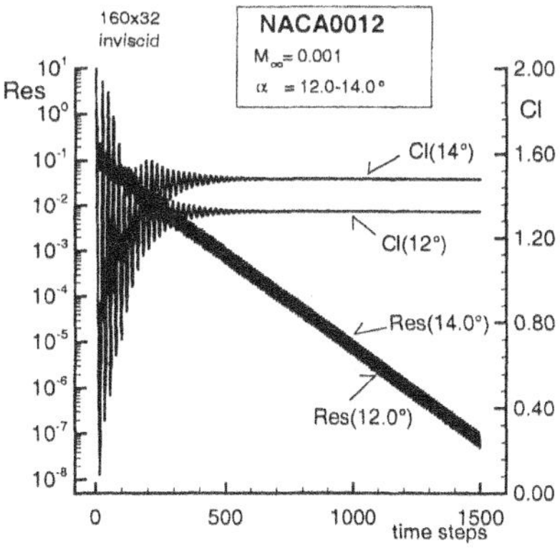

Fig. 9: Convergence for inviscid, incompressible flow around NACA 0012 airfoil at $\alpha = 12° - 14°$

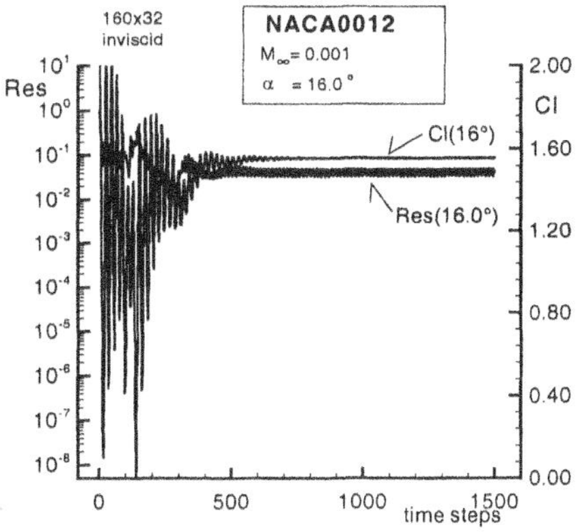

Fig. 10: Convergence for inviscid, incompressible flow around NACA 0012 airfoil at $\alpha = 16°$

Parallel and Adaptive Finite Element Techniques for Flow Simulation

J. Stiller[1], W. Wienken[2], U. Fladrich[3], R. Grundmann[1], and W. E. Nagel[3]

[1] TU Dresden, Institut für Luft- und Raumfahrttechnik (ILR)
[2] TU Dresden, Institut für Strömungsmechanik (ISM)
[3] TU Dresden, Zentrum für Hochleistungsrechnen (ZHR)
D-01062 Dresden, Germany

Summary

We present the software library MG which provides an interface for the parallel implementation of adaptive flow solvers using unstructured grids. The excellent scalability on distributed memory architectures is demonstrated. Current applications include finite element solvers for incompressible as well as compressible flows. As an example we present large eddy simulations of the turbulent channel flow at $Re_\tau = 180$ and the flow past a square cylinder at $Re = 22,000$ using an SUPG FEM for the compressible Navier-Stokes equations.

1 Introduction

The development of accurate, scalable, solution-adaptive solvers is a cornerstone on the way toward the realistic simulation of problems involving a broad spectrum of scales, as typical for turbulent flows. Since most present (and future) high performance computers rely on distributed memory architectures it is not surprising that virtually every modern CFD code is parallelized. Prominent examples are the DLR FLOWer and TAU codes [1], NPARC's Wind code [3], or PETSc-FUN3D [5]. Moreover, most commercial CFD codes are now parallelized as well.

The integration of adaptive and parallelization techniques, however, involves a new dimension of algorithmic complexity. Not only more flexible and intricate data structures are required. As adaptation changes the local grid density in an unpredictable manner, it also creates the need for dynamic load balancing. So far, only a few research projects have addressed the problem of scalability of parallel adaptive solvers using three-dimensional unstructured grids [2,4,11,14]. The MG library [14] developed by the authors intends to provide a lightweight but highly scalable interface for implementing unstructured flow solvers. In Section 2 we shortly discuss the features and the design of MG. Further the aspects of scalability on parallel computer systems are analyzed in some more detail. In Section 3 we present a finite element Navier-Stokes solver based on MG and preliminary results of its application to large-eddy simulation (LES) of turbulent flows.

2 The MG Library

2.1 Features

The objective of developing MG is to provide a software library which combines the following features within one package:

- data structures and procedures for handling unstructured grids
- stable and fast methods for grid adaptation
- transfer operators for parallel multigrid solvers
- dynamic partitioning and load balancing of multilevel grids
- support for high-order (spectral element) methods

The current implementation is restricted to tetrahedral cells and does not include grid generation and visualization facilities. On the other hand it offers fully integrated and very efficient tools for parallelization, grid adaptation and multigrid methods. Furthermore, the recently added spectral element library provides the basis for high-order methods that can be applied to direct numerical simulation (DNS) and large-eddy simulation (LES) of complex turbulent flows.

2.2 Scalability

The scalability of a flow solver is of crucial importance for large-scale simulations. Especially on massively parallel systems a high efficiency of all subsystems: grid adaptation and partitioning, implicit solvers and explicit updates, is mandatory.

The scalability of parallel grid adaptation has been studied extensively in [14]. Under quite general assumptions the cost (wall clock time) of one adaption cycle can be estimated as

$$T = \mathcal{O}(L + M_1 \log(P/P_1) + M),$$

where L is the number of grid levels, M_1, M are the sizes, and P_1, P the numbers of coarse and fine grid partitions, respectively. Neglecting the first contribution, the resulting parallel efficiency is

$$E(P) = \frac{T_1}{PT} < \left(\frac{C}{M}\log(P/P_1) + 1\right)^{-1}$$

where C is a problem dependent constant. An immediate consequence of this estimate is that the parallel efficiency of a fixed size problem will break down inevitably for large P. On the other hand, scalability (in terms of a non or slowly growing wall clock time) could be achieved if the ratio P/P_1 remains bounded and P is chosen such that a constant partition size M is maintained. It should also be noticed that the actual efficiency is affected by several other factors including load balance and possible bottlenecks of the implementation and of the hardware platform. The key features of the algorithm provided with MG are as follows:

1. It is recursive, i.e., fine grid partitions result from coarse grid partitions by subdivision (Fig. 1). These are independent from each other and, hence, this process is easily parallelized.
2. It is multi-objective, i.e., on a given grid level the partitioning method considers the impact on each subsequent grid level. For this purpose the multi-constraint version of MeTiS [7] is used. As demonstrated in [13] this approach leads to a well balanced distribution of all grid levels.
3. All substeps (refinement pattern completion, partitioning, adaptation, and data remapping) are fully parallelized.

Altogether these features allow for almost optimal scalability up to several hundred processors (see Figure 2(a)). The parallelization of the other solver components is less critical than grid adaptation. As an example Fig. 2(b) depicts the excellent scalability of the finite element Navier-Stokes solver described in the following section.

3 Application to FEM for Compressible Flow

The applications implemented on top of MG include finite element based solvers for advection-diffusion problems, incompressible and compressible Navier-Stokes equations as well as magnetohydrodynamic problems at low magnetic Reynolds numbers. In the following we focus on the compressible Navier-Stokes solver and its application to the large-eddy simulation of subsonic flows.

3.1 Numerical Model

The filtered Navier-Stokes equations for a perfect gas can be written as

$$\frac{\partial \bar{\mathbf{U}}}{\partial t} + \frac{\partial \bar{\mathbf{F}}_i}{\partial x_i} = \frac{\partial \bar{\mathbf{D}}_i}{\partial x_i}$$

where $\bar{\mathbf{U}}$ are the resolved conservative variables and

$$\bar{\mathbf{F}}_i = \mathbf{F}_i(\bar{\mathbf{U}}) + \mathbf{F}_i^{sgs}, \quad \bar{\mathbf{D}}_i = \mathbf{D}_i(\bar{\mathbf{U}}) + \mathbf{D}_i^{sgs}$$

are the advective and diffusive fluxes with superscript *sgs* denoting the unresolved (subgrid-scale) contribution. The SGS contribution of diffusive fluxes can be neglected in most situations [17]. The unresolved advective fluxes are modeled using the simplest reliable approach consisting of the trace-free Smagorinsky model with van Driest damping and the Reynolds analogy for the SGS heat flux. Furthermore, the isotropic part of the SGS stress can be neglected for moderate turbulence Mach numbers [19].

The equations are discretized using the Streamline-upwind/Petrov-Galerkin (SUPG) method developed by Hughes and coworkers [12] and further improved by Jansen et al. [6]. The basic idea of SUPG consists in adding a solution-dependent "upstream-weight" to the Galerkin test functions. In contrast to artificial viscosity methods this approach is consistent with the exact filtered equations. Using linear shape functions it leads to a second order upwind scheme. For time integration of the resulting system of ordinary differential equations we use the standard 4-stage Runge-Kutta method combined with a block Jacobi iteration for resolving the consistent mass matrix.

3.2 Results

The numerical model was implemented on top of MG. Using the tools provided by the library, parallelization and grid adaptation required only a few lines of code. An example that illustrates the capabilities of the solver is given in Fig. 3. It shows a snapshot of the computed flow (Mach number distribution) and the adapted grid in comparison to an Schlieren photograph by A. C. Charters [15]. The computational grid consisting of about 6×10^5 nodes was adapted and redistributed over 16 processors after every tenth time step (at a Courant number of 0.1). Though the Reynolds number was considerably smaller in the computation, the agreement of the resolved flow structures with the experiment is remarkable. A quantitative analysis of the results and performance of this application is in progress and will be subject of a forthcoming paper.

In the remaining part of this section we discuss the application to the large-eddy simulation of the turbulent flow in a plane channel and the flow past a square cylinder. In these preliminary computational experiments we deliberately did not use the grid adaption facilities, but focused on the validation of the method itself. A more detailed description and further examples are provided in [18].

Turbulent Channel Flow. The turbulent flow in a plane channel at Reynolds number $Re_\tau = 180$ (based on the shear velocity and the channel half height δ) and a mean Mach number of 0.1 was considered. To fit the relevant structures, a computational domain of dimension $4\pi\delta \times 2\delta \times \delta$ was chosen. For the homogeneous flow and spanwise directions periodic boundary conditions were imposed. Unfortunately, the available grid generator was not capable of producing an unstructured two-periodic grid. Therefore, an equidistant quasi-structured tetrahedral grid consisting of $202 \times 64 \times 50$ nodes and approximately 3.9 million cells was used. In the subgrid-scale model the Smagorinsky constant was set to $C_s = 0.1$ and a Prandtl number of $Pr_t = 0.7$ was assumed.

In Fig. 4 the computed mean velocity and velocity fluctuations are compared to DNS data of Kim et al. [8]. The overestimation of centerline velocity indicates that the wall normal mesh spacing ($\Delta y^+ = 5.625$) is still too large for computing the wall shear stress accurately. Apart from this, the quality of results matches that of typical second order finite difference approximations.

Flow past a Square Cylinder. The second example is devoted to the flow past a square cylinder at a Reynolds number of 21,400. This flow was studied experimentally by Lyn et al. [10, 9]. Moreover it was also subject of various numerical studies summarized in [16]. For our study, the computational domain was fixed to $20D \times 14D \times 4D$ with the cylinder of side length D placed at a distance of $5D$ from the inlet. Laminar inflow, isothermal no-slip cylinder wall and weakly reflecting outflow conditions were imposed. In the subgrid model $C_s = 0.1$ and $Pr_t = 0.9$ were assumed. The computational domain was discretized using an unstructured grid consisting of approximately 268,000 nodes and 1.29 million tetrahedral cells. The global mesh spacing of $\Delta x = D/5$ was refined to a resolution of $D/20$ near the cylinder. Though this grid is pretty coarse for a large-eddy simulation at such Reynolds number the results are very promising. In Fig. 5(a) the characteristic flow parameters are compared to the experimental data of Lyn & Rodi and numerical results from [16]. The Strouhal number Sr almost perfectly agrees with the experiment while the drag coefficient, c_d, and the average length of the recirculation zone, L_f, exhibit a difference of less than ten percent. The standard deviations of c_d and the lift coefficient c_l fit well in the available numerical data. The non vanishing lift coefficient indicates, however, that the averaging time was still insufficient for removing the statistical uncertainty. Figures 5(b–g) demonstrate the good agreement of the mean velocity and fluctuations to experiment at the given position. Finally, Fig. 5(h) shows a typical isobar surface revealing the instantaneous vortex structure.

4 Conclusion

The presented MG library provides a highly scalable interface for parallel and adaptive flow solvers using unstructured grids. A finite element Navier-Stokes solver was implemented on top of MG and successfully applied to perform LES of turbulent flows in a channel and past a square cylinder. Though not a reference method, the considered linear SUPG-FEM appears to be suitable for application-oriented LES.

References

[1] P. Aumann, H. Barnewitz, H. Schwarten, K. Becker, R. Heinrich, B. Roll, M. Galle, N. Kroll, Th. Gerhold, D. Schamborn, M. Franke: "MEGAFLOW: Parallel complete aircraft CFD". Parallel Comp. **27**, 2001, pp. 415-440.

[2] P. Bastian, K. Birken, K. Johannsen,S. Lang, N. Neuss, H. Rentz-Reichert, C. Wieners: "UG – A Flexible Software Toolbox for Solving Partial Differential Equations". Computing and Visualization in Science **1**, No. 1, 1997, pp. 27-40.

[3] M.S. Fisher, M. Mani, D. Stookesberry: "Parallel processing with the Wind CFD code at Boeing". Parallel Comp. **27**, 2001, pp. 441-456.

[4] J.E. Flaherty, R.M. Loy, C. Özturan, M.S. Shephard, B.K. Szymanski, J.D. Teresco, H.L. Ziantz: "Parallel structure and dynamic load balancing for adaptive finite element computation". Appl. Num. Math **26**, 1998, pp. 241-263.

[5] W.D. Gropp, D.K. Kaushik, D.E. Keyes, B.F. Smith: "High-performance parallel implicit CFD". Parallel Comp. **27**, 2001, pp. 441-456.

[6] K.E. Jansen, S.S. Collis, C. Whiting, F. Shakib: "A better consistency method for low-order stabilized finite element methods". Comp. Meth. Appl. Mech. Engrg. **174**, 1999, 153-170.

[7] G. Karypis, V. Kumar: "Multilevel Algorithms for Multi-Constraint Graph Partitioning", Univ. Minnesota, Dep. Computer Science, TR 98-019, 1998.
[8] J. Kim, P. Moin, R. Moser: "Turbulence Statistics in Fully Developed Channel Flow at Low Reynolds Number". J. Fluid Mech. **177**, 1987, pp. 133-166.
[9] D. Lyn, S. Einav, W. Rodi, J. Park: "A Laser-Doppler Velocimetry Study of Ensemble-Averaged Characteristics of the Turbulent Wake of a Square Cylinder". J. Fluid Mech. **304**, 1995, pp. 285-319.
[10] D. Lyn, W. Rodi: "The Flapping Shear Layer Formed by Flow Separation from the Forward Corner of a Square Cylinder". J. Fluid Mech. **267**, 1994, pp. 353-376.
[11] L. Oliker, R. Biswas: "PLUM: Parallel Load Balancing for Adaptive Unstructured Meshes". NAS Rep. 97-020, NASA Ames Research Center, 1997.
[12] F. Shakib, T.J.R. Hughes, Z. Johan: "A new Finite Element Formulation for Computational Fluid Dynamics: X. The Compressible Euler and Navier-Stokes Equations". Comp. Meth. Appl. Mech. Engrg. **89**, 1991, 141-219.
[13] J. Stiller, K. Boryczko ,W.E. Nagel: "A New Approach for Parallel Multigrid Adaption". In B. Hendrickson et al. (Eds.): Proc. 9th SIAtheM Conf. on Parallel Processing for Sci. Comp., SIAM, 1999, ISBN 8-89871-435-4
[14] J. Stiller, W.E. Nagel, U. Fladrich: "Scalability of Parallelel Multigrid Adaption". In E. Dick et al. (Eds.): Multigrid Methods VI. Springer 2000, pp. 228-234.
[15] M. Van Dyke: "An Album of Fluid Motion", Parabolic Press, 1982.
[16] P. Voke: "Flow Past a Square Cylinder: Test Case LES2". In J. Challet et al. (Eds.): Direct and Large Eddy Simulation II. Kluwer, 1997.
[17] B. Vreman, B. Geurts, H. Kuerten: "Subgrid-modelling in LES of compressible flow". Appl. Sci. Research **54**, 1995, pp. 191-203.
[18] W. Wienken: "Eine Finite-Element-Methode für die Large-Eddy-Simulation und ein darauf basierendes Verfahren zur Bestimmung des Kavitationsbeginns". PHD thesis, TU Dresden, Fakultät Maschinenwesen, Jan 2003.
[19] T.A. Zang, R.B. Dahlburg, J.P. Dahlburg: "Direct and large-eddy simulations of three-dimensional compressible Navier-Stokes turbulence". Phys. Fluids A **4**, 1992, pp. 127-140.

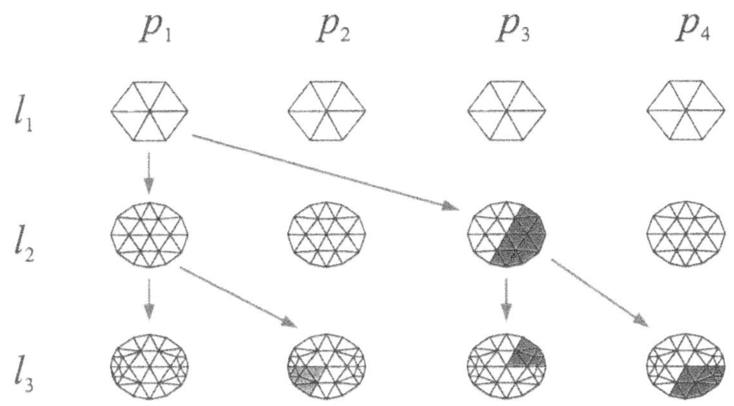

Figure 1 Mulitilevel grid partitioning

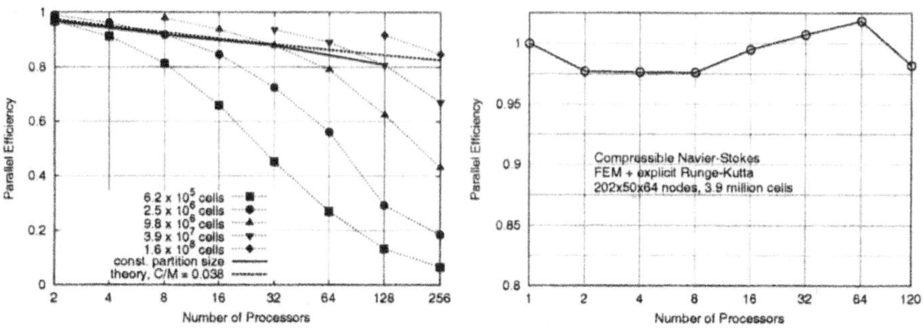

(a) Local grid adaptation (Cray T3E-1200)

(b) Compressible Navier-Stokes (SGI O3800)

Figure 2 Scalability

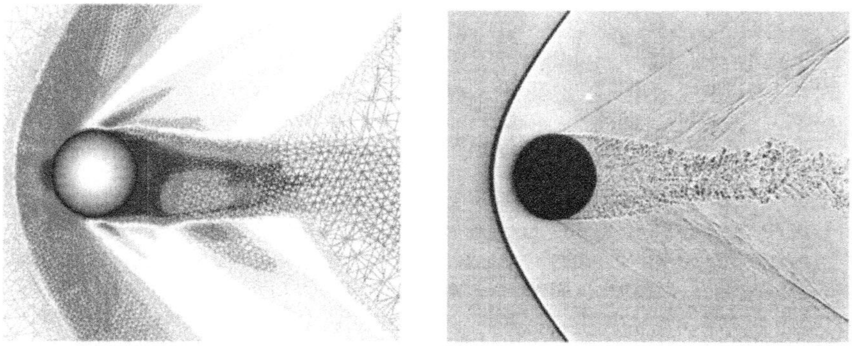

Computation, $Re_d = 2000$ (Mach number coloring)

Experiment, $Re_d \approx 4.2 \times 10^5$ [15]

Figure 3 Flow past a sphere at $Ma = 1.53$

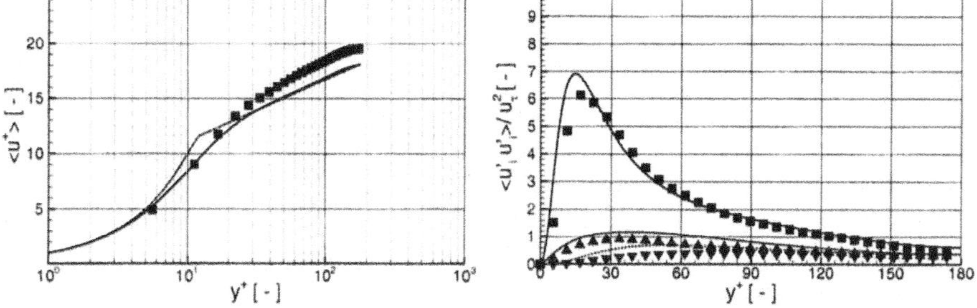

(a) Mean velocity

(b) Velocity fluctuations

Figure 4 Turbulent channel flow. Symbols DNS [8], solid lines LES

	LES	exp. [10]	min [16]	max [16]
Sr	1.31	1.32	1.30	1.61
c_d	2.30	2.1	2.03	2.79
c'_d	0.14	-	0.12	0.36
c_l	−0.12	-	−0.125	0.04
c'_l	1.07	-	1.01	1.68
L_f	1.32	1.38	1.02	1.61

(a) Comparison of flow parameters

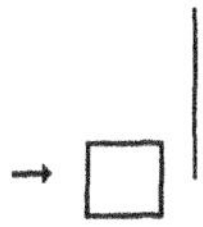

(b) Reference position

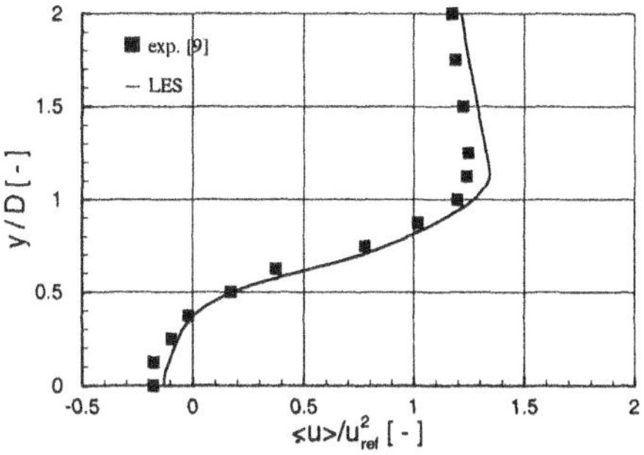

(c) Mean velocity in flow direction

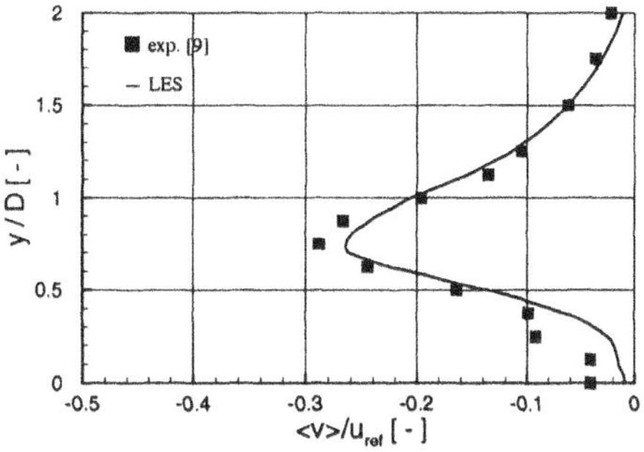

(d) Mean cross flow velocity

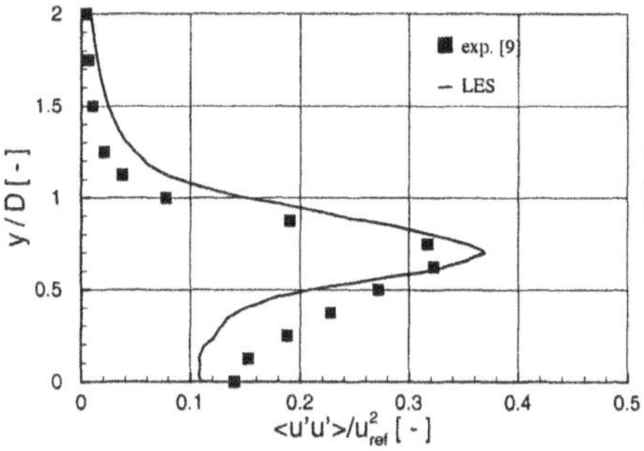

(e) Streamwise velocity fluctuations

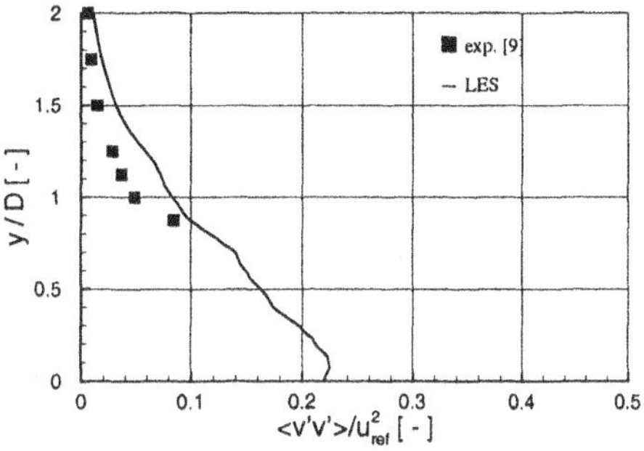

(f) Cross flow velocity fluctuations

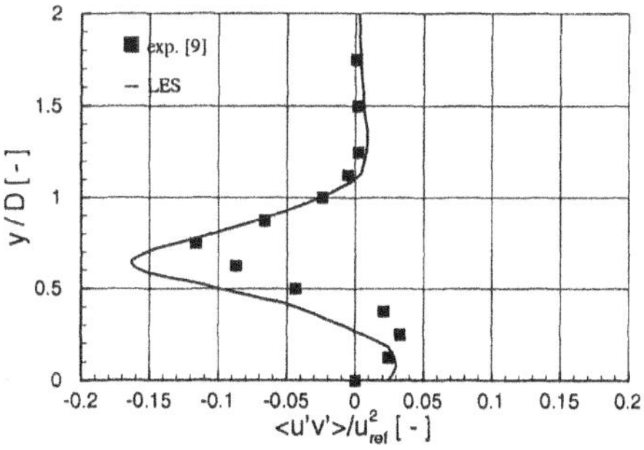

(g) Reynolds stress

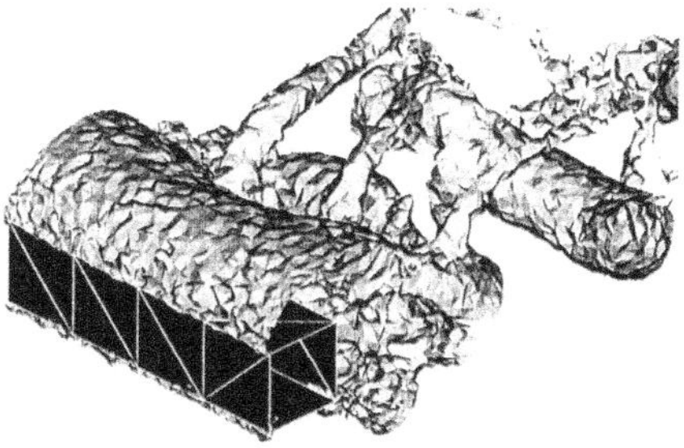

(h) Typical isobar surfaces

Figure 5 Flow past square cylinder

Automatic Feature-Based Sampling of Native CAD Data for Surface Grid Generation

Frank Deister[1], Udo Tremel[1], Ernst Heinrich Hirschel[2], and Herbert Rieger[1]

[1] EADS Deutschland GmbH, Military Aircraft, P.O. Box 80 11 60, D-81663 München, Germany, Frank.Deister@m.eads.net

[2] Universität Stuttgart, Institut für Aero- und Gasdynamik, Pfaffenwaldring 21, D-70550 Stuttgart, Germany

Summary

An automatic feature-based sampling of native CAD data (NASA-IGES, STEP, FLITE) is presented together with applications. The user only specifies a minimal and maximal allowed length and a maximal allowed curvature angle for the complete geometry or rather parts of it. Now locally the curvature of the CAD curve or rather CAD surface is computed, which defines a local length depending on the input parameters. This local length cannot be larger than the length of the corresponding CAD curve or rather CAD surface. Therewith, a length distribution is obtained fully automatically for the complete geometry. A locally refined Cartesian background mesh (octree data-structure) is constructed based on the length distribution, which stores and smoothes the local lengths. Afterwards the triangulation is generated by an advancing front triangulator: the local resolution of the surface triangulation depends on the mesh size specification, which is given by the locally refined Cartesian mesh. The sampling program takes benefit of the object-oriented design and implementation (programming language C++) and is parallized for computer systems with distributed memory using message passing (MPI).

1 Introduction

Grid generation is still one of the bottlenecks in application-oriented numerical fluid mechanics. The geometrical complexity of a realistic configuration makes lengthly and costly grid generation necessary, [1]. Particularly, this holds for surface mesh generation, which requires a strong user interaction with regard to time and expertise. The presented sampling of native CAD data appears to have the potential to overcome this: it is fully automatic without any user interventions. User-specified sources prescribing a local mesh size are avoided completely. The application (fighter aircraft with more than 2700 NURBS surfaces) in Chapter 6 reveals the ability to handle highly complex geometries, which show differences in characteristic length of many orders of magnitudes. Supported geometry import formats are the industrial relevant data exchange formats NASA-IGES, [2], STEP, [3], and the traditional FLITE, [4] format. Various parametric curve and surface types represented by spline composite curves and tensor-product surfaces such as Ferguson, Bezier or NURBS are available.

The surface triangulation is generated by an EADS-inhouse advancing front triangulator, [5], which is developed in collaboration with the University of Wales, Swansea, U.K. The triangulator takes strong advantage of the object-oriented design and implementation: all specific point, curve and surface objects are derived from one abstract interface respectively. Therefore, all different types share a common interface that algorithms can work with. In this way, the advancing front triangulation algorithm is independent of the underlying specific mathematical surface definition.

The sampling program and the surface triangulator are parallelized for computer systems with distributed memory using message passing (MPI). On this account, automatic surface generation starting from native CAD data is very fast: the turn-around time for the fighter aircraft (2.3 million triangles) is less than 15 minutes on a PC-cluster with 32 XEON processors (2 GHz).

2 Sampling of Native CAD Curves

The sampling of CAD curves are controlled by three sampling parameters, which are specified by the user: minimal arc length L_{min}, maximal arc length L_{max} and maximal curvature angle α_{max}. Now a CAD curve is subdivided into consecutive curve segments applying the following three sampling criteria: the arc length of a curve segment must not be smaller than the specified minimal arc length L_{min}. Conversely the curve segment length must be smaller then the maximal arc length L_{max}. Last but not least, the curvature angle must be smaller than the maximal curvature angle α_{max}. The last sampling criterion is only applied, if the arc length is larger than the minimal arc length L_{min}. The curvature angle is taken as the angle between the tangential vectors at the two end points of a curve segment.

If the CAD curve is larger than the minimal arc length L_{min}, the arc length and curvature angle is investigated using a forward stepping algorithm: marching along the CAD curve with sufficient small steps Δs, the arc length L and curvature angle α are calculated repeatedly. As soon as a sampling criterion is violated, an interval nesting algorithm is applied. With that the location on the CAD curve is computed, where the arc length L or rather the curvature angle α matches the corresponding sampling parameter L_{max} or rather α_{max}. Anyway, the CAD curve is subdivided at that location. The algorithm is illustrated in Figure 1(a). Because of the object-oriented program design and implementation, the algorithm is independent of the specific mathematical CAD curve definition, [5]: it works only on one common abstract interface for CAD curves.

All curve segments are approximated by straight lines. Finally, for each straight line a bounding box is determined: the bounding box is called raster box, because it controls the local resolution of the later Cartesian background mesh, Chapter 4. Figure 1(b) shows the raster boxes for a CAD curve around the leading edge of the ONERA M6 wing. The entire CAD curve is enclosed by the raster boxes. Besides, the final Cartesian grid is visible in the plane cutting the CAD curve.

3 Sampling of Native CAD Surfaces

The sampling of CAD surfaces requires the same three user specified sampling parameters L_{min}, L_{max} and α_{max}, which are used also for curve sampling. Because only the part inside a trimmed CAD surface is considered, a scan-line algorithm from computer graphics is applied, [6] and [7]. As first step, the trimming CAD curves are approximated by sequences of straight lines. For this, they are sampled as described in Chapter 2. The resulting straight lines in physical (Cartesian) space are transformed to the (u-v)-parameter space of the CAD surface, in which the remaining computation takes place. The approximated trimming curves are checked for intersection using an ADT approach (alternating digital tree, [8]). If an intersection occurs, the corresponding CAD curve is sampled again with smaller values for the sampling parameters.

The second step consists of computing the stencil point distribution, where the local surface curvature will be investigated later. For both u- and v-direction equally distributed iso-curves (probes) of the CAD surface are sampled applying again the curve sampling algorithm from Chapter 2. But this time the curvature angle is defined as the angle between the surface normal vectors at the end points of a curve segment. The end points of the evaluated curve segments are taken as the sought stencil points. For each direction, the final stencil point distribution is extracted from these probes. In Figure 2 the final stencil point distribution in u- and v-direction are represented by the circles.

Now the scan-line algorithm is applied separately for the u- and v-iso-curves, which are defined by the stencil points (third step). It identifies that part of the iso-curves, which are inside the trimmed surface or rather inside the polygon constituted by the sequences of straight lines. The demarcations are drawn as squares in Figure 2. However, these inner curve parts are sampled again with the approach presented in Chapter 2. Afterwards, the computed sample lengths are related to the stencil points located inside the trimmed surface (circles drawn with thick lines in Figure 2). If the number of stencil points inside the trimmed surface is less than a user defined minimum, then the stencil is doubled and the scan-line algorithm is repeated. In the case of bad local parameterization of the CAD surface alternative strategies to continue the sampling process are incorporated.

Every stencil point inside the trimmed surface gets a raster box according to the stored sample length. Additional raster boxes are created, if the distance between two stencil points is larger than their sample lengths. In this way, the trimmed surface is enclosed by raster boxes completely, which are used for the generation of the Cartesian background mesh, Chapter 4.

4 Cartesian Background Mesh

The locally refined Cartesian background mesh specifies the mesh size required by the surface mesh generator. It bases on the hierarchical octree-data structure, Figure 3, which describes the connectivity between the Cartesian cells, [9], [10] and

[8]. The raster boxes along the CAD curves and CAD surfaces, Chapter 2 and 3, determine the local resolution of the Cartesian background mesh: all Cartesian cells are identified, which intersect a raster box. These Cartesian cells must not be larger than the current raster box. If a Cartesian cell is larger, then it is recursively subdivided into children cells until a child cell is smaller than the raster box. Besides, the sample length of the corresponding curve segment is stored in every intersected Cartesian cell. At the end, the Cartesian background mesh is smoothed due to an one-level difference rule: it is not allowed, that two neighbouring Cartesian cells differ by more than one refinement level. The stored sample lengths are prolongated through the Cartesian mesh. The rate of change between adjacent Cartesian cells is limited by an user-defined slope. In this way, a smoothed sample length distribution is achieved in the complete flow domain. Moreover, the user is able to control the rate of coarsening of the triangulation by these slope parameter.

During generation of the surface triangle mesh, the Cartesian background mesh specifies the local mesh size. Now the mesh size is required for a point. First, the Cartesian cell is identified applying the hierarchical octree data-structure, which encloses this point. Then the sought mesh size is interpolated linearly using the sample length and its gradient (least square method), which are stored in the Cartesian cell. It is to emphasis, that the smoothed mesh size specification is also available in space and therefore usable by a tetrahedral mesh generator.

5 Parallelization

The parallelization of the sampling process is incorporated via a pipelining approach: the loop over all topological curves and faces to be sampled is executed in parallel. The parallel algorithm bases on the message passing programming model (MPI). In Figure 4 the architecture of the parallel system is outlined. Initially, the available processes are divided into three groups. One process has the role of the distributor, one of the collector and all others are assigned to the worker group. The distributor process initializes the complete geometry definition and has then the task to extract and send sub-geometries to the next worker waiting for data. When the worker process has sampled the received sub-geometry, i.e. computed all the raster boxes enclosing the CAD curve or CAD surface, these raster boxes are delivered to the distributor. At the end of the pipeline this distributor receives all the raster boxes and generates the Cartesian background mesh and prolongates the sample lengths. Therefore, currently the operations for the Cartesian background mesh are implemented sequentially. The parallelization of these operations will be done soon.

6 Applications

The presented concept of fully automatic surface mesh generation is demonstrated with a fighter aircraft. The CAD description is imported via the STEP format and consists of 6788 NURBS curves and 2749 NURBS surfaces. The only input parameters specified by the user are the minimal arc length L_{min} (4 mm for the fighter and

0.5 mm for the vane sensor), the maximal arc length L_{max} (50 mm for the fighter and 4 mm for the vane sensor) and the maximal curvature angle α_{max} (15°). It is to emphasize, that not a single source is specified. In parallel the sampling takes less than 2 minutes on a cluster with 16 XEON-processors, Table 1. Additional 3:13 minutes are needed to prolongate the sampled lengths sequentially. Altogether, the wall clock time for the complete program (CAD import, sampling, Cartesian mesh, octree output) was 6:14 minutes.

The parallel surface mesh generator triangulates the fighter aircraft in less than 7 minutes. Here, the computation is done on a cluster with 32 XEON processors (2 GHz). Details are provided in [5]. The resulting surface mesh consists of more than 2.3 million triangles. Figure 5 shows the front view of the fighter aircraft together with several stores. The vane sensors are located at the lower side of the aircraft slightly behind the radome, as can be seen in the lower right-hand side of Figure 5.

More details of the surface triangulation are presented in Figure 6. Especially in Figure 6(a) the curvature-based refinement appears. The exhaust nozzle is completed by a thin ring. Here the underlying CAD surface has a quadratic shape. Although the ring is defined by two circular trimming curves, the narrow ring is also well resolved automatically, Figure 6(b).

7 Conclusion

A fully automatic and parallel feature-based sampling of native CAD data has been developed. The local curvature and characteristic length are investigated along CAD curves and inside trimmed CAD surfaces in order to define local sample lengths. A locally refined Cartesian background mesh (octree data-structure) is constructed to prolongate and therewith to smooth the sample lengths. During the surface mesh generation, the Cartesian background mesh serves as the mesh size specification. The object-oriented design and implementation has strong advantages. In particular, the sampling algorithms are independent of the underlying specific mathematical curve or surface definition. Due to parallelization for distributed memory systems, the turn-around time for surface mesh generation around a complex fighter aircraft (2788 NURBS surfaces, 2.3 million triangles) is reduces to less than 15 minutes.

References

[1] E.H. Hirschel, W. Schwarz: "Mesh Generation for Aerospace CFD Applications". *Surveys on Mathematics for Industry,* Volume 4, 1995: pp. 249–265.

[2] NIST (National Institute of Standards and Technology): "Initial Graphics Exchange Specification (IGES)". Version 5.3, 1990.

[3] ISO (International Organization for Standardization): "Standard for the Exchange of Product model data (STEP)". ISO 10303, Geneva.

[4] Computational Dynamics Research (CDR): "FLITE-3D: User Manual". Innovation Center, University College, Singleton Park, Swansea SA2 8PP, U.K.

[5] L. Fornasier, F. Deister, O. Hassan, U. Tremel, N.P. Weatherill: "Robust and Efficient Generation of Unstructured Surface Grids about Geometrically Complex Configurations Using 'Real-Design' CAD Data". AIAA-Paper 2003-805, 2003.
[6] L.A. Piegl, A.M. Richard: "Tessellating trimmed NURBS surfaces". Computer Aided Design, Volume 27, No. 1, 1995: pp. 16–26.
[7] J.D. Foley, A. van Dam, S.K. Feiner, J.F. Hughes: "Computer Graphics - Principles and Practice", Addison-Wesley Publishing C., Reading, Massachusetts, 2nd edition, 1996.
[8] M.J. Aftosmis: "Solution Adaptive Cartesian Grid Methods for Aerodynamic Flows with Complex Geometries". in *Computational Fluid Dynamics,* Lecture Series 1997-02, von Karman Institute for Fluid Dynamics, 1997.
[9] F. Deister: "Selbstorganisierendes hybrid-kartesisches Netzverfahren zur Berechnung von Strömungen um komplexe Konfigurationen". Dissertation, Universität Stuttgart, Germany, 2002.
[10] F. Deister, E.H. Hirschel: "Self-Organizing Hybrid Cartesian Grid/Solution System with Multigrid". AIAA-Paper 2002-0112.

List of Tables

Table 1 Parallel performance for fighter aircraft (Cpu-time, 16 XEON-processors, 2 GHz).

SAMPLING OF 6788 CURVES	3.67 sec.
SAMPLING OF 2749 SURFACES	1:22 min.
TOTAL TIME SAMPLING (INCLUDING ALL)	1:59 min.
LENGTH PROLONGATION (not parallel)	2:38 min.
GRADIENT CALCULATION (not parallel)	21.7 sec.
TOTAL TIME PROLONGATION (not parallel, 1.4 million cells)	3:13 min.
TOTAL WALL CLOCK TIME (CAD IMPORT UP TO OCTREE OUTPUT)	6:14 min.

List of Figures

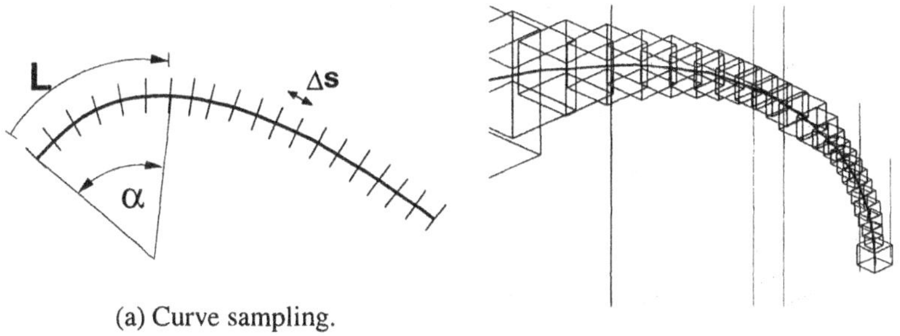

(a) Curve sampling.

(b) Raster boxes.

Figure 1 Sampling of curves.

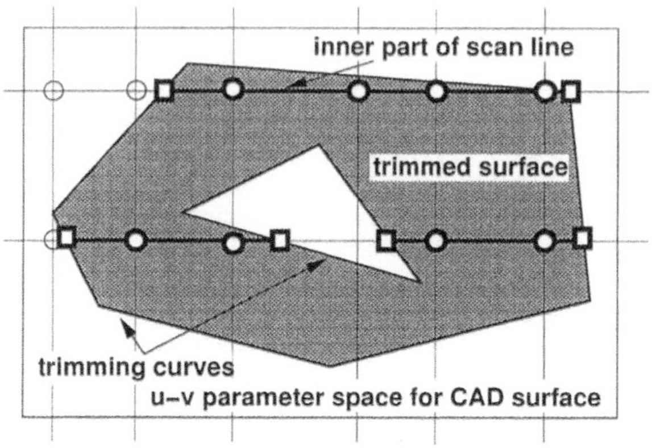

Figure 2 Sampling of trimmed surface.

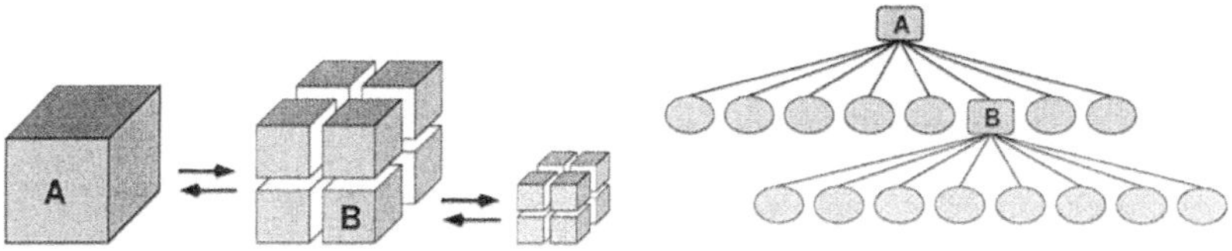

(a) Recursively subdivided cells.

(b) Octree representation.

Figure 3 Hierarchical octree data-structure.

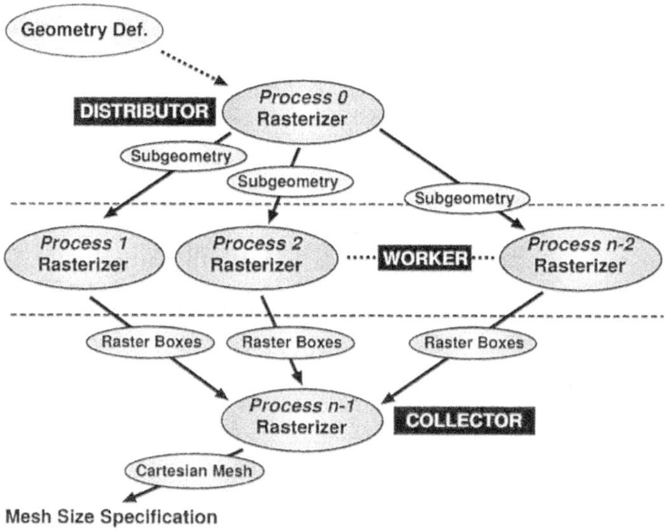

Figure 4 Architecture of the parallel sampling program.

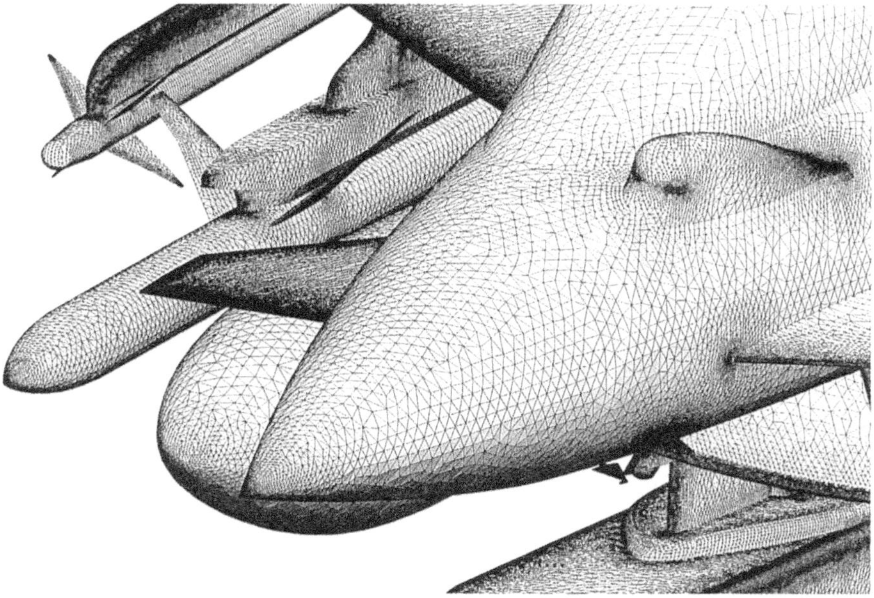

Figure 5 Front view of fighter aircraft, 2.3 million surface triangles.

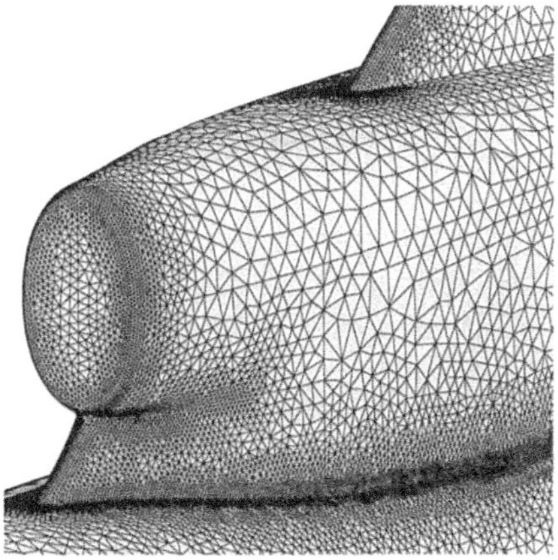

(a) Surface mesh near fin.

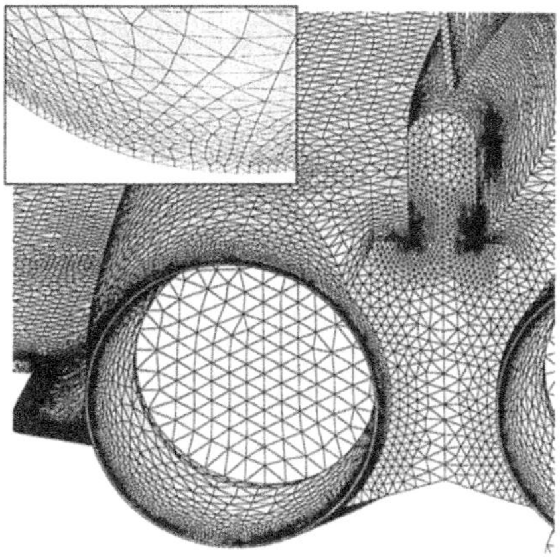

(b) Surface mesh near exhaust.

Figure 6 Details of surface mesh for fighter aircraft.

Post-Parametrization of CAD-Geometries Using Freeform Deformation and Grid Generation Techniques

Arno Ronzheimer

DLR, Institute of Aerodynamics and Flow Technology,
Lilienthalplatz 7, D-38108 Braunschweig, Germany
Arno.Ronzheimer@dlr.de

Summary

A post-parametrization method, based on freeform deformation techniques, has been established to simplify and to automate the parametrization of shapes from CAD-models, to be used for aerodynamic design. To create the necessary deformation lattices, structured grid generation techniques of a grid generation system, developed at DLR, are utilised. As this grid generation system has the salient feature to store and to replay a sequence of processes with different parameter settings, a deformation of the shape, given by polygonal curves and surfaces, can be controlled quantitatively. In two examples the applicability of the method for aerodynamic wing design and detailed design of a wing tip is demonstrated.

1 Introduction

Numerical methods in computational fluid dynamics have been continuously improved and have become a reliable tool in industry during the complete process of aircraft development. Particularly unstructured CFD-methods have gained a degree of automation and reliability to perform routinely flow analyses on geometries, having nearly any detail of a final aircraft.

However, to improve an aircraft design in a late phase, especially when aerodynamic interference effects occur, which may degrade the promised performance of the aircraft, the turnaround time for further design studies is mainly governed by geometrical complexity and the effort to construct alternate geometry models within a CAD-System. Presently CAD-models of the outer shape of an aircraft may contain several hundred panels to be used for grid generation. If all panels form a watertight shape, unstructured grid generation methods are able to perform the grid generation task highly reliable and automated [2].

To eliminate this lag of automation in geometry modelling for redesign, freeform deformation techniques, which are extensively used in the area of soft object animation, have been adapted, similar to the approach described in [8]. However, to gain a higher flexibility in applying freeform deformation to a great variety of forthcoming tasks, the freeform deformation method, described in Chapter 2, has been implemented into the grid generation system MegaCads [1]. MegaCads provides import and export of geometry data and the present functionalities are used to create the set-up for a specific deformation task.

2 Principle of Freeform Deformation

The classical method of freeform deformation has been proposed by Sederberg and Parry [10] in 1986. Extensions to this method have been made by Griessmair and Purgathofer [6] in 1989 and by Coquillart [3] in 1990.

The first step of freeform deformation is to map a given object, defined by polygonal curves and surfaces with vertices $\vec{P}_{Object}$ into a parametric volume $\vec{P}(u,v,w)$. In the present implementation a trivariate B-spline volume, given by

$$\vec{P}(u,v,w)=\sum_{i=0}^{n_u}\sum_{j=0}^{n_v}\sum_{k=0}^{n_w}N_{i,m_u}(u)N_{j,m_v}(v)N_{k,m_w}(w)\vec{Q}_{i,j,k} \tag{1}$$

with B-spline basis functions $N_{i,m_u}(v)$, $N_{j,m_v}(v)$, $N_{k,m_w}(w)$ of degree m_u, m_v, m_w, defined on knot vectors $\{0,\ldots,0,u_{m_u+1},\ldots,u_{n_u},1,\ldots,1\}$, $\{0,\ldots,0,v_{m_v+1},\ldots,v_{n_v},1,\ldots,1\}$ and $\{0,\ldots,0,w_{m_w+1},\ldots,w_{n_w},1,\ldots,1\}$ is used. The basis functions are calculated recursively:

$$\begin{aligned} N_{i,1}(u) &= \begin{cases} 1 & (u_i \le u < u_{i+1}) \\ 0 & otherwise \end{cases} \\ N_{i,m_u}(u) &= \frac{u-u_i}{u_{i+m_u-1}-u_i}N_{i,m_u-1}(u)+\frac{u_{i+m_u}-u}{u_{i+m_u}-u_{i+1}}N_{i+1,m_u-1}(u). \end{aligned} \tag{2}$$

The definitions of $N_{j,m_v}(v)$ and $N_{k,m_w}(w)$ are similar.

The control points $\vec{Q}_{i,j,k}$ define the initial deformation lattice, that encloses the initial object, as shown in Figure 1. Each vertex point $\vec{P}_{Object}$ of the initial object is mapped into the B-spline volume by solving for the parametric co-ordinates (u_O, v_O, w_O) of $\vec{P}_{Object}$ from eq. (1), using a Newton method. In a second step the deformation of the initial object is achieved by re-mapping the previously calculated parametric co-ordinates (u_O, v_O, w_O) to yield new vertex points:

$$\vec{P}_{NewObject}(u_O,v_O,w_O)=\sum_{i=0}^{n_u}\sum_{j=0}^{n_v}\sum_{k=0}^{n_w}N_{i,m_u}(u_O)N_{j,m_v}(v_O)N_{k,m_w}(w_O)\vec{R}_{i,j,k}\,. \tag{3}$$

Eq. (3) is mainly identical to eq. (1), but with a different control point lattice $\vec{R}_{i,j,k}$, which determines finally the deformation of the object.

3 Control of Deformation

As the present method of freeform deformation is an indirect method, where the surrounding space is deformed, rather than the object itself, proper deformation lattices have to be provided, to achieve a desired geometrical variation of an object. For this purpose the DLR interactive structured grid generation system MegaCads [1] is applied.

As MegaCads was initially intended to generate grids around geometries given by cross-section-curves, several CAD-functionalities have been implemented to create curves and surfaces, to translate rotate and intersect curves and surfaces, and to project points onto curves and surfaces. For grid generation customary techniques are available to distribute points along curves, to create surface and volume grids using bilinear and trilinear interpolation techniques. Therefore the described deformation method has been implemented into MegaCads, using the present structures and functionalities to construct adequate deformation lattices.

In Figure 2 freeform deformation is applied to a structured surface grid around the ONERA M6-wing [9]. For this purpose two hexahedral grids of size $6 \times 2 \times 1$ cells have been generated, which correspond to the wing planform.

Besides the prescribed functionalities of MegaCads the underlying replay concept allows to repeat a complete construction sequence in batch mode with defined parameters. These parameters are used among other things to scale the length of curves and lines. This is applied for the M6-wing case and demonstrated in Figure 3, showing 2-dimensional grid sections which are used to build the 3-dimensional initial and final deformation lattice in further construction steps. As MegaCads supports arithmetics and the definition of dependencies between parameters, finally a control of the deformation incorporating certain geometrical constraints is enabled.

The manual effort to construct the prescribed deformation mechanism for the M6-wing using MegaCads is conceivably small and is already done in a few hours. Beyond this small effort, a geometrical variant by changing one or more parameters can be generated in seconds.

4 Application of Freeform Deformation

The basic question is now, if the freeform deformation method, which at first is used to animate imaginary creatures, is applicable in aerodynamic design. In the above deformation example a variation of a single parameter curls up a curved bump on the wing surface, what in matter of an improvement of the aerodynamic performance appears to be devious. However, the present example of the M6-wing deformation has been used as a test case for aerodynamic shape optimisation.

For this purpose the batch version of MegaCads has been integrated into the SynapsPointer® optimisation framework [4], which manages the complete optimisation procedure. The described deformation of the M6-wing surface grid with those parameters shown in Figure 3, represents the geometry generator. To avoid expensive computations, parameters at the tip of the wing $p_1 ... p_{11}$ have been linked to corresponding parameters of the root section. Additionally a parameter p_{12}, which determines a twist of the wing, has also been incorporated into the deformation lattice.

The optimisation procedure is completed by a volume grid generator, which generates a structured O-O type grid around the deformed surface grid, and for the solution of the Euler-equations the DLR-FLOWER code [7] has been integrated. The optimisation algorithm, which is also integrated into the optimisation framework, is based on a conjugated gradient method and parameters from the optimisation algorithm are routed to the replay file of MegaCads via a template. The objective of the optimisation is to minimise the inviscid drag coefficient c_D, which is extracted from the output of the flow solver and passed to the optimization algorithm. Although there was an adjoint method available [5], gradients have been computed explicitly during optimisation.

The resulting surface grid of the optimised wing and the appropriate parameters, which have been obtained after 85 optimisation loops, are shown in Figure 4. As all flow solutions have been computed for a target lift coefficient $c_L = 0.3$ and an onflow Mach number $M_\infty=0.84$ the resulting surface is cambered near the trailing edge. Along the leading edge the thickness is decreased. In Figure 5 lines of constant Mach number on the upper surface and pressure distributions at a spanwise cut are compared for initial and optimised wing. The typical λ-type shock on the upper surface of the basic M6-wing geometry has vanished and the level of suction pressure has been reduced on the upper surface after optimisation. More significant is the fact, that the inviscid drag has been reduced more than 40%. But despite the fact, that parameters have been linked to preserve thickness, from Figure 5 a small reduction can be observed.

5 Parametrization of CAD-based Shapes

In the above example freeform deformation has been applied successfully to parameterise an elementary geometry. However, due to the simplicity of the freeform deformation method, there should be no obstacle to adopt freeform deformation for more complex design problems. As there is only one restriction, which limits the application of freeform deformation to points of polygonal curves and polygonal surfaces, either knot-points or collocation points or discretized representations of spline curves and spline surfaces can be treated.

Ideally the CAD-conversion engine of the CentaurSoft™ hybrid grid generation system [2] provides a suitable re-discretisation of CAD-data, which are already cleaned and are representing a watertight domain for grid generation. Therefore an interface for importing and exporting CentaurSoft™ geometry data has been integrated into MegaCads to enable freeform deformation of CAD-data. As far as the topology of the CAD-model is preserved, whereas the geometry may have changed by freeform deformation, the exported geometry data can be processed immediately by the CentaurSoft™ intrinsic grid generation, which runs automatically in batch mode. Thus the main prerequisites are given for the set-up of shape optimisation loops to improve the design of complex aircraft CAD-models. This is demonstrated for a CAD-model of a transport aircraft in cruise configuration in Figure 6. In this case freeform deformation is applied to a detail, namely the wing tip. Figure 7 shows possible geometrical variants, which have been generated using the same MegaCads replay file, but with different parameter settings.

6 Conclusion

The freeform deformation method provides a great potential in aerodynamic shape design. However, the flexibility and applicability of this method to parameterise a given geometry retrospectively is at first enabled by a combination with the structured grid generation system MegaCads. MegaCads manages import and export of geometrical data, supports the construction of necessary deformation lattices and provides the control of deformation through the underlying replay concept.

The presented examples have demonstrated, that routinely applications of the freeform deformation method to a variety of aerodynamic design problems are established.

References

[1] O. Brodersen, A. Ronzheimer, R. Ziegler, T. Kunert, J. Wild, M. Hepperle: "Aerodynamic Applications using MegaCads", 6th Intern. Conf. on Numerical Grid Generation, Ed.: M.Cross et.al., London, 1998, pp. 793-802.

[2] CentaurSoft™: http://www.centaursoft.com.

[3] S. Coquillart: "Extended Free-Form Deformation: A Sculpting Tool for 3D Geometric Modeling", Proceedings of SIGGRAPH '90, Dallas, Texas, Aug. 6-10, 1990.

[4] O. Frommann: "Conflicting Criteria Handling in Multiobjective Optimization Using the Principles of Fuzzy-Logic", AIAA-98-2370, 1998.

[5] N. Gauger, J. Brezillon: "Aerodynamic Shape Optimization Using Adjoint Method", Journal of Aero. Soc. of India, Vol 54, No. 3, 2002

[6] J. Griessmair, W. Purgathofer: "Deformation of Solids with Trivariate B-Splines", EUROGRAPHICS '89, Elsevier Science Publishers (North Holland), 1989, pp.137-148.

[7] N. Kroll, C.-C. Rossow, K. Becker, F. Thiele: "MEGAFLOW – A Numerical Flow Simulation System", 21st ICAS Congress, Melbourne, 1998.

[8] J. A. Samareh: "A Novel Shape P Approach ", NASA TM-1999-209116, NASA Langley Research Center, Hampton, VA 23681.

[9] V. Schmitt, F. Charpin: "Pressure Distributions on the ONERA-M6-Wing at transonic Mach Numbers", AGARD Advisory Report No. 138, May 1979.

[10] T. W. Sederberg, S. R. Parry: „Freeform Deformation of Solid Geometric Models", Proceedings of SIGGRAPH' 86, Dallas, Texas, Aug. 18-22, 1986.

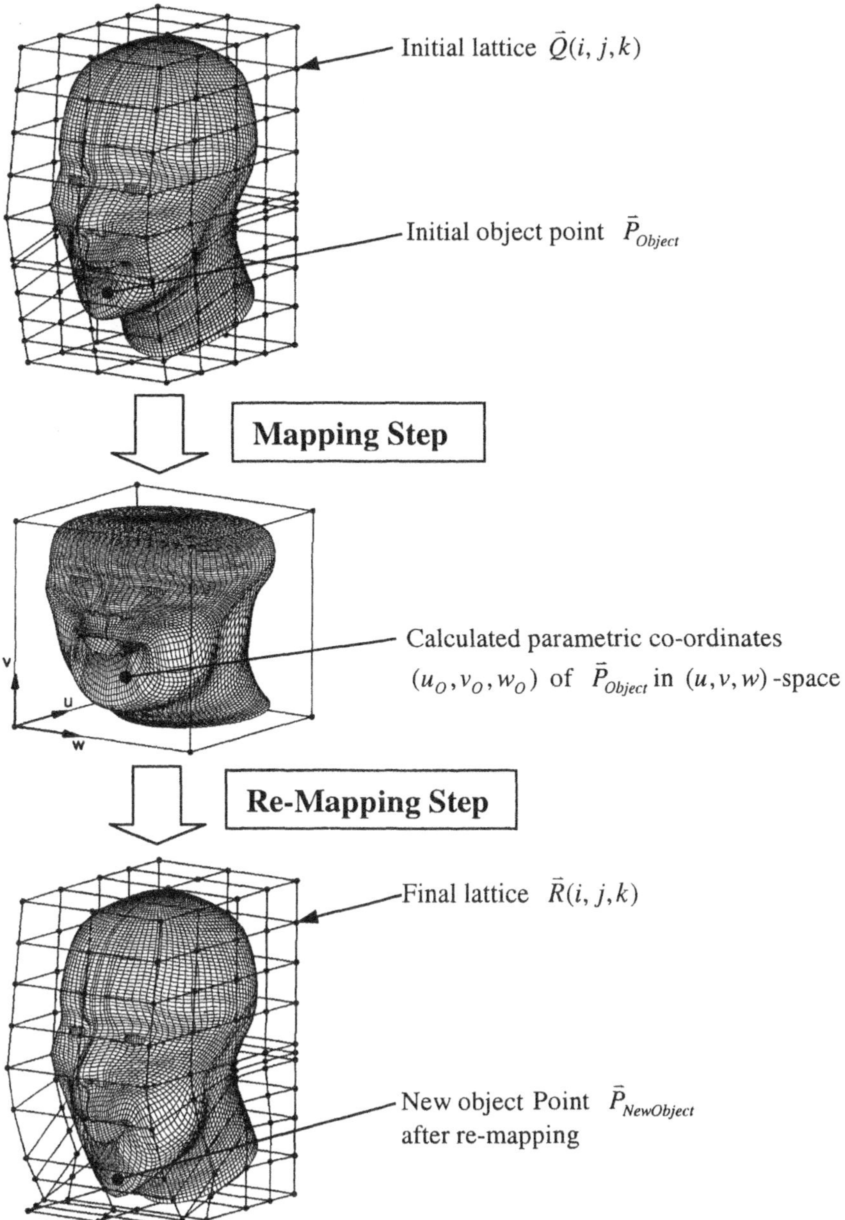

Figure 1 Principle steps of freeform deformation

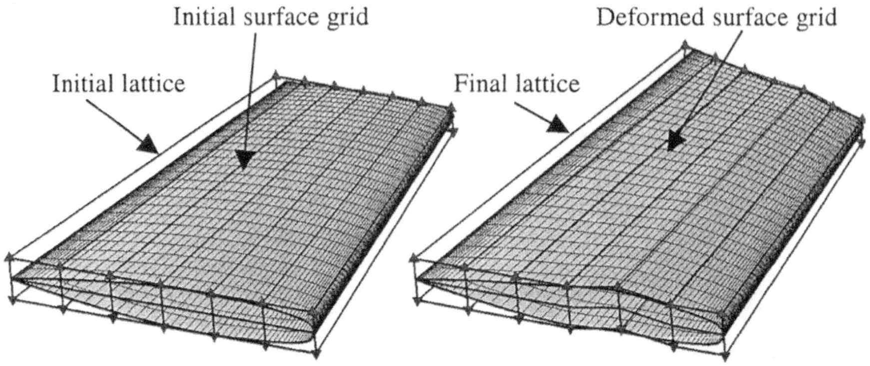

Figure 2 Deformation lattices for ONERA M6-wing

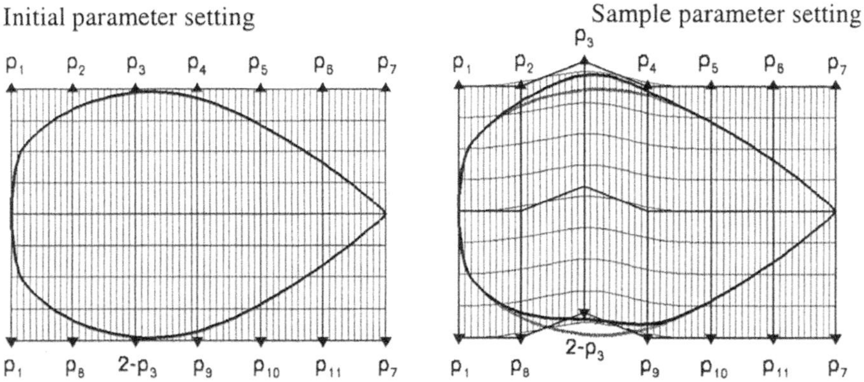

Figure 3 Deformation of 2-d root section with parameter adjustments

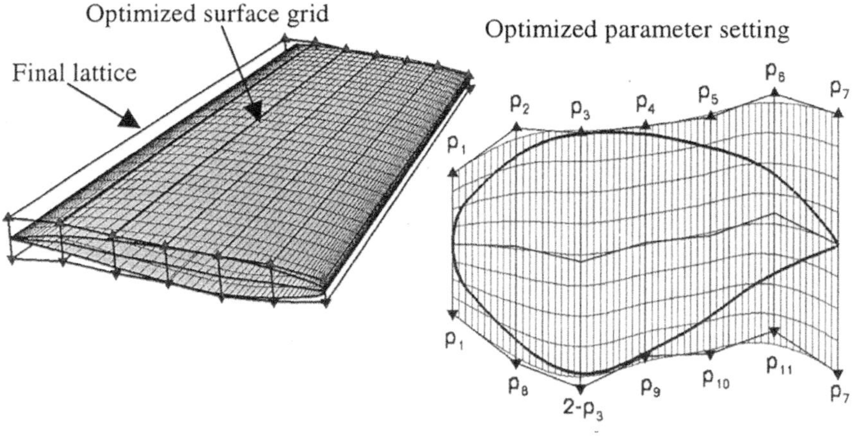

Figure 4 Optimised M6-wing geometry

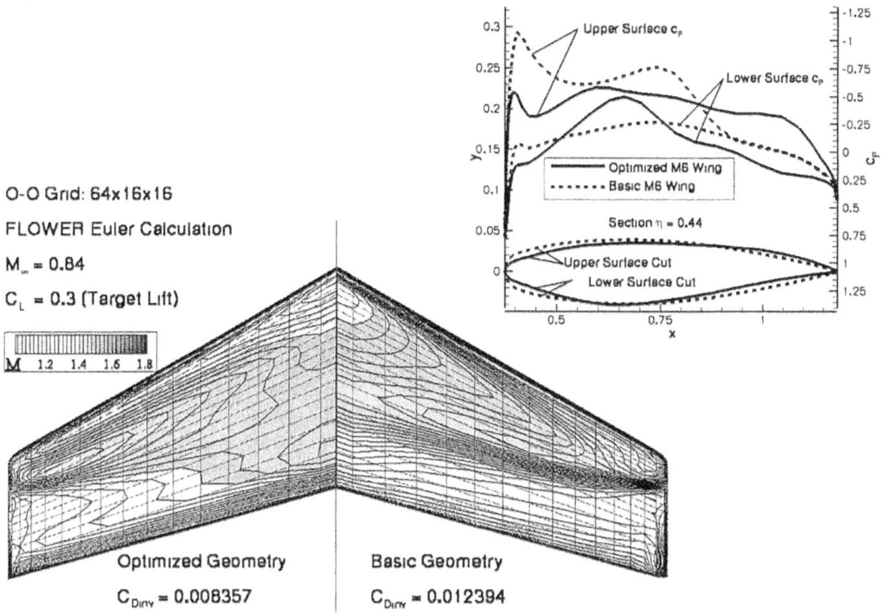

Figure 5 CFD-results of basic and optimized M6-wing geometry

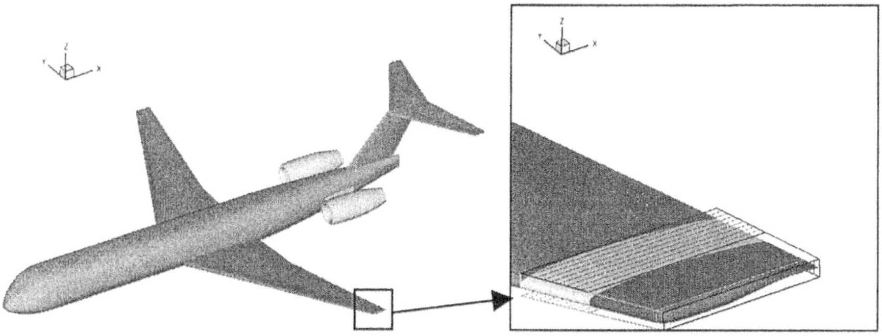

Figure 6 CAD geometry with wing tip prepared for deformation

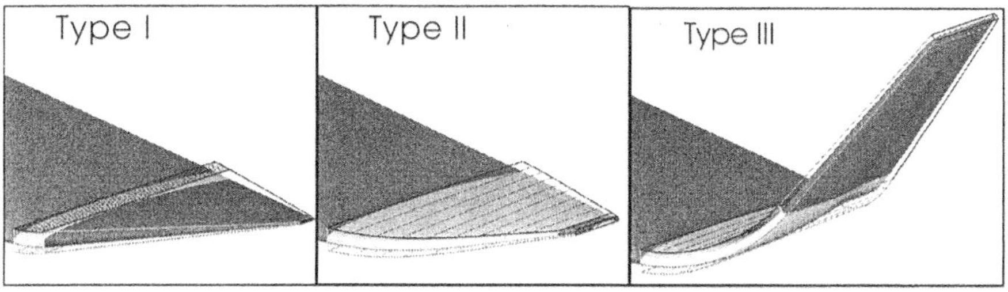

Figure 7 Geometrical variations of wing tip

Low Mach-Number Preconditioning for the DLR-TAU Code and Application to High-Lift Flows

Stefan Melber and Ralf Heinrich

German Aerospace Center (DLR), Institute of Aerodynamics and Flow Technology,
Lilienthalplatz 7, D-38108 Braunschweig, Germany

Summary

This report presents first results for the application of the DLR TAU-code with respect to low Mach-number preconditioning for complex high-lift configurations. It comprises the numerical simulation of the viscous compressible flow around simple test cases and a transport aircraft high-lift configuration with engine.

1 Introduction

The efficient design of a transport aircraft high-lift configuration with respect to low speed take-off/landing performance and handling qualities represents a complex aerodynamic problem. The flow field around the wing with deployed high-lift devices and engines at high incidences is characterized by the co-existence of complex flow phenomena. In addition there are some critical areas with respect to the flow topology like the engine/pylon/wing intersection. Aside from established experimental investigations of such aerodynamic effects in wind-tunnels or flight tests, the application of computational fluid dynamics (CFD) gains increasing importance for the analysis and improvement of high-lift aircraft configurations ([6] - [8], [11] and [12]).

A problem linked with high-lift flows is typically a mix of compressible and incompressible flow regions. The difficulty in solving the compressible equations for low Mach-numbers is associated with the large disparity of the acoustic wave speed and the waves convected with the fluid speed, which results in a growing stiffness of the equations to be solved and a misbalance of the artificial dissipation. This leads to very poor convergence and accuracy properties with decreasing Mach-number. The application of preconditioning promises to bypass these problems by changing the eigenvalues of the system of compressible flow equations to remove this disparity in areas of low Mach-numbers [10]. A preconditioning technique based on the work of Choi and Merkle [1] has been implemented in the TAU-code. A guideline for the implementation used here can be found in [10].

Before applying the preconditioned solver to complex 3D applications simple test cases have been performed first to test the implementation and the potential of preconditioning. Encouraged by the results we applyed preconditioning to the vortex dominated flow for two different onflow conditions for a complex high-lift configuration. An improvement in the convergence ratio could be demonstrated and especially an improved resolution of flow features like vortices could be found.

2 Flow Solution Method and Preconditioning

The Reynolds-averaged Navier-Stokes equations (RANS) are solved by the DLR hybrid unstructured flow solver TAU, which is based on a three-dimensional finite volume scheme. The governing equations are solved on a dual grid, which, together with the edge-based data structure, allows to run the code on any type of cells. The solver is part of the MEGAFLOW-project [5].

A multi-step Runge-Kutta scheme is used for time integration. Space discretization is realized with a Roe- or AUSM-type 2nd-order upwind scheme, or by employing a central method with scalar dissipation, which is used for the present study. The viscous fluxes are discretized using central differences. In order to accelerate the convergence to steady state, local-time stepping, residual smoothing, and a multigrid technique based on agglomeration of the dual-grid volumes are employed. For parallel computations a domain-decomposition is used providing a subset of dual grids. All numerical simulations in this paper are computed by using the one-equation Spalart-Allmaras turbulence model.

In order to bypass the problems associated to low Mach-number flows, a preconditioning method has been implemented in the TAU code. As guideline reference [10] has been used. In the following some "simple" applications are presented which have been used to check on one side the implementation and on the other to test the potential of preconditioning for high-lift applications.

A simple numerical experiment has been performed to test the implementation of the preconditioning algorithm. The inviscid flow around a NACA0012 airfoil has been calculated for a Mach-number range from 0.1 down to 10^{-3}. For all computations the angle of attack is set to $\alpha = 2.0^o$. The drag which is zero for an inviscid, subsonic flow, can be used to estimate the accuracy of the numerical solution. Table 1 summarizes results for the simulation with and without preconditioning. For the three Mach-numbers taken into account the number of multigrid cycles and the final drag for a convergence of 7 orders of magnitude with respect to the L2-norm of the density residual is printed. The convergence and accuracy become unacceptable without using preconditioning for Mach-numbers approaching zero. For preconditioning, the convergence as well as the final drag is almost Mach-number independent. Even for the best solution without preconditioning ($M = 0.1$), the drag-coefficient for all solutions using preconditioning is improved by a factor of 2.6. Another well suited value for measuring the numerical quality of a scheme is the total pressure loss coefficient $cp_{tot,max} = 2\ (p_{tot} - p_{tot,\infty})_{max}/(\rho_\infty \mathbf{v}_\infty^2))$. For Mach-numbers ranging from 10^{-5} to 0.6 the maximum value of the coefficient on the profile is plotted in Fig. 1a with and without preconditioning. Using preconditioning, it becomes visible, that the coefficient remains almost constant in the whole subsonic region. This shows, that the quality of the numerical solution is independent of the size of the Mach-number for subsonic flows. A similar behavior has been achieved for viscous high-lift flows. The flow around a three-element airfoil has been simulated for Mach-numbers from 10^{-3} to 0.1, a Reynolds-number of $3.52 \cdot 10^6$ and an angle of attack of $\alpha = 20.18^o$. Fig. 1b shows the convergence of the total drag–coefficient without preconditioning (only for $M = 0.1$; the solutions for $M < 0.1$ are not visible in the range up to 3000 multigrid cycles!) and for preconditioning. Again the preconditioned solution is almost Mach-number independent for the whole range of test cases.

Another example showing the accuracy improvements of the preconditioned solver is the convection of a vortex. Using analytical functions observed from a data-fit of experimental data [13] a vortex has been prescribed on the inflow boundary of a cartesian mesh. The vortex core radii is resolved by 4 cells. Fig. 2 shows the evolution of the pressure coefficient on the center line of the vortex and profiles of axial- and tangential velocity as well as the pressure coefficient distribution in a plane 50 core radii behind the inflow boundary. As shown in [2] the numerical solutions should be close to the analytical functions, if the Euler equations are solved on a sufficiently fine mesh. Here we select a relatively coarse mesh to point out the advantages of preconditioning. It becomes visible, especially for the pressure coefficient, that the preconditioned solution is closer to the analytical solution.

The results shown underline the advantages of preconditioning for nearly incompressible flows compared to standard schemes solving the compressible equations. The advantage of preconditioning compared to algorithms solving the incompressible equations is, that applications including incompressible and compressible regions are enabled. This is especially important for high-lift applications.

3 Numerical Results

The DLR ALVAST wing/fuselage wind-tunnel model [4] with deployed slat and flaps in take-off configuration including the UHBR engine is selected as the baseline high-lift configuration, Fig. 3a. The geometry specifications are similar to an AIRBUS A320. The considered UHBR-engine simulator represents an advanced engine concept, with a very high bypass ratio in order to improve the propulsion efficiency of the engine.

The hybrid unstructured grids are generated with the code system CENTAUR [3]. The grids consist of two parts as depicted in Figure 3b: a quasi-structured prismatic cell layer is utilized in order to achieve an appropriate resolution of the viscous effects inside the boundary layer. In contrast to this, tetrahedral cells are used to fill the outer domain of the flow-field.

3.1 Vortex dominated flow - Peniche

Especially at high angles of attack differences in the lift-polar can be found for half and full model testing results. Although the semi-span model should reach the same results as a full-model for symmetric incoming flow.

The symmetry condition of the half model is approximated by mounting the model on the wind tunnel floor or ceiling. As a consequence, the model would be positioned partially in the boundary-layer of the wind-tunnel walls. To prevent this undesired effect the half-model is mounted on an extension of the fuselage in the symmetry-plane, a so-called peniche. This peniche can be on the other hand a source of the differences between semi- and full-span measurements.

With numerical methods the peniche influence and the effects referring to the wing-root aerodynamics has been investigated in [7] by considering the isolated (half-)fuselage of the ALVAST configuration, because the fuselage is most directly influenced by the peniche, and a high grid resolution is achievable with a limited overall effort. More details and results can be found in [7]. Here the influence of preconditioning on the vortex-resolution for this case should be demonstrated.

Figure 4 displays a cut of the flow field normal to the fuselage-axis on the same grid with and without preconditioning. The flow computations were performed for a free stream Mach-number of $M_\infty = 0.22$ and a Reynolds-number of $Re_\infty = 17 \cdot 10^6$. In the Figures the vortex above the fuselage can be seen, which is a result of the flow separation behind the equator of the fuselage at higher angles of attack. Clearly visible is an increased flow-resolution and vortex strength using preconditioning.

3.2 Aircraft with running Engine - Nacelle-Vortex

The simulation of the flow around realistic high-lift aircraft configurations with a high level of detail is one of the near future objectives of CFD in aeronautical applications. In this context, the capabilities of the TAU-code with respect to the topic of engine/airframe integration have been investigated on the ALVAST high-lift configuration with engines in [9], Fig. 3a. Here the applicability of preconditioning for such a complex case should be demonstrated.

The ALVAST with UHBR-engine was computed at $\alpha = 12^o$, $M_\infty = 0.22$ and $Re_\infty = 2 \cdot 10^6$ in take-off configuration and engine-setting of about 90% maximum thrust. To enable a realistic numerical simulation the jet effect is modeled numerically by special inflow and outflow conditions, neglecting the inner parts of the engine [9]. Such complex configurations are accompanied by three-dimensional flow features: As an example in Figure 5a the area of wing/pylon/engine junction is depicted, which shows vortex shedding from the nacelle trailing edge, rolling up and crossing the slat and wing upper-side. This vortex is called "nacelle-vortex". A more detailed analysis can be found in [9].

The effect of preconditioning on the calculation quality for this complex configuration is shown in Figure 5 on the nacelle vortex with and without preconditioning. The increased vortex strength, the stronger shear layer and a slightly different path of the vortex over the wing using preconditioning is clearly visible.

3.3 Static Aircraft with running Engine

Another application of low Mach-number preconditioning is a static aircraft with running engine. This condition can be found at warming up the engines of an aircraft before take-off or at the moment when full power is turned on on the runway. These are critical cases for the design of an aircraft because of the possibility of absorbing particles or the hazard for vehicles or persons in front of the engines. A related problem is the usage of thrust reversers at the landing phase, especially at lower rolling speeds. In this case the recirculation of hot exhaust gases into the engine intake or the absorbing of particles is a problem which must be considered in the design to prevent power loss or damages to the engines.

From CFD-side these cases are complicated to simulate because of the big areas of nearly zero flow with its resulting stiffness of the flow equations. The usage of preconditioning promises to improve the solution quality and effectiveness for such cases. The ALVAST with UHBR-engine as shown in the previous section was used to demonstrate the capabilities of the TAU-code for such type of flow problems.

In Figure 6a the ALVAST configuration is shown with running engine calculated with preconditioning at $Ma_\infty = 0$. The critical area in front of the engine is bounded by an iso-surface of velocity, here for example $|V| = 20\ m/s$. Behind the engine the interaction of the jet and the thrust gate can be seen. In Figure 6b a cut through the engine including the Mach-number distribution is depicted. The streamlines in this plane clarify the path of the flow in front and behind the engine. Behind the engine the jet-boundary with its oscillations is depicted.

4 Conclusion

This paper shows the capabilities of the hybrid unstructured DLR TAU-code with respect to preconditioning at low Mach-numbers especially for complex high-lift configurations. For nearly incompressible flows Mach-number independent convergence and accuracy properties can be obtained. Besides a faster convergence of global quantities like drag and lift, an improved solution quality could be shown. This helps to save computational costs compared to the code without preconditioning, what is especially important for complex high-lift configurations. In that context the improved resolution of vortices could be shown for high-lift configurations with running engine.

References

[1] Choi, Y.-H., Merkle, C. L.: *The Application of Preconditioning to Viscous Flows.* Journal of Computational Physics 105, 1993, pp. 207-223.

[2] Heinrich, R.: *Numerical Simulation of Wake-Vortex Encounters Using the Chimera-Technique.* in Notes on Numerical Fluid Mechanics, Volume 77, pp. 74 - 81., Springer Verlag, Berlin, Heidelberg, New York, 2002.

[3] Kallinderis, Y.: *Hybrid Grids and Their Applications.* Handbook of Grid Generation, CRC Press, Boca Raton, London, New York, Washington D.C., pp. 25-1 - 25-18, 1999.

[4] Kiock, R.: *The ALVAST Model of DLR.* DLR IB 129-96/22, 1996.

[5] Kroll, N.; Rossow, C.-C.; Schwamborn, D.; DLR Insitute of Aerodynamics and Flow Technology; Becker, K.; Airbus-Deutschland; Heller, G.; Fairchild-Donier: *MEGAFLOW - A Numerical Flow Simulation Tool for Transport Aircraft Design.* 23st ICAS Congress, Toronto, ICAS 2002-1.10.5, 2002.

[6] Mavriplis, D.J.: *Parallel Performance Investigations of an Unstructured Mesh Navier-Stokes Solver.* ICASE Report No. 2000-13, March 2000.
[7] Melber, S.; Wild, J.; Rudnik, R.: *Overview about the HLRS Projet "Numerical High Lift Research" (NHLRes).* DLR IB 124-2002/15, 2002.
[8] Melber, S.: *3D RANS Simulations for High Lift Analysis of Transport Aircraft Configurations.* Notes on numerical fluid mechanics, Volume 77, Springer Verlag, Berlin, Heidelberg, New York, 2002, pp 27-34.
[9] Melber, S.: *3D RANS-Simulations for High-Lift Transport Aircraft Configurations with Engines.* DLR IB 124-2002/27, 2002.
[10] Radespiel, R.; Turkel, E.; Kroll, N.: *Assessment of Preconditioning Methods.* DLR FB 95/29, 1995.
[11] Rogers, S.E.; Roth, K.; Cao, V.Hoa; Slotnick, J.P.; Whilock, M.; Nash, S.M.; Baker, M.D.: *Computation of viscous Flow for a Boeing 777 Aircraft in Landing Conf.* AIAA paper 2000-4221, 2000.
[12] Rudnik, R.; Melber, S.; Ronzheimer, A.; Brodersen, O.: *Three-Dimensional Navier-Stokes Simulations for Transport Aircraft High Lift Configurations.* Journal of Aircraft, Vol. 38, pp.895-903, 2001.
[13] Struijs, R.; Jonville, G.; Darraq, D.: *Inviscid computation of the effect of wake vortices on a model airplane.* WAVENC-TN16, 1999.

Table 1 Multigrid cycles, drag for a convergence of 7 orders of mag. (invis. NACA0012).

	no prec.		preconditioning	
	MG-cycles	drag	MG-cycles	drag
Ma = 0.1	381	0.000763	184	0.000289
Ma = 0.01	1497	0.005384	184	0.000288
Ma = 0.001	5432	0.042612	184	0.000288

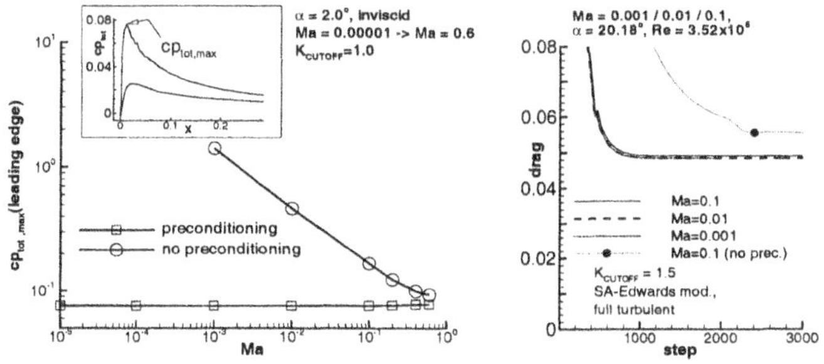

(a) (b)

Figure 1 (a) Total pressure loss coefficient for inviscid NACA0012. (b) Drag convergence for 3 element high lift airfoil.

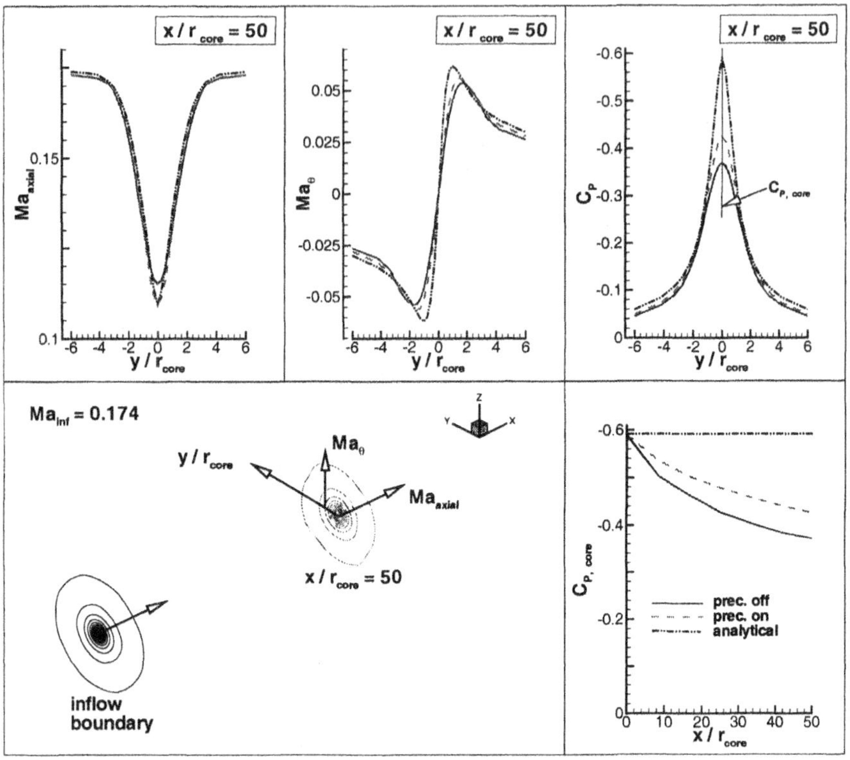

Figure 2 Convection of a vortex.

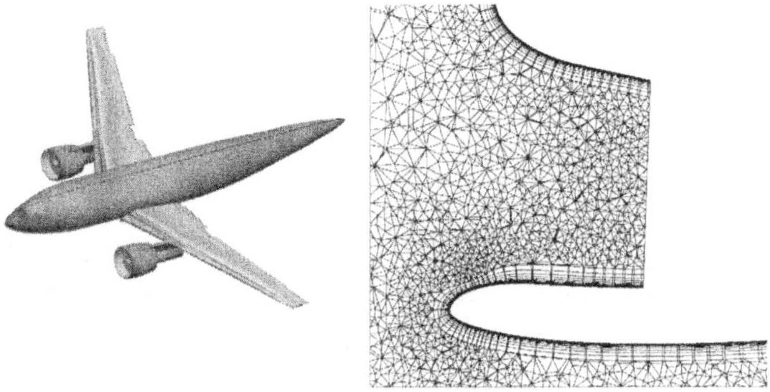

(a) (b)

Figure 3 (a) ALVAST high lift configuration with UHBR-engine, pressure coefficient on the surface. (b) Hybrid unstructured grid in a cut through UHBR-engine near fan-intake.

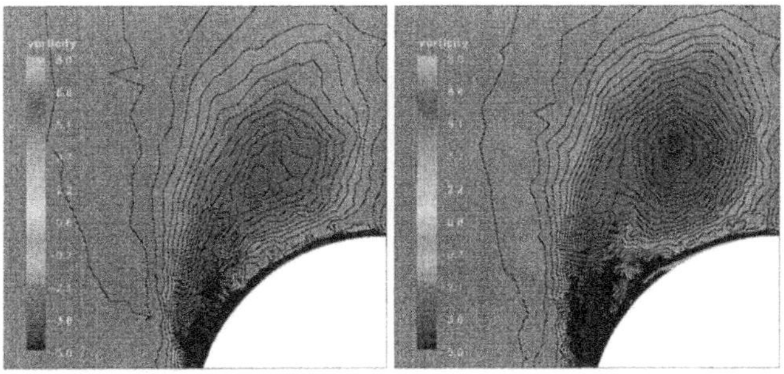

(a) (b)

Figure 4 Vorticity in a plane normal to the fuselage-axis for half-fuselage without peniche, $\alpha = 20.99^o$, (a) preconditioning off, (b) preconditioning on.

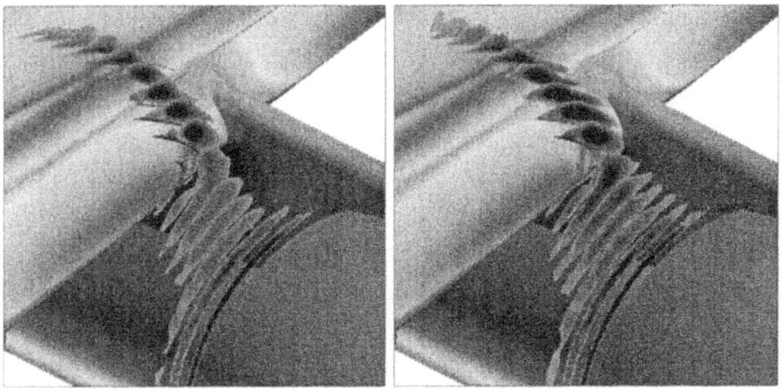

(a) (b)

Figure 5 ALVAST high lift configuration with UHBR-engine, pressure coefficient on the surface and vorticity cuts of nacelle vortex: (a) preconditioning off, (b) preconditioning on.

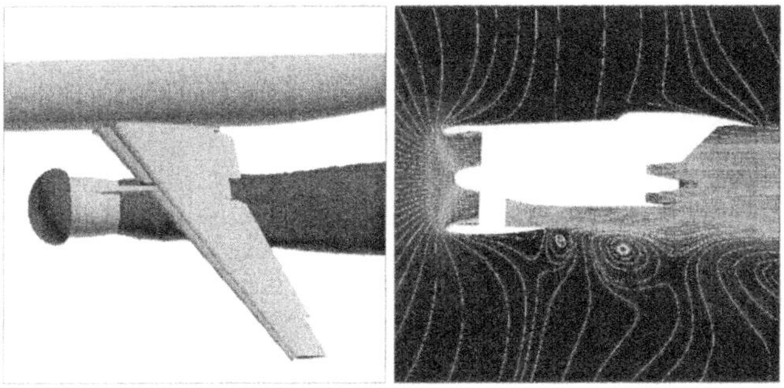

(a) (b)

Figure 6 Computation of ALVAST high-lift configuration with UHBR-engine, static aircraft and running engine: (a) iso-surface $|\boldsymbol{V}| = 20\ m/s$, (b) cut through engine, Ma-distribution and streamlines in this plane.

Improvement of upwind schemes with the Least Square method in the DLR TAU Code

M. Widhalm and C. C. Rossow

DLR, Institute of Aerodynamics and Flow Technology,
Lilienthalplatz 7, D-38108 Braunschweig, Germany
markus.widhalm@dlr.de, cord.rossow@dlr.de

Summary

Accuracy investigations for upwind schemes in the framework of a node based and cell centered discretisation for the Euler and Navier Stokes equation on unstructured hybrid grids are presented. A crucial factor for the accuracy of the upwind discretisation is the used computation of gradients. The Least-Square approach will be investigated in comparison to the well known Green-Gauss algorithm. Numerical results are obtained for subsonic inviscid and subsonic turbulent flows around airfoils. The frequently used central scheme for applications of subsonic to transonic flows will be used as a reference for the results. In case of upwind discretisations the cell centered formulation in combination with the Least-Square reconstruction provides the most accurate results.

1 Governing equations

The compressible Navier-Stokes equations can be written in conservative form as

$$\frac{\partial}{\partial t}\int\limits_{V} \vec{w} dV + \int\limits_{\partial V} \bar{\bar{F}}_i \bullet \vec{n} dS + \int\limits_{\partial V} \bar{\bar{F}}_v \bullet \vec{n} dS = 0 \tag{1}$$

for a domain V with the boundary ∂V. The vector $\vec{w}$ contains the conservative variables for mass, momentum and energy. $\bar{\bar{F}}_i$ represents the inviscid, $\bar{\bar{F}}_v$ the viscous flux density tensor and $\vec{n}$ is the face vector outwards of the boundary ∂V.

2 Solution Algorithm

The accuracy investigation were carried out with the DLR Navier-Stokes Code TAU [5]. The basic solver uses a node based dual mesh with an edge based data structure, created in a preprocessing step. The governing equations are discretised with a Finite Volume method on unstructured hybrid meshes. An explicit time stepping scheme with central or upwind space discretisation is the basic solution algorithm. Convergence acceleration is achieved with local time stepping, residual smoothing [3], and an agglomeration based multigrid technique [7], [3].

2.1 Finite Volume Discretisation

node based The Finite Volume discretisation follows an egde based node centered discretisation as described for the TAU code. The control volume V_i is given by a median dual mesh approach. It is computed using the centroids of the neighbouring cells and the edge midpoints as depicted in Figure 1.

cell centered As an alternative a cell centered approach, Figure 2, has been implemented. The control volumes are defined by the primary grid connectivity, and the flow quantities are stored in the cell center. The edge based structure is used again which leads to a similar computing, except for boundaries. Additionally, the grid coordinates of the face centers, used in the flux computation, are stored. At this time, the presented implementation does not support all functionalities of the standard node based TAU code.

2.2 Flux Computation

The inviscid fluxes are computed on the faces of each control volume using the flux-difference technique of Roe [8]. The fluxes are calculated from data on either side of the face as:

$$\vec{F_{lr}} = \frac{1}{2}\left(\vec{F_l} + \vec{F_r}\right) - \frac{1}{2} \mid \bar{\bar{A}}(\vec{w_l}, \vec{w_r}) \mid (\vec{w_r} - \vec{w_l}). \tag{2}$$

The vector $\vec{F_{lr}}$ represents the flux density vector normal to a cell interface, and $\bar{\bar{A}}$ is the corresponding flux Jacobian. The lower indices l and r denote variables to the left and right of the cell face. The scheme as written in equation (2) is of first order in space, where the solution is constant over the control volume. To achieve second order accuracy, the solution is assumed to be piecewise linear in the control volume. For the flux computation the primitive variables are reconstructed by extrapolation to all faces of the control volumes:

$$\begin{aligned} \phi_{face,r} &= \phi_{node,r} + \frac{1}{2}\psi\vec{\nabla}\phi_r\left(\vec{l_l} - \vec{l_r}\right), \quad \text{and} \\ \phi_{face,l} &= \phi_{node,l} + \frac{1}{2}\psi\vec{\nabla}\phi_l\left(\vec{l_r} - \vec{l_l}\right). \end{aligned} \tag{3}$$

Since in the cell centered scheme the coordinates of the face centers are available, here the geometrically correct extrapolation of variables at the cell face can be used:

$$\begin{aligned} \phi_{face,r} &= \phi_{node,r} + \psi\vec{\nabla}\phi_r\left(\vec{l}_{face} - \vec{l_r}\right), \quad \text{and} \\ \phi_{face,l} &= \phi_{node,l} + \psi\vec{\nabla}\phi_l\left(\vec{l}_{face} - \vec{l_l}\right). \end{aligned} \tag{4}$$

The value ϕ at the face is computed with the gradients $\vec{\nabla}\phi$ at the node at each side of the face, and $\vec{l}$ is the location vector of the corresponding points. The Limiter ψ, as proposed in [9], reduces the scheme to first order at discontinuities to avoid oscillations of the gradient.

3 Reconstruction schemes

Green-Gauss reconstruction The default reconstruction in the TAU Code is the Green-Gauss reconstruction. To obtain the gradient of a variable ϕ the Green-Gauss theorem reads

$$\int_V \vec{\nabla}\phi dV = \int_{\partial V} \phi \cdot \vec{n} dS. \tag{5}$$

The use of a discrete version of the above equation leads to

$$\vec{\nabla}\phi = \frac{1}{V}\sum_{i}^{N}\frac{1}{2}\left(\phi_i + \phi_0\right)\vec{n}dS, \tag{6}$$

with N denoting the number of surrounding points. The errors introduced by this method can be assessed by specifying a linear variation of a variable on a given grid. The method gives reasonable errors on single element grids. In case of a hybrid grid, however, Figures 3 and 4, significant errors were detected in the vicinity of the interface between hexahedral and prism elements as seen in Figure 5. In a node based scheme at the interfaces distortions in the dual mesh are critical for computing with the Green-Gauss approach, while with cell centered scheme, especially when the fluxes are evaluated exact at the face centers, element distortions are not severe.

Least-Square reconstruction The Least-Square approach is based upon a first order Taylor series approximation for each surrounding edge of node P0, Figure 1, ([1], [6]). This represents an over-determined system of linear equations, $Ax = b$. To solve this system a Gram-Schmidt process is used by decomposing the matrix A into a product of an orthogonal matrix Q, and an upper triangular matrix R, [4].

$$\begin{bmatrix} w_1 \Delta x_{01} & w_1 \Delta y_{01} \\ w_2 \Delta x_{02} & w_2 \Delta y_{02} \\ \vdots & \vdots \\ w_N \Delta x_{0N} & w_N \Delta y_{0N} \end{bmatrix} \begin{pmatrix} \frac{\partial \phi}{\partial x} \\ \frac{\partial \phi}{\partial y} \end{pmatrix} = \begin{pmatrix} w_1\left(\phi_1 - \phi_0\right) \\ w_2\left(\phi_2 - \phi_0\right) \\ \vdots \\ w_N\left(\phi_N - \phi_0\right) \end{pmatrix}, \quad x \cong R^{-1}Q^T b \tag{7}$$

These equations are written with a weigthing factor w_i providing freedom in its choice. These weighting coefficients were chosen as a function of the grid metrics, which leads to $w_i = 1.0/|\vec{l_i} - \vec{l_0}|^t$ for values of $t = 0, 1, 2$, with $\vec{l_i} - \vec{l_0}$ as the length of the edge. The highest robustness with all computations were reached with $t = 1$. The linear Least-Square method gives exact gradients for linear functions on meshes with arbitrary cells, Figure 5. For the discretisation of viscous fluxes, which are higher order terms, the stencils which are obtained on quadrilateral grids get decoupled, see [6], [2]. This occures near solid boundaries in hybrid viscous grids, where hexahedral or prisms are used to resolve the boundary layer flow. This decoupling allows spurios solution modes and is undesirable. It can be prevented by using the Least Square approach for reconstructing the primitive variables only. When gradients are required elsewhere, as in the production terms for the turbulence models, Green-Gauss is used.

4 Numerical Results

Accuracy Investigation To compare the order of accuracy associated with the gradient computation using Green-Gauss and Least-Square in a node based scheme, a grid refinement study with an arbitrary non-cartesian mesh, e.g. a body fitted grid around an expansion ramp, Figures 6 and 7, is done. By specifying a nonlinear function, $\phi = x^2 + y^2$, on the grids with freestream at all boundaries and focusing on the first derivative, we can measure the error between the exact and numerical gradient evaluated on the grids. By plotting the logarithm of the L2-norm of the error against the logarithm of the characteristic cell size, $1/\sqrt{N}$, one can infer the order of the truncation error from the slope of the plot, Figure 8. For the structured grids the slope for Green-Gauss and Least-Square is 1.93 and 1.94, respectively and indicates second order accuracy. For unstructured grids the resolved order for the Green-Gauss reconstruction is below one and should therefore not be used for a node based dual mesh. Least-Square holds a slope of 1.3, but still needs further investigation for improvement.

Inviscid Flow For subsonic flows the total pressure loss can be used for the accuracy investigation of a numerical scheme. In case of an inviscid flow without shocks it has to be exactly zero. The total pressure loss is qualified to be a good indicator for the discretisation error.
RAE2822 airfoil The test case presents a subsonic flow at an inflow Mach number of 0.5 and the solution is obtained without any limiters. The meshes contain 10618 and 31704 points as depicted in Figures 9,10. The left graph of Figure 11 shows the total pressure loss on the wall. The Least-Square reconstruction is sligthly better than Green-Gauss. The convergence history is shown in Figure 12 (left). The comparison of the node centered (nb) and cell centered (cc) scheme can be seen in the right part of Figure 11 and 12. On unstructured grids, the cell centered scheme benefits in reduction of pressure loss. The Mach number distribution displayed in Figure 12 (right) shows that the node based scheme creates a spurious entropy layer, which is reduced for the cell centered approach. Drag prediction is more accurate with the cell centered scheme, seen in Table 4. The drag is halfed when node based and cell centered schemes are compared in case of Least-Square.

node based dual metric				cell centered	
	central	upwind gg	upwind ls	upwind gg	upwind ls
lift	0.6617	0.6576	0.6591	0.6590	0.6696
drag	0.0036	0.0046	0.0041	0.0028	0.0022
pitching-moment	-0.2547	-0.2530	-0.2538	-0.2546	-02595

Turbulent Flow *Airfoil with blunt trailing edge*
In viscous flow computations, the pressure drag is very sensitive to numerical errors, and any inaccuracy in the discretisation can produce spurious additional drag. The selected grid, Figures 13 and 14, contains arbitrary formed elements and is particularly suitable for this accuracy investigation. The inflow Mach number is 0.2 and

the Reynolds number is 6.0 million. The computations performed with Least-Square are with and without limiters. Due to stability reasons the Green-Gauss calculation could be carried out only with limiters. As can be seen in Figure 15, for a node based scheme the Least-Square predicts lower pressure drag than the Green Gauss reconstruction, but still is not competitive to the central scheme. In a cell centered scheme, Figure 16, the Least-Square reconstruction shows a significant improvement and is even better, if computed without limiter, than the central scheme. The improvement with Green-Gauss is moderate and again not comparable to the central scheme.

5 Conclusion

The accuracy investigation has shown for an upwind discretisation in a node based scheme, that the computation of gradients does not fullfill the accuracy requirements for drag prediction. Especially Green-Gauss is not suitable on hybrid grids with arbitrary cells. Compared with Green-Gauss the Least-Square reconstruction offers improvement but is not sufficient. For the cell centered discretisation both reconstructions have increased accuracy in conjunction with the correct flux approximation. However, in the cell centered scheme Green-Gauss is not competitive with the central scheme and is therefore not recommended on unstructured hybrid grids. The drag prediction benefits at the most from the Least-Square approach. Considering the memory costs in comparison of a node based dual mesh and cell centered scheme, the cell centered scheme needs higher memory storage. Even with this aspect a cell centered approach should be considered further, because it seems to be more suitable for upwind schemes than a node based scheme.

References

[1] Anderson W. Kyle, Bonhaus Daryl L.: An Implicit Upwind Algorithm for Computing Turbulent Flows on Unstructured Grids. Computers Fluids Vol. **23**, No.1, pp. 1-21, 1994

[2] Coirier W. J.: An adaptively-refined, cartesian, cell based scheme for the Euler and Navier-Stokes equations. PhD thesis, University of Michigan, USA, 1994

[3] Galle M.: A Method for the Numerical Simulation of Compressible Viscous Flows on Hybrid Grids. DLR-FB 99-04, 1999

[4] Golub G., van Loan C.: Matrix Computations. The Johns Hopkins University Press, 1991

[5] Gerhold T., Friedrichs O., Evans J., Galle M.: Calculation of Complex three-dimensional configurations employing the DLR TAU Code. AIAA-97-0167, 1997

[6] Haselbacher A., Blazek J.: On the Accurate and Efficient Discretisation of the Navier-Stokes Equations on Mixed Grids. AIAA Paper 99-33552, 1999

[7] Mavriplis D. J., Jameson A., Martinelli L.: Multigrid Solution of the Navier-Stokes euqations on triangular meshes. ICASE-Report, No. 89-35, 1989

[8] Roe P. L.: Approximate Riemann Solvers, Parameter Vectors, and Difference Schemes. Journal of Computational Physics 43, pp. 357-372, 1981

[9] Venkatakrishnan V.: On the Accuracy of Limiters and Convergence to Steady State Solutions. AIAA 93-0880

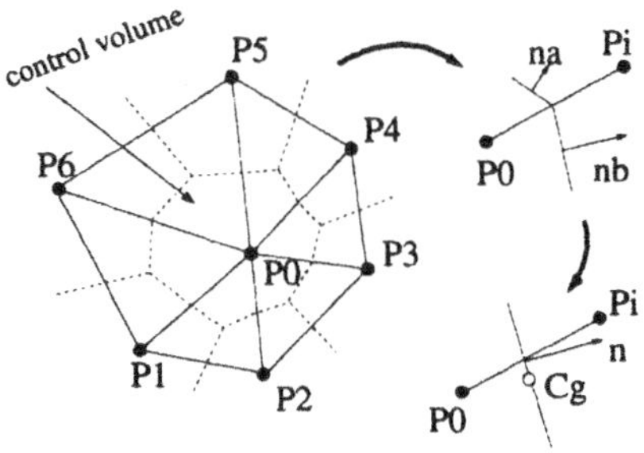

Figure 1 Representative median dual cell for node based scheme

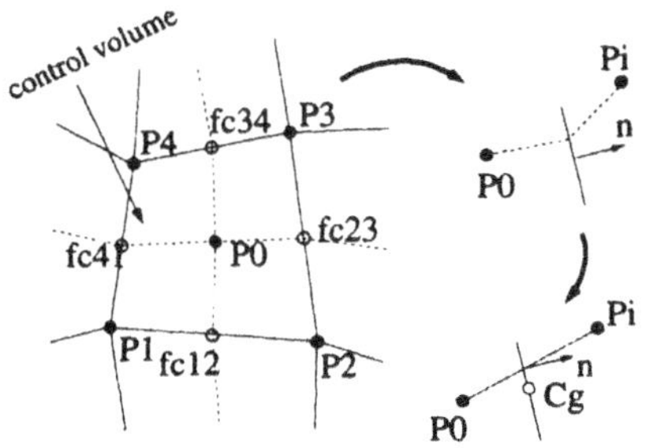

Figure 2 Control volume for cell centered scheme

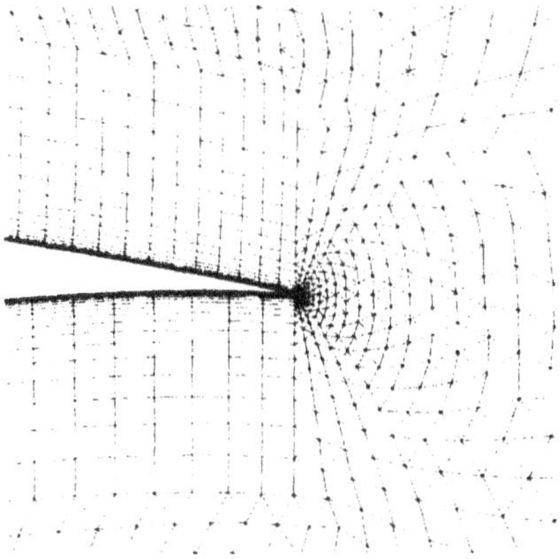

Figure 3 Primary grid at the interface between hexahedrals and prisms

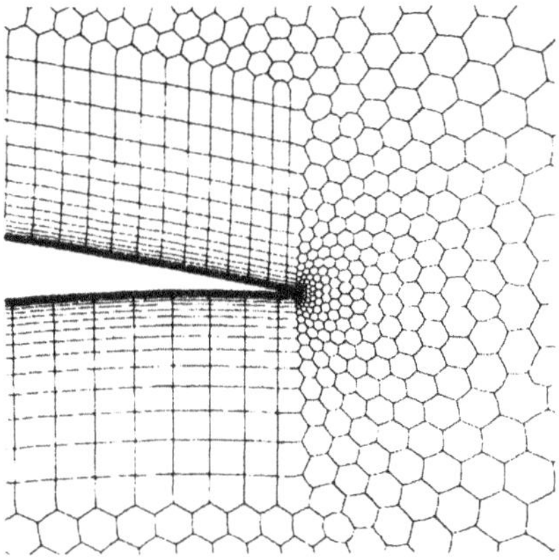

Figure 4 Dual grid at the interface between hexahedrals and prisms

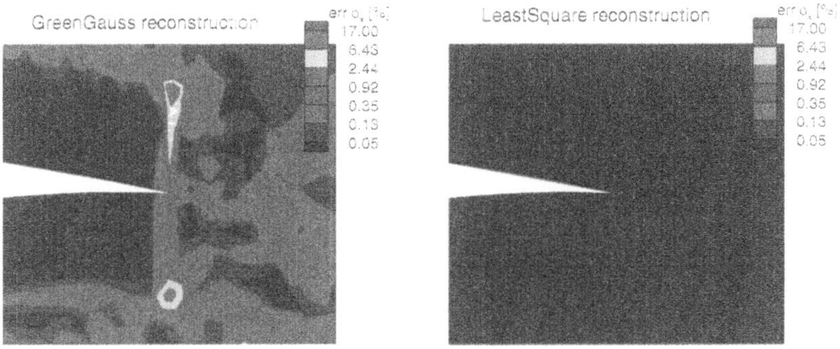

Figure 5 Percentage error of the Green-Gauss and Least-Square reconstruction of the magnitude of the first derivative of a linear function

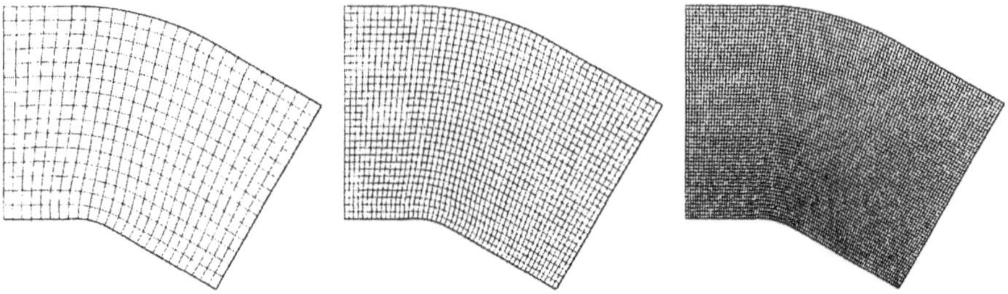

Figure 6 Structured coarse, medium and fine expansion ramp grids with 832, 3162 and 12322 points

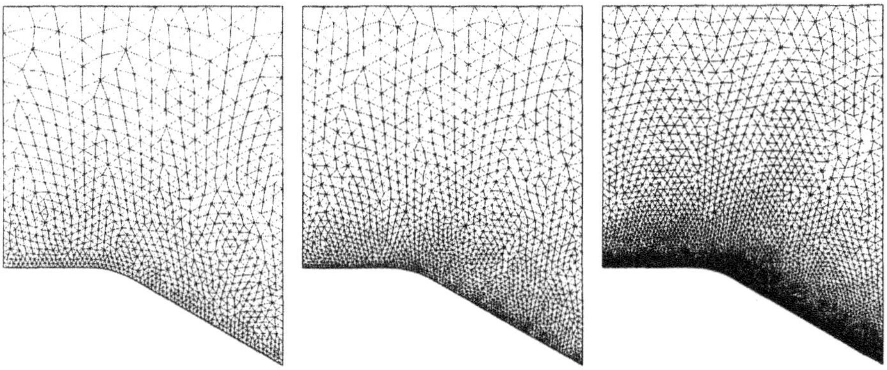

Figure 7 Unstructured coarse, medium and fine expansion ramp grids with 1848, 3212 and 7050 points

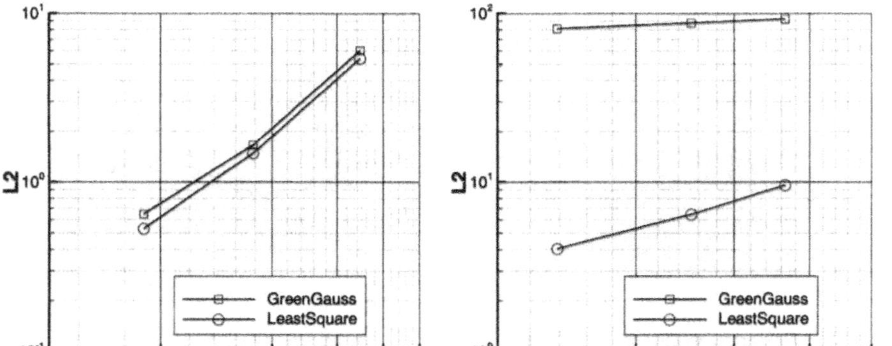

Figure 8 L2 norms of the error of gradient computation on expansion ramp grids ,left structured - right unstructured grids, for node based scheme

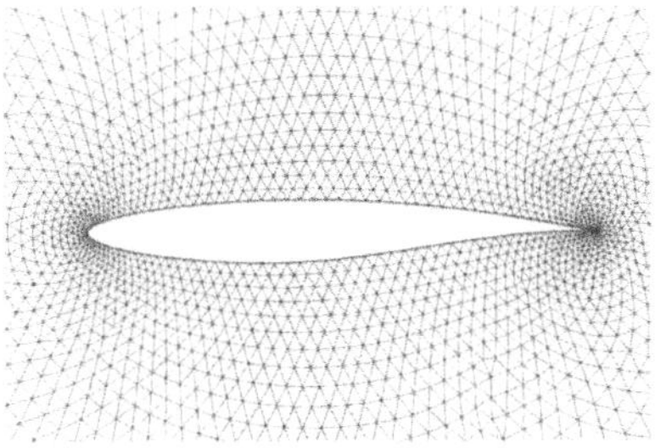

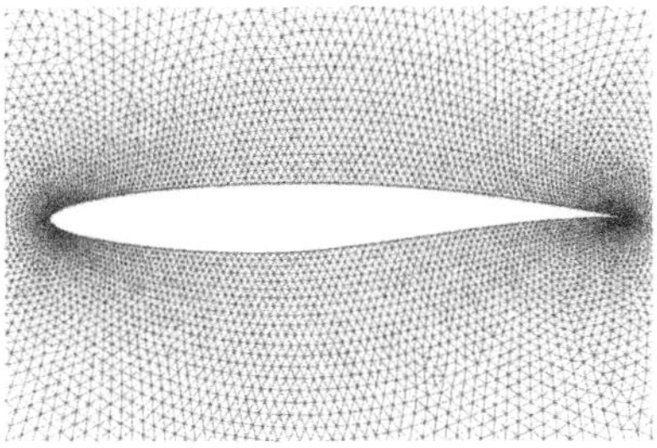

Figure 9 Close view of the coarse grid, RAE2822 airfoil

Figure 10 Close view of the medium grid, RAE2822 airfoil

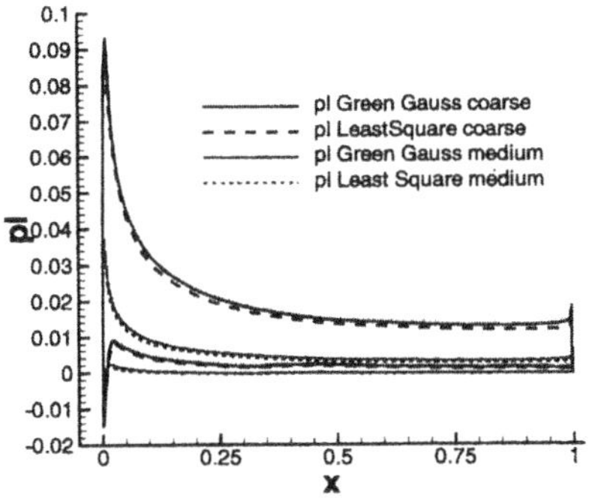

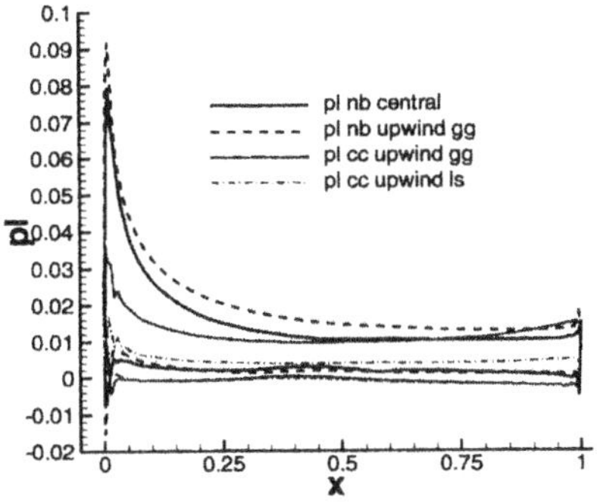

Figure 11 Comparison of total pressure losses (pl) from different reconstructions in a node based (nb) discr. - left, and comparison of upwind vs. central schemes in the node based and cell centered (cc) discr. - right, for the coarse grid, RAE2822 airfoil, M = 0.5, $\alpha = 2.8^o$

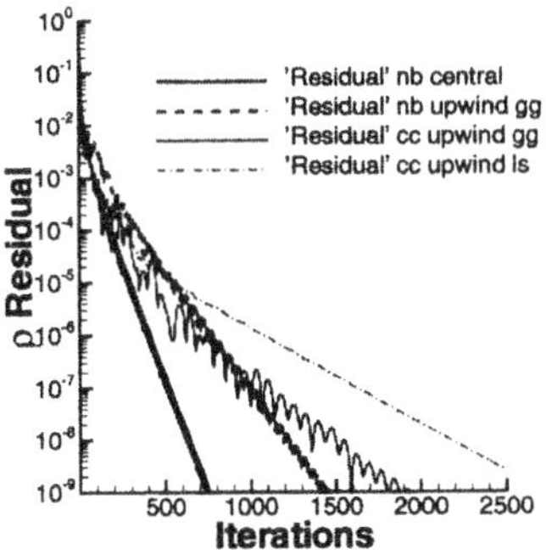

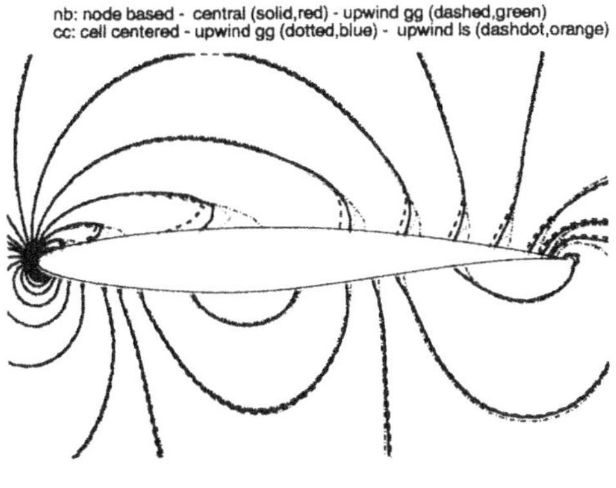

Figure 12 Convergence history (left) and Mach number contours (right) for the coarse grid, RAE2822 airfoil, M = 0.5, $\alpha = 2.8^o$, nb: node based, cc: cell centered, gg: GreenGauss, ls: LeastSquare

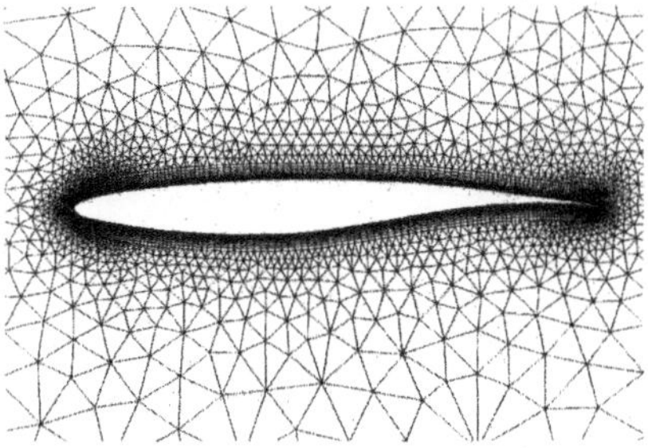

Figure 13 Near field view of the grid around an airfoil with blunt trailing edge

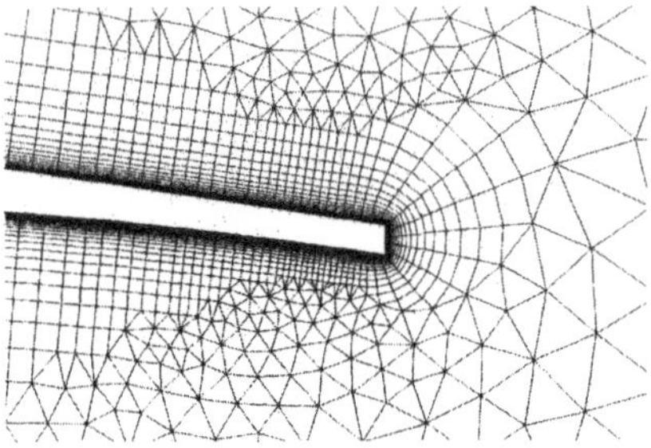

Figure 14 Close up view from the trailing edge of the grid

Lift coefficient C_l vs. pressure drag coefficient C_{dp} for an airfoil with blunt trailing edge at M = 0.2, Re = 6 Mill.

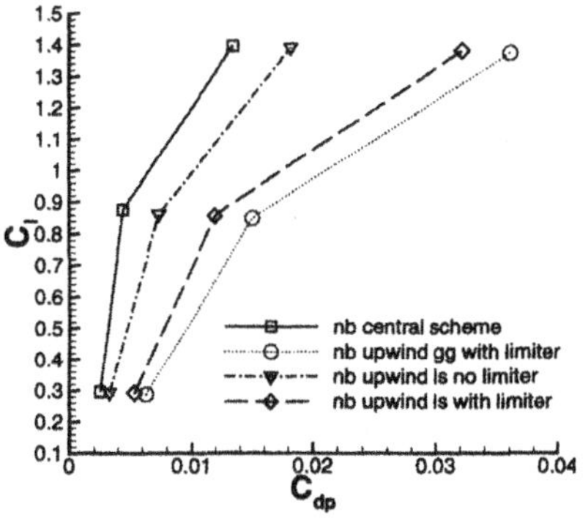

Figure 15 Comparisons of results for the Central scheme and the Upwind scheme with Green Gauss (gg) and Least Square (ls) reconstruction for the node based discretisation (nb)

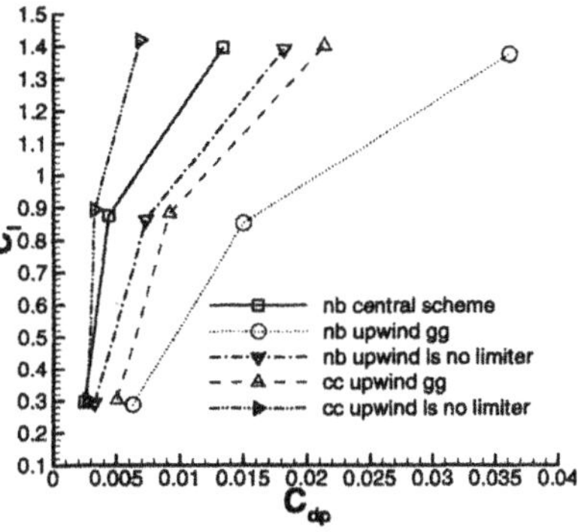

Figure 16 Comparisons of results for Central scheme and the Upwind scheme in a node based (nb) and cell centered (cc) discretisation

Development and Application of the Thin Oil Film Technique for Skin Friction Measurements in the Short-Duration Hypersonic Wind Tunnel

E. SCHÜLEIN
German Aerospace Center DLR, Institute of Aerodynamics and Flow Technology
Bunsenstr. 10, 37073 Göttingen, Germany, erich.schuelein@dlr.de

Summary

The utilization of image acquisition, image processing and skin friction computations has been developed to support skin friction measurements at the Ludwieg-tube facility at DLR. Verification of the oil-film-interferometry technique with analytical tests at different flows with given shear stress distributions showed its advantages and handicaps for skin friction measurements in 3D gradient flows. The abilities of the thin oil film technique with its application in a short duration wind tunnel were demonstrated on some examples. The comparison of experimental results obtained by different optical skin friction methods demonstrated good agreement.

1 Introduction

Skin friction measurements are very important for understanding many types of flow fields, and valuable for the validation of CFD turbulence models. The difficulties of shear stress measurements with existing methods in interacting flows or in flows with thin boundary layers are well known. They are mainly related to the intrusiveness of the measuring device and the corresponding flow disturbance. The oil-film-interferometry technique, developed in [10], overcomes many of these limitations, since it provides non-intrusive measurements. It is based on the relationship between the thinning of an oil film, placed on the test surface and exposed to the flow at the local surface shear τ, which is for 3D flows given in the following form [10]:

$$\tau = \frac{2\mu}{nh^2 t}\int_0^{s_0} nh\, ds\,, \quad \text{or} \quad \tau = \frac{2\mu I}{t} \quad \text{for } I = \frac{1}{nh^2}\int_0^{s_0} nh\, ds \tag{1}$$

where t is the run time, μ is the oil viscosity; s, h and n are the oil film dimensions corresponding to Figure 1 (left). The rate of thinning of the oil film in time is determined by using optical interference created when an incident light beam is partially reflected at the oil and test surfaces as shown in Figure 1 (right). This method, originally used in low-speed flows, was later extensively modified for high-speed 3D gradient flows. The most progressive approaches have been developed in the last ten years and can be commonly named *Global Interferometry Skin Friction* technique *GISF* [1, 3, 4, 6, 8]. The global shear stress distributions can be obtained by this technique during a single run. The size of the testing surface is limited only by the performances of the light source and the camera. The modern GISF-methods for 3D gradient flows can be split into: a) *One-Image-Method*, which requires for the calculation of the skin friction only one interferogram and the total run time t [1, 10], and b) *Two-Image-Method*, where only the time interval Δt is important and at least two images obtained during the run are needed [3, 4, 6, 8]. The 1-Image-Method is simpler, but requires the exact run time statement, equation (1). This can lead to measurement errors especially in short-duration wind tunnels. A typical equation for the 2-Image-Method is given for example in [4]:

$$\tau = \frac{2\mu}{nh^2}\frac{d}{dt}\int_0^{s_o} nh\, ds \tag{2}$$

The differences of 1- and 2-Image-Methods are not really significant. It can be shown that the techniques of different methods can be converted into each other by usage of the *equivalent* run time assuming that the local wall shear is uniform in time. For example, according to (1) the condition of $\tau(t)$ = const requests that the relation $I(s_o)/t$ is also constant. The equivalent run time t_{eq} can consequently be calculated if I_1 and I_2 at two different times t_1 and t_2 are known:

$$t_{1eq} = t_1 + \frac{I_2 t_1 - I_1 t_2}{I_1 - I_2} \tag{3}$$

This equivalent time enables usage of equation (1) provided, that at least two interferograms obtained during the run are available, but the exact run time is still unknown.

2 The Super- & Hypersonic Ludwieg Tube Facility at DLR

The presented GISF-meter was primarily developed for the DLR Göttingen Ludwieg-tube facility (DNW-RWG) which covers a Mach number range of $3 \leq M \leq 7$ and a unit Reynolds number range of $5\times10^6 \leq Re_1 \leq 80\times10^6$ m^{-1}. Figure 2 shows the general arrangement of the wind tunnel with the main features and flow conditions for the two test-legs A and B. The peculiarity of a Ludwieg-tube facility is the usage of a long tube as pressure reservoir, which is closed at one end and has a gate valve attached to the other end, followed by a supersonic nozzle, the test section and a dump tank. After opening the gate valve, the flow is started by an expansion wave, which travels to the closed tube end, is being reflected there and travels back. As long as the expansion wave has not come back to the nozzle throat, test gas flows out at constant stagnation conditions through the nozzle and the test section into the dump tank. The tube length of 80 meters results in a run time of over 0.3 seconds. The low operation costs, relatively large test section, and the good optical access predestine this facility for optical experimental methods, like the thin-oil-film technique.

3 Verification of GISF-methods with analytical tests

To conduct the optical skin friction measurements at the Ludwieg-tube at DLR a software application for image acquisition, image processing and skin friction computations was developed. The best known GISF-methods are adapted and implemented in this application. To check the limits of GISF-techniques in a short-duration wind tunnel, different 2D and 3D flows with given shear stress distributions

were investigated. The known shear stress is used to simulate the oil-film deformation in time under conditions, which are similar to the Ludwieg-tube conditions. These tests include simulations of adverse and favourable wall shear stress gradients as well as convergence and divergence of the surface streamlines in different combinations, which are typical for 3-D interacting flows.
In one of these tests, for example, the increasing shear stress in the flow with a gradient of 5 Pa/mm was assumed, so that the shear stress increases over a length of 10 mm from 50 up to 100 Pa. The corresponding sequence of interferograms, which describes the deformation of the oil film, has been analytically simulated with the thin-oil-film equation for the total run time of about 300 ms by 30 frames per second. The evaluation of these interferograms by GISF-meter with different implemented methods and comparison of the given shear stress distribution with the obtained results have been demonstrated the limits of simplified approaches [3, 5] and good results for more general methods [1, 4, 6, 10], which are based on the 3D thin-oil-film equation. The simplified methods give acceptable results only near the leading edge of the oil film. Results of all other enforced analytical tests confirm these conclusions.

4 GISF-measurements at different interacting flows

4.1 3-D wake flow behind a vertical fin on a flat plate

The first application of GISF-technique in the Ludwieg-tube is the investigation of the 3D wake flow behind a vertical fin on a flat plate at Mach 5 (Figure 3). The test model with the interaction zone, which was investigated, is shown on the top sketch. The image of oil-flow visualization and the corresponding sketch of the limiting streamlines show the properties of the flow at the area of interest. A typical fringe pattern obtained by the oil film interferometry shows the influence of the highly 3D shear gradients on the oil film thickness distribution: the thinning of the oil film along the reattachment line R and the oil accumulation near the separation lines S. The corresponding skin friction distribution in a cross section of the fin wake (Figure 4) demonstrates the same properties by the high shear level along the reattachment line and reduction of shear at the flow separation. The small imprecision of the fin orientation along the flow explains the asymmetrical distribution of the skin friction in the wake flow because of the lateral displacement of the central reattachment line with respect to the coordinate $Z = 0$. The comparison of results at different GISF-methods shows its good agreement and demonstrates the confidence of all these methods.

4.2 Incident shock / turbulent boundary layer interaction flow

The investigation at 2D incident shock / turbulent boundary layer interaction flow was enforced in the Ludwieg tube at Mach 5 [2, 9]. Some flow cases were studied with different shock generator angles. The experimental data include shadowgraphs, velocity profile measurements upstream and downstream of the interaction as well as optical skin friction and heat transfer measurements. Figure 5 shows shadowgraphs for the interacting flows with a small (10°) and a large separation zone (14°). The comparison of experimental skin friction distributions in these cases

(Figure 6) obtained by GISF-method (10° and 14°) and from the velocity profile data (10°) shows a good agreement of different skin friction techniques.

4.3 Longitudinal vortices in supersonic turbulent separated flow at 20° ramp

In most 2D separated supersonic boundary layers the flow shows tracks of longitudinal vortices behind the reattachment lines. These vortices are very similar to the ones investigated by Görtler. The presented oil-flow-visualization image and the corresponding sketch of the limiting streamlines on the wall (Figure 7) show the structure of these vortices and demonstrate its tracks downstream of the reattachment line for a 20° compression corner turbulent flow at Mach 3 [7]. The cross sectional distribution of the measured skin friction at different distances from the leading edge of the ramp (Figure 8) shows behind the reattachment line fluctuations of about ±20% of the skin friction values. The fluctuations will get smaller with increasing distance from the separated zone, but still remain notable, also at the farthest investigated distance.

Conclusions

A software application for image acquisition, image processing and skin friction computations has been developed for optical skin friction measurements in the Ludwieg tube at DLR. The efficiency of known GISF-methods for measurements under specific conditions of a short duration wind tunnel has been analytically investigated for different 2D and 3D flows with given shear stress distributions. The tests showed good results for all GISF-methods, that are based on the general thin-oil-film equation. The simplified methods showed restrictions and can only be used in combination with the array-oil-drop technology. It was shown that evaluation techniques of different GISF-methods could be converted into each other by usage of the equivalent run time. The abilities of the GISF-technique and its application in the Ludwieg tube were demonstrated on some examples. The comparison of the results obtained by different optical skin friction methods demonstrated good agreement and the reliability of all these methods.

References

[1] Garrison T.J. and Ackman M. Development of a Global Interferometer Skin Friction Meter. *AIAA-96-1968*, AIAA, 1996.

[2] Fedorova N.N., Fedorchenko I.A., Schülein E. Experimental Investigation and Numerical Simulation of the Impinging Oblique Shock Wave / Turbulent Boundary Layer Interaction at M=5. *ZAMM*, Berlin, 81, Suppl. 3, 2001, pp. 773-774.

[3] Kornilov V.I., Pavlov A.A., Shpak S.I. On the Techniques of Skin Friction Measurement Using Optical Method. *Proceedings of Inernational Conference on the Methods of Aerophysical Research,* Novosibirsk, Russia, 1992, Part 1, pp. 71-74.

[4] Maksimov A.I., Pavlov A.A., Shevchenko A.M. Development of the Optical Skin Friction Measurement Technique for Supersonic Gradient Flows. *Proceedings of Inernational Conference on the Methods of Aerophysical Research,* Novosibirsk, Russia, 1994, Part II, pp. 172-177.

[5] Monson D.J. and Higuchi H. Skin Friction Measurements by a Dual-Laser-Beam Interferometer Technique. *AIAA J.*, Vol. 19, No. 6, 1981, pp. 739-744.

[6] Naughton J.W. and Brown J.L. Surface Interferometric Skin-Friction Measurement Technique. *AIAA-96-2183*, AIAA, 1996.

[7] Schülein E. Experimentelle Untersuchungen zur Längswirbelbildung in turbulenten Überschallströmungen mit Ablösungen. *DLR IB 224 – 2002 A 12*, Göttingen, 2002.

[8] Schülein E., Koch S., Rosemann H. Skin Friction Measurement and Transition Detection Techniques for the Ludwieg-Tubes at DLR. *AGARD-CP-601*, AGARD, 1998, pp. 23-1–23-9.

[9] Schülein E., Krogmann P., Stanewsky E. Documentation of two-dimensional impinging shock / turbulent boundary layer interaction flow. *DLR IB 223 - 96 A 49*, Göttingen, 1996, 69 p.

[10] Tanner, L., and Blows, L., A Study of the Motion of Oil Films on Surfaces in Air Flow, with Application to the Measurement of Skin Friction. *Journal of Physics, E:Scientific Instruments*, Vol. 9, March 1976, pp. 194-202.

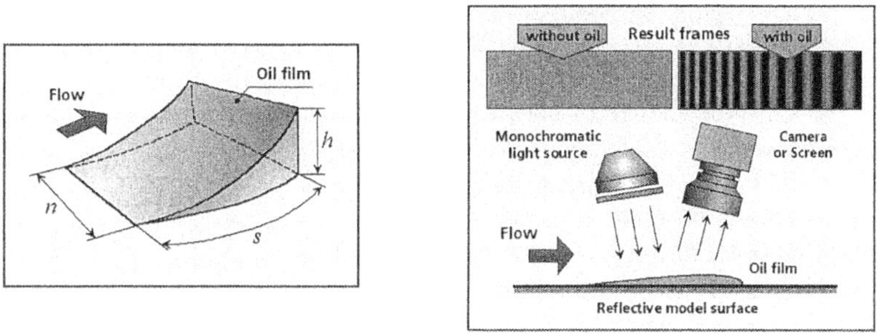

Figure 1 Principle of oil-film interferometry

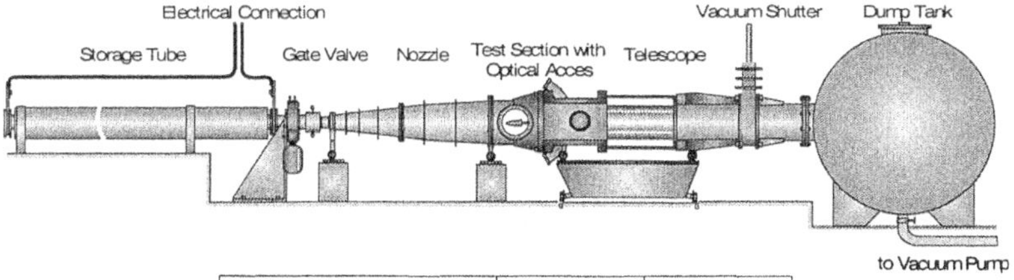

Performance	Tube A	Tube B
Max. Stag. Pressure, bar	15	40
Max. Stag. Temperature, °C	20	400
Mach Number	2.79 to 4.65	5.0 to 6.9
Unit Reynolds Number, $10^6 m^{-1}$	5 to 60	5 to 50
Run Time, s	0.4	0.3
Test Section	0.5 x 0.5 m²	ø 0.5 m

Figure 2 Sketch of the Ludwieg Tube Facility

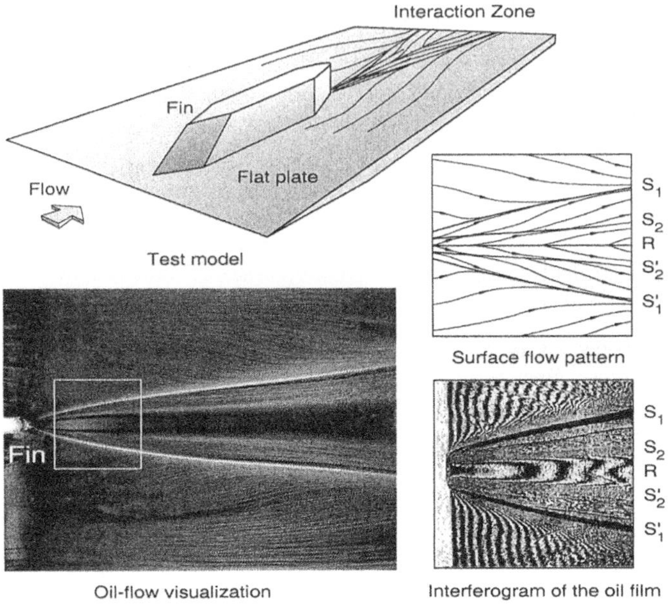

Figure 3 3D wake flow behind a vertical fin on a flat plate

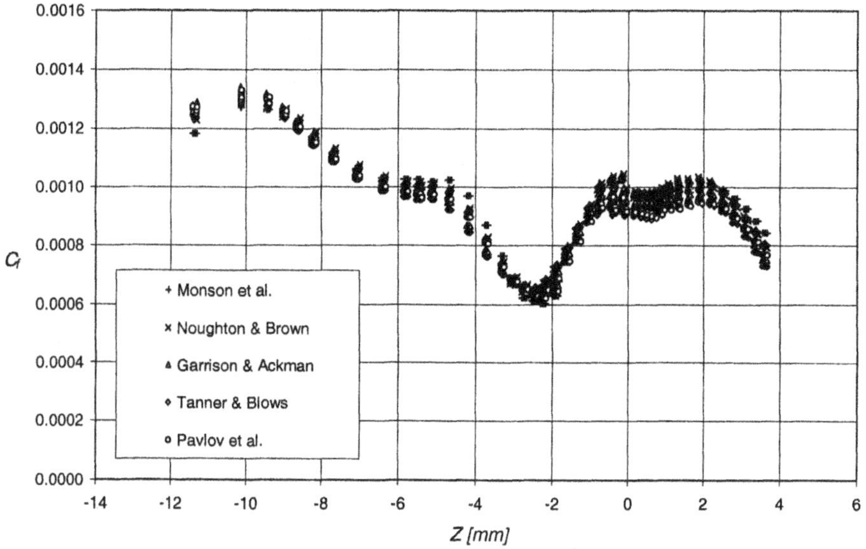

Figure 4 Skin friction distributions in a cross section of the fin wake, obtained by different GISF-methods. The X-distance from the trailing edge of the fin is about 55-58 mm.

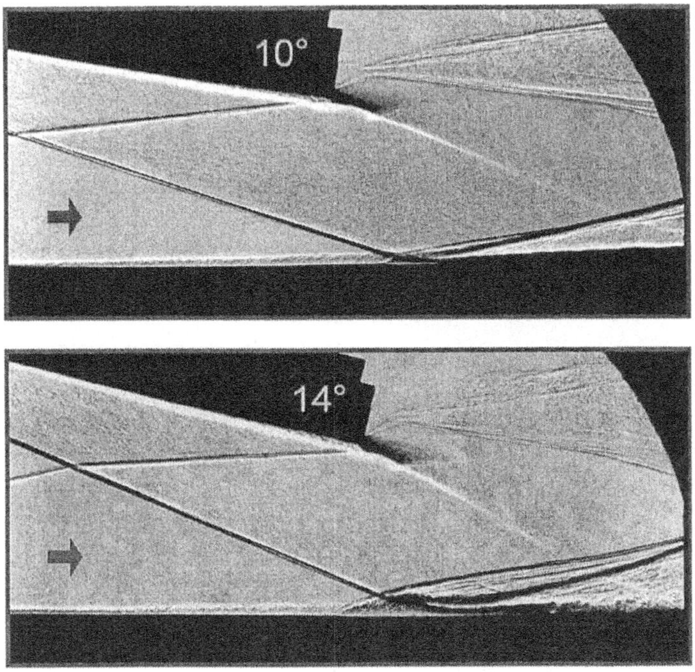

Figure 5 Shadowgraphs of 2D incident shock / turbulent boundary layer interaction flow at Mach 5 (10° and 14° indicate shock generator angle)

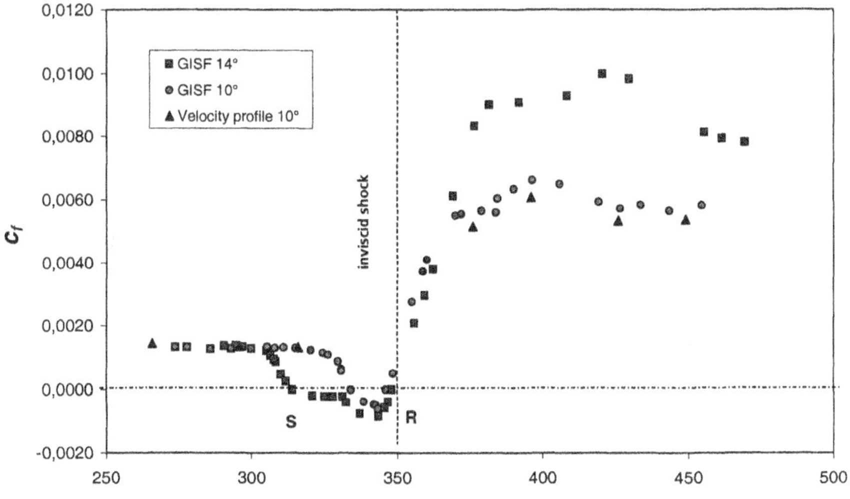

Figure 6 Skin friction distributions in 2D incident shock / turbulent boundary layer interaction flow at Mach 5

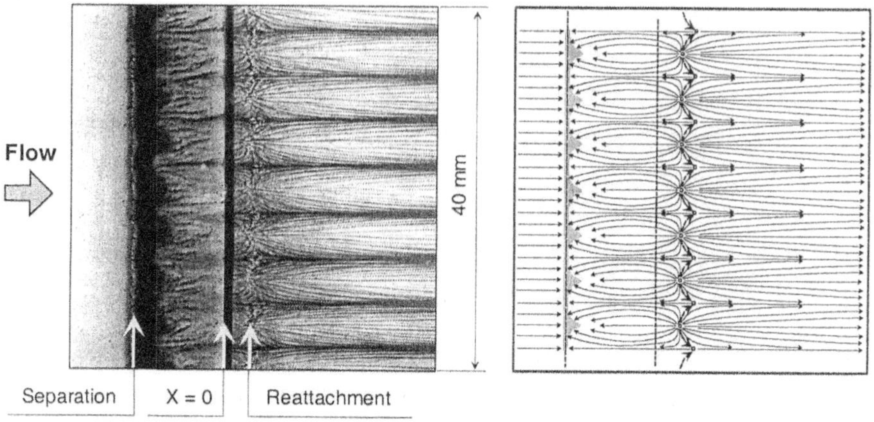

Figure 7 Oil-flow visualization and sketch of limiting streamlines of a 20° compression corner flow at Mach 3

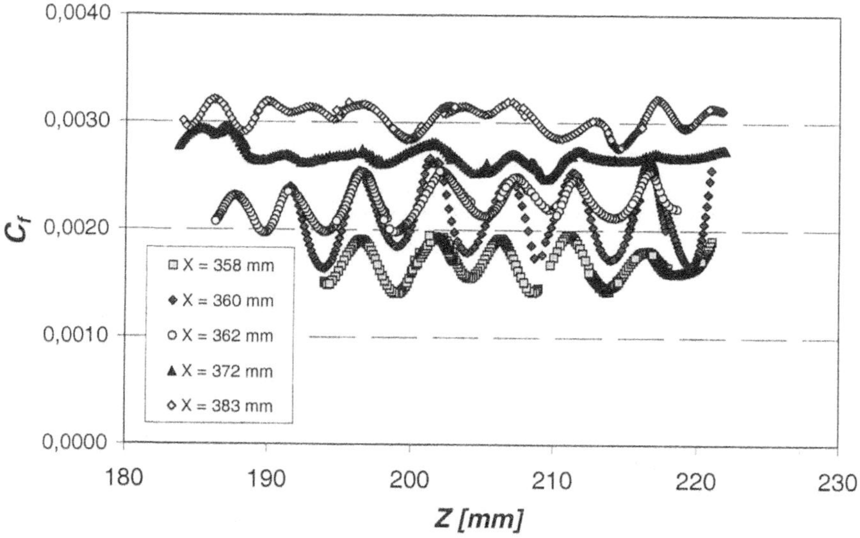

Figure 8 Skin friction distribution in cross sections of a 20° compression corner flow at Mach 3

Experimental and Semi-Empirical Study of Supersonic Separated Ramp Flow

J. Müller, R. Mümmler, W. Staudacher
Universität der Bundeswehr München, Institut für Luftfahrttechnik,
Trisonischer Windkanal München – TWM
Werner-Heisenberg-Weg 39, D 85577 Neubiberg, Germany

Summary

This paper reports about a method to predict the position of separation in a supersonic flat plate – 24° ramp flow. The investigation is restricted to a Mach number range of 2.2 to 2.9 and unit Reynolds numbers between $29\cdot10^6\ m^{-1}$ and $58\cdot10^6\ m^{-1}$. An average boundary layer thickness of 13 mm is representative. Separation location was determined by laser Doppler anemometer and compared with pressure measurements. The entire data set, taken at the Trisonic Windtunnel Munich (TWM), was analyzed to derive a mathematical description of the physical dependencies, based on empirical parameters.

1 Introduction

To determine the line of separation and reattachment in supersonic ramp flows, different techniques were applied during the last fourty years. An easy and cheap method is the oil flow visualization technique [2]. The qualitative results can be quite accurate, but an influence of accumulated oil cannot be avoided.
A classic way to reveal separated flow is to find the three points of inflection in the pressure distribution. Yet being a reliable indicator, there is no physical correlation between the local pressure and the location of separation. This gap of information can be closed, using an additional interference free method like LDA. As reported in [2], the locations of separation and reattachment are characterized by a change of sign of the local shear stress and the appropriate flow reversal in immediate proximity to the surface can be detected by LDA. In conjunction with the pressure distributions a discussion follows how to predict separation location as a function of the characteristic flow parameters. The mathematical approach closely follows a study published by Settles [4] and Green [1] to predict the length scale of upstream influence.

2 Test setup

The *T*risonic *W*indtunnel *M*unich (TWM) is an intermittent blow down type wind tunnel. Test section dimensions are 300 x 675 mm^2. Typical test conditions are for stagnation temperature and model temperature 293 K and turbulence intensity $\approx$ 1% (measured by LDA). Due to the short run times (15 sec – 80 sec) adiabatic model conditions are given.

The model is a flat plate-ramp configuration (Figure 1). The flat plate length is 1265 mm, provided with a sharp leading edge. A 24 degree ramp with a length of 90 mm is attached downstream. The span of the whole configuration is 300 mm, identical to the width of the wind tunnel. Side wall boundary layer thickness is 60 mm.
The transition to turbulent boundary layer was calculated to occur about 15 mm downstream of the leading edge. Just ahead of the ramp an average boundary layer thickness of 13 mm can be assumed.

3 Measurement techniques

3.1 Laser-Doppler-anemometer (LDA)

The 2d-LDA at TWM consists of a 6 Watt Argon-Ion laser (Coherent), a high power transmitter box, fiber optic cables and transmitter and receiver optics (DANTEC). It is operated in forward scatter mode. The tracer particles, diameter $\approx 0.8\ \mu m$, are injected by a Pallas AGF 5.0 liquid aerosol generator.

3.2 Pressure measurement technique

Along the centerline of the model 63 taps for static pressure-measurements are aligned. The longitudinal spacing is 3 mm. The origin of the coordinate system ($x = 0$, $z = 0$) is the plate-ramp junction. The pressures are digitized by two ESP-48 electronic pressure scanners.

4 Experimental results

To find separation and reattachment by LDA, the flow reversal locations on the model surface must be detected. Hence the measurements must be taken immediately adjacent to the surface [2].
A summary of the LDA-results is given in Figure 2. It shows separation points for a Mach number variation between M = 2.2 and 2.9. A constant Reynolds number of $Re = 41 \cdot 10^6 m^{-1}$ was selected for all Mach numbers. Additionally the results for upper/lower wind tunnel operating Reynolds number are displayed. The picture confirms the well known shrinking of the separation bubble with increasing Mach and Reynolds number [4].
Plotting the static pressure versus x-coordinate the characteristic three points of inflection curves are found, e.g. Figure 3. Introducing the coordinates of separation due to LDA the corresponding pressure levels can be emphasized. The same procedure is applied for the pressure gradient plots. Obviously separation occurs downstream of the maximum (see Figure 3).
Keeping the unit Reynolds number constant, higher Mach numbers result in lower separation pressures, Figure 4. The pressure gradient reveals quite a constant value.

Figure 5 shows the location of separation pressure and pressure gradient. For increasing total pressure and constant Mach number the separation pressure increases significantly with a slight downstream shift of separation location.
In contradiction, increasing Mach number with constant Reynolds number (different from constant total pressure) results in a pronounced downstream shift of separation.
To judge the influence of Mach and Reynolds number on separation pressure, pressure gradient and location, we must recall that the Reynolds number is adjusted by appropriate choice of the total pressure for a given Mach number, e.g. for a constant Mach number of 2.5 an increase of Re from $29 \cdot 10^6 m^{-1}$ to $58 \cdot 10^6 m^{-1}$ is achieved by increasing the total pressure from 2.87 to 5 bar, see Figure 6. Now regarding the Mach number variation for Re = $41 \cdot 10^6 m^{-1}$ the total pressure is 3.5 bar for M = 2.2 and 5 bar for M = 2.9. Obviously, the Mach number variation results in a total pressure difference of 1.5 bar, compared to 2.13 bar in case of Reynolds number variation.
The boundary layer thickness, given in the following chapters, was determined by LDA at a reference coordinate of x = -80 mm. Taking

$$\delta = 0.3839\left(\frac{\nu}{u}\right)^{1/5}(1265mm + x)^{4/5}$$, ν: kinematic viscosity, u: free stream velocity,

a good agreement with the experimental results could be achieved. Hence this equation was selected for the calculation of the boundary layer thickness at the position of the first pressure rise due to interaction.

4.1 Results of test parameter variation

Increasing total pressure, constant Mach number results in:

a) Increase of static pressure
b) Decrease of boundary layer thickness, δ = 13.24 mm (M = 2.2, p_{tot} = 3.5 bar), δ = 12.36 mm (M = 2.2, p_{tot} = 5 bar)
c) An increase of 1.5 bar (at M = 2.2) increases the Reynolds number from $41 \cdot 10^6 m^{-1}$ to $58 \cdot 10^6 m^{-1}$, shifting the location of separation 2.8 mm downstream (from –40.3 to –37.5 mm)
d) The static pressure gradient for separation (same parameters as for c)) increases from 5 to 7.69 mbar/mm

Increasing total pressure, constant Reynolds number

This constellation can be achieved only by increasing the Mach number. The effects are as follows:

a) Decrease of static pressure
b) Increase of boundary layer thickness (for Re = $41 \cdot 10^6 m^{-1}$, p_{tot} = 3.5 bar and M = 2.2 the boundary layer thickness is 13.24 mm. At M = 2.9, p_{tot} = 5 bar, it comes to 13.48 mm)
c) For Re = $41 \cdot 10^6 m^{-1}$, M = 2.2 an increase of 1.5 bar total pressure increases the Mach number to 2.9 and the separation location moves downstream from –40.3 to –18.8 mm, Δx = 21.5 mm

d) The static pressure gradient for separation (same parameters as for c)) decreases from 5 to 3.7 mbar/mm

Constant total pressure, increasing Mach number (= decreasing Reynolds number)

For constant total pressure, only a distinct Mach – Reynolds number combination is possible. As pointed out, an increase of Mach number is directly linked with a decreasing Reynolds number and a shrinking separation bubble, see Figure 5. Hence a constant total pressure and increasing Mach number results in:

a) Decrease of static pressure
b) Increase of boundary layer thickness
c) Upstream shift of separation location
d) Static pressure gradient decreases

4.2 Semi-empirical approach

As mentioned above, the location of upstream influence is assumed to be a function of initial Mach and Reynolds number, ramp angle, the overall pressure gradient and undisturbed boundary layer thickness. Further investigations [1] reveal that due to the dimensional consistency the length scale of the interaction L_m at fixed ramp angle α and Mach number M will be a function of boundary layer thickness δ and unit Reynolds number Re, both raised to fractional powers. According to this approach we find:

$$L_m = f(\mathrm{Re}, M, \alpha, \delta)$$

We now discuss the applicability of this approach for a prediction of separation location. δ represents the boundary layer thickness at the position of the first pressure rise due to the interaction. The Reynolds number is normalized to 1 meter to avoid the dependency of δ. For constant Mach number and ramp angle a combination of x_{sep}, Re and δ must now satisfy the dimensional analysis, leading to:

$$x_{sep} = \delta^{a+1}\,\mathrm{Re}^{a}\,C\,,$$

C being a constant value. Experimental results were presented in chapter 4.1 and show an opposite tendency of boundary layer thickness and Reynolds number. If we want to introduce the experimental data, we must use the equation as follows:

$$|x_{sep1}| - |x_{sep2}| = \left[\delta_1^{a+1}\,\mathrm{Re}_1^{a} - \delta_2^{a+1}\,\mathrm{Re}_2^{a}\right]\cdot C$$

The constant can be substituted by:

$$C = \frac{|x_{sep1}|}{\delta_1^{a+1}\,\mathrm{Re}_1^{a}}$$

Taking x_{sep1} = -40.26 mm, x_{sep2} = -37.5 mm, δ_1 = 13.24 mm; δ_2 = 12.36 mm; $\mathrm{Re}_1 = 41\cdot10^6\mathrm{m}^{-1}$; $\mathrm{Re}_2 = 58\cdot10^6\mathrm{m}^{-1}$, we obtain:

$$a = \log_{\left[\frac{\delta_2}{\delta_1}\frac{\mathrm{Re}_2}{\mathrm{Re}_1}\right]}\left(\frac{|x_{sep2}|}{|x_{sep1}|}\frac{\delta_1}{\delta_2}\right) = -0.00785$$

We now extend the discussion on the Mach number variation with constant Reynolds number. Obviously the separation length decreases when increasing the Mach number. This is expressed by the following approach:

$$x_{sep} = \delta^{a+1} \mathrm{Re}^a M^b C_M$$

According to the experiment, x_{sep3} = -40.26 mm, x_{sep4} = -18.8 mm, δ_3 = 13.24 mm; δ_4 = 13.48 mm; $\mathrm{Re}_3 = 41 \cdot 10^6 \mathrm{m}^{-1}$; M_3 = 2.2; M_4 = 2.9 and substituting C like above leads to

$$b = -2.821 \text{ and } C_M = 31.192$$

$$\left| x_{sep} \right| = 31.192 \cdot \delta^{0.99125} \mathrm{Re}^{-0.00785} M^{-2.821}$$

Referencing the equation to the boundary layer thickness gives

$$\left| \frac{x_{sep}}{\delta} \right| = C_M \cdot \mathrm{Re}_\delta^a \cdot M^b = 31.192 \cdot \mathrm{Re}_\delta^{-0.00785} \cdot M^{-2.821}$$

5 Conclusions

Due to the required formulation, Figure 2 transforms to Figure 7. Obviously the downstream shift of separation with increasing Re has vanished. Moreover, between M = 2.2 and 2.6 x_{sep}/δ is quite unaffected by the Reynolds number. This is reflected by the power a = -0.00785 which effectively damps the Reynolds number influence (see Figure 7).

The power b = -2.821 sufficiently reproduces the experimentally found Mach number dependency. The most significant difference between calculated and experimental results is found for M = 2.7, 2.8 and 2.9, showing an increasing x_{sep}/δ with decreasing Reynolds number. Due to the numeric approach this effect is not reproduced.

A closer look at the test conditions shows that an increasing Mach number with decreasing Reynolds number leads to decreasing static pressures, finally resulting in wind tunnel blockage. The flow field is now characterized by a detached normal shock wave and thus must carefully be differentiated from the undisturbed flow case. However, immediately before blockage an increasing upstream shift of separation can occur, superimposing the pure separation effects.

It must be pointed out, that the equations are derived using experimental data composed of pressure and LDA measurements. To introduce local flow conditions the most downstream boundary layer thickness, still unaffected by separation is taken. The problem associated is to determine the upstream influence location, the accuracy of which depends on the spacing of the pressure ports. A method proposed by Settles [3] takes the intersection of the tangent on the first inflection point of the first inflection point of the pressure curve with the undisturbed pressure level. Due to the pure mathematical definition there remains inaccuracy. A better technique is to calculate the pressure gradient to identify the coordinate where a certain level is exceeded. Of course this method can produce

offsets, caused e.g. by improper pressure level criteria or inadequate spacing of the pressure ports.
As the calculation is performed using experimental data, the equations must be understood as a good tool to predict separation in the Mach and Reynolds number range of this study. The applicability beyond this range could not be verified.
All equations and results are based on a constant ramp angle of 24 degree. To introduce the ramp angle effect an experimental approach like $e^{C\alpha}$ could work. This was proposed by Law, Rosko and Thomke and Settles to find the position of upstream influence.

Acknowledgements

The financial support of the Deutsche Forschungsgemeinschaft (DFG) in the frame of collaborative research center SFB 255, "Transatmosphärische Flugsysteme, part A5: Shock/Boundary-layer interactions" is gratefully acknowledged.

References

[1] J.E. Green. "Interactions Between Shock Waves and Boundary Layers". Progress in Aerospace Sciences, Vol. 11, 1970, Pergamon Press, pp. 235-340.

[2] J. Müller, R. Mümmler and W. Staudacher. "Comparison of some measurement techniques for shock-induced boundary layer separation". In Aerospace Science and Technology 5, 2001, pp. 383-395.

[3] G.S. Settles. "An Experimental Study of Compressible Turbulent Boundary Layer Separation an High Reynolds Numbers". PhD. Thesis, Dept. of Aerospace and Mechanical Sciences, Princeton University, 1975.

[4] G.S. Settles, J.J. Perkins, S.M. Bogdonoff. "Upstream Influence Scaling of 2D & 3D Shock/Turbulent Boundary Layer Interactions at Compression Corners". AIAA 19th Aerospace Sciences Meeting, 1981.

Figures

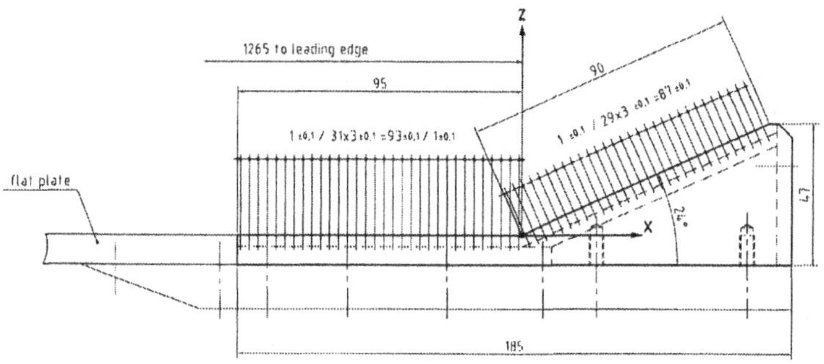

Figure 1 Test model, dimensions in mm

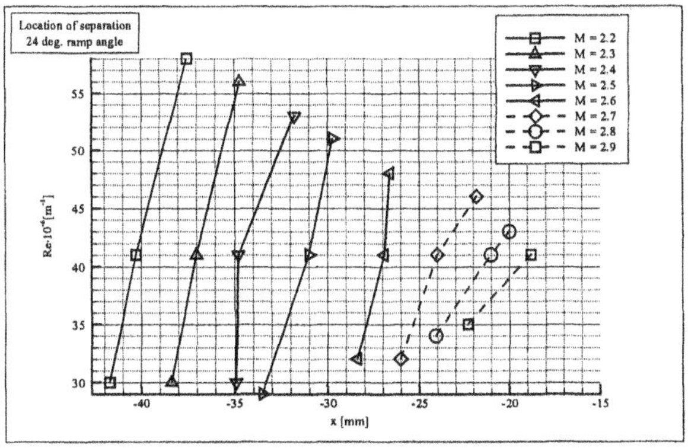

Figure 2 Location of separation measured by LDA

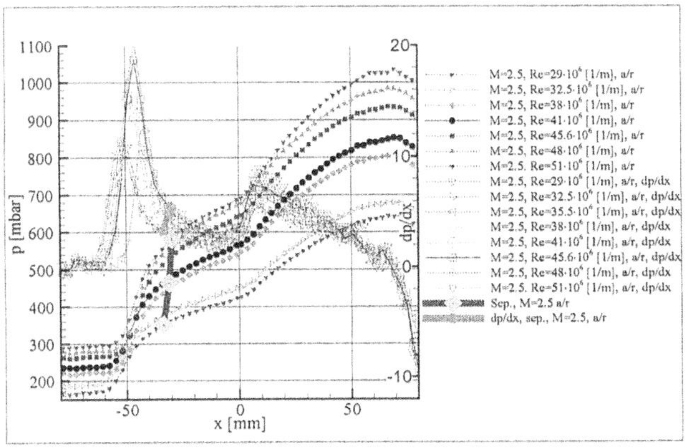

Figure 3 M = 2.5, static pressure, pressure gradient, sep. by LDA

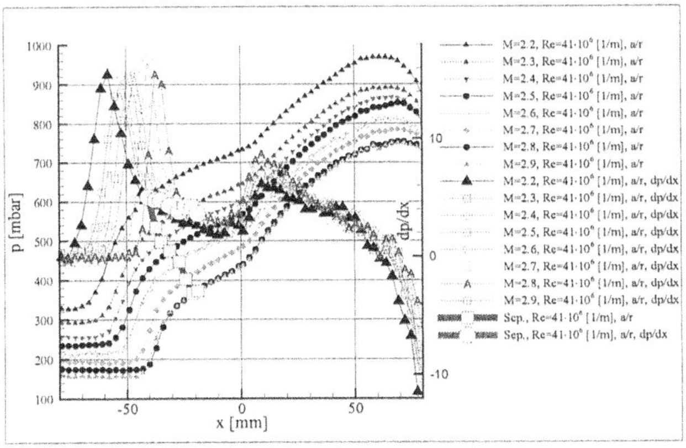

Figure 4 Re = $41 \cdot 10^6 m^{-1}$, static pressure, pressure gradient, sep. by LDA

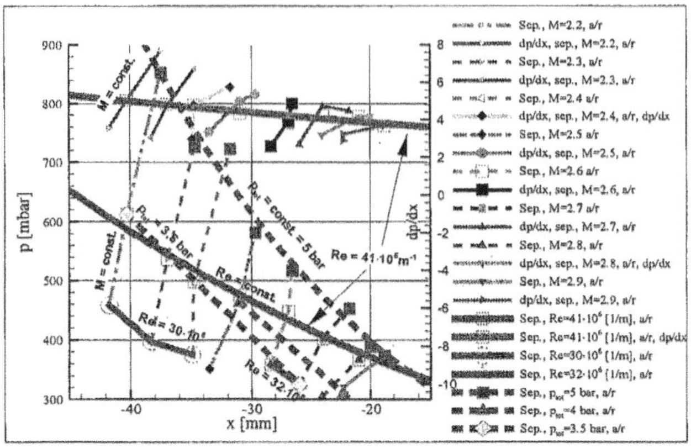

Figure 5 Characteristics of separation

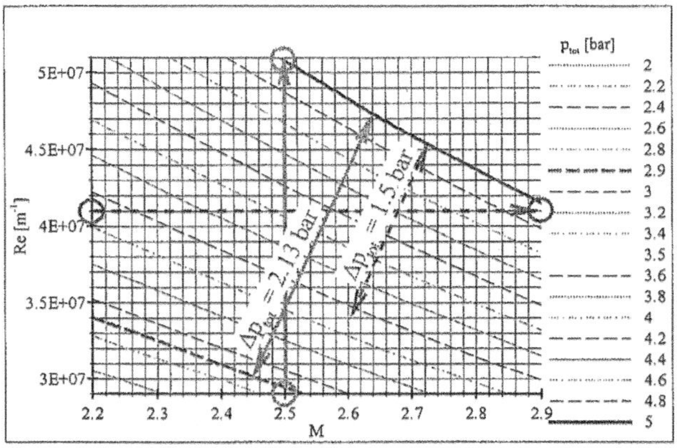

Figure 6 Total pressure influence on Mach and Reynolds number

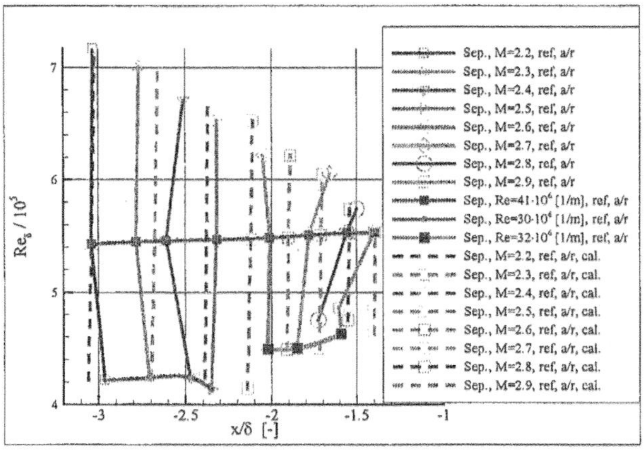

Figure 7 Characteristics of separation, referenced on boundary layer thickness at upstream interaction location

In-Flight and Wind Tunnel Investigations of Natural Transition on a Laminar Flow Airfoil

I. Peltzer and W. Nitsche

Institut für Luft- und Raumfahrt, Technische Universität Berlin, Marchstr.12, 10587 Berlin, Germany
Inken.Peltzer@tu-berlin.de

Summary

The temporal and spatial development of natural Tollmien-Schlichting (TS) instabilities in the linear and weakly nonlinear stage were investigated in extensive flight as well as in wind tunnel experiments on a laminar airfoil. Two-dimensional TS waves dominate in the early linear stage of the boundary layer flow, but small three-dimensional disturbances are already existent. These individual and spatial independent wave packets are amplified later than the two-dimensional waves and they dominate the late stage of the TS development. The experiments were carried out on a laminar profile performed as a glove for a sailplane, employing different surface sensor arrays (surface hot-wire sensors, piezo sensors, microphones).

1 Introduction

The experimental investigation described in this paper deals with the laminar turbulent boundary layer transition on an airfoil. There are three major aspects of the transition process: receptivity, linear stability and nonlinear breakdown [2]. The first aspect (receptivity) is devoted to the generation of instability waves (Tollmien-Schlichting (TS) waves) in the region from the airfoil's leading edge to the chord position where instability waves are amplified. The process of generating the TS waves depends strongly on the respective flow conditions. Therefore, it is necessary to investigate the transition under real conditions in flight experiments. The second stage of transition (linear stability) corresponds to the development and amplification of the instability waves. Experimental investigations on the amplification and the spatial propagation of these instability waves require spatial and temporal measurements with high-resolution surface sensor arrays.
Based on experiences gained in the Deutsche Forschungsgemeinschaft -funded university group research project (see e.g. [3]) this paper deals with in-flight and wind tunnel measurements of the temporal and spatial propagation of TS instabilities on a laminar wing glove. Arrays of different types of surface sensors (piezo sensors, surface hot-wires and microphones) were employed. The subsequent explanations focus on measurements to investigate the spanwise distribution of natural TS waves at their early linear as well as their weakly nonlinear stage.

2 Experimental Set-up

For the in-flight measurements a laminar wing glove was used, which was developed and certified for the Grob G103 TWIN II two-seater sailplane at the Berlin Institute for Aeronautics and Astronautics. Figure 1 shows the sailplane equipped with the measuring glove. The measuring glove has a two-dimensional (2D) central part of 1.0 m span and a chord length of 1.22 m. One special feature of the measuring glove employed is its exchangeable wing segment. Depending on the experiment conducted, this segment can be equipped with surface sensors or actuators without any changes in the supporting structure of the measuring glove. Furthermore the segment can be employed both, during flight and on the experimental wing in the wind tunnel.
The measurement system is located in the equipment support underneath the glove. A Prandtl tube, which is also attached underneath the glove, is used to obtain the freestream velocity. The pressure differences between two pressure taps on the glove ($\Delta p_{x1,x2} : (x_1/c = 37.2\%, x_2/c = 52\%)$ and $\Delta p_{x3,x4} : (x_3/c = 2.8\%, x_4/c = 8.5\%)$) are measured to monitor the angle of attack of the glove. A thermocouple and a manometer are used to measure the air temperature and absolute pressure, respectively. These parameters are continuously recorded to obtain the freestream boundary conditions. The flight tests were carried out at flow velocities of 23 m/s to 28 m/s ($Re_c \approx 2 \cdot 10^6 - 3 \cdot 10^6$).
The wind tunnel experiments were conducted in the laminar wind tunnel at Stuttgart University. This wind tunnel has a very low turbulence level ($Tu = 1.2 \cdot 10^{-4} - 5 \cdot 10^{-4}$) and is consequently suitable in particular for investigations on boundary layer transition.

2.1 Measurement System

A miniaturised and modularly arranged measuring system especially developed for in-flight measurements is employed in this case for in-flight and wind tunnel experiments. The multi-channel measurement data acquisition system (MEDAS) was described in Suttan [6]. One part of the measurement system is situated underneath the glove and contains the wing-side part of MEDAS as well as the air data acquisition and other electronic devices (e.g. anemometer bridges, amplifiers). The measuring computer in the cockpit includes the board part of MEDAS and makes it possible to control the measurement system at the wing during flight. Communication and data transfer for storage between the cockpit and the wing glove are carried out via fiber optical transmission.

2.2 Sensors and Sensor Positions

Three wing segments equipped with different sensor arrays at different chord positions were employed. On the first segment 16 surface hot-wire sensors are arranged spanwise at a chord length of 50 % (set-up A, not shown). This highly sensitive wall sensor was developed for measuring weakest wall shear stress fluctuations. A hot wire (5μm) is welded flush to the surface above a thin slot (0.075 mm to 0.1 mm) of a printed circuit board (copper-laminated). This arrangement produces a reduced heat-flux into the structure. Therefore, a high signal-to-noise ratio is achieved [4]. Modularly arranged hot-wire bridges were designed especially for in-flight measurements to be able to operate a high number of surface hot-wire sensors during flight in constant-temperature mode [1].

The second sensor set-up (set-up B, Fig. 2) employs a spanwise row consisting of 16 surface hot-wire sensors at 33 % of the chord length. The sensors were employed for measuring minor disturbances in the early linear amplification stage of TS waves. Furthermore a 74 piezo sensor array with four spanwise rows of 16 sensors each and additional sensors arranged in streamwise direction are used. The first spanwise piezo sensor row is located at 46.3 % and the fourth row at 49.6 % of the chord length. There are 14 mm between each row and 7 mm between the spanwise sensors. The piezo sensor array is made from a 28 μm thin polarised plastic foil (Polyvinylidenfluoride), coated with a metal layer on both sides. The sensor material produces small electric charges caused by pressure and temperature fluctuations. Due to the closed foil on the wing surface, the sensors produce negligible roughness. The signal-to-noise ratio can be increased using the pyro effect [5]. The small temperature gradient between the glove and the flow needed is generated by a heating layer underneath the sensors.

The third segment is equipped with very small wall-microphones (set-up C). 16 microphones are installed in three spanwise rows downstream, so that their positions and the distance between the sensors are equivalent to the segment with the piezo sensors. Figure 3 shows the principle sketch of the construction. The microphones are mounted underneath the surface with a 0.3 mm drilled hole. They have a sensitivity of -63 dB, a frequency range of 20 Hz to 12000 Hz and are 6.3 mm in diameter and 4.3 mm high.

3 Measurements of Natural Tollmien-Schlichting Waves

3.1 Measurements with a Surface Hot-Wire Sensor Array

Results of wind tunnel measurements of set-up A are to be seen in Figure 4. In Figure 4a) time traces of the spanwise sensor array at $x/c = 50\%$ are shown. Typical TS wave packets can be identified in the signals. The contour plot (Fig. 4b) shows 15 ms of these signals. Spanwisely coherent wave packets with a span of 20 mm to 30 mm can be observed. Obviously, wave packets occur individually, that means the three-dimensional (3D) portions dominate the local TS structure, which illustrates the later amplification stage.

The spanwise sensor arrays allow a spatial time-resolved data acquisition. The data can be transformed into the frequency and spanwise wavenumber domain, using the Fourier analysis [1]. The wavenumber represents the inverse spanwise wavelength and is normalized by the lateral sensor spacing ($2{\cdot}\Delta z = 14\,mm$). A wavenumber of zero reflects the 2D part of TS waves and a wavenumber of one corresponds to a spanwise Nyquist wavelength of 14 mm.

Figure 4c) shows the frequency wavenumber spectrum of the time traces. The maximum of the amplitude values is in the range of the TS frequency of the glove (300 Hz to 900 Hz) close to wavenumber zero, which illustrates that the instabilities are threedimensional with small oblique angles. Higher wavenumbers and the higher harmonic frequencies also occur in the spectrum, indicating the beginning nonlinear stage of transition.

Figure 5 shows the TS instabilities in a very early amplification stage measured in flight experiments with the surface hot-wire sensors at 33 % chord length (set-up B). A representative part of time traces, the contour plot and the frequency wavenumber spectrum is displayed again. The range of the amplitudes is ten times lower, compared to the signals of the case described before. Also, there is no significant phase shift between adjacent sensors at the measured span position. This clearly illustrates the 2D propagation of the early stage of TS wave excitation. However, a view to the frequency wavenumber spectrum shows amplitudes at higher wavenumbers. That means small 3D disturbances are already existent.

3.2 Measurements with a Piezo Sensor Array

Figure 6 shows typical time traces of the piezo sensors in the three rows, positioned at chord lengths of 46 %, 47 % and 48 %, respectively (wind tunnel measurements). For the signal analysis (using FFT and inverse FFT) the time trace data were filtered to emphasize the relevant signal components (200 Hz to 1800 Hz). An excitation of the wave trains from one row to the next can be observed. The contour plot (Fig. 7 of these signals clearly indicates the phase of the spanwise adjacent sensors. The first row shows a 2D spanwise distribution of instabilities (in phase) but the third row shows oblique wave trains and individual spatial independent wave packets. The angle of such wave trains rises with increasing chord length.

3.3 Measurements with a Microphone Array

Figure 8 displays time traces of the microphones of in-flight measurements, which were installed at the same position as the piezo sensors. With this technique an excitation of individual wave packets can also be observed clearly. The area of the contour plot marked with A illustrates a slightly oblique wave train and the temporal staggering of this structure from one row to another. The development of a wave packet from a 2D wave formed to an individual 3D wave packet can be seen in the area marked B.

Figure 9 gives the power spectra for one respective sensor of a particular row at different flight velocities. The first chart reflects the spectra of a sensor of the first row at velocities of 23.6 m/s, 25.0 m/s and 27.8 m/s, respectively. The angle of attack of the wing decreases with increasing flight velocities. Consequently, the sensors are situated in an increasingly stable area of the boundary layer at increasing velocities. Basically, each spectrum clearly shows an increase within the TS frequency band typical for the wing glove. At 25.0 m/s the spectrum illustrates a second hump,

which reflects the excitation of the higher harmonic frequencies. At the lower velocity of 23.6 m/s the spectrum of each row has risen so much, that the curve of the higher harmonic frequencies disappears. At higher velocities the higher harmonic frequencies are not yet excited. The spectra of the rear sensors downstream are on a higher level than the spectra of the first row. These spectra all point to the existence of TS instabilities in various stages of amplification.
Consequently, TS instabilities can also be recorded using microphones, which would significantly simplify the measuring technique for in-flight experiments and offers a convenient alternative. The microphone array described above combines some advantages of both piezo and surface hot-wire sensors. The passive measuring system facilitates a large number of sensors, taking up very little space, as is the case with piezo sensors. The signal-to-noise ratio in the case of the microphones is better than that for the piezo sensors. However, in contrast to the piezo sensors the microphones can produce additional roughness, caused by the drilled holes.

4 Conclusions

Temporal and spatial high-resolution measurements of Tollmien-Schlichting (TS) waves on a laminar wing glove during flight and in wind tunnel were successfully carried out to investigate the early and later stages of the development of TS waves. For the measurements different types of surface sensor arrays (surface hot-wires, piezo sensors and microphones) in dependence on their specific characteristics were employed. Results show that, in the early linear amplification stage, primarily 2D TS waves develop, but 3D disturbances are also observed. As the TS instabilities are in a higher amplification stage, the individual and spatial independent wave packets dominate the boundary layer flow. Furthermore, the TS instabilities measured in flight and in wind tunnel behave similarly. The measurement techniques used and their results obtained can be useful to explore and understand the spatial TS wave development in boundary layers on a laminar wing.

Acknowledgements

The financial support on this research project by Deutsche Forschungsgemeinschaft (DFG), reference Ni 282/12-2, is gratefully acknowledged. Special thanks go to the Institute of Aerodynamics and Gasdynamics, Stuttgart University, where the wind tunnel measurements were carried out.

References

[1] Baumann, M.: Aktive Dämpfung von Tollmien–Schlichting Wellen in einer Flügelgrenzschicht. In: Fortschritt–Berichte VDI, Reihe 7, VDI Verlag, Düsseldorf (1999) Dissertation an der Technischen Universität Berlin.

[2] Kachanov, Y.: Physical Mechanisms of Laminar-Boundary-Layer Transition. Annu. Rev. Fluid Mech. **26** (1994) 411–482

[3] Nitsche, W., Suttan, J., Becker, S., Erb, P., Kloker, M., Stemmer, C.: Experimental and Numerical Investigations of Controlled Transition in Low-Speed Free Flight. Aerosp. Sci. Technol. **5** (2001) 245–255
[4] Sturzebecher, D., Anders, S., Nitsche, W.: The Surface Hot Wire as a Means of Measuring Mean and Fluctuating Wall Shear Stresses. Exp. in Fluids **31** (2001) 294–301
[5] Sturzebecher, D., Nitsche, W.: Visualization of the Spatial-Temporal Instability Wave Development in a Laminar Boundary Layer by Means of a Heated PVDF Sensor Array. In Körner, H., Hilbig, R., eds.: New Results in Numerical and Experimental Fluid Mechanics. Volume 60 of NNFM., Vieweg Verlag, Braunschweig (1997) 335–342
[6] Suttan, J.: Entwicklung und Einsatz von Multisensor–Piezofolienarrays an Laminarflügeln zur zeitlich und flächig hochauflösenden Messung von dynamischen Wandkräften in der Transition. In: Fortschritt–Berichte VDI, Reihe 8, VDI Verlag, Düsseldorf (1999) Dissertation an der Technischen Universität Berlin.

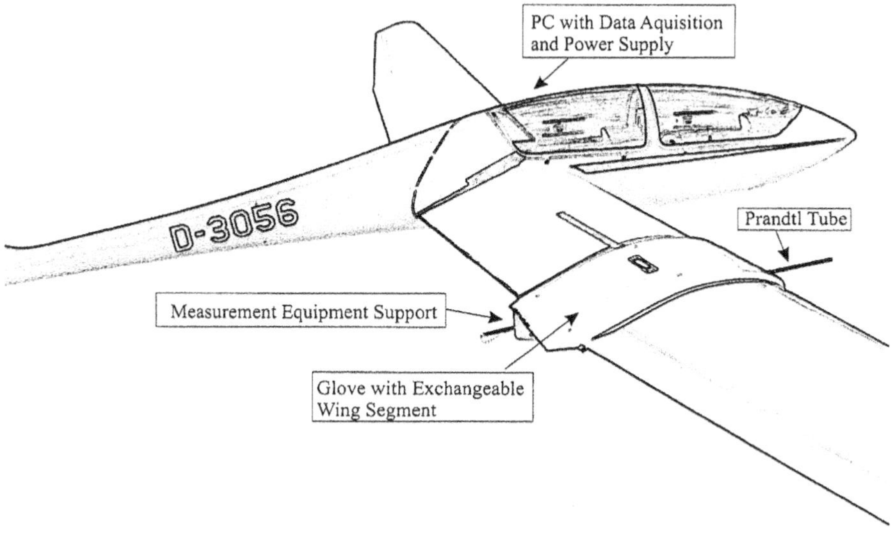

Figure 1 Sailplane with glove

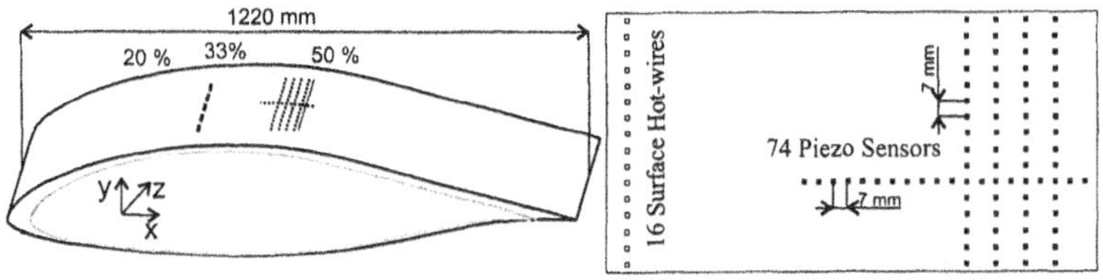

Figure 2 Sketch of set-up B

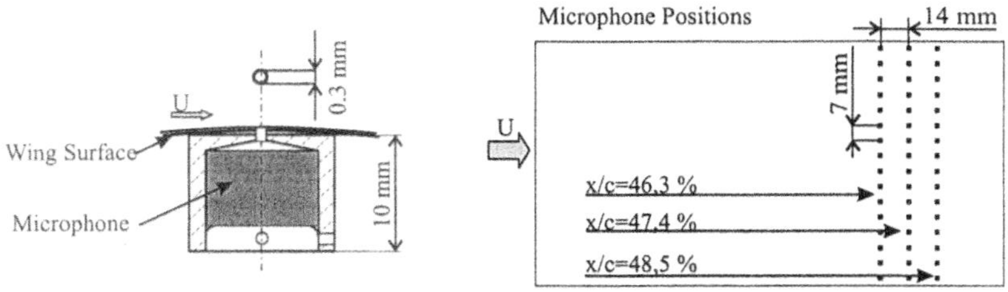

Figure 3 Sketch of the microphone set-up and microphone positions (set-up C)

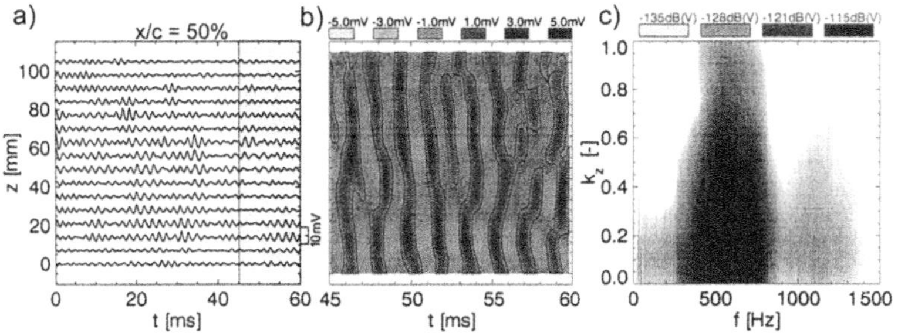

Figure 4 Time traces, contour plot and frequency wave number spectrum of spanwise hot wire sensors (wind tunnel, x/c=50 %, u=23.6 m/s)

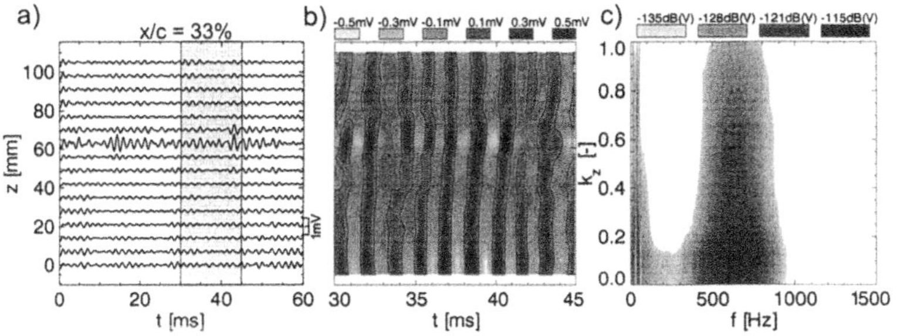

Figure 5 Time traces, contour plot and frequency wavenumber spectrum of spanwise hot wire sensors (in-flight, x/c=33 %, u=22.2 m/s)

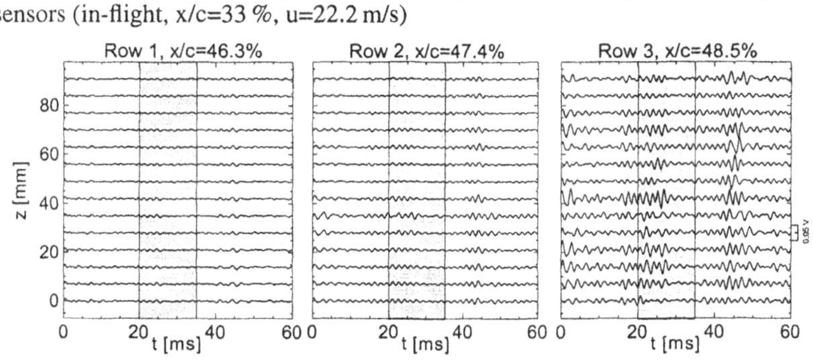

Figure 6 Time traces of piezo sensors (wind tunnel, u=23.6 m/s)

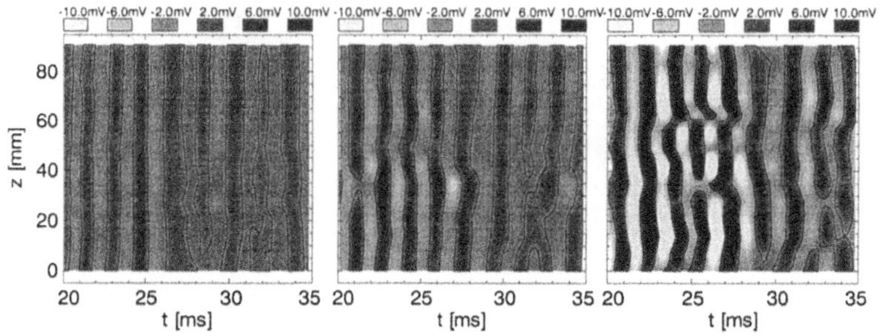

Figure 7 Contour plots of piezo sensors (wind tunnel, u=23.6 m/s)

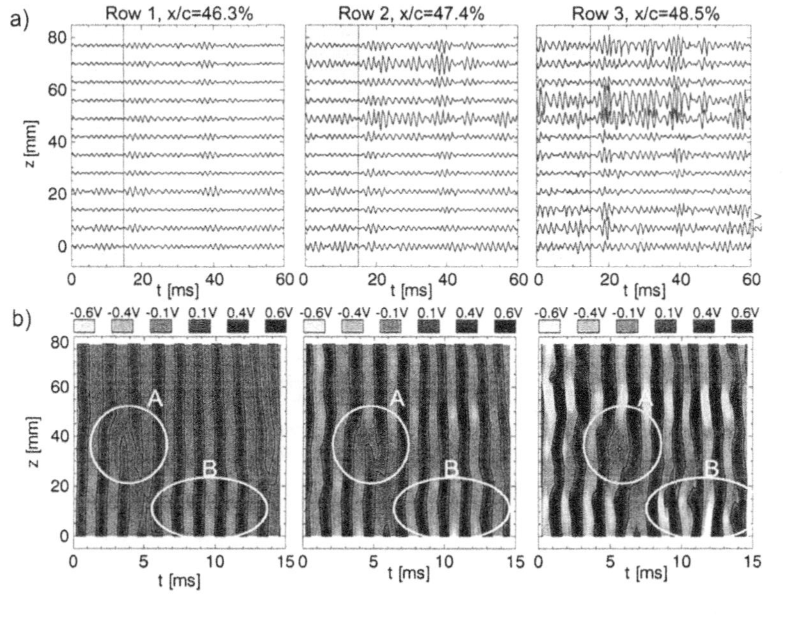

Figure 8 Time traces and contour plots of the microphones (in-flight, u=23.6 m/s)

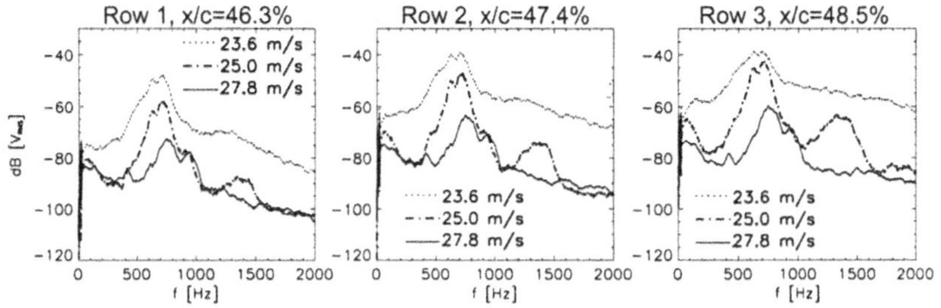

Figure 9 Spectra of a microphone per row at different velocities (in-flight)

Investigation of Wall and Support Interference Using Two TWIG Models of Different Scale

Hartmut Holst[1], Markus Jacobs[2], and Sieghard Tusche[2]

[1] German Aerospace Centre DLR, Bunsenstrasse 10, 37073 Göttingen, Germany,
hartmut.holst@dlr.de
markus.jacobs@dnw.aero, sieghard.tusche@dnw.aero,
WWW home page: http://www.dlr.de, http://www.dnw.aero
[2] German Dutch Wind Tunnels DNW, Bunsenstraße 10, 37073 Göttingen, Germany

Summary

The flow about a 3D model in a wind tunnel is influenced by the walls and by the model suspension. This has to be accounted for by corrections. The transonic tunnel of DNW Göttingen is equipped with a two-dimensional adaptive test section of 1×1 m^2, with flexible top and bottom walls, for 2D and for 3D measurements. For 3D models, the Wedemeyer/Lamarche method is taken for wall adjustment, and this adaptation method produces zero wall interference on the test section centreline. For off-centreline points, residuals can be determined by a wall pressure method, and averages can be taken for corrections. Two models of different scale were used together with 3 different types of support. Nominally, interference-free results were available from NLR-HST. Comparison of present results with the reference data shows the quality of the results, and also the limits of model size. The influence of the different supports was identified.

1 Introduction

The flow in a test section differs from free flight, and it is necessary to determine the influence of the walls - not present in free flight - to correct the test data. For straight walls, the correction of force and flow field parameters can be done according to AGARDographs 109 [3] and 336 [2]. The transonic facility of DNW-TWG at Göttingen is equipped with a twodimensional adaptive test section. By application of the Wedemeyer/Lamarche method - also known as VKI method - [7], [8], wall shapes can be found, for which the wall interferences all along the centreline of the test section are zero. This wall adaptation method is very fast, and it converges in a single iterative step [5]. The spanwise increase of wall interference is not removed, but the wall interferences can be computed by a wall pressure method everywhere in the test section, [1], [4]. Using appropriate spanwise averages of Mach number correction ΔM and angle of attack correction $\Delta\alpha$, the correction of force coefficients and flow variables can be performed. For the application of the wall adaptation procedure, only top and bottom centreline pressures are needed, and of course the wall shapes of the flexible top and bottom walls. For the computation of the residual wall

interferences however, the full wall pressure distribution is necessary, and therefore each of the four walls is equipped with three lines of pressure orifices.
By the application of the wall pressure method, the success of the adaptation procedure can be checked. This is of major importance for the transonic facility of TWG at DNW in Göttingen, because the relative model size can be increased in a wind tunnel with adaptive walls without significant loss in the quality of results. Therefore, the test on the two TWIG models - **T**ransonic **W**all **I**nterference **G**enerator - have been performed, to investigate the wall interferences. In addition, it was desired to determine the influence of the rear support.

List of Symbols

α	angle of attack
c	chord
c_D	drag coefficient
c_L	lift coefficient
c_m	pitching moment coefficient
c_p	pressure coefficient
$\Delta\alpha$	angle of attack correction
ΔM	Mach number correction
M	Mach number
u	induced velocity
U_∞	undisturbed flow velocity
v_n	velocity normal to the wall
x,y,z	cartesian coordinates

Subscripts

c/4	quarter chord
D	drag
L	lift
m	pitching moment
min	minimum
support	...referring to support of model

2 Tunnel and test section

The transonic wind tunnel of DNW Göttingen is a continuously working tunnel. It is equipped with 3 exchangeable test sections for supersonic (Laval nozzle), transonic (perforated) and subsonic measurements; the latter is the twodimensional adaptive test section for threedimensional tests, top and bottom walls are flexible. The 4- (or 8-) stage axial compressor has an electrical power of 12 MW, and for transonic measurements in the perforated test section an auxiliary suction plant is available. For flexible handling of the different components, test sections, model support and nozzles are equipped with an air cushion transport system.
The adaptive test section is shown in Fig. 1. Together with the nozzle it has been shifted upstream, to allow the model installation. The test section is 4.51 m long, and the cross section is 1 × 1 m^2. The wall shapes of top and bottom walls can be adjusted by 22 and 20 jacks respectively. The top wall centreline is equipped with 25 pressure orifices, the bottom wall has 23 pressure orifices. In total there are 207 pressure orifices. There are two different model positions: For twodimensional measurements in the middle (left window in fig.1, closed), and for threedimensional measurements at x = 3.28 m. The reason for this is that the 3D model has to be mounted by a rear sting, and the longer the rear sting, the smaller is the achievable angle of attack.

3 Models and supports

The models used here are TWIG/NL300 - see Fig. 2 - and TWIG/D750. The first model has a spanwidth of 0.3 m, the latter of 0.75 m. Except the scale, the models are identical. The wing has sharp edges, and TWIG is a low wing configuration. The reference data for comparison have been produced in the high speed tunnel HST of NLR Amsterdam. The geometrical relations of the two models in the two different tunnels are shown in Fig. 3. Force measurements with the larger model TWIG/D750 have been performed only in the smaller tunnel TWG, and the small model TWIG/NL300 has been used in both tunnels.
Three different types of rear support have been used for the measurements with the small model as it is shown in Fig. 4. The support configurations no. 1 and no. 2 are using the NLR device, equipped with a joint for adjusting the angle of attack, the horizontal boom is 104 mm in diameter. For conf. 1 the joint stays straight, the angle of attack is changed by the circular arc support of TWG, which can be seen in Fig. 5, and the model stays on the tunnel centreline. Using conf. 2, the angle of attack is adjusted by the joint of the NLR device, and the TWG circular arc support is only used to keep the model on the tunnel centreline, by changing its vertical position. The small rear sting normally used in TWG, shown as support configuration no. 3, has a diameter of about 30 mm. For the bigger model TWIG/D750, also the small rear sting had been used, referred to as conf. A, and an aluminium shell, approximating the geometry of conf. 1, referred to as conf. B.

4 Wall adaptation method

The wall adaptation method is described in detail in Refs. [7] and [5]. The top and bottom flexible walls are adjusted to a wall shape, for which the wall interferences on the test section centreline are zero, the sidewalls remain parallel. The information needed is the wall pressures on top and bottom wall centrelines and the wall shapes. The method converges in a single iterative step, as will be shown. Only for higher subsonic Mach numbers, more than one step is necessary when starting from aerodynamically straight walls. But the normal case is the determination of a new wall shape starting from a "neighbouring" Mach number or angle of attack. The Wedemeyer/Lamarche method is valid for subsonic compressible flows, when the Prandtl-Glauert approximation applies. Supersonic regions in the flow field might not extend to the walls.

5 Wall pressure method for interference computation

The wall pressure distribution is being measured on three lines per wall by a psi-system, and wall shapes of the flexible top and bottom walls are known by the position of the jacks for wall adjustment. So all boundary conditions at the test section walls are known: According to the relationship $u/U_\infty = -c_p/2$, the longitudinal velocity is known, and the velocity normal to the wall v_n/U_∞ corresponds to the

slope of the wall. It should be mentioned that the sidewalls remain parallel, and that the growth of boundary layer displacement thickness is compensated by starting with divergent top and bottom walls (about 2mm/m). This divergent initial wall shape is called *aerodynamically straight*; it is not necessary to start from aerodynamically straight walls, any wall shape can be taken, and the method converges in one single step.
To enable the integration of the wall interferences according to Green's theorem - which allows to transform the integration over the volume of the test section to an integration over the surface - [4], the test section walls are subdivided into a large number of panels, and an interpolation procedure supplies all panels with data for c_p and v_n, see Fig. 6.

6 Computation of wall interference

Wall interference results for the bigger model are presented in Fig. 7 to Fig. 10, a comparison of adapted walls and aerodynamically straight walls. The effect of wall adaptation is shown for the different incidence angles $\alpha_{support} = -1.49^0$, 8.41^0, 9.58^0 and 11.82^0. For a support angle of 10.00^0 there was - caused by bending of rear sting due to normal force - a correction of 1.82^0, resulting in an angle of 11.82^0 . For the Mach number correction it is observed that at the reference point of the model $(x - x_{c/4}) = 0$, ΔM is brought to zero by wall adaptation, see Fig. 7. The spanwise increase of ΔM remains, Fig. 8. The test section ends at $(x - x_{c/4}) = 1.23m$, where the wall interference result starts to be incorrect due to the proximity of the end of the test section. The same observation is made for the angle of attack correction $\Delta\alpha$ in Fig. 9 and Fig. 10 respectively. Here the necessity of bending back the flexible walls - to stay attached to the rest of the tunnel circuit - introduces inaccuracies at the end of the test section. Nevertheless, at the location of the model the wall interferences are remarkably reduced by wall adaptation. But it is very clear that the model is located in an inhomogeneous flow, so that spanwise averages have to be taken for the corrections.

7 Results

In Fig. 11 it can be seen that after wall adaptation and correction with residuals, the lift for the big model conf. B agrees quite reasonable with the reference results from HST. For the small model conf. 1, the not adapted case corrected according to [2] fits even better to the reference data than the adapted wall case, for which the wall interferences along the test section centreline are zero. For the drag in Fig. 12 it is seen that the results from the big model do not agree in any case with the reference results, whereas perfect agreement is achieved for the small model. The convergence behaviour is demonstrated by the computation of ΔM at the reference point of the model using the wall pressure method. After a single step, the final result is reached in Fig. 13. Typical wall shapes are shown in Fig. 14. The smaller rear sting of conf. 3 causes a smaller outward movement of top and bottom walls. Fig. 15 shows the influence of Reynolds number, and the influence of the upwash correction on the reference results. In Fig. 16 a comparison of the three different supports used for the small model is made. At higher angles of attack, the lift is larger for the small rear sting of conf. 3. Finally, in Fig. 17 a comparison of drag coefficients is made. At one Reynolds number, a $\Delta c_{D_{min}}$ as difference between the NLR device of conf. 2 and the TWG small sting could be identified. When adding this value to the NLR reference data, a comparison could be made with data at about the same Reynolds number. Good agreement was found. The pitching moment is shown in Fig. 18, and again the agreement was good for the small model, whereas the inhomogeneities of wall interference cause differences to the reference results for the big model. For more details see [6].

8 Conclusion

For the big model TWIG/D750 it has been seen that there are residual wall interferences after wall adaptation. The residuals are inhomogeneous over the model, so that corrections have to be performed using spanwise averages. After correction, the lift coefficients agree quite well, but the drag is not satisfactory after correction. This is probably due to the proximity of the test section end, relatively closer to the trailing edge of the big model than for the smaller. For the smaller model, three different support configurations have been used. It is clear that increasing residuals towards the physical end of the test section are not to be expected, because the model is relatively not so close to the end. Also spanwise inhomogeneities are not to be expected, because of the smaller model size (only 0.30 m instead of 0.75m in a test section of 1 m× 1 m). The lift results agree perfect with the reference data from the high speed tunnel HST of NLR. A difference in drag associated with the boom of conf. 1 and conf. 2 in comparison with conf. 3 was found, which has to be taken into account (for corrections). The Wedemeyer/Lamarche method converges in a single step. The wall interferences along the centreline are reduced to zero, as expected.

References

[1] Ashill, P.R., Weeks, D.J.: "A Method for Determining Wall-Interference Corrections in Solid Walled Tunnels from Measurements of Static Pressure at the Wall." AGARD CP335, 1982, Paper no, 1.

[2] Ewald, B.F.R. (editor): "Wind Tunnel Wall Corrections." AGARDograph AGARD-AG-336, 1998.

[3] Garner, H.G., Rogers, E.W.E., Acum, W.E.A., Maskell, E.C.: "Subsonic Wind Tunnel Wall Corrections." AGARDograph AGARD-AG-109, 1966.

[4] Holst, H.: "Procedure for Determination of Three-Dimensional Wind Tunnel Wall Interferences and Wall Adaptation in Compressible Subsonic Flow Using Measured Wall Pressures." DLR-FB 91-09, Göttingen, Germany, 1991.

[5] Holst, H.: "Realtime Computation of Wall Shapes in a Two-Dimensional Adaptive Test Section." ICAS-94-3.8.1, Anaheim, California, USA, 1994, pp. 2738-2753.

[6] Holst, H.: "Wall interference and support interference for the models TWIG/D750 and TWIG/NL300 in TWG." DLR-IB 224 2002 A 07, Göttingen, Germany, 2002.

[7] Lamarche, L., Wedemeyer, E.: "Minimization of Wall Interference for Three-Dimensional Models with Two-Dimensional Wall Adaptation." Technical Note 149, von Kármán Institute for Fluid Dynamics, Brussels, Belgium, 1984.

[8] Wedemeyer, E.: "Adaptive Wind Tunnel Walls: Technology and Applications. (Section 4: Testing of 3-D Models in 2-D Adaptive Wall Test Sections.)" AGARD-AR-269, 1990.

Figure 1: The adaptive test section of DNW/TWG in Göttingen.

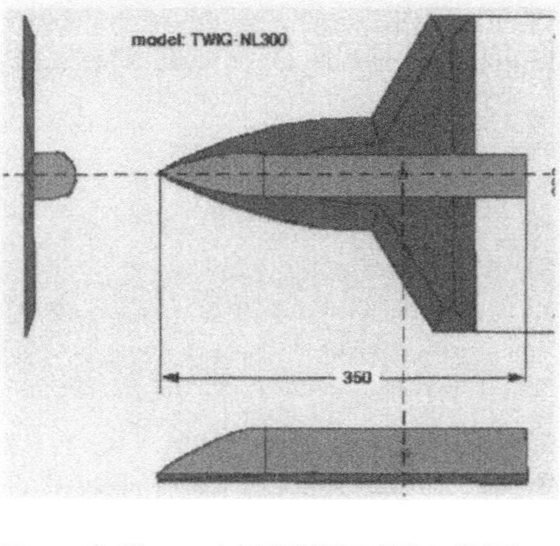

Figure 2: The model TWIG/NL300 in TWG.

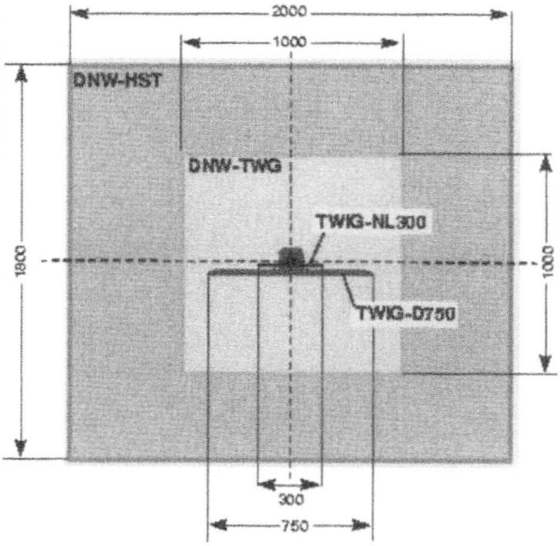

Figure 3: Geometrical relations of the two TWIG models in TWG and in HST.

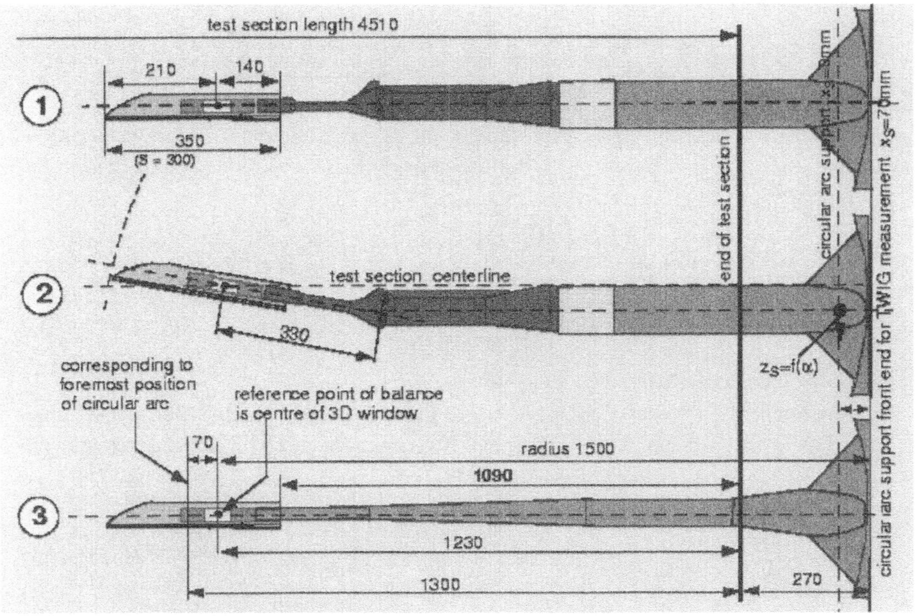

Figure 4: Sketch of the three different support configurations.

Figure 5: The model TWIG/NL300 in the transonic tunnel TWG.

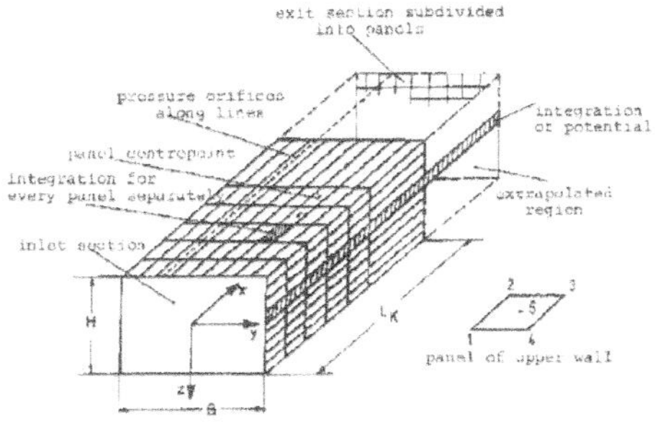

Figure 6: Subdivision of tunnel surface for integration of wall interferences, when using the wall pressure method.

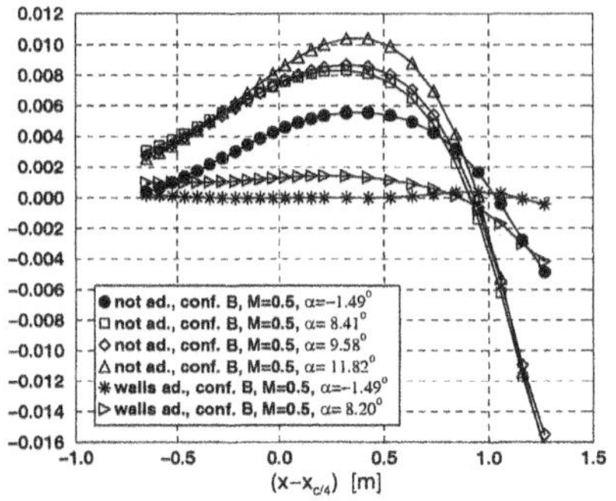

Figure 7: Blockage interference along test section centreline for model TWIG/D750.

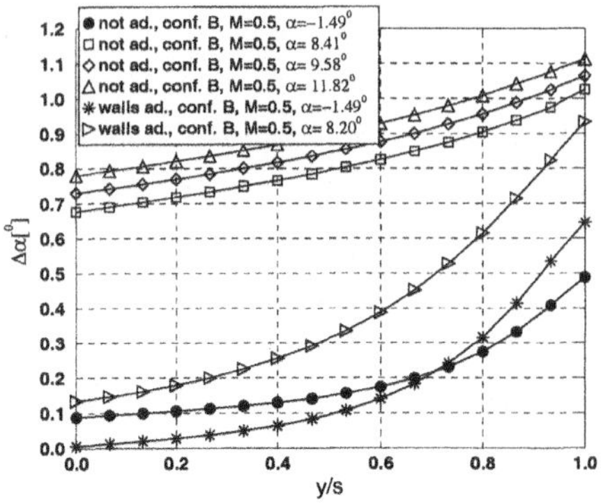

Figure 10: Angle of attack correction along swept quarter chord for model TWIG/D750.

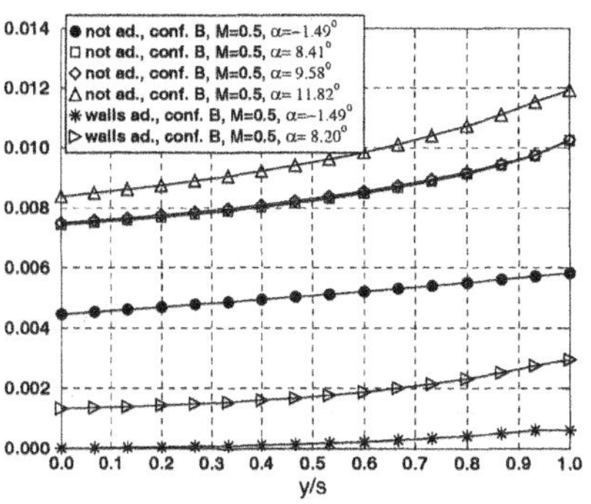

Figure 8: Blockage interference along swept quarter chord for model TWIG/D750.

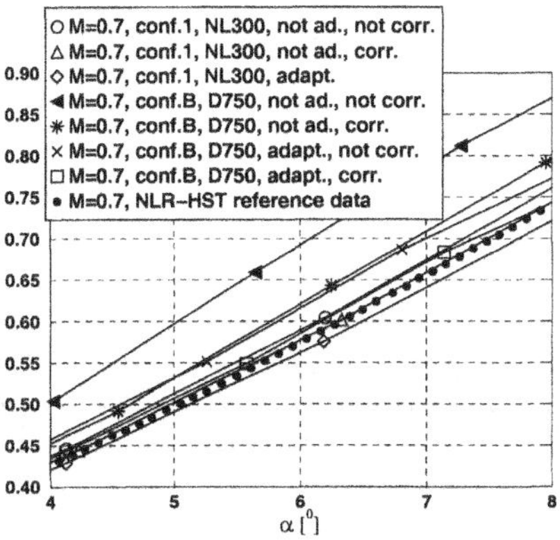

Figure 11: Comparison of c_L for the two TWIG models with conf. "B" and "1".

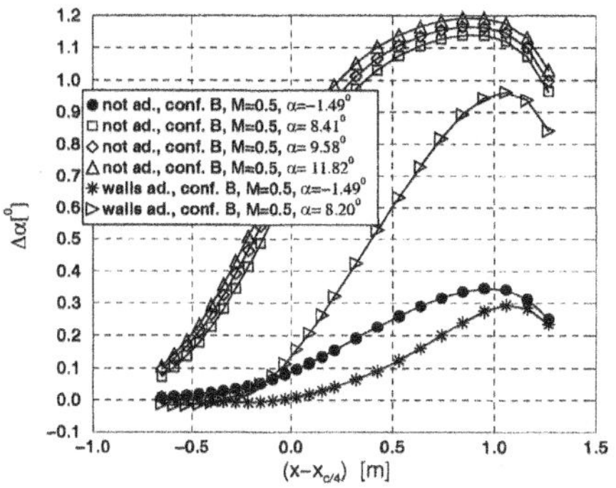

Figure 9: Angle of attack correction along centreline for model TWIG/D750.

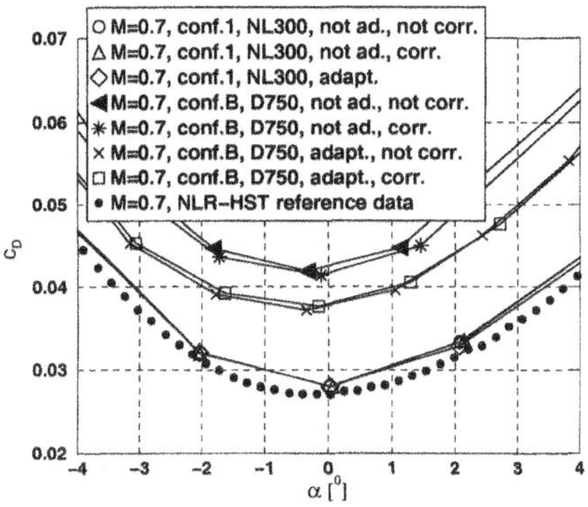

Figure 12: Comparison of c_D for the two TWIG models, conf. "B" and "1".

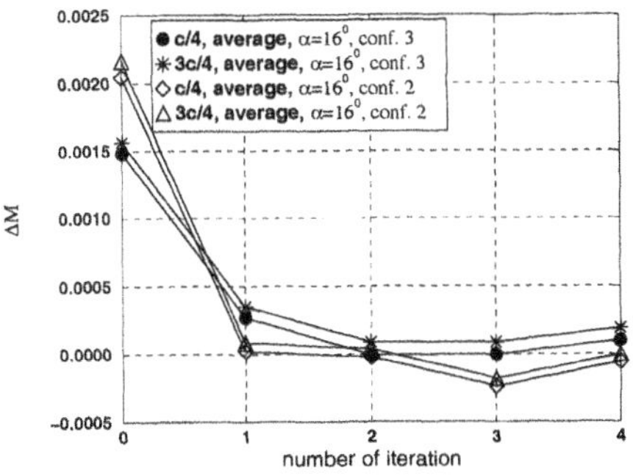

Figure 13: Convergence behaviour of adaptation procedure, averages along quarter chord line and 3/4-chord.

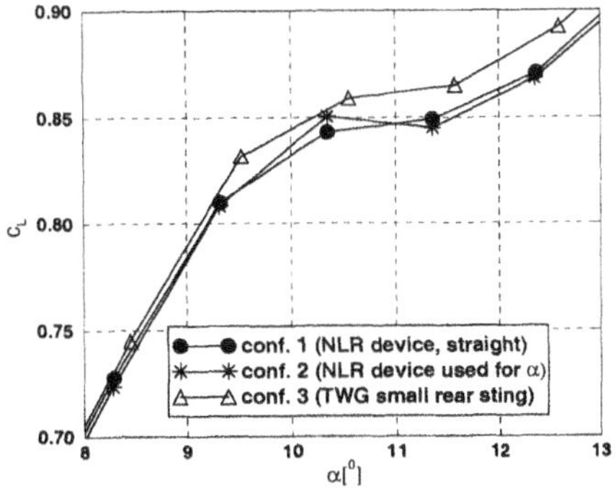

Figure 16: Direct comparison of lift results, for the 3 supports used for TWIG/NL300, at M=0.8, walls adapted.

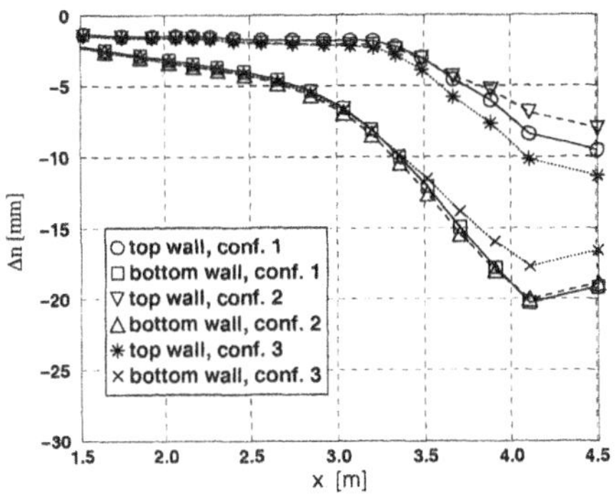

Figure 14: Examples of wall shapes for the different support configurations used at $\alpha \approx 17.5^0$.

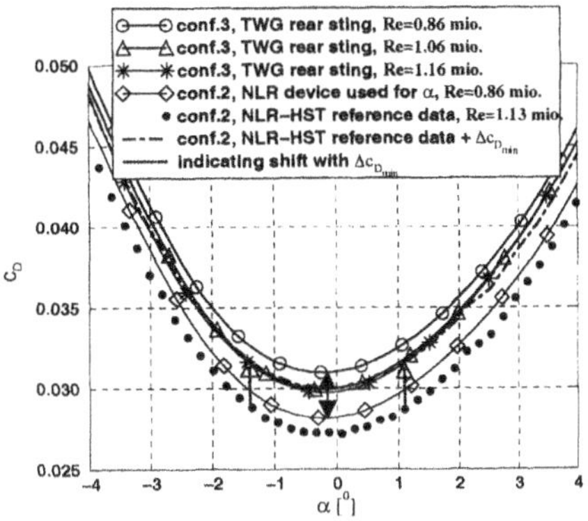

Figure 17: Comparison of drag results, identification of difference $c_{D_{min}}$.

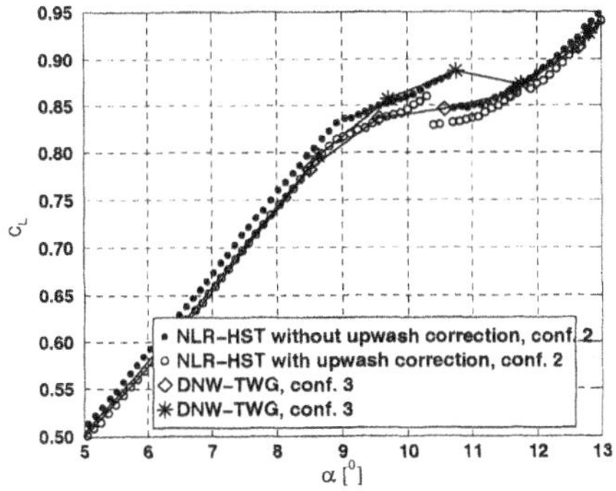

Figure 15: Comparison of lift results, influence of Reynolds number.

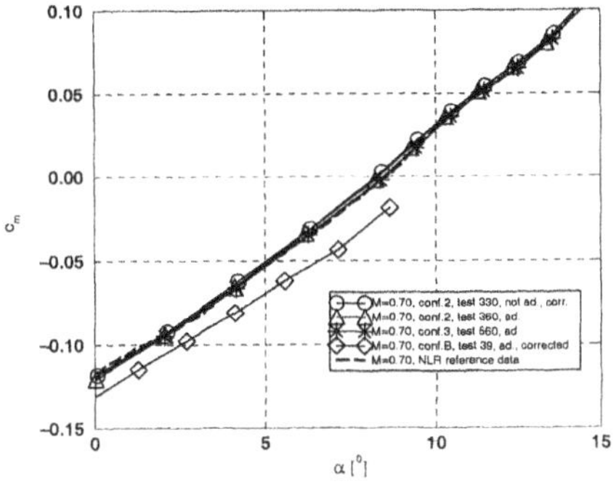

Figure 18: Comparison of pitching moment.

Experimental Investigations on the Adaptation Accuracy of Adaptive Slotted Wind Tunnel Test Sections Using Selected Test Bodies

Oliver Meyer, Stefan Blume, and Wolfgang Nitsche

TU Berlin, Institute of Aero and Astronautics (ILR), Marchstr. 12-14, 10587 Berlin, Germany
email: Oliver.Meyer@TU-Berlin.de,
WWW: http://www.aero.tu-berlin.de

Summary

The technique of flexible, adaptive walls to reduce wall interferences in wind tunnels is still restricted to small research wind tunnels due to its mechanical complexity. However, a very promising alternative is the use of adaptive slots in test section walls. This paper presents the latest experimental results on the effectiveness of adaptive slots. The results presented focus on detailed adaptation investigations under high subsonic flow conditions. The 2D slot adaptations of an airfoil (CAST7) and the 3D adaptation of a body of revolution (ETB) show in both test sections (2D-adaptive and slotted) matching results within an experimental error of about $\pm 3\%$. The deviation can be minimized significantly by applying a method developed to correct for slot adaptations.

1 Introduction

1.1 The Adaptive Slot Idea

The technique of adaptive slots can be understood as a combination of the conventional passive slotted wall and the flexible wall method. The flexible wall method delivers superior wall interference reductions in wind tunnel test sections but with the disadvantage of being a mechanically costly alternative. Therefore, this method until now was been restricted to small research wind tunnels, [12], [2]. A much simpler and more popular technique for industrial large scale wind tunnels is the implementation of passive slots, but this results in poorer adaptation capabilities. The current idea is to provide these longitudinal slots with flexible, adaptive bottoms, [11], [10]. With this method, only small slots in the solid wall are deflected and this grants the opportunity to apply a superior adaptation strategy even to large scale test sections while using all four test section walls. Even fully 3D adaptations become feasible with a lower complexity than shown by other alternatives comparatively, [7], [5]. The premise is to use common adaptation procedures, [1], [6] for flexible walls and convert the calculated, interference reducing wall shapes to the particular slot arrangements via the area rule, Figure 1, p. 5. Then the flow in an adaptive slotted test section depends on particular slot-typical parameters such as the number, the cross-sectional distribution, and the relative width of the slots.

1.2 Slot Adaptations – A Brief Outline of the Status Quo

Various numerical [3], [8], [11] and some experimental investigations [8], [9] were performed to get an overall impression of the capability of adaptive slots to account for low interference flow in adaptive test sections. The numerical work has focused on the parametric study using single and multi-slot arrangements under high subsonic 2D and 3D flow conditions. These investigations showed the performance of adaptive slots, even for complete 3D adaptations. However, it was also observed that the adaptation quality decreases slightly after applying particular slot parameters, [3]. The following experimental work should validate the numerical investigation and, in particular, investigate the source of the decreasing performance of adaptive slots and further – if possible – yield a correction for the adaptation deviation.

Figure 2 (a), p. 6, depicts some exemplary numerical results for the ETB (ETW Transition Body). The pressure distributions are quite close together for the 2D reference and the different single slot adaptations, but it is also obvious from the magnified section, that the pressures are slightly lower for the slot adaptations and decrease further using smaller relative slot widths $(b/t)_{min}$. The offset from the free stream calculation can be attributed to a poor 3D adaptation in the 2D test section and is irrelevant for this adaptation investigation.

After preliminary experimental tests showed a similar behavior, the following experimental investigations dealt mainly with detailed in-slot measurements to find the cause of the deviating adaptation accuracy using slender slots, [7]. The mass flows in different cross sections for different slot geometries were determined by measuring the velocity inside the slots, they were then compared with the equivalent 2D mass flows. A clear tendency towards decreasing mass flow in the slot compared to the 2D wall deflection depending on the relative slot width and the Mach Number was observed as shown in Figure 2 (b), p. 6. These results are now of major importance: the actual decreased mass flow in the slot can be estimated for different relative slot widths and different Mach Numbers, and it can even be used for multi-slot arrangements by simply adding the amounts of not accommodated mass flow of each slot. This mass flow must be added to the main test section flow – depending on the test section parameters – to get an estimation of an increased mass flow and hence increased velocity and decreased pressure in the main test section, i.e. at the test body. Finally an appropriate correction can be performed.

The following slot adaptations of a CAST7 airfoil and a body of revolution (ETB) will show the capability of adaptive slots in 2D and 3D flows. The above mentioned method will be applied where applicable to correct for particular slot-dependent deviations.

2 Test Set-Ups

The following experiments are performed under high subsonic flow conditions in the transonic wind tunnel at the TU Berlin. The adaptation accuracy of adaptive slots is determined by comparative measurements in a newly built slotted test section, [7], and the 2D adaptive test section of the ILR, [4]. Both, pressure and Laser-2-Focus flow field measurements are utilized to assess the adaptation quality of adaptive slots. The wall deflections are calculated according to the procedures of Amecke [1] (2D) and Holst [6] (3D). The wall shapes are then transferred to the slots using the area rule. The width of the adaptive slots is constantly 40mm, the resulting relative slot widths $(b/t)_{min}$ are defined by the actual slot deflections (slot depths).

The **CAST7 airfoil** is tested under two angles of attack (α) and three Mach Numbers ($M_\infty = 0.35, 0.5, 0.6$). The $\alpha = -3°$ case represents the minimum lift condition with slot deflections only into the solid test section walls, whereas the $\alpha = 1°$ test case requires a slot inlay to account for a modelled slot deflection into the test section as shown in Figure 3, p. 6. The displayed set-ups and slot deflections are to scale. The gray shaded area represents the measured flow field around the airfoil. The blockage ratio is in both cases approximately 8% in the test sections with non-deflected walls.

The **ETB** is tested under similar conditions as the airfoil but without angle-of-attack variation ($\alpha = 0°$). The measured area was reduced to one half of the model due to the symmetry of the set-up, Figure 3, p. 6. The measured flow field covers the area of the strongest velocity gradients along the test body, i.e. from the stagnation area to the suction peak. The blockage ratio for this test set-up is approximately 5.4%.

3 Test Results

3.1 Velocity Fields

Examples of measured flow fields for $M_\infty = 0.6$ are shown in Figure 4 (a), p. 7. The left column lists the measured flow fields in the 2D test section and the center column the corresponding flow field in the slotted test section. As these flow fields look very much alike, the right column depicts the difference of both flow fields in percentage of the local Mach Numbers.

CAST7 at $\alpha = -3°$: The local velocities around the airfoil differ in both test sections between -1.7% and 1.2% and result in an average error of $\Delta M = 0.45\%$ which is well within the measurement accuracy of the L2F measurement technique. The minimum relative slot widths are $(b/t)_{min} = 3.8$ for the upper slot and $(b/t)_{min} = 1.9$ for the lower slot. According to Figure 2 (b), p. 6, the actual mass flow in the slot – compared to the 2D mass flow – is reduced by 5% and 10%, respectively. This results theoretically in a main test section mass flow increase of about $\Delta M = 0.52\%$. Hence, it appears senseless to correct within the measurement scatter for this minor amount in this particular test.

CAST7 at $\alpha = 1°$: Here the flow velocity in the slotted test section is on average 1.8% higher than in the 2D test section. Deviations occur locally at about $\Delta M = 2.8\%$. This can now be attributed to the boundary layer effect in the slot because the relative slot width is about $(b/t)_{min} = 1$ for a relatively large upper slot (see Figure 3, p. 6). According to Figure 2 (b), p. 6 the mass flow in the slot is reduced by 10% in the upper slot and 5% in the lower slot ($(b/t)_{min} = 3.4$). Due to the larger upper slot area the flow in the main test section is affected more significantly, namely $\Delta M = 0.9\%$. Hence, the average deviation in the slotted test section can be reduced to $\Delta M = 0.9\%$. The corrected flow field is shown in Figure 4 (b), p. 7.

ETB: The ETB adaptation is performed with the 3D procedure for 2D adaptive walls [6]. Even this more complex flow field can be adapted to within an average deviation in both test sections of $\Delta M = 0.58\%$. The theoretical slot correction yields a deviation of $\Delta M = 0.59\%$ but further corrections appear purposeless within this minor scatter. The accurate results can also be attributed to large relative slot widths of $(b/t)_{min} = 2.2$ for this adaptation, but the trend of slightly higher velocities in the slotted test section is again visible.

The results of the lower free stream Mach Numbers not shown yield in all cases equivalent deviations.

3.2 Pressure Distributions

The pressure distributions which correspond to the flow fields are shown in Figure 5. Both, the adapted and unadapted cases are compared. The unadapted cases in both test sections show a minor offset for which the adapted cases are corrected, because this offset can not be attributed to the actual adaptation.

CAST7 at $\alpha = -3°$: With the offset correction applied, both adapted curves show perfect match, which reinforces the flow field results discussed above. Even the numerical curves show good coincidence with the experiment, the minor difference can be attributed to slightly different wall shapes.

CAST7 at $\alpha = 1°$: This test case shows again the successful slot adaptation. The slot adaptation is corrected for the zero offset of the unadapted case and additionally for the -0.9% obtained from the slot correction. Finally the slot adapted curve is very close to the 2D reference. Also the numerical curve shows a good match, better on the lower side of the airfoil, poorer at the suction peak.

ETB: Again, the c_p curves show perfect correspondence after the offset correction. Hence, 3D adaptations also appear feasible as the flow field investigations already showed.

4 Conclusion

The comparative experiments in the 2D and the slotted adaptive test section showed that adaptive slots can be used as a successful adaptation technique for 2D and 3D flows. Firstly, the experimental results reinforce the observations from the numerical spadework but, more importantly, show that corrections for particular slot typical deviations are successfully applicable. Slot adaptations can therefore be performed within a deviation of $\pm 1\%$ from the 2D adaptation.

References

[1] J. Amecke: "Direkte Berechnung von Wandinterferenzen und Wandadaption bei zweidimensionaler Strömung in Windkanälen mit geschlossenen Wänden", DFVLR-FB 85-62, 1985.
[2] B.R.F. Ewald: "Wind Tunnel Wall Corrections", AGARD-AG-336, 1998.
[3] I. Futterer: "Numerische Untersuchung zum Einsatz adaptiver Schlitze zur Minimierung von Windkanalwandinterferenzen bei kompressiblen Unterschallströmungen", Dissertation, TU Berlin, 2000.
[4] U. Ganzer, Y. Igeta, J. Ziemann: "Design and Operation of TU Berlin Wind Tunnel with Adaptable Walls", ICAS Paper 84-2.1.1, 1984.
[5] A. Heddergott, D. Kuczka, E. Wedemeyer: "The adaptive Rubber Test Section of the DFVLR Göttingen", IEEE 85CH2210-3, 1985, pp.154-156.
[6] H. Holst: "Procedure for Determination of Three-dimensional Wind Tunnel Wall Interferences and Wall Adaptation in Compressible Subsonic Flow Using Measured Wall Pressures", DLR-FB-91-09, 1991.
[7] O. Meyer, S. Blume, W. Nitsche: "Experimental Investigations on the Adaptation Accuracy of Adaptive Slots in Wind Tunnel Test Section Walls", ICAS 2002-3.7.2, Toronto, Canada, 2002.
[8] O. Meyer, I. Futterer, W. Nitsche: "Numerische und experimentelle Untersuchungen zum Einsatz adaptiver Schlitze in Windkanalmeßstrecken", DGLR-JT2000-91, Dresden, 2000.
[9] O. Meyer, I. Futterer, W. Nitsche: "Numerical and Experimental Investigations on the Reduction of Wind Tunnel Wall Interference by Means of Adaptive Slots", Aeronautical Journal, Vol. 105, Nr 1052, London, 2001, pp.571-580.
[10] W. Nitsche, I. Wallbruch, J. Quest: "Einsatzmöglichkeiten adaptiver Schlitze zur Verringerung von Wandinterferenzen", DGLR-JT97-123, München, 1997.
[11] J. Quest, W. Nitsche, A. Mignosi: "Adaptive Slots: An Alternative Concept to Reduce Wall Interferences", AGARD-CP-585, 1996.
[12] S.W.D. Wolf: "Adaptive Wall Technology for Improved Wind Tunnel Testing Techniques - a Review", Prog. Aerospace Sci., Vol.31, 1995, pp.85-136.

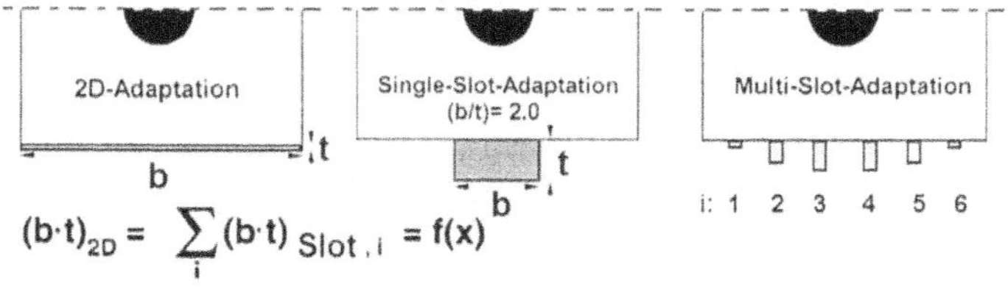

Figure 1 Area rule to convert wall adaptations into slot adaptations

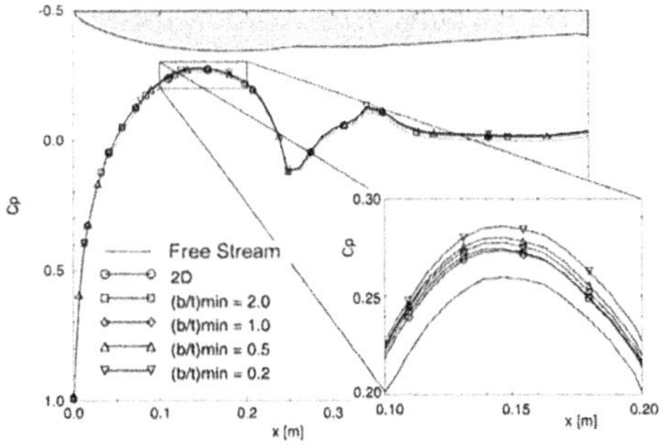

(a) ETB, Pressure Distributions for $M_\infty = 0.7$ (num.) [3]

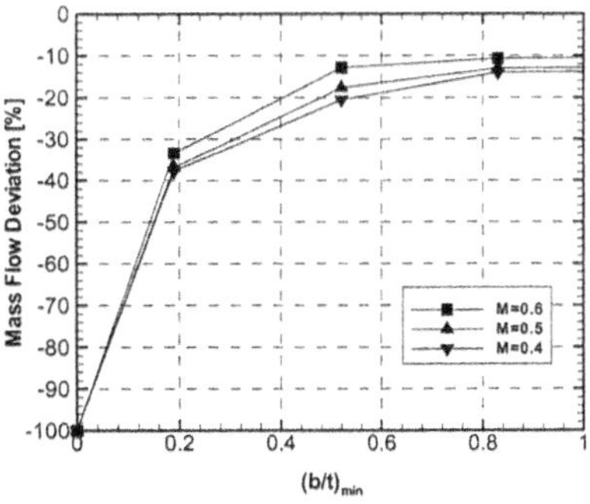

(b) Slot Mass Flow vs 2D Mass Flow (from Exp.) [7]

Figure 2 Effect and Magnitude of Reduced Slot Mass Flow

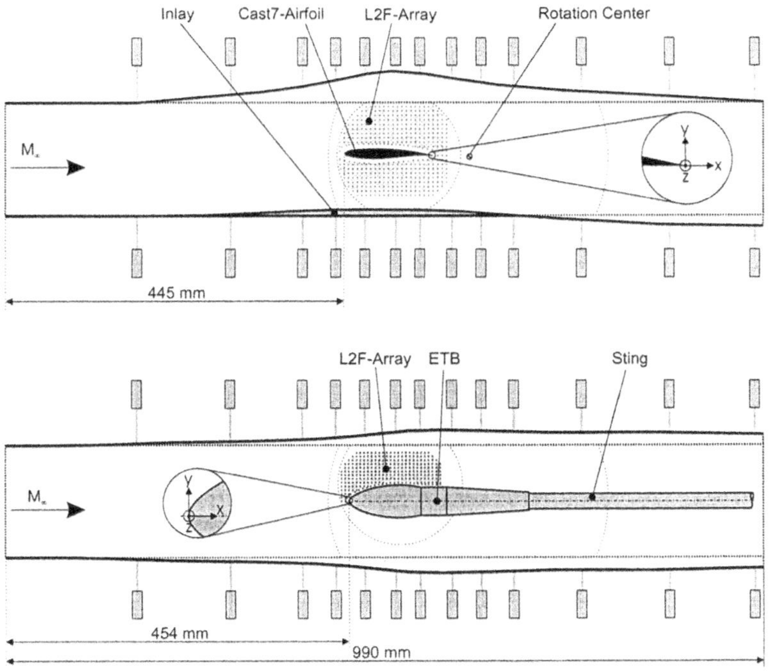

Figure 3 Test Set-Ups for CAST7 airfoil, $\alpha = 1°$ (upper) and ETB (lower)

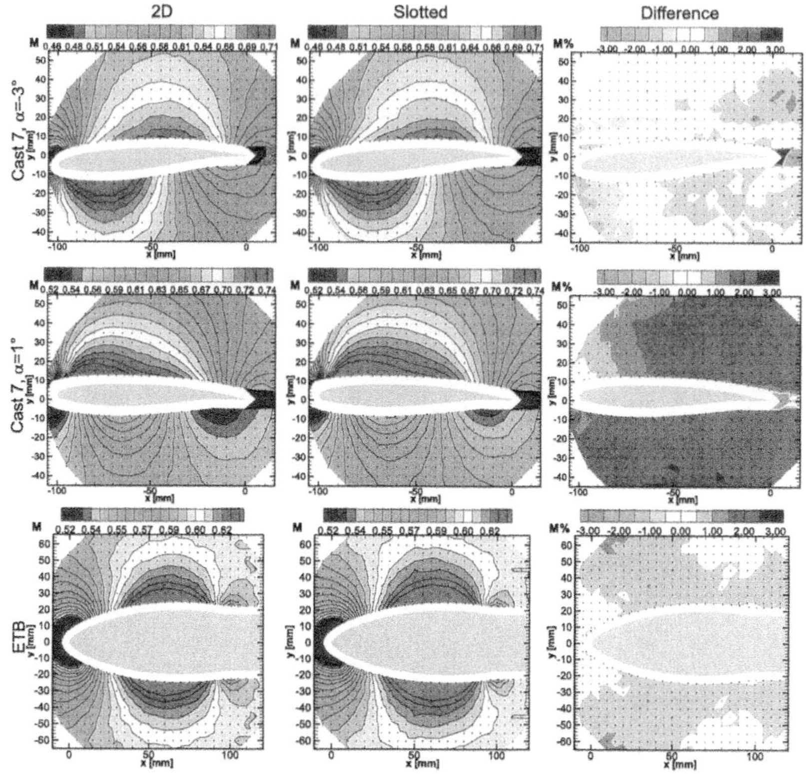

(a) Flow Fields at $M_\infty = 0.6$: CAST7, $\alpha = -3°, 1°$ and ETB (from Top to Bottom)

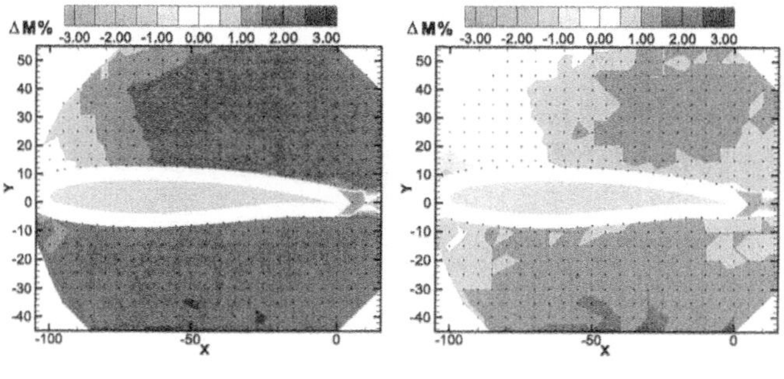

(b) Applied Correction to CAST7 Airfoil Flow at $\alpha = 1°$

Figure 4 Flow Field Investigations

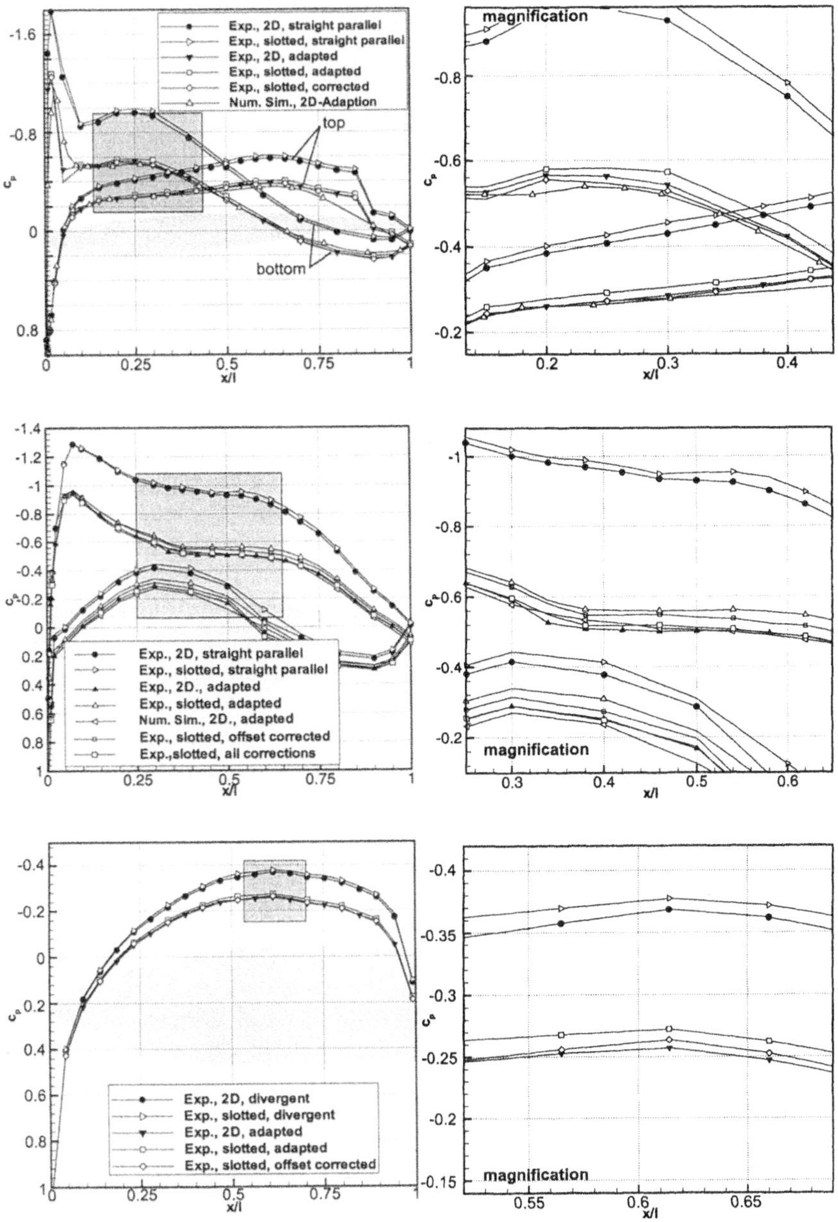

Figure 5 Pressure Distributions at $M_\infty = 0.6$: CAST7, $\alpha = -3^\circ, 1^\circ$ and ETB

Flutter Stability of a Transonic Compressor Rotor - Application and Comparison of Different Numerical Methods

Ralf Kemme[1], Gerhard Kahl[2] and Stefan Schmitt[3]

[1] DLR Institute of Aeroelasticity, Bunsenstr. 10, D-37073 Göttingen, Germany
[2] MTU Aero Engines, P.O. Box 500640,D-80976 Munich, Germany
[3] DLR Institute of Propulsion Technology, Linder Höhe 10, D-51147 Porz-Wahn, Germany

Summary

This paper investigates the flutter mechanism of a high loaded transonic compressor rotor by applying different numerical methods. For predictions of the stability in the frequency domain, harmonic fluid forces are calculated with a linearized Euler method and with a nonlinear time domain Navier-Stokes solver for forced motion. To include nonlinear effects in the stability prediction, the time domain method is applied in a fluid-structure coupled mode of operation. The results of this 3D viscous aeroelastic computations in the time domain for the whole annulus of the rotor are compared to the frequency domain solutions.

The comparative study shows that shock-boundary layer interactions significantly influence the stability of the compressor rotor. While for higher reduced frequencies the rotor still behaves linearly, for lower reduced frequencies limit cycle oscillations (LCO) due to shock-boundary layer and shock-shock interaction occur.

1 Introduction

The development of modern aircraft engine compressors with increased pressure ratio and reduced weight has led to highly loaded stages with transonic inflow. As a result, the design engineers of completed engines have often encountered severe aeroelastic problems that are difficult and costly to solve. Therefore, tools are needed which can be used during the design process to correctly predict the aeroelastic behaviour of the blading.

In the classical approach, the aeroelastic problem is divided into separate steps of aerodynamic and structural computations. The mode shapes of the structural dynamic model are used to generate harmonic motions for the calculation of airloads. The aeroelastic problem is then solved by using these airloads in the structural dynamic problem, where the aeroelastic stability has to be determined. By neglecting the additional mass and stiffness effects of the fluid force on the structure, the aeroelastic stability can be determined by examining the aerodynamic damping - this comprises the so called energy method. If these added mass and stiffness effects are significant and cannot be neglected, an eigenvalue problem has to be solved. As typical examples of the frequency domain method, the publications of Platzer and Carta [13] and Försching [4] are mentioned.

To include nonlinearities in both the structure and/or the aerodynamics, aeroelastic equations can be accurately solved in the time domain. To avoid the development of completely new methods for solving the fluid and the structural equations simultaneausly, time staggered algorithms for coupling of existing codes have been frequently used. A principle insight into the numerical stability and accuracy of different coupling techniques is given by Giles [5] and Piperno [12].

Since the first papers about time domain methods have been published, several methods for solving the aeroelastic equations in the time domain have been developed. Most of them are restricted to 2D or inviscid problems due to the large computational effort required. In 1996, Marshall and Imregun [9] published a work showing 3D viscous whole annulus aeroelastic time domain calculations which are the only aeroelastic computations of this complexity so far.

It should be noted here that the very time consuming computations of time domain methods are only justified if nonlinear effects significantly influence the stability. Grüber and Carstens [6] applied a 2D nonlinear Navier-Stokes-Solver to compute the stability depending on the oscillation amplitude of a cascade. They showed that strong nonlinearities could occur which are dominated by viscous effects. More recently Carstens and Belz [2] and Bendiksen [1] impressively showed the nonlinear effects on the stability of a 2D cascade in transonic flow. Vahdati [16] investigated the stall flutter of a transonic fan by applying different prediction methods and concluded that viscous effects are significant for the stability prediction, however, the aeroelastic behaviour of the fan was predominantly linear.

The main objective of this paper is to apply and compare different numerical methods to the determination of the stability of a highly loaded compressor rotor in transonic flow. Particular attention is directed to the application of a time domain method for 3D viscous computations on the complete rotor annulus.

2 Basic Aeroelastic Equations and Prediction Methods

The dynamic aeroelastic model of the compressor rotor is derived from a finite element model of the rotor which is coupled with the fluid forces. By assuming a perfectly stiff rotor disk, all blades could be modeled seperately. Since only the low frequency eigenmodes of the blades are prone to flutter, the structural system may be reduced to only a few modes. It is convenient to transform the governing aeroelastic equation to modal coordinates, which can be written in matrix form as

$$\mathbf{M}\ddot{\mathbf{q}}(t) + \mathbf{D}\dot{\mathbf{q}}(t) + \mathbf{K}\mathbf{q}(t) = \mathbf{f}\left(\mathbf{q}, \dot{\mathbf{q}}, \ddot{\mathbf{q}}\right) \tag{1}$$

Here, $\mathbf{q}(t)$ is the vector of modal coordinates and $\mathbf{M}$, $\mathbf{D}$ and $\mathbf{K}$ denote the diagonal matrices of generalized mass, damping and stiffness coefficients. For aeroelastic investigations, the right-hand side vector $\mathbf{f}\left(\mathbf{q}, \dot{\mathbf{q}}, \ddot{\mathbf{q}}\right)$ represents the generalised nonlinear aerodynamic forces acting on the structural system.

2.1 Unsteady airloads

Time dependent airloads are needed to complete the aeroelastic model of equation (1). Classical aeroelastic methods linearize the vector $\mathbf{f}$ with respect to $\mathbf{q}$ and its time derivatives. The additional assumption of a harmonic time dependence of the displacement vector then allows a decoupled computation of the unsteady airloads for given eigenmodes. The aeroelastic eigenmodes in a perfectly tuned cascade are traveling waves where the N blades vibrate with constant and equal amplitudes and a constant interblade phase angle

$$\sigma_k = \frac{2\pi k}{N} \qquad for \qquad k = 0, ..., N-1. \tag{2}$$

For nonlinear aeroelastic investigations, equation (1) is solved for each timestep, thus airloads for each time step are required.

In this study the computations of the unsteady airloads were performed by applying two different methods:

Time Linearized Euler Method Lin3d

Lin3d is a 3D inviscid linearized Euler method. It is incorporated as standard aeroelastic application in the design cycle for turbomachinery components at MTU Aero Engines. It has proven to fit the needs of the designer by providing the required information quickly, robustly and within reasonable accuracy, as long as viscous or nonlinear effects are negligible. The basic idea of the linearization is that the unsteady flow for small blade amplitudes is a small perturbation of the steady state flow. By neglecting higher order terms in the equations derived under this assumption, the unsteady flow equations become linear. For every arbitrary traveling wave mode of a cascade, the stability can be determined for given frequencies. Lin3d features deforming meshes, non reflecting boundary conditions, forced response calculations for generalised forces due to up- and downstream disturbances and the possibility of calculation of acoustic modes. A more detailed description of the method can be found in the papers by Kahl and Klose [7] and Kahl [8].

Nonlinear Navier-Stokes-Method TRACE

The second method applied for the current study is the TRACE code of DLR, (Nürnberger et.al.[11]). It is focused on turbomachinery applications and has been validated against a wide range of test cases. This method solves the three-dimensional nonlinear Reynolds-averaged Navier-Stokes equations in the time domain. For turbulence closure here the one-equation turbulence model of Spalart and Allmaras [15] is used in combination with wall functions. The spatial discretisation of the convective fluxes is obtained from Roe's upwind scheme in combination with MUSCL extrapolation. For time integration here a dual time-stepping technique is used which allows time steps of reasonable size. Parallel computing is applied based on the decomposition of the computational domain that is meshed by a structured multi-block grid.

The aeroelastic module of the code (Schmitt et.al.[14]) can handle arbitrary blade mode shapes using an algebraic mesh deformation technique. Two modes of operation are possible: the blade vibration can either be imposed as a harmonic blade motion or calculated using a fluid-structure-coupling approach. By prescribing harmonic blade motion unsteady time harmonic airloads can be calculated and used for aeroelastic analysis in the frequency domain. To compute airloads for all possible traveling wave modes, multipassage computations are performed.

The second mode of operation includes nonlinear effects in the aeroelastic analysis. In the following sections, the aeroelastic computations are explained in more detail.

2.2 Frequency domain methods

Assuming a linear structural behaviour and a linear dependence of the unsteady airloads on the small blade vibration amplitudes, time t can be removed from equation (1). This is achieved by inserting

$$\bar{\mathbf{q}} = \hat{\bar{\mathbf{q}}} e^{j\bar{\lambda}t} \quad \text{and} \quad \bar{\mathbf{f}} = \bar{\mathbf{A}}\bar{\mathbf{q}} \tag{3}$$

into (1), where $\bar{\mathbf{q}}$ and $\bar{\mathbf{f}}$ now denote the complex displacement and force vector and $\hat{\bar{\mathbf{q}}}$ and $\bar{\mathbf{A}}$ signify the displacement amplitude or aerodynamic influence coefficient matrix, respectively. The resulting equations can be cast into the form of a quadratic eigenvalue problem. The complex eigenvalues $\bar{\lambda}$ that constitue the solution of this equation are composed of the eigenfrequency ω and the damping coefficient Λ for each traveling wave and blade mode, with

$$\omega = \Re\{\bar{\lambda}\} \quad \text{and} \quad \Lambda = 2\pi\frac{\Im\{\bar{\lambda}\}}{\Re\{\bar{\lambda}\}} \tag{4}$$

Note that solution (4) is strictly valid only if the solution provides harmonic oscillations, thus if the damping coefficient is zero. However, for small values of the damping coefficient, we assume that the harmonic fluid forces remain unaffected.

If the airloads do not change the vibration frequency ω significantly and coupling between blade modes is negligible, it is convenient to determine the stability of the rotor just by computing the harmonic airloads. With the unsteady aerodynamic work per cycle of oscillation W_{aero}, where a positive value of W_{aero} implies energy transferred from the flow to the oscillating airfoil, the aerodynamic damping coefficient Λ can be calculated using

$$\Lambda = \frac{-W_{aero}}{2KE_{max}}. \tag{5}$$

In this equation, KE_{max} denotes the maximum kinetic energy of the oscillating blades. This approach is called the energy method or work per cycle method.

2.3 Time domain method

The aeroelastic investigations in the time domain are carried out using a fluid-structure coupling approach incorporated in the TRACE code. In this approach the fluid equations are coupled to a linear structural model in generalized co-ordinates. The structural dynamic problem is solved using the integration method of Newmark [10]. To calculate the displacement vector for the next time step within the Newmark method, the airloads are required for the unknown timelevel. As the unsteady airloads depend on the displacement vector and its time derivatives, a time staggered algorithm is used here to solve this problem. The inevitable energy error arising from such a staggered algorithm is minimized by using an optimized algorithm proposed by Farhat [3]. This method has been proven to be very accurate, robust and computationally efficient.

To initialize a computation in the time domain, the flow solver starts from the steady state flow solution with a velocity impulse of all modes given to one blade. This initialization procedure guarantees that all possible traveling wave modes are excited. In the further computation, the blades are free to oscillate in the flow, so that after a sufficiently long time, all damped modes will have decayed and only unstable or undamped modes will exist in the solution. The computed time series can be analyzed using modal identification techniques to gain detailed information about aeroelastic modes, frequencies and damping coefficients, if required.

3 Numerical Results

The numerical results presented below have all been computed for the highly loaded rotor of a low pressure compressor. The flow field at design conditions is characterized by a complex system of shocks interacting with the boundary layer of the blades. In **Fig. 2** the mach number distribution in a blade to blade section at 90% span is presented, which shows the complex nature of the flow. Aeroelastic computations with the TRACE code have been performed on the whole annulus of the compressor rotor to gain damping results for all possible traveling wave modes. The mesh for these computations consists of about 1.5 million grid points and is shown in **Fig. 1**. In all computations the effect of the tip clearence vortex has been neglected.

Unsteady aerodynamic coefficients were calculated for the first five eigenmodes but this study focuses on the first bending mode, **Fig. 3**, because of its low aerodynamic damping for small interblade phase angles σ. **Fig. 4** shows the aerodynamic damping for the first eigenmode as function of the interblade phase angle. The results of the TRACE code are compared with the data of Lin3d. Although the overall behaviour of the damping predicted by the viscous method is similar to that of the linearized Euler procedure, the TRACE code yields a lower damping over the whole range of interblade phase angles. It is assumed that the differences between the two procedures are due to viscous effects which obviously lower the aerodynamic damping for the present configuration. Additionally, in **Fig. 4** the results of a viscous computation of just a part of the whole annulus consisting of 4 blades are plotted (black circles). Although the traveling wave modes that are represented by this sectional approach are not exactly part of the traveling waves for the complete rotor, results fit well into the curve drawn by the full rotor damping coefficients. For time consuming parametric studies this could be a convenient approach of saving computational time by modeling the aeroelastic properties of the compressor rotor with just 4 blades.

For interblade phase angles near $\sigma = 0^o$, the TRACE code predicts a negative damping for the first bending mode. This will cause self excited vibrations if the structural damping of the blades is not large enough to damp the aerodynamic excitation. Given the fairly high reduced frequency of $\omega^* = 1.7$ based on full chord length, this is a rather surprising result. To ensure that this result is not an artifact due to a coarse computational mesh, the calculation of the aerodynamic coefficient for $\sigma = 0$ was repeated with a finer grid yielding an almost identical damping value (**Fig. 4**, diamonds).

The differences between the results of the two codes are emphasized by comparing the local aerodynamic damping instead of the global damping **Fig. 5**. The local damping coefficient identifies those regions on the blade surface which contribute either to a stable or to an unstable vibration. **Fig. 5** shows that both methods predict similar stability contours on the pressure side of the blades up to approximately thirty percent of span, but exhibit strong discrepancies in the tip region. The linearized Euler method (right-hand side) predicts a small unstable area close to the leading edge but a locally stable oscillation in the vicinity of the shock and the rearward part of the blade. The Navier-Stokes code, on the contrary, shows unstable and stable areas around the shock and a large zone of instability after the shock. Obviously, the pronounced shock-boundary layer interaction leads to this different stability behaviour in the tip region of the blades. This interaction induces a much stronger shock movement and hence has a large impact on the global stability of the rotor.

3.1 Flutter investigations in the time domain

The aeroelastic investigations in the time domain were performed on the whole annulus mesh of **Fig. 1** taking into account the first five eigenmodes per blade. With 17 blades on the rotor, the aeroelastic model consists of 85 degrees of freedom which enter the system (1) of dynamic equations. Since the unstable aeroelastic modes are not known a priori, the flutter investigations were started following the procedure described in section 2.3.

The first result, presented in **Fig. 6**, shows the time series of three blades for the first bending mode over many periods of vibration. The free vibrations of the blade assembly are started with the initial deflection of blade 0. Due to the aerodynamic coupling, the adjacent blades are also starting to oscillate by taking vibrational energy from the initially deflected blade. It is hard to judge the flutter susceptibility of the blades solely from the time series, therefore the blade vibrations are decomposed into traveling wave modes by performing a Fourier analysis in time and space. By comparing the amplitudes of two adjacent time periods, the logarithmic decrement of each traveling wave mode was determined. It should be stressed that this approach assumes a time slice linear behaviour of the system and implies no strong variations in vibration frequency of all traveling wave modes. Although this identification technique yields only approximate results, it clearly shows the time dependent vibration properties of the system.

Fig. 7 shows the aerodynamic damping for the first eigenmode as function of the interblade phase angle. In this figure, results for frequency domain and time domain methods using the TRACE code are plotted. The results obtained by modal decomposition of the time series fit the frequency domain solution perfectly. It should be mentioned that the logarithmic decrement does not change over time, thus identifying a linear system. The figure illustrates that the rotor behaves aeroelastically linear at this operating point, even though a complex flow structure with nonlinearities like shocks and shock boundary layer interaction is present.

3.2 Flutter investigation in the time domain for lower reduced frequencies

This section shows that for certain operating conditions the rotor does not show linear vibration behaviour and that this nonlinear behaviour can be identified using the described methods in the time domain. Nonlinear behaviour occurs when the reduced frequency for the first bending mode is set to $\omega^* = 0.85$ by decreasing the modal stiffness. For computational reasons, the simulation was performed on only 2 blade channels with five modes per blade. The used aeroelastic system comprises traveling wave modes of $\sigma = 0^o$ and $\sigma = 180^o$.

The time history of blade 0 (**Fig. 8**) displays the classical displacement characteristic of limit cycle oscillations (LCO), where the vibration amplitude of the blade first increases and after some periods levels off to a constant value. The aeroelastic mode for this phenomena is the traveling wave mode of $\sigma = 180\,^o$. The time dependent damping coefficient for this mode, decomposed using the method described before, is presented in **Fig. 9**. The damping coefficient represents no longer a constant line in this graph but varies with time and hence with vibration amplitude. It should be emphasized that this behaviour occurs even for small amplitude levels. Compared to results obtained using the frequency domain method (dashed line), the damping coefficient indicates always unstable behaviour whereas the frequency method result indicates stability. The discrepancies emerge due to strong passage shock movement which interacts with the boundary layer and with the oblique shock on the suction side, dependent on the vibration amplitude. The results show that neglecting the nonlinear aeroelastic behaviour for this operating condition would result in erroneous results, which would lead to a flutter problem with probable blade failure. In the design cycle of a compressor, a costly redesign would result. This scenario can be avoided by using adequate nonlinear aeroelastic analysis, as shown above.

4 Conclusion

A comparative study of aeroelastic stability prediction methods has been presented here. For predictions in the frequency domain, the computation of the time harmonic unsteady airloads were performed using either a linearized inviscid method (Lin3d) or a nonlinear viscous code (TRACE). The results gained were compared with solutions of the nonlinear TRACE code in its fluid-structure-coupled mode of operation which corresponds to a real time domain aeroelastic simulation. All methods were used to compute the flutter stability of a highly loaded transonic low pressure compressor rotor. The flow field in the rotor at design conditions is characterized by a complex system of shocks interacting with the boundary layer of the blades. Hence nonlinear aeroelastic behaviour of the rotor blades was expected.

In the first part of this study, the stability of the compressor rotor was computed at design conditions. Although the overall behaviour of the damping predicted by the viscous method is similar to that of the linearized Euler procedure, the TRACE code yields a lower damping over the whole range of interblade phase angles. The most striking result of the viscous aerodynamic method is that a negative damping for an interblade phase angle of $\sigma = 0$ was predicted, contrary to predictions of the inviscid method. These discrepancies are explained by a strong shock boundary layer interaction resulting in greater shock movement, which could not be predicted with an inviscid method. Although it could be expected that this phenomena leads to nonlinear aeroelastic behaviour, computations in the time domain on the whole annulus of the rotor show no significant nonlinearity.

In the second part of this study, it was shown that the aeroelastic behaviour for lower reduced frequencies was governed by nonlinear unsteady flow phenomena. This is deducted from a time domain analysis, where the oscillation of the rotor blades resulted in limit cycle oscillation (LCO).

References

[1] Bendiksen, O. O., *Transonic Bending Flutter in Rotor and Cascades* Proceedings of the 9th International Symposium on Unsteady Aerodynamics, Aeroacoustics and Aeroelasticity of Turbomachines (ISUAAAT), Lyon, September 4-8, 2000, Presses Universitaires de Grenoble, pp. 791-802, 2001

[2] Carstens, V., Belz, J., *Numerical Investigation of Nonlinear Fluid-Structure Interaction in Vibrating Compressor Blades,* Journal of Turbomachinery, Vol 123, No 2, 2001, pp 402-408

[3] Farhat, C., Lesoinne, M., *Higher-Order Staggered and Subiteration Free Algorithms for Coupled Dynamic Aeroelasticity Problems,* AIAA Paper 98-0516, 1998

[4] Försching, H., *Aeroelastic Stability of Cascades in Turbomachinery* Prog. Aerospace Sci., Vol. 30, pp. 213-266, 1994

[5] Giles, M., *Stability and Accuracy of Numerical Boundary Conditions in Aeroelastic Analysis,* COSMASE (Computation of Sciences, Methods and Algorithms an Supercomputing for Engineering), Short Course on Fluid-Structure-Interaction, May 2–5, KTH, Stockholm, Sweden, 1995

[6] Grüber, B., Carstens, V., *The Impact of Viscous Effects on the Aerodynamic Damping of Vibrating Transonic Compressor Blades - A Numerical Study* Journal of Turbomachinery, Vol. 123, No. 2, pp. 409-417, 2001

[7] Kahl, G., Klose, A., *Computation of Time Linearized Transonic Flow in Oscillating Cascades,* ASME Paper, 93-GT-269, 1993

[8] Kahl, G., *Application of the Time-Linearized Euler Method to Flutter and Forced Response Calculations,* ASME Paper, 95-GT-123, 1995

[9] Marshall, J.G., Imregun, M., *An Analysis of the Aeroelastic Behaviour of a Typical Fan-Blade with Emphasis on the Flutter Mechanism,* ASME Paper 96-GT-78, 1996

[10] Newmark, N.M., *A Method of Computation for Structural Dynamics,* A.S.C.M., Journal of Engineering Mechanics Division, Vol. 85, pp. 67-94, 1959

[11] Nürnberger, D., Eulitz, F., Schmitt, S., Zachcial, A., *Recent Progress in the Numerical Simulation of Unsteady Viscous Multistage Turbomachinery Flow* Proceedings of the

15th International Symposium on Air Breathing Engines, ISABE Paper 2001-1081, 3-7 September, Bangalore, India, 2001

[12] Piperno, S., Farhat, C., *Partitioned procederures for transient solution of coupled aeroelastic problems Part II: energy transfer analysis and three-dimensional applications* Comp. Methods Appl. Mech. Engrg. Vol. 190, pp. 3147-3170, 2001

[13] Platzer, M.F., Carta, F.O., *Aeroelasticity in Axial-Flow Turbomachines, Volume 1: Unsteady Turbomachinery Aerodynamics. Volume 2: Structural Dynamics and Aeroelasticity AGARD, AGARDograph* 298, 1987/1988

[14] Schmitt, S., Eulitz, F., Nürnberger, D., Carstens, V., Belz, J., *Simulation of Propfan Forced Response Using a Direct Fluid-Structure Coupling Method,* Proceedings of the 4th European Conference on Turbomachinery,Fluid Dynamics and Thermodynamics, March 20-23, 2001, Firenze, Italy, pp 1023-1033

[15] Spalart, P.R., Allmaras, S.R., *A One-Equation Turbulence Model for Aerodynamic Flows,* AIAA-Paper 92-0439, 1992

[16] Vahdati, M., Sayma, A. I., Marshall, J. G., Imregun, M., *Mechanisms and Prediction Methods for Fan Blade Stall Flutter* Journal of Propulsion and Power, Vol. 17, No. 5, pp. 1100-1108, 2001

Figure 1 Mesh for 17-bladed rotor

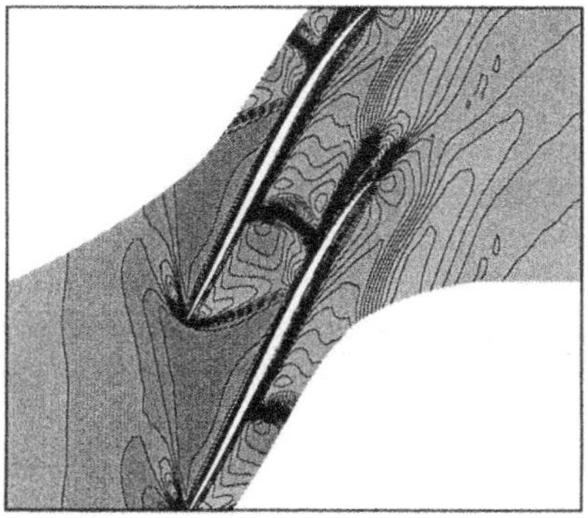

Figure 2 Mach distribution of steady flow solution

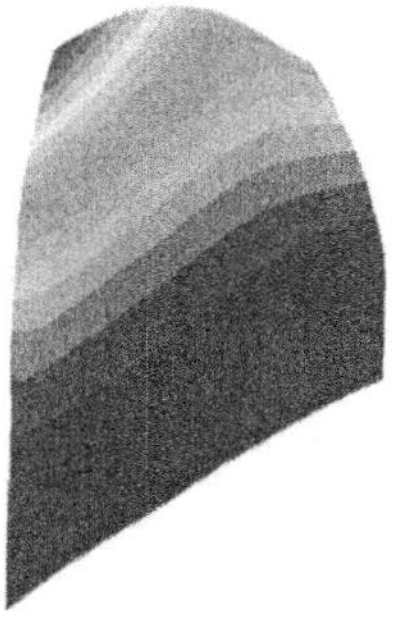

Figure 3 Deflection contours of first bending mode

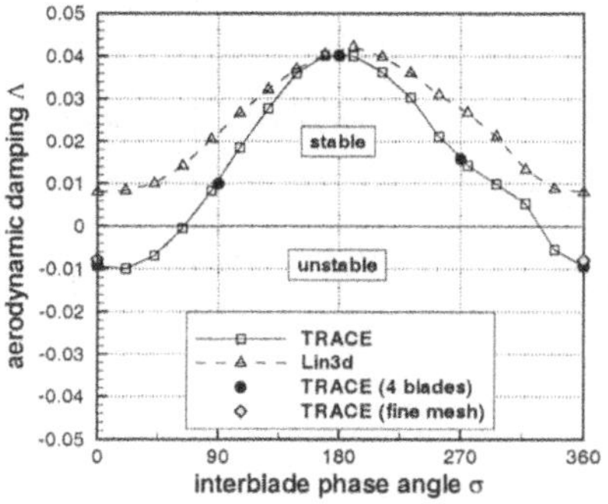

Figure 4 Aerodynamic damping of first bending mode, $\omega^* = 1.7$

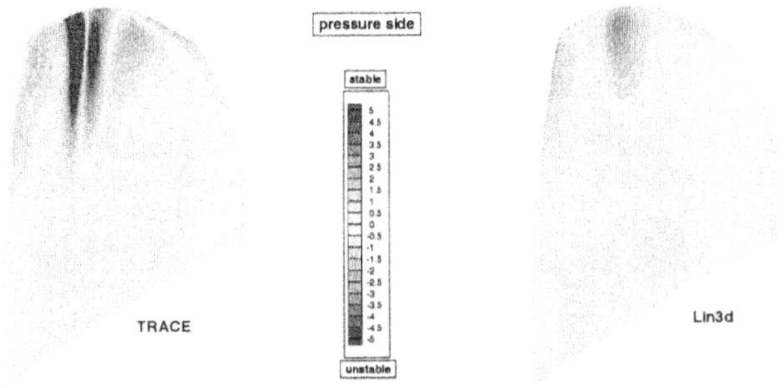

Figure 5 Comparison of the local aerodynamic damping coefficent, $\sigma = 0^o$, $\omega^* = 1.7$

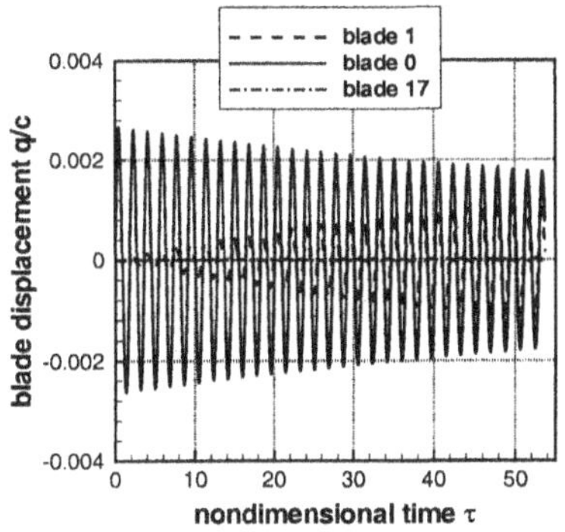

Figure 6 Time series of three blades, first bending mode, $\omega^* = 1.7$

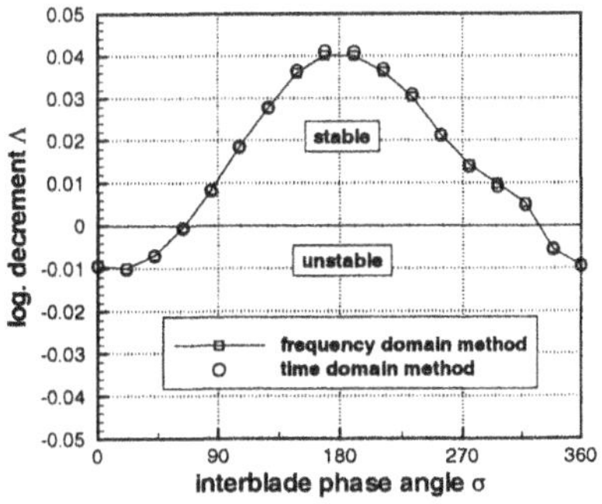

Figure 7 Comparison of damping results for the first bending mode, $\omega^* = 1.7$

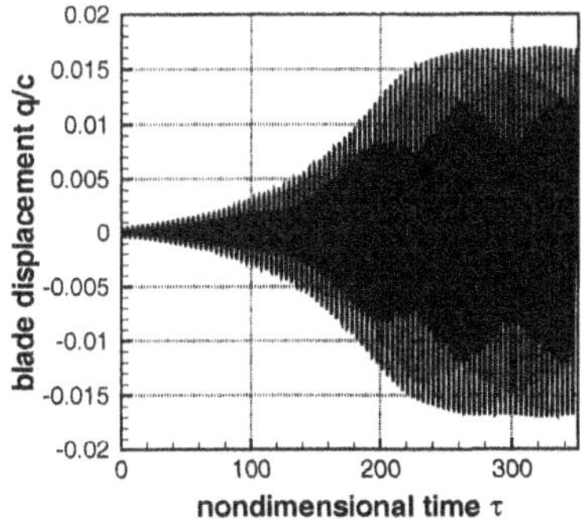

Figure 8 Time series of one blade, first bending mode, $\omega^* = 0.85$

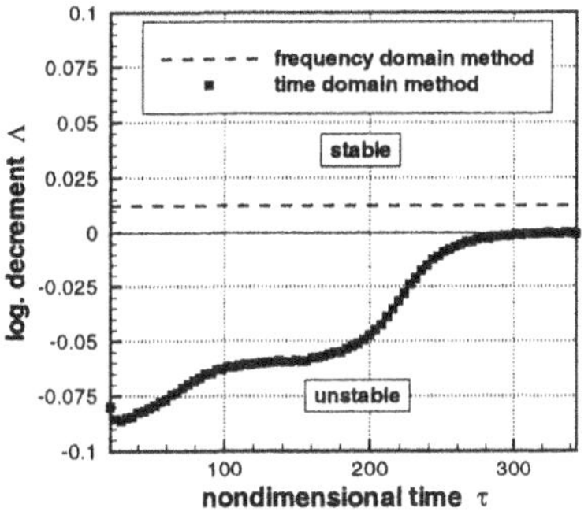

Figure 9 Time dependence of damping coefficient for the first bending mode, $\sigma = 180^o$

Shock-Vortex Interaction in a Transonic Turbocompressor

Willy H. Hofmann, Josef Ballmann

RWTH Aachen, Mechanics Laboratory,
Templergraben 64,
52062 Aachen, Germany,
willy.hofmann@lufmech.rwth-aachen.de,
WWW home page: http://www.lufmech.rwth-aachen.de

Summary

Tip-vortices on axial compressor rotor blades develop due to the pressure difference between suction and pressure side and roll up above the suction side. On a transonic rotor blade the leading edge thickness causes a bow shock which changes over to an expansion wave on the suction side. On the pressure side the shock is being intensified and crosses the passage hitting the neighbouring blade on its suction side. The tip-vortex has to cross the passage shock. The shock-vortex interaction leads to a sudden and strong deceleration of the flow, a rapid expansion of the vortex and, as a consequence, to a strong change in the vorticity distribution of the vortex. Depending on the intensity of the shock-vortex interaction, vortex breakdown occurs and possibly causes compressor stall.

1 Introduction

Large efficiency decrease occurs in the tip region of transonic rotors. In this region losses are caused mainly by blade tip-vortices which roll up above the suction sides of the blades [2]. The tip-vortices arise due to the pressure difference between suction and pressure side of the blade. The pressure difference induces a jet-like flow through the tip gap and causes the vortex formation. The vortex trajectory starts at the tip of the leading edge with a feeding sheet along the tip edge and propagates into the passage. The rolling-up forms a stream-wise slender vortex with a wake-like axial velocity profile [14, 3, 4]. Clearance flow and tip vortex influence on the overall peformance has been the subject of several recent papers, which mainly focus on blade tip design and gap height [14].

On a transonic rotor blade the leading edge thickness causes a bow shock which changes over to an expansion wave on the suction side. On the pressure side within the enclosed part of the bladed passage the shock is being intensified and crosses the passage hitting the neighbouring blade on its suction side. The aforementioned tip-vortex has to cross this so-called passage shock. The shock-vortex interaction leads to a sudden and strong deceleration of the flow, a rapid expansion of the vortex and, as a consequence, to a strong change in the vorticity distribution of the vortex (Fig. 2). Important flow parameters influencing the breakdown are the adverse pressure gradient in the stream-wise direction and the swirl intensity of the vortex.

More details on shock induced breakdown of slender vortices are given by Rizetta [9], and Thomer [13].

Schlechtriem and Lötzerich[10] detected tip-clearance vortex breakdown in a transonic axial compressor rotor downstream of the shock-vortex interaction. They conjectured that a blockage resulting from the breakdown has an important impact on stall and might even induce compressor surge. In order to clarify these phenomena numerical flow simulations are presented for the NASA rotor 37 which is described in [6, 8].

2 Geometry

The NASA rotor 37 was designed as a low aspect ratio transonic inlet stage for a core compressor to study the effects of aspect ratio and loading level on the performance of speed stages. The rotor design pressure ratio is 2.106 at a mass flow rate of 20.19 kg/s with a measured choking mass flow of 20.93 kg/s. At the design wheel speed of 17,188.65 rpm, the relative Mach number at the inlet is 1.13 at the hub and 1.48 at the tip, giving a nominal tip speed of 454 m/s. The rotor has 36 blades with an aspect ratio of 1.19, hub to tip ratio of 0.70, and a tip solidity of 1.288. The running tip clearance was estimated to be 0.356 mm (0.5% span). Details of the rotor aerodynamic design were reported by Reid and Moore[8], overall aerodynamic performance of the stage by Moore and Reid[6].

The experiment was performed in the single stage transonic compressor facility at the NASA Lewis Research Center. The rotor was tested by itself to avoid the interaction effects generated by the presence of an upstream inlet guide vane or a downstream stator blade row. Radial distributions of static and total pressure, total temperature, and flow angle were measured at two axial stations, designated as station 1 and 4 (Fig. 3). Using the blade hub leading edge as the origin z=0, station 1 is located at z=-4.19 cm and station 4 at z=10.19 cm. The flow field was also surveyed within the rotor at 20% axial chord downstream from the leading edge, named as station 2, at 30, 50, 70, 90, and 95% span. Recently, the team responsible for the experimental investigation of NASA Rotor 37 raised some doubts about the validity of the experimental data near the hub[11]. However, it is believed that the experimental data is quite accurate near the tip.

3 Computational Mesh

Structured multi-block grids are employed to deal with the geometric complexity of a turbo-compressor (Fig. 1). For this, the freely-available grid generator MegaCads, part of the MEGAFLOW project, was chosen [5]. It is an interactive grid generation code (no automation) with a few simple CAD functions for structured multi-block Navier-Stokes grids for high Reynolds-number flows. The resulting algebraic grids can be smoothed using elliptic techniques. The whole design is done in a parametric fashion and each design step is stored in a script file.

The surrounding area near the blade is discretized with an O-type body fitting grid. Hence orthogonal or nearly orthogonal cells along the surface are ensured. Outside this domain in the main flow region an H-type grid is provided. The tip gap is gridded with an H-grid embedded into an O-grid. In the endwall region near the blade tip, point-clustering is more refined in span-wise direction (Fig. 1(b)). The tip clearance gap is fully gridded and the blade tip is modelled with a rounding off like the actual airfoil tip shape (Fig. 1(c)). The inlet and exit of the computational domain were chosen to coincide with stations 1 and 4.

Used grid consists of 65 points span-wise for the blade, of which 25 are located at the upper last 2.5% span. Curvatures at leading and trailing edge are approximated with 17 points to resolve sufficiently the flow incidence. More details of the discretization can be found in Table 1.

In order to evaluate the viscous fluxes at the walls by applying the no-slip and adiabatic wall condition without a wall function method, an important parameter of grid quality is the wall distance of the first node away from the wall. The minimum grid spacing for the grid presented in this work is chosen such that dimensionless normal wall distance $y^+ \leq 1$.

4 Numerical Method

The computations are performed for three-dimensional flow using the commercial code NUMECA FINE Turbo [7]. The Favre-Reynolds-averaged Navier-Stokes equations are discretized using a cell-centered explicit finite volume scheme after Jameson in a relative coordinate system which is rotating with the reference frame. Time integration is performed using a four step Runge-Kutta scheme. To stabilize the time-marching scheme, scalar eigenvalue-based second and fourth order difference smoothing operators are used. To speed up convergence to steady state, local time stepping, residual smoothing, and multigrid techniques are applied. In the present work the two-equation k–ϵ turbulence model referred to as Yang/Shih model is used. Time correct unsteady calculations are performed employing the dual time stepping method.

4.1 Boundary Condition

At solid boundaries no-slip and no-heat transfer conditions are imposed. At the inflow boundary, flow angle, total pressure and total temperature are prescribed according to the experimental data. At subsonic outlet, static pressure is prescribed at the hub and its radial variation is calculated based on the radial equilibrium equation. At periodic boundaries from blade passage to adjacent blade passages, periodic flow conditions are imposed.

5 Results

The current study concentrates on operating conditions at design speed. Mass flow is varied by changing the outflow pressure such that steady state is followed by unsteady behaviour. Calculated mass flow rate was normalized by the choking flow rate calculated by the respective code. Similarly, the experimental mass flow rate was normalized by the experimental choking flow rate. The computational flow rate amounts to 20.82 kg/s, which lies within the experimental uncertanties 20.93 ± 0.14 kg/s.

For the calculations of the overall performance we followed the procedure proposed in [1], adapted to the discretization in the inflow and outflow stations (stations 1 and 4 in [1]). Figure 4 shows the overall pressure ratio $P_{t,4}/P_{t,ref}$ versus normalized mass flow rate of experimentally obtained data in comparison to numerically predicted values. Calculated operating conditions are denoted with symbols and interpolated with spline functions. Calculated steady state pressure ratios are in good agreement with the experimental data. The peak pressure rise is located at the same normalized flow rate of about $\dot{m}/\dot{m}_{max} = 0.915$. Diminishing the mass flow rate lower 0.912 by increasing the back pressure no convergence of the steady computations was found. We will come back to this point.

For a more detailed analysis, the measured distributions of relative Mach number are compared with computations (Fig. 7) on blade-to-blade surfaces of revolution at 20% chord downstream of the leading edge (position indicated in Fig. 6(a)). Therefore, experimental data is available for the respective flow conditions close to stall ($\dot{m}/\dot{m}_{max} = 0.925$). Figure 7 shows the relative Mach number M_{rel} distribution from pressure side to suction side at 70, 90, and 95% span. The numerical flow field prediction agrees favorably with the experiment at all positions, especially at the outer span positions. The passage shock position as well as the post- and pre-shock conditions are captured very well.

Figure 6(a) illustrates on an S1-surface at 90% span the relative Mach number distribution as well as isolines with increments of 0.1. Downstream of the passage shock, a roundish, low Mach number region exists which results from the interaction of the passage shock with the tip-clearance vortex. Taking the shroud as 100% span, the vortex core is located in an S1-plane at about 98.5% span (Fig. 6(b)) which is twice the tip-gap height below the blade tip. The Mach number contours show a trough which identifies the tip clearance vortex. The vortex induces local deformation of the shock surface. Comparing Figure 6(b) and Figure 6(c), the latter corresponds to operating condition closer to stall, the passage shock is moving upstream, the shock surface goes baggy, and the low Mach number sink increases because of an intensified shock-vortex interaction.

A deeper insight into the tip clearance vortex is gained by considering cutting planes normal to the vortex axis upstream of the passage shock (cut B, indicated in Fig. 6(b)). Distributions of flow quantities are depicted in Figure 6. The vortex axis is determined by connecting density minima within the vortex core in two successive cuts. Due to the fact that the vortex flow field is not rotationally symmetrical and in order to compare different operating conditions, data are extracted along a line, which represent nearly one major extension direction of the vortex profile (Fig. 8(a)). Within the vortex core, density minimum and pressure minimum (not shown) are nearly congruent (Fig. 8(a)). The distributions exhibit almost mirror symmetry. In comparison of different operating conditions, density and pressure distributions are shifted to lower values (Fig. 8(b)). The vortex core radius is found by observing the locations of the maximum and minimum of the circumferential velocity W_φ in the cutting plane. The density minimum coincides with the minimum of W_φ, the vortex core radius measures 1.5 mm in x-direction independently of the three different operating conditions as presented in Figure 8(c). For comparison only, the radial tip gap height measures about 0.4 mm, or 0.5 % span. Raising exit pressure causes increasing circumferential velocity of the vortex.

As mentioned above steady state computations can be performed till a normalized mass flow rate of about $\dot{m}/\dot{m}_{max}$ = .912. Therefore, for the case of lower mass flow rate computations have been conducted in the unsteady mode. treated in an unsteady way. Overall pressure ratio of every second calculated time step is marked in Figure 4. Figure 5 shows the time-dependent function constituting the exit pressure at the outlet boundary condition which follows a ramp. The starting value is identical to the one of the last converged steady state computation. Within a period of 2 milliseconds (ms) the pressure is increased about 0.6 %. Thereafter the pressure is kept constant. Within the first 2 ms overall pressure ratio is slightly unchanged. Only a small raise is visible because disturbances at the outlet need time to reach the inlet. Mass flow rate variation is also very small. After that, the changes of pressure ratio and of mass flow rate increase step by step with time till finally the computations does not converge anymore.

Figure 9 shows vortex limiting 3D streamlines for converged steady calculation. The aforementioned low Mach number regions as visible in Figures 6(b) and 6(c) can be identified in Figures 9(b) and 9(c) as recirculation zones within the vortex. For mass flow rates lower than 0.912 the unsteady computations yield an increase of the vortex recirculation zone (Fig 10), which provokes a blockage, enlarges in streamwise and radial direction and migrates upstream in the same way the shock surface does. Finally the vortex breaks down and inhibits the primary flow field. Vortex breakdown depends on the vorticity, shock conditions, and the angle enclosed by the vortex trajectory and the shock surface. Results support the conjecture that compressor stall is strongly influenced by tip vortex breakdown.

6 Conclusions

In the present study, an existing commercial 3D, RANS flow-solver has been used to simulate the stationary and unsteady flow field within the transonic axial compressor rotor NASA rotor 37. Numerical results are shown for various operating conditions using a structured multi-block computational grid. The tip clearance gap is fully gridded and the blade tip is modelled with a rounding off more like the actual airfoil tip shape. The turbulent viscosity has been estimated by the use of the two equation turbulence model $k - \epsilon$ with enhancements by Yang and Shih. Quantitatively, the whole passage flow field as well as the vortex flow are sufficiently predicted.

Tip clearance vortices can be classified as slender with a wake-like axial velocity profile. Vortex density distributions are shifted to lower levels changing operating conditions along the speed line, whereas circumferential velocity increases.

When the tip-vortex approaches the passage shock, vortex core flow decelerates and the vorticity distribution changes abruptly. The shock surface receives a dent and, downstream, a roundish, bubble-like region of very low Mach numbers exists, which is the vortex recirculation area. Increasing back pressure, the passage shock moves upstream and shock-vortex interactions are enforced. Blockage in terms of recirculation increases. Finally the vortex breaks down and inhibits the primary flow field. From this point of view, vortex breakdown is joined by stall in an axial compressor rotor.

Future work will emphasize on the analysis of different circumferential rotor speeds and different modified blade tip shapes in order to influence the tip vortex and the passage shock position.

Acknowledgements

This work is sponsored at RWTH Aachen by the Deutsche Forschungsgemeinschaft within the frame of the Transfer-Project "Tip Leakage Vortex - Passage Shock - Interaction concerning Transonic Compressor Rotors and their Impact onto Surge Conditions" with industrial partnership of Rolls-Royce Germany, Dahlewitz near Berlin.

References

[1] Davis, R. L., Delaney, R. A., Denton, J. D ., Giles, M, B., Strazisar, A. J., and Wisler, D. C.: "CFD Code Assessment in Turbomachinery - Author's Information Package". The American Society of Mechanical Engineers Turbomachinery Committee, 1993.

[2] Gerolymos. G., Vallet I., : "Tip-Clearance and Secondary Flows in a Transonic Compressor Rotor". ASME Paper 98-GT-366, Presented at the International Gas Turbine & Aeroengine Congress & Exhibition, Stockholm, Sweden, June, 2-5, 1998.

[3] Hofmann, W., Ballmann J., : "Tip-Vortices in Transonic Axial-Compressors". Notes on Numerical Fluid Mechanics, Springer, Vol. 77, pp. 341-349, 2002.

[4] Hofmann, W., Ballmann J., : "Tip Clearance Vortex Development and Shock-Vortex-Interaction in a Transonic Axial Compressor Rotor". AIAA Paper 2002-0083, 2002.

[5] Kroll, N., Rossow, C.-C., Becker, K., and Thiele, F. : "The MEGAFLOW project". Aerospace Science and Technology, Vol. 4, 2000, pp. 223-237.

[6] Moore, R.D., Reid, L., : " Performance of Single-stage Axial-Flow Transonic Compressor Rotor and Stator Aspect Ratios of 1.19 and 1.26, Respectively, and with Design Pressure Ratio of 2.05". NASA TP 1659, 1980.

[7] "FINE^{TM}Numeca's Flow Integrated Environment". User Manual FINE^{TM}, April 2000.

[8] Reid, L., Moore, R.D., : "Design and Overall Performance of Four Highly Loaded, High Speed Inlet Stages for an Advanced High-Pressure Ratio Core Compressor". NASA TP 1337, 1978.

[9] Rizzetta, D.P., : "Numerical Simulation of Vortex-Induced Oblique Shock-Wave Distortion". AIAA Journal, Vol. 35, No. 1, pp. 209-211, 1997.

[10] Schlechtriem, S., Lötzerich, M., : "Breakdown of Tip Leakage Vortices in Compressors at Flow Conditions Close to Stall". ASME Paper No. 97-GT-41, 1997.

[11] Shabbir, A., Celestina, M.L., Adamczyk, J.J., and Strazisar, A.J., : "The effect of Hub Leakage on 2 High Speed Axial Flow Compressor Rotors". ASME Paper 97-GT-346, 1997.

[12] Strazisar, A. J., : "Data Report and Data Diskette for NASA Transonic Compressor Rotor 37". NASA Lewis Research Center, Cleveland, OH, 1994.

[13] Thomer, O., Krause, E., Schröder, W., Meinke, M., : "Computational Study of Normal and Oblique Shock-Vortex Interactions". ECCOMAS 2000, Barcelona 11-14, September, 2000.

[14] Van Zante, D.E., Strazisar, A.J., Wood, J.R., Hathaway, M.D., Okiishi, T.H.: "Recommendations for achieving accurate numerical simulation of tip clearance flows in transonic compressor rotors". ASME Journal of turbomachinery, 122(4): pp. 733-742, 2000.

Table 1 Computational grid summary.

Points, total (Mio.)	1.26	Pitch-wise, blade-to-blade	77
Span-wise, blade	65	Pitch-wise, tip gap	21
Span-wise, tip gap	17	Stream-wise	255
		Stream-wise, blade	85

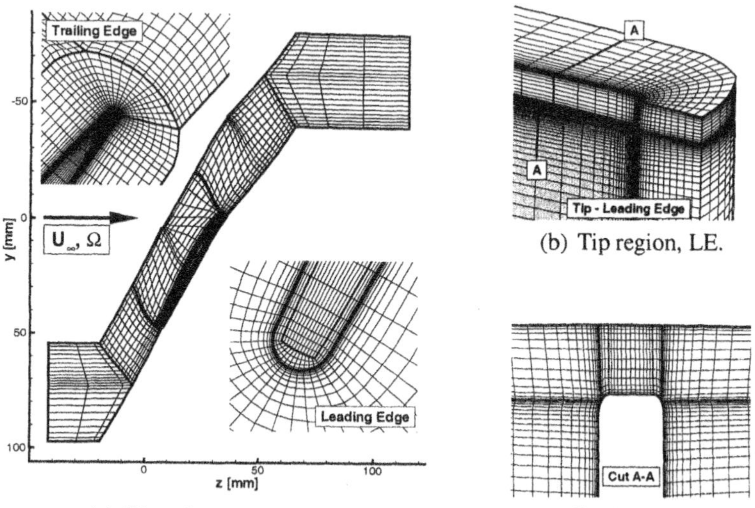

(a) S1 surface, near casing. (b) Tip region, LE. (c) Cutting-section.

Figure 1 Computational grid.

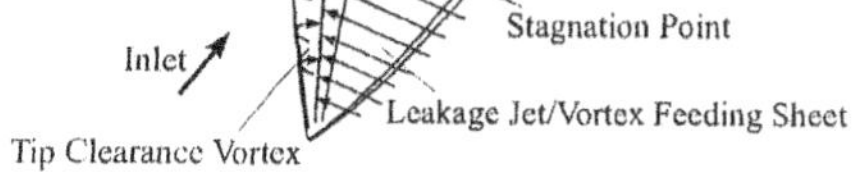

Figure 2 Shock-vortex interaction in a cmpressor blade duct, schematic view.

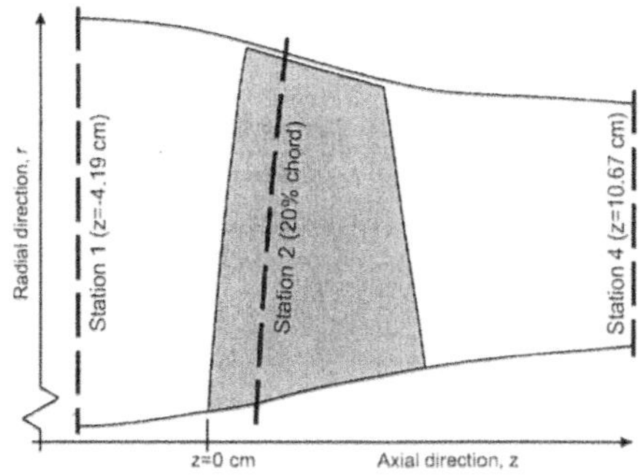

Figure 3 Measurement stations, meridional view, according to [1].

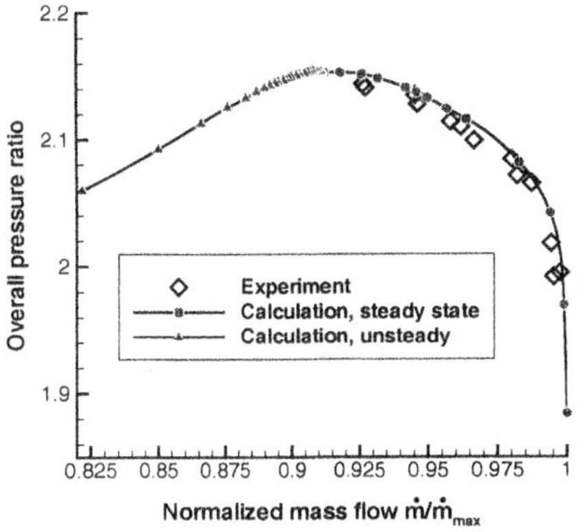

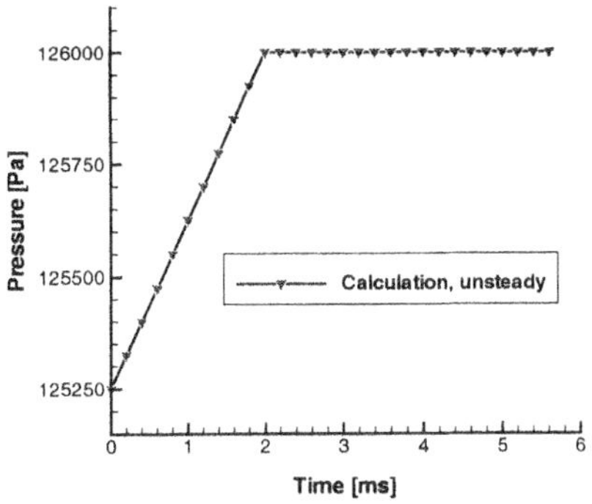

Figure 4 Overall pressure ratio, 100% speed line.

Figure 5 Outflow boundary: time history of exit pressure raise at hub.

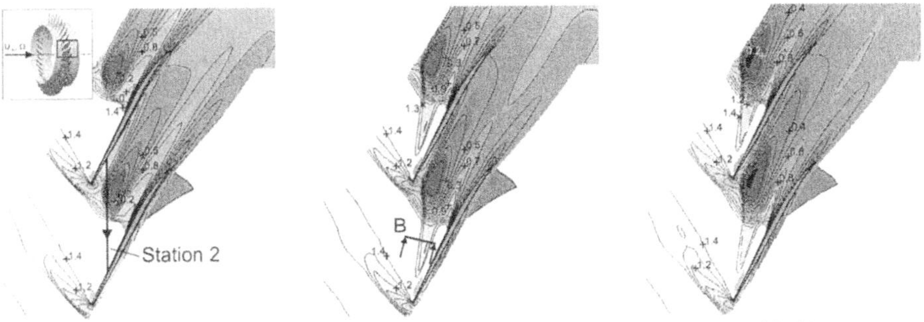

(a) S1-95% span. (b) S1-98.5% span. (c) S1-98.5% span.

Figure 6 Distribution of M_{rel} on S1: (a), (b) $\dot{m}/\dot{m}_{max} = .925$; (c) $\dot{m}/\dot{m}_{max} = .912$.

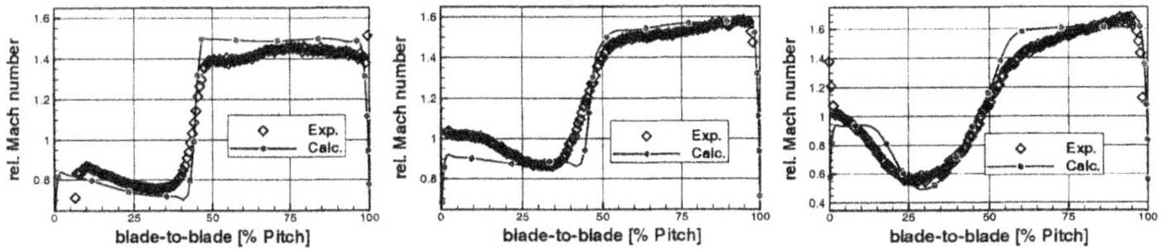

(a) 70% span. (b) 90% span. (c) 95% span.

Figure 7 Relative Mach number M_{rel}, Station 2, PS-to-SS, $\dot{m}/\dot{m}_{max} = .925$.

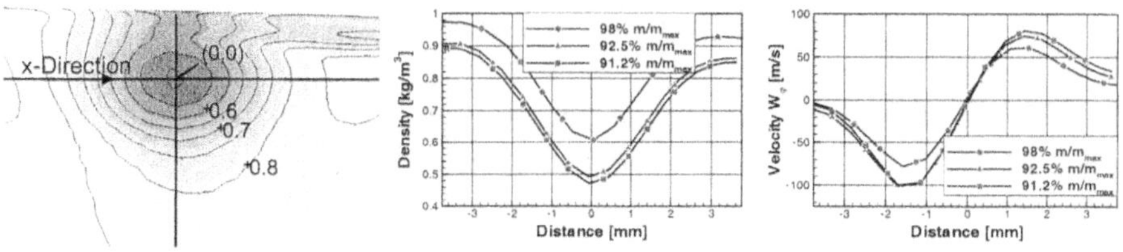

(a) (b) (c)

Figure 8 Density distribution in cut B (see Fig.6(b)) (a) for $\dot{m}/\dot{m}_{max} = .925$; density profile (b) and profile of velocity W_φ (c) across the vortex in x-direction.

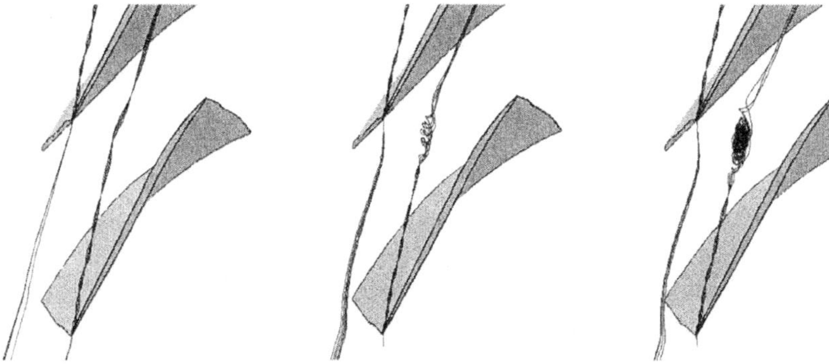

(a) $\dot{m}/\dot{m}_{max} = .980$. (b) $\dot{m}/\dot{m}_{max} = .925$. (c) $\dot{m}/\dot{m}_{max} = .912$.

Figure 9 Vortex limiting 3D streamlines, steady state calculation

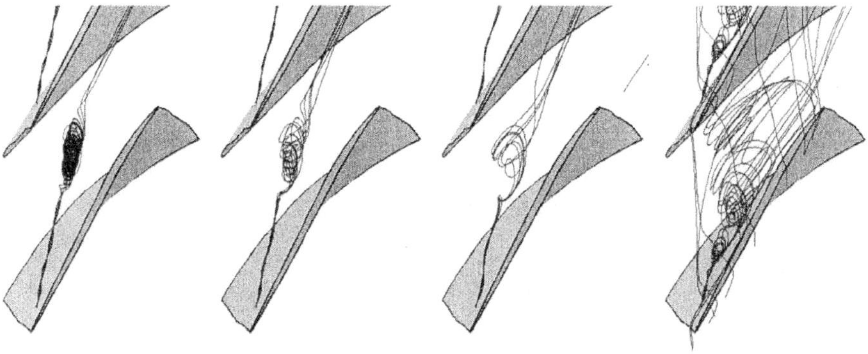

(a) $\dot{m}/\dot{m}_{max} = .902$, $time = 3.0\ ms$. (b) $\dot{m}/\dot{m}_{max} = .897$, $time = 3.8\ ms$. (c) $\dot{m}/\dot{m}_{max} = .890$, $time = 4.6\ ms$. (d) $\dot{m}/\dot{m}_{max} = .866$, $time = 5.4\ ms$.

Figure 10 Instantaneous vortex limiting 3D streamlines, unsteady calculation.

Second-Moment Closure Modelling of Swirl-Induced Separation

R. Jester-Zürker and S. Jakirlić

Fachgebiet Strömungslehre und Aerodynamik,
Darmstadt University of Technology, Petersenstr. 30, 64287 Darmstadt, Germany

Summary

A Second-Moment Closure (SMC) study of an isothermal, incompressible swirling flow in three generic combustor configurations featuring different inflow structure with respect to the circumferential velocity type (both configurations with annular and central swirling jet were considered), combustor confinement (in terms of expansion ratio ER) and swirl intensity (S) was performed. The study focuses primarily on the flow separation induced by an imposed swirl, the phenomena termed as the *vortex breakdown*, but also on the flow relaxation in the far field. Both the basic high-Reynolds number SMC model, modified to account for the non-linearities in the pressure scrambling and dissipation processes, and a low-Reynolds number SMC model, accounting separately for the viscous and non-viscous wall blockage, were applied. Inflow conditions are computationally generated, rather than prescribed. In the course of the inflow data generation, a number of the "equilibrium" swirling flows in the concentric annulus and pipe geometries, from which coaxial and central swirling jets expand into a combustion chamber, are computed. The SMC results reproduced all important mean flow and turbulent features in a good agreement with available experimental and LES data.

1 Introduction

A mechanism being most responsible for the flame stabilization in a combustor configuration is the flow reversal generated locally in the flow core. This flow phenomena is a consequence of an intensive action of the angular momentum, being of sufficiently high intensity, entering the combustor. This swirl-induced bubble represents the outcome of the transition from the non-swirling (or weakly swirling), so-called, supercritical flow situation, to the subcritical flow state, with respect to the reverse flow in the core. This sudden change of the flow structure corresponds to the well-known *vortex breakdown* phenomena (see, e.g., Escudier and Keller, 1985). In the practically relevant combustion chamber geometries, being the subject of the present work, flow is recovered (that is non-recirculating) after passing through the subcritical regime in the region of the backflow. This leads to the conclusion that the flow in the combustor is locally subcritical. It is even more pronounced if the reactive flow in combustor is observed. An increase in the fluid viscosity due to the combustion process can cause flow acceleration in the streamwise direction, contributing significantly to the re-establishment of the supercritical flow situation. The

adverse pressure gradient, being an outcome of the sudden expansion, supports additionally the generation of a large free recirculation zone. The joint adverse pressure gradient (dilatational effects) and additional straining due to presence of the swirl motion (secondary shearing and streamline curvature effects) cause the weakening of the stress anisotropy, being especially pronounced in the flow core, Jakirlić et al. (2002).
The boundary conditions of the flows considered correspond to the experimental configurations investigated by Roback and Johnson (1983), So et al. (1984) and Nejad et al. (1989). The Roback and Johnson's experimental data base was recently enriched by the high quality LES results (Pierce and Moin, 1998a). The corresponding Reynolds numbers, expansion ratios and swirl intensities are given in Table 1. The stated Reynolds numbers are related to the incoming flow. All three flow configurations are presented schematically in Fig. 1, illustrating clearly the differences in the structure of swirling inflow. The variability in the inflow conditions increases significantly the sensitivity of the prediction to the quality of the turbulence models.

2 Computational method

The combustor flows considered are in fact three-dimensional, for which all three velocity components and all six Reynolds stress components must be accounted for. However, the Reynolds-averaged continuity and momentum equations are solved with assumption of axisymmetry ($\partial/\partial\varphi = 0$):

$$\frac{1}{r}\frac{\partial(rU_j)}{\partial x_j} = 0\,; \quad \frac{1}{r}\frac{\partial(rU_jU_i)}{\partial x_j} = \frac{1}{r}\frac{\partial}{\partial x_j}\left[r\left(\nu\frac{\partial U_i}{\partial x_j} - \overline{u_iu_j}\right)\right] - \frac{1}{\rho}\frac{\partial P}{\partial x_i} + S_{U_i}$$

where the vector $x_i(r,z)$ denotes the coordinate directions and $U_i(U_r, U_\varphi, U_z)$ ($\equiv (V, W, U)$) stands for the mean velocity vector. The source terms S_{U_i} arise from the coordinate transformation (see e.g., Jakirlic et al., 2000, for further details about governing equations written in the cylindrical coordinate frame).

Numerical framework All computations were performed by a computer code based on a finite-volume numerical method for solving the Reynolds Averaged Navier-Stokes (RANS) equations in orthogonal coordinate systems with a collocated variable arrangement. The velocity-pressure coupling is ensured by the pressure correction method based on the well known SIMPLE algorithm. The equations are linearized and solved sequentially using an iterative ILU method. Diffusion fluxes are approximated by central differences (CD). For discretization of the convective fluxes, a blended upwind-central differencing scheme, implemented in the so-called deferred-correction manner was used. The value of the blending factor was typically between 0.7 (for Second-Moment Closure models) and 1.0 (for $k-\varepsilon$ models) for all variables. The solution domain used has, typically, a shape of the suddenly expanding axisymmetric pipe geometry, with a length L and a height $R = D/2$, D being a combustor diameter. The dimensions of the solution domain and final sizes of the numerical grids used for high and low-Reynolds number model computations are given in Table 1. Coarser, but also refined grids were applied in the course of the grid independence study (see e.g., Fig. 2). The grids were squeezed in the shear layer and near-wall regions, providing the dimensionless wall distance y^+ of the numerical node closest to the wall being ≥ 15, when using high-Reynolds models, i.e. ≤ 0.5, if low-Re models were applied. Standard wall functions were used for high-Reynolds number model computations. It is common to compute combustor flows in a steady manner, if RANS models are used to describe turbulence. The so-called "temporal underelaxation" was applied here to strenghten additionally the diagonal dominance of the coefficient matrix, implying actually the unsteady computations were performed. However, such an unsteady calculation ended up, as expected, in a steady solution.

Turbulence models Computations were performed with the low-Re-number Reynolds-stress model (Hanjalic and Jakirlic - HJ low-Re SMC, 1998) and with the basic high-Re-number Reynolds-stress model (Gibson and Launder - GLm, 1978) modified to account for the non-linearities in:

• the slow part of the pressure scrambling process, $\Phi_{ij,1}$ (Speziale et al. - SSG, 1991) and

• the dissipation correlation, ε_{ij} (Hallbäck et al.-HGJ, 1990).

The non-linear parts of the model formulations for $\Phi_{ij,1}$ and ε_{ij} can be easily coupled:

$$\Phi_{ij,1} - \varepsilon_{ij} = -C_1 \varepsilon a_{ij} - (C_1' + \alpha)\varepsilon \left(a_{ik}a_{kj} - \frac{1}{3}A_2\delta_{ij} \right) - \left[(1 - f_s)\frac{2}{3}\delta_{ij}\varepsilon + f_s \frac{\overline{u_i u_j}}{k}\varepsilon \right]$$

where $a_{ij} = \overline{u_i u_j}/k - 2/3\delta_{ij}$, $A_2 = a_{ij}a_{ji}$, $f_s = 1 + \alpha(A_2/2 - 2/3)$ and $\alpha = 3/4$. This most general model for the slow part is quadratic, because the Cayley-Hamilton theorem allows higher powers of a_{ij} to be expressed in terms of a_{ij} and a_{ij}^2. The DNS database for fully developed channel flow enables the evaluation of the profiles of the model coefficients C_1 and C_1' (see Hanjalic and Jakirlic, 2002, for more details). Fig. 3 shows the coefficients evaluated in such a way for different combinations of the $\Phi_{ij,1}$-components. The coefficient C_1 reaches the standard high-Reynolds number value $1.7 - 1.8$ away from the wall, in a good agreement with the most of the model proposals (not shown here). The evaluated coefficient C_1' is negative, apart from very close to the wall, showing in fact an opposite behaviour from C_1. Far from a wall the value of C_1' is close to that proposed by Speziale et al. (1991), $C_1' = -1.05$, at least in the logarithmic region. The HGJ model is developed for homogeneous flows, meaning that the dissipation rate ε represents actually the homogeneous dissipation, which differs from the total dissipation rate only very close to the solid walls ($y^+ \leq 15 - 20$). The non-linear model formulation by Hallbäck et al. captured the dissipation anisotropy in good agreement with the DNS data of the flows being affected by separation (not shown here) and swirling effects, Fig. 4. The Rotta's isotropic model ($\varepsilon_{ij} = 2/3\delta_{ij}\varepsilon$) is also shown for comparison.

In addition, all flow cases considered are solved using the standard high-Reynolds number $k - \varepsilon$ model and its low-Reynolds number extension due to Launder and Sharma (1974).

Inflow conditions The specification of the inflow boundary conditions in the case of swirling flows deserves special attention. The inlet cross-section of the computational domain traditionally coincidences with the first available measurement location, usually being situated in the interior of the combustor. In this work the inlet boundary was located prior to expansion and the inflow data were generated by doing a separate computation of the fully developed annular swirling flow, imposing a uniform profile of the fictitious azimuthal pressure gradient (that is azimuthal body force), in line with the proposal of Pierce and Moin (1998b). Fig. 5 shows the mean circumferential velocity W and corresponding turbulent shear stress components $\overline{uw}$ and $\overline{vw}$ across the concentric annulus of the Roback and Johnson's model combustor obtained by such a computational procedure.
The low-Reynolds number Second-Moment closure model due to Hanjalic and Jakirlic (1998), tested in a range of the flow situations featuring different mean flow and turbulent phenomena, was used for the inflow data generation. Correct capturing of the highly anisotropic near-wall region of the fluid streams in the concentric annulus and central pipe was very important keeping in mind that these wall layers unify in the combustor, passing into a curved shear layer.

3 Results and discussion

Fig. 6 displays the streamline pattern obtained by the GLm model illustrating clearly the mean flow structure, being characterized by an annular recirculation region (corner bubble) and a large free separation region in the flow core. A short wake region between the inner and annular streams passes into a large-eddy shear region between both recirculation zones. The most intensive turbulence production occurs just in this flow region bordering both the central and corner bubbles. Whereas the annular swirling jet separates at the sharp edge of the sudden expansion generating the corner bubble, which ends at the wall building the reattachment region, both the separation and reattachment points of the large recirculation zone are situated at the combustor centerline i.e. in the free stream. Proper prediction of the separation onset and reattaching length of such a free bubble represents a special challenge for the statistical turbulence models. Fig. 7 shows evolution of the mean axial velocity profiles for two of three model combustors investigated. Whereas the joint effects of the high expansion ratio ($ER = 2.1$) and swirling inflow in the case of the Roback and Johnson's combustor contributed to the suprisingly good cut off of the standard $k - \varepsilon$ model, Fig. 7 upper (in spite of the inherent, well-known weaknesses by treating swirling and separation separately), a somewhat milder flow expansion ($ER = 1.5$, Nejad et al. case), by comparable swirl intensity (see Table 1), causes the free separation zone being totally missed, Fig. 7 lower. Both the onset of the free separation zone and its length are correctly predicted by the GLm model described above. Although no free separation occurs in the third flow configuration (due to a small non-swirling central jet) examined by So et al., a similar misrepresentation of the axial flow by using the standard linear $k - \varepsilon$ model is documented (not shown here). Another consequence of isotropic eddy viscosity is the so-called "solid-body rotation" form of the tangential velocity profile, Fig. 8, which is always obtained, independent of the initial velocity profile. One of the reasons for such a behaviour is the simple linear relationship between Reynolds-stress and mean rate of strain tensors, implying essentially that the eddy viscosity is isotropic. But it is well known that swirl causes a strong anisotropy of both the stress and dissipation tensors, as well as a highly anisotropic eddy viscosity. Fig. 9 displays the evolution of the streamwise turbulence intensity and the shear stress component along the Nejad's et al. flow. The highest values of these turbulence quantities are clearly located in the regions of the strong mean shear, where the most intensive interaction between the mean and reversed flow regions and turbulence takes place. Both phenomena, the separation due to sudden expansion and strong swirl in the entrance region are associated with the adverse pressure gradient effects, which, as already known, act toward isotropic state. It leads to the weakening of the stress anisotropy, which is most pronounced in the combustor core. Such a behaviour is closely connected to the creation of a large free recirculation zone. The profiles of the two reminding diagonal stress components (not shown here) are similar shaped indicating a high production rate around the combustor center in the entrance region, being the consequence of the strong secondary shearing ($\partial W/\partial r$) and streamline curvature effects ($-W/r$). The shear stresses are somewhat undepredicted in the regions coinciding with the shear layers, representing well known weakness of the statistical turbulence model. However, it is not of the crucial importance in this flow class, keeping in mind that the flows are strongly dominated by pressure gradient and angular momentum.

4 Conclusions

Three swirling flow configurations akin to those encountered in combustion chambers, varying in the structure of the swirling inflow, were investigated numerically. The variability in the inflow conditions significantly influences the prediction results, according to the turbulence models used. Second-moment closure models, providing information about all stress components and containing exact terms for dealing with the swirling and rotation effects in the stress equation, are inherently capable of capturing most of the phenomena mentioned. Indeed, in contrast to the k-ε models, in all flow cases reported here the second-moment closures produced superior results.

Acknowledgements This work has been financially supported by the Deutsche Forschungsgemeinschaft (DFG) through the grant SFB 568.

References

[1] Eggels, J.G.M., Boersma, B.J., and Nieuwstadt, F.T.M. (1994): Report Lab. for Aero- und Hydrodynamics, Delft University of Technology, Delft, The Netherlands, October
[2] Escudier, M.P., and Keller, J.J. (1985): *AIAA J.*, **23(1)**, pp. 111-116
[3] Gibson, M.M.. and Launder, B.E. (1978): *J. Fluid Mech.*, **86**, pp. 491-511

[4] Hallbäck, M., Groth, J., and Johansson, A.V. (1990): *Physics of Fluids*, A2 (10), pp. 1859-1866
[5] Hanjalić, K., and Jakirlić, S. (1998): *Computers and Fluids*, **27**, pp. 137-156
[6] Hanjalić, K., and Jakirlić, S. (2002): In *Closure Strategies for Turbulent and Transitional Flows*, B.E. Launder and N.H. Sandham (Eds.), Cambridge University Press, Cambridge, UK, pp. 47-101
[7] Jakirlić, S., Volkert, J., Pascal, H., Hanjalić, K., and Tropea, C. (2000): *Int. J. of Heat and Fluid Flow*, **21(5)**, pp. 627-639
[8] Jakirlić, S., Hanjalić, K., and Tropea, C. (2002): *AIAA J.*, **40(10)**, pp. 1984-1996
[9] Launder, B.E. and Sharma, B.I. (1974): *Letters in Heat and Mass Transfer*, **1**, p 131
[10] Nejad., A.S., Vanka, S.P., Favaloro, S.C., Samimy, M. and Langenfeld, C. (1989): *ASME J. Eng. for Gas Turbines and Power*, **111**, pp. 36-45
[11] Pierce, C.D., and Moin, P. (1998a): *AIAA Paper* 98-2892
[12] Pierce, C.D., and Moin, P. (1998b): *AIAA J.*, **36(7)**, pp. 1325-1327
[13] Roback, R., and Johnson, B.V. (1983): NASA CR 168252
[14] So, R.M., Ahmed, S.A., and Mongia, H.C. (1984): NASA CR 3832
[15] Speziale, C.G., Sarkar, S., and Gatski, T.B. (1991): *J. Fluid Mech.*, **227**, pp. 245-272

Table 1 Characteristics of the flow configurations considered and sizes of the numerical grid used for high and low-Reynolds number model computations

Test case	Re_S	ER	S	L, m	R, m	Grid size, CV's High-Re/Low-Re
Roback & Johnson (1983)	47500	2.1	0.41	0.775	0.061	$200x70/210x120$
So et al. (1984)	83000	1.0	2.24	0.3567	0.0625	$250x70/300x130$
Nejad et al. (1989)	125000	1.5	0.5	0.762	0.0508	$240x70/280x120$

a) **b)** **c)**

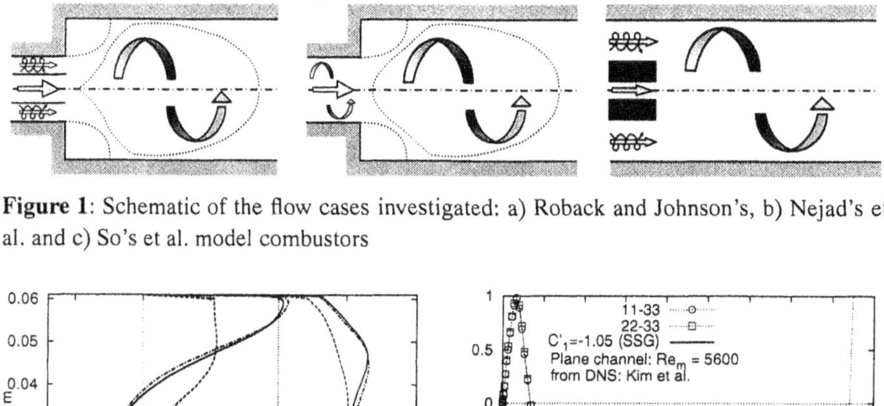

Figure 1: Schematic of the flow cases investigated: a) Roback and Johnson's, b) Nejad's et al. and c) So's et al. model combustors

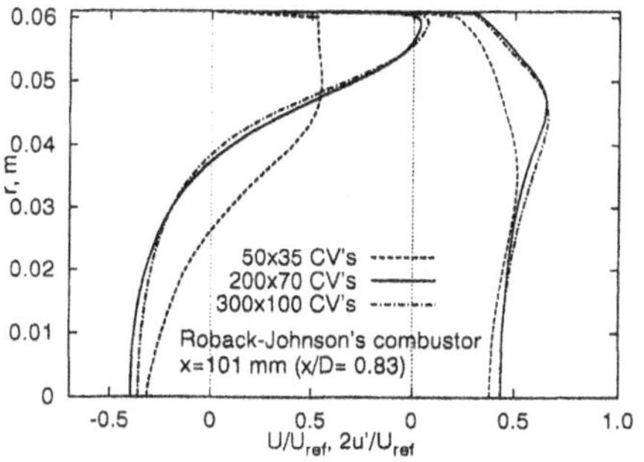

Figure 2: Effect of grid refinement: axial velocity and streamwise turbulence intensity

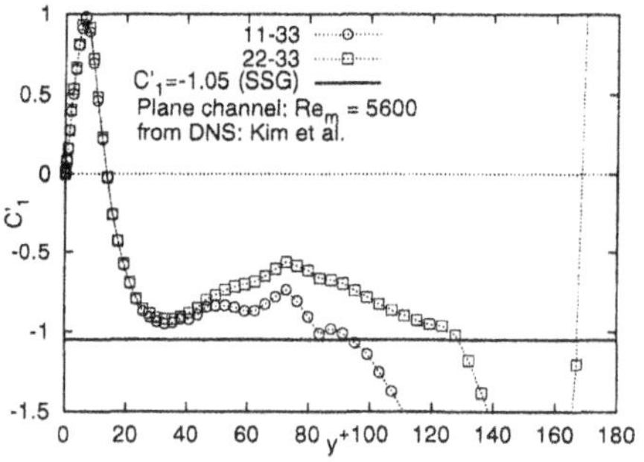

Figure 3: Profile of C_1' coefficient in the fully developed channel flow

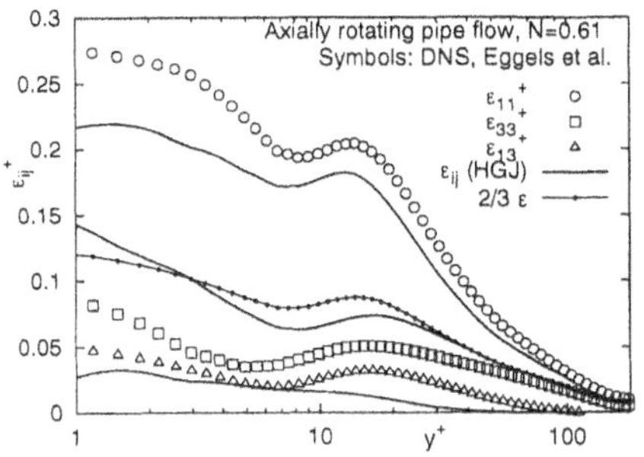

Figure 4: Dissipation correlation components obtained by the HGJ model in the swirling flow in an axially rotating pipe

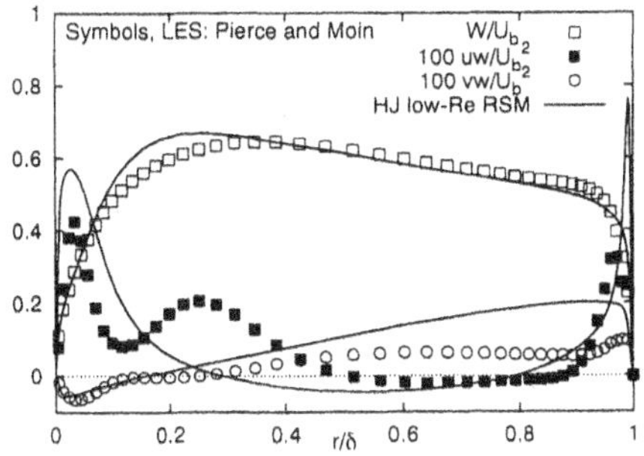

Figure 5: Mean swirling velocity and shear stress profiles across the annular swirling flow of the Roback and Johnson's combustor

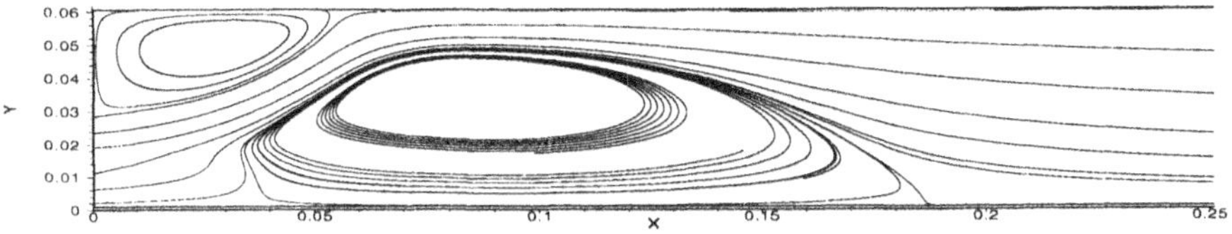

Figure 6: Computationally obtained streamline pattern (Roback and Johnson's case)

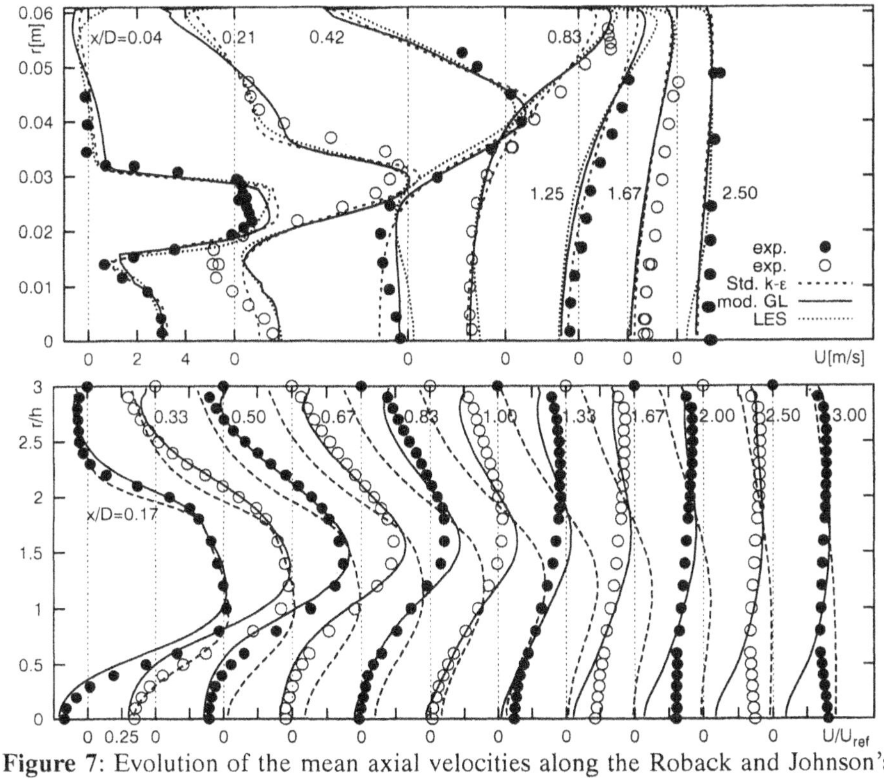

Figure 7: Evolution of the mean axial velocities along the Roback and Johnson's (upper) and Nejad's et al. (lower) combustor configurations

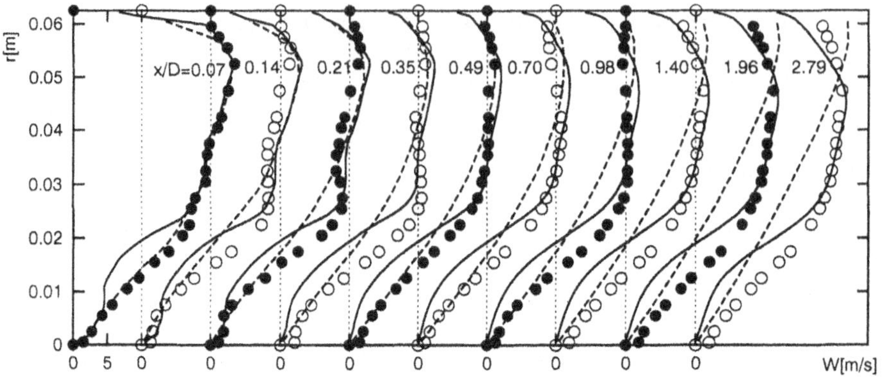

Figure 8: Evolution of the mean circumferential velocities along the So's et al. combustor configuration. Key as Fig. 7

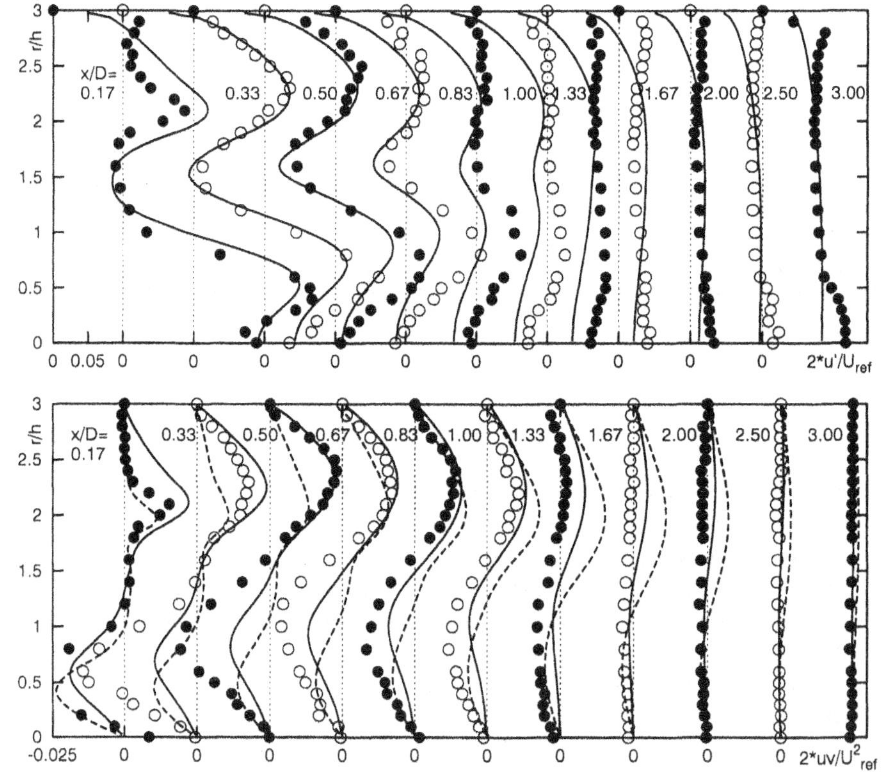

Figure 9: Evolution of the streamwise turbulence intensity (upper) and the shear stress component (lower) along the Nejad's et al. combustor configuration. Key as Fig. 7

Influence of transonic Flow on the Thrust Determination of TPS during Wind Tunnel measurements

H. Frhr. von Geyr, C.-C. Rossow, H. Hoheisel

German Aerospace Center DLR, Institute of Aerodynamics and Flow Technology,
Lilienthalplatz 7, 38108 Braunschweig, Germany, heiko.vgeyr@dlr.de

Summary

The drag determination of powered configurations equipped with Turbine-Powered-Simulators (TPS) at transonic speeds in the wind tunnel is basically the problem of correct net thrust determination, since neither drag nor thrust can be measured directly. For thrust evaluation the Standard-Bookkeeping-Method (SBM) is applied, which computes the gross thrust from measured quantities and coefficients, which are obtained during static engine calibration. In wind tunnel experiments, peculiar drag behaviour was observed in terms of the effect of power settings and the influence of different bypass ratios (BPR). Within this contribution analysis using CFD are performed to demonstrate the influence of external flow on the gross thrust for an isolated TPS. From the obtained results the shortcomings of the SBM at transonic external flow are identified and a thrust corrections to the SBM are derived. The application of the corrections to the measured Lilienthal-polars stress the necessity of a valid bookkeeping procedure for transonic Mach numbers.

1 Introduction

Civil air traffic will, according to the opinion of experts, further grow within the future. Environmental and economical demands are the key drivers for the necessity to substantially reduce the specific fuel consumption of an aircraft. Besides the effort spend by the aircraft manufactures to further reduce the aerodynamic drag, the largest potential is found in the increase of the engine's bypass ratio. The growing outer nacelle diameter with increasing BPR, limitations on the landing gear height, and the demand for sufficient nacelle ground clearance require a close coupling between wing and engine for the underwing installation of high-bypass-ratio engines [7]. Therefore, jet propulsion system integration becomes of major importance, since otherwise the gain on engine efficiency can be compensated by the increase of the installation drag [2].

The determination of engine installation effects with respect to influences on the aerodynamic drag is one of the key objectives of wind tunnel measurements. The key problem of drag measurements of a TPS-driven configuration in the wind tunnel is based on accurate determination of the net thrust, since neither of the components can be measured directly. Therefore the net thrust has to be determined accurately, which requires information about the gross thrust present for given wind tunnel conditions and engine power setting. The presently applied method is the Standard-Bookkeeping-Method [6], which defines correlation coefficients by static engine calibration to be used for wind tunnel gross thrust determination. The main assumption behind the SBM is, that these coefficients as pure functions of the fan nozzle pressure ratio (FPR) are independent of the external flow. The main objective of the current contribution is to outline the influence of the external flow on the resulting force contributing to the gross thrust. The presented results are based on numerical computations of isolated TPS and will underline the necessity of thrust corrections of the SBM to accurately determine the net thrust at transonic conditions [5]. This is underlined by the application of the corrections to drag polars obtained for the TPS-powered ALVAST configuration [10] during the ENIFAIR project [2].

2 Applied Methods and Test Devices

Numerical Methods and Test Devices The flow solver used for the this study is the DLR in-house flow solver FLOWer [8]. This code solves the time averaged Navier-Stokes equations on a structural multiblock grid using a finite-volume cell-vertex formulation. Space discretisation is realised through a central scheme with artificial viscosity. Time integration is carried out by an explicit hybrid multistage scheme. Local time stepping, implicit residual smoothing and multigrid techniques are used to accelerate the convergence to steady state. The two-equation k-ω model of Wilcox considers turbulence effects. The generation of the structured multi-block grids has been performed with the grid generator MegaCads [1].

The DLR-ALVAST configuration represents a twin-engine AIRBUS A320 like geometry with respect to fuselage and wing planeform, scaled by a factor of 1:10. The model can be equipped with three different TPS-units, a Turbo-Fan (TF) engine (BPR=5), a Very-High-Bypass-Ratio engine (VHBR) (BPR=9.2) and a Ultra-High-Bypass-Ratio (UHBR) engine (BPR=15.7) [4] and was installed in the ONERA S1MA wind tunnel as a half model for measurements. The cruise condition is defined by a Mach number of $Ma = 0.75$ and Start-Of-Cruise (SOC) power settings, for which the TF and VHBR simulators are operating at choked conditions.

The Standard-Bookkeeping-Method In the wind tunnel measurement the net thrust F_N is defined as the change of momentum between station *E0* at infinity upstream of the engine and the stations *E19, E9* at infinity downstream of the engine. The corresponding control volume is sketched by the dashed lines in Figure 1. The thrust domain is defined by the intake stream tube, the flow through the engine, and the post exit streamtubes, which preserve the fan and core mass flows. Hence, the momentum balance leads to the following expression for the wind tunnel balance force X_W:

$$X_W = (F_G - \dot{m}_{Fan} \cdot v_0) - (\phi_{Nacelle} + \phi_{Pre} + \phi_{Post}) ,$$
$$\text{with } F_N = (F_G - \dot{m}_{Fan} \cdot v_0) \ , D = (\phi_{Nacelle} + \phi_{Pre} + \phi_{Post}) \ . \qquad (1)$$

The aerodynamic drag D consists of surface forces along the external nacelle surface $\phi_{Nacelle}$, the pre-entry forces ϕ_{Pre} of the intake streamtube, and the post-exit forces ϕ_{Post} of the representative fan and core jet streamtubes. The net thrust definition F_N consists of the gross thrust F_G and the inlet momentum, where the gross thrust is found to be:

$$F_G = \dot{m}_{Fan} \cdot v_{Fan,19} + \dot{m}_{Core} \cdot v_{Core,9} \tag{2}$$

Thus, the presented definitions allow a formally correct force accounting to the thrust and the drag domain. In order to determine the gross thrust, the unknown velocities at *E19*, *E9* and the unknown fan mass flow are computed using the fan velocity coefficient C_v and the discharge coefficient C_D, obtained by static engine calibration. These coefficients correlate the particular quantity with the quantity at isentropic conditions. Hence, the gross thrust is obtain by [6]:

$$F_G = \dot{m}_{Fan,id} \cdot C_D \cdot v_{Fan,19,id} \cdot C_v + \dot{m}_{Core} \cdot v_{Core,9,id}$$
$$C_v = \frac{X_C - \dot{m}_{Core} \cdot v_{Core,id}}{\dot{m}_{Fan} \cdot v_{Fan,id}} \text{ and } C_D = \frac{\dot{m}_{Fan}}{\dot{m}_{Fan,id}} \tag{3}$$

At calibration conditions no external flow is present and therefore the calibration balance force X_C is equal to the the gross thrust at calibration condition, which leads to the definitions of C_v.

3 Results

Application of the SBM At Through-Flow-Nacelle (TFN) conditions no effective propulsion force is generated. The drag difference between the installed configuration at TFN and the uninstalled wing-body (WB) configuration defines the installation drag $\Delta C_{D,I} = C_{D,TFN} - C_{D,WB}$. Figure 2 shows the measured Lilienthal-Polars and the installation drag for all configurations at $Ma = 0.75$. Compared to the WB-configuration the drag of all installed configurations is increased and this shift becomes more distinct with increasing BPR. When the engines are running at SOC-conditions, thrust determination becomes important. Figure 3 summarizes the polars for the installed configurations and shows in the right part the drag changes due to the jet effect. The latter is defined as $\Delta C_{D,J} = C_{D,SOC} - C_{D.TFN}$. Compared to TFN conditions the drag at SOC appears to reduce, independent of the installed engine and nearly no BPR influence is visible. Based on previous results and from well-established knowledge this behaviour is not understandable. With respect to the above outlined SBM, the most likely explanation for this peculiar behaviour is the inaccuracy of the bookkeeping procedure and here in particular a possible influence of the external flow in the correlation coefficients.

Influence of the freestream Mach number The capability of the numerical method to capture local engine installation effects in terms of power effect and the configurational influence has been demonstrated in [3] and [9]. In order to investigate the influence of the freestream Mach number on the thrust development, numerical computations of different isolated simulators have been performed for different power settings. Within this contribution the focus is put on the TF and exemplary results will be discussed for the SOC-conditions. The Mach number is stepwise decreased from $Ma = 0.75$ towards the calibration condition of $Ma = 0.0$. Here the computations are performed using a pre-conditioned system of the governing equations, available as a feature of the flow solver. Figure 4 compares the jet development between $Ma = 0.75$ and $Ma = 0.34$ by showing ISO-Mach contours overlayed with profiles of the axial velocity component in several sections downstream of the nozzle exits. From the velocity profiles it is clearly seen, that the fan jet spreading at $Ma = 0.34$ is more distinct than for the high speed case. This already indicates, that a difference of the gross thrust contributing forces between calibration and wind tunnel conditions must be expected. These forces are identified considering the dotted control volume of Figure 1 for calibration conditions. Due to the missing external flow in the calibration chamber and absence of the inlet momentum, the calibration balance force consists of forces on the fan and core nozzle exit stations (E18, E8) and those forces, which are created on the core cowling between the stations E18 and E8. The additional friction forces of the bellmouth are subtracted from the measured balance force.

Putting the focus now on the question, how the thrust related forces are affected by the fan jet, Figure 5 shows the influence of the external flow on the radial pressure distribution of the fan nozzle exit station. Due to the choked nozzle flow conditions at SOC the fan jet reaches supersonic speed in this station. Nevertheless, the external flow alters the pressure distribution close to the nacelle trailing edge. Here, two effects are interacting. First, the flow recompression close the trailing edge of the nacelle increases with increasing Mach number as shown in Figure 6, while at $Ma = 0.0$ the trailing edge pressure is close to the ambient static pressure. Second, the different fan jet spreading results in lower fan jet velocities for $Ma = 0.75$ and associated higher static pressure. As indicated by the ISO-Mach contours in Figure 4, the pressure development along the core cowling between stations E18 and E8 is significantly affected by the different fan jet development, as shown in Figure 7. Figure 8 quantifies the observed effects by showing the changes of the force components relative to the calibration condition as a function of the freestream Mach number. The influence of the external flow speed is now clearly visible in the increase of the deviation from the calibration forces when the Mach number increases. The summation of all contributing forces is presented in Figure 9 as a function of the freestream Mach number. Here the gross thrust shows the same non-linear increase for higher Mach numbers, leading to a gross thrust ratio of 1.85% at $Ma = 0.75$. For low Mach numbers up to $Ma = 0.25$ the deviation of the calibration gross thrust appears to be lower than 0.15%. This result, which is also found for the VHBR simulator, leads to the conclusion, that the SBM is valid only in the low speed regime, since here the correlation coefficients are functions of the fan nozzle pressure ratio only.

As being indicated in the pressure distributions, external flow interacts with the fan jet leading to an increase of the static pressure at the fan and core nozzle exit stations as the Mach number increases. Especially for unchoked nozzle conditions this effect reduces the mass flows, as long as the fan and turbine working point remains the same. In order to compensate the losses of mass flow and additionally retain the fan pressure ratio, the TPS is driven at slightly increased rotational speed compared to the calibration condition. The corresponding gross thrust ratio for SOC conditions is marked with the filled symbol in Figure 9. As expected, the gross thrust further increases compared to the calibration thrust. The gross thrust correction is now derived by evaluating C_v for calibration conditions and for the conditions at $Ma = 0.75$ and applying equation 3 for both cases. Due to the choked fan and core conditions the discharge coefficient remains unaffected by the external flow. Evaluating the ratio of the corresponding net thrusts, following equation 1 leads to the corrected Lilienthal-Polars as presented in Figure 10. Compared to the results of the SBM the corrected polars now show a drag increase when switching from TFN to SOC conditions. In agreement to the TFN results the aerodynamic drag increases proportional to the BPR. The validity of the applied correction is underlined by the comparison of numerically computed polars for the installed ALVAST configuration, taken from the AIRDATA project [9],with the corrected polars of the SBM. The BPR influence in the corrected polars and in the numerical computational results, which are based on the correct thrust/drag decomposition, are qualitatively in good agreement.

4 Conclusion

The drag determination of the ALVAST configuration with different powered TPS-units, measured at transonic conditions in the wind tunnel, showed an unexpected drag behaviour with respect to the bypass ratio and power settings of the engine. Drag extraction from measured database provided by the Standard-Bookkeeping-Method, using informations of a static TPS calibration. A general formulation of the net thrust definition proves on the one hand the formal correctness of the SBM, but on the other hand indicates the shortcomings of the SBM procedure, applied to powered configurations at transonic external flow conditions. With the use of CFD as a complementary tool the dependence of the gross thrust from the external Mach number is analysed. The results obtained for the isolated TF at SOC conditions show a clear dependence of the gross thrust on the external flow. When increasing the Mach number the jet spreading decreases and the recompression of the flow on the outer nacelle gets more distinct. Both effects cause a pressure increase on the fan and core nozzle exit stations and alter the core cowling forces. The resulting calibration balance force is therefore depending on the external Mach number and not only on the fan nozzle pressure ratio. The calibration force as a function of the external Mach number shows a non-linear increase with a growing Mach number. This leads to two conclusions. First, the SBM method as applied provides a valid net thrust for Mach numbers below $Ma = 0.25$. Second, for transonic Mach numbers the SBM underestimates the gross thrust, so that a accurate thrust/drag force decomposition is not possible. The application of thrust correction to Lilienthal-polars, extracted by SBM from measurements, leads with respect to power and BPR effects to a drag behaviour, which is consistent with common knowledge. The results clearly underline the necessity of a correct bookkeeping method. Thrust determination at transonic wind tunnel conditions has to consider the effect of the external flow and therefore static calibration should be combined with a calibration procedure of the isolated TPS in presence of external flow in the wind tunnel.

References

[1] Brodersen, O.; Hepperle, M.; Ronzheimer,A.; Rossow, C.C.; Schöning, B.
The parametric grid generation system MegaCads
5th International Conference on Numerical Grid Generation in Computational Field Simulations, pp.353-362, Mississippi, 1996, NSF, B.K.

[2] Burgsmüller, W.; Hoheisel, H.
ENIFAIR-EU Research into Engine Integration On Future Transport Aircraft
Air&Space Europe, Vol.2, No.2, 2000

[3] Frhr. von Geyr, H; Brodersen, O.; Melber, S.
Engine Integration on transport aircraft - The European research activities on Propulsion Airframe Integration
Proceedings of the ECCOMAS conference 2001, Swansea Wales, UK, 4-7 Sept., 2001

[4] Godard, J.L.; Piat, J.F.
Performing High Speed tests with ENIFAIR Powered Nacelles in the S1MA Wind Tunnel
Contribution to the Workshop on EU-Research on Aerodynamic Engine/Aircraft Integration for Transport Aircraft, DLR Braunschweig, 26-27th, September, 2000

[5] Hoheisel, H.
Engine Thrust Determination for Wind Tunnel Experiments
Contribution to the Workshop on EU-Research on Aerodynamic Engine/Aircraft Integration for Transport Aircraft, DLR Braunschweig, 26-27th, September, 2000

[6] Huhnd, M.
Introduction of the TPS-Technique in the DNW-NWB
German Aerospace Center DLR, IB 129-99/8, 1999

[7] Hurez, A.
Recent Progress on Powerplant / Airframe Integration at Aerospatiale Matra Airbus
Contribution to the Workshop on EU-Research on Aerodynamic Engine/Aircraft Integration for Transport Aircraft, DLR Braunschweig, 26-27th, September, 2000

[8] Kroll, N.; Rossow, C.C., Becker, K., Thiele, F.
MEGAFLOW A Numerical Flow Simulation
ICAS-98-2.7.4, 1998

[9] Laban, M.
Aircraft Drag and Thrust Analysis (AIRDATA) Project overview and Key results
Contribution to the Workshop on EU-Research on Aerodynamic Engine/Aircraft Integration for Transport Aircraft, DLR Braunschweig, 26-27th, September, 2000

[10] Puffert-Meißner, W.
The development and Use of Model Hardware Contribution to the Workshop on EU-Research on Aerodynamic Engine/Aircraft Integration for Transport Aircraft
DLR Braunschweig, 26-27th, September, 2000

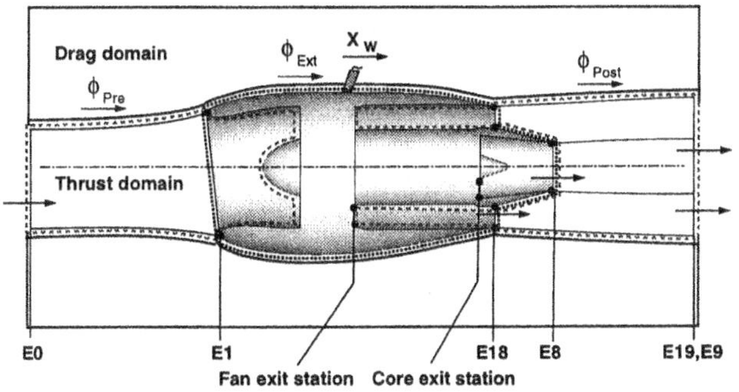

Figure 1: Control volume for thrust/drag decomposition

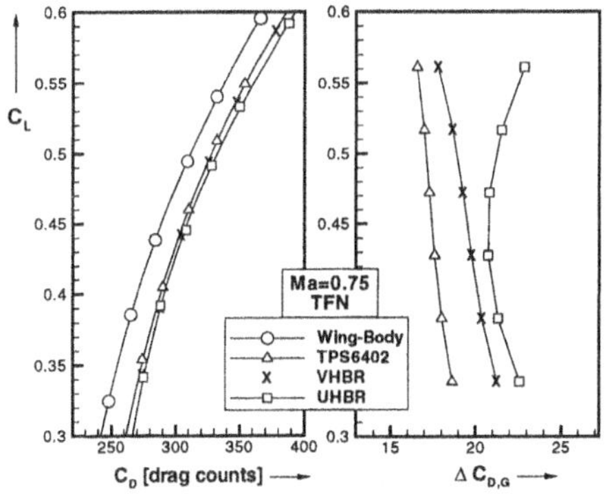

Figure 2: ALVAST Lilienthal-Polars at TFN

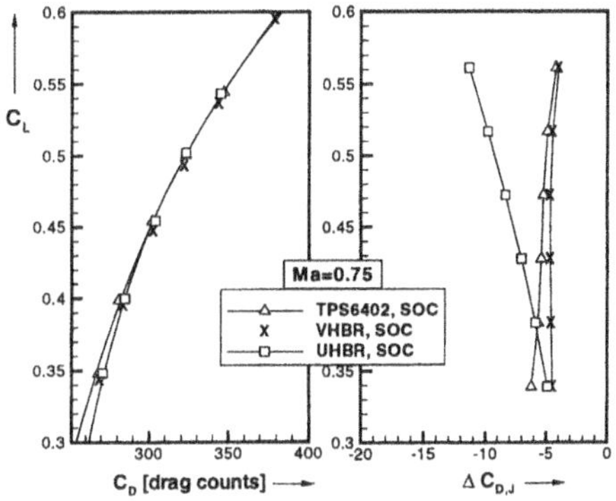

Figure 3: ALVAST Lilienthal-Polars at SOC

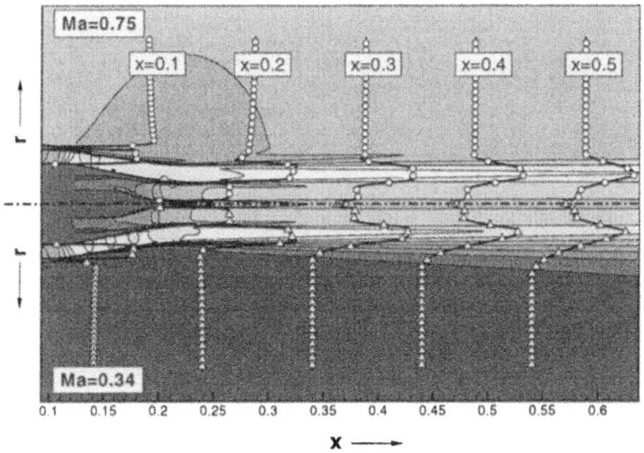

Figure 4: Jet development depending on the external Mach number

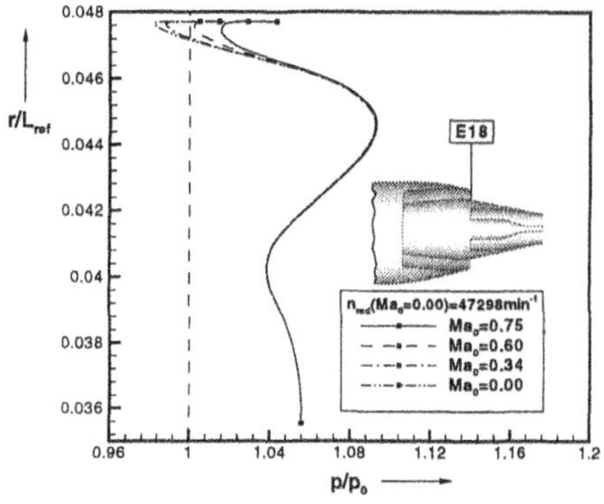

Figure 5: Mach number influence on fan nozzle exit pressure

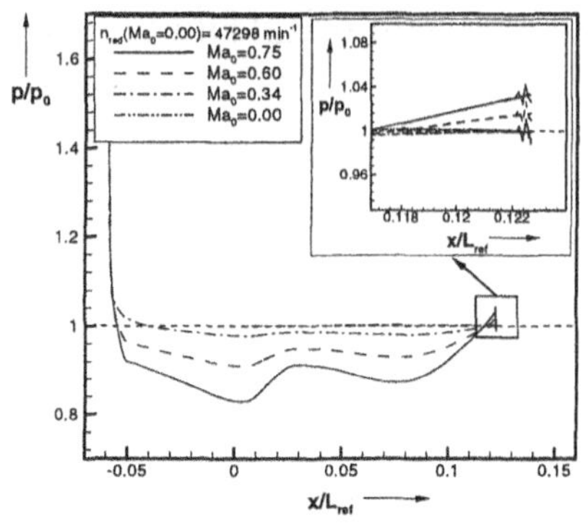

Figure 6: Mach number influence on nacelle pressure

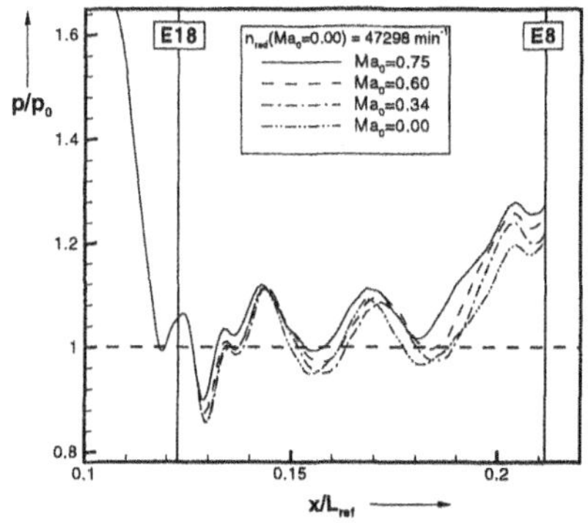

Figure 7: Effect of different fan jet developments on cowling pressure

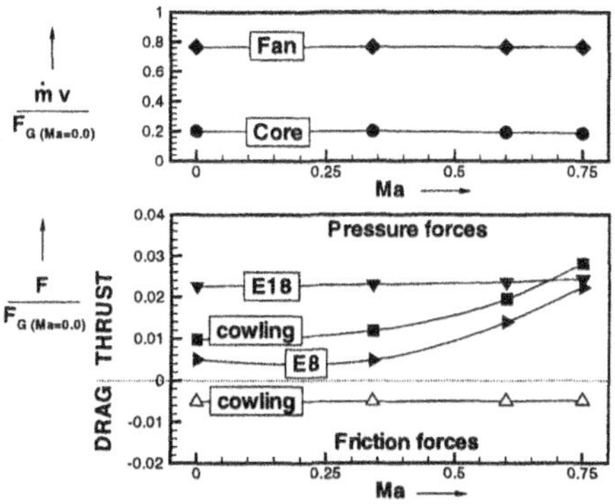

Figure 8: Mach number influence on gross thrust components

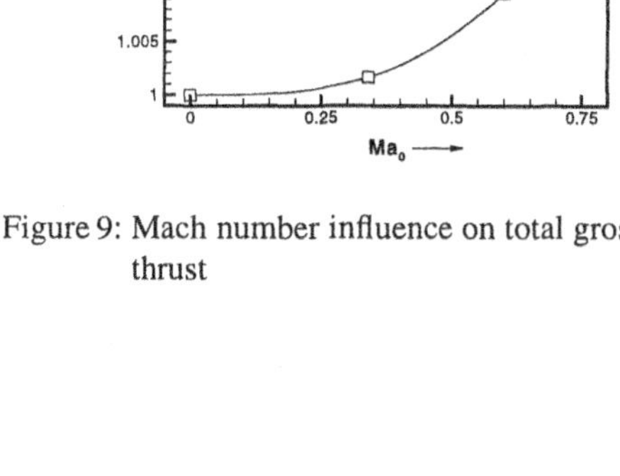

Figure 9: Mach number influence on total gross thrust

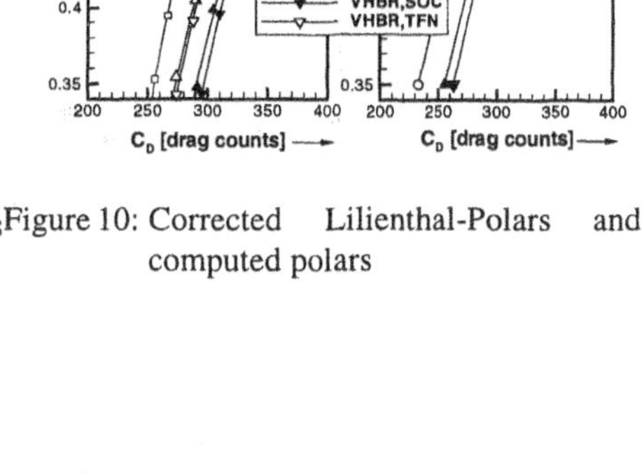

Figure 10: Corrected Lilienthal-Polars and computed polars

Notes on Numerical Fluid Mechanics and Multidisciplinary Design

Available Volumes

Volume 86: S. Wagner, M. Kloker, U. Rist (eds.): Recent Results in Laminar-Turbulent Transition - Selected numerical and experimental contributions from the DFG priority programme 'Transition' in Germany. ISBN 3-540-40490-2

Volume 85: N. G. Barton, J. Periaux (eds.): Coupling of Fluids, Structures and Waves in Aeronautics - Proceedings of a French-Australian Workshop in Melbourne, Australia 3-6 December 2001. ISBN 3-540-40222-5

Volume 84: J. Ballmann (ed.): Flow Modulation and Fluid-Structure Interaction at Airplane Wings - Research Results of the Collaborative Research Center SFB 401 at RWTH Aachen, University of Technology, Aachen, Germany. ISBN 3-540-40209-8

Volume 83: L. Davidson, D. Cokljat, J. Fröhlich, M. A. Leschziner, C. Mellen, W. Rodi (eds.): LESFOIL: Large Eddy Simulation of Flow Around a High Lift Airfoil - Results of the Project LESFOIL supported by the European Union 1998 - 2001. ISBN 3-540-00533-1

Volume 82: E. H. Hirschel (ed.): Numerical Flow Simulation III - CNRS-DFG Collaborative Research Programme, Results 2000-2002. ISBN 3-540-44130-1

Volume 81: W. Haase, V. Selmin, B. Winzell (eds.): Progress in Computational Flow-Structure Interaction - Results of the Project UNSI, supported by the European Union 1998-2000. ISBN 3-540-43902-1

Volume 80: E. Stanewsky, J. Délery, J. Fulker, P. de Matteis (eds.): Drag Reduction by Shock and Boundary Layer Control - Results of the Project EUROSHOCK II, supported by the European Union 1996-1999. ISBN 3-540-43317-1

Volume 79: B. Schulte-Werning, R. Grégoire, A. Malfatti, G. Matschke (eds.): TRANSAERO - A European Initiative on Transient Aerodynamics for Railway System Optimisation. ISBN 3-540-43316-3

Volume 78: M. Hafez, K. Morinishi, J. Periaux (eds.): Computational Fluid Dynamics for the 21st Century. Proceedings of a Symposium Honoring Prof. Satofuka on the Occasion of his 60th Birthday, Kyoto, Japan, 15-17 July 2000. ISBN 3-540-42053-3

Volume 77: S. Wagner, U. Rist, H.-J. Heinemann, R. Hilbig (eds.): New Results in Numerical and Experimental Fluid Mechanics III. Contributions to the 12th STAB/DGLR Symposium Stuttgart, Germany 2000. ISBN 3-540-42696-5

Volume 76: P. Thiede (ed.): Aerodynamic Drag Reduction Technologies. Proceedings of the CEAS/DragNet European Drag Reduction Conference, 19-21 June 2000, Potsdam, Germany. ISBN 3-540-41911-X

Volume 75: E.H. Hirschel (ed.): Numerical Flow Simulation II. CNRS-DFG Collaborative Research Programme, Results 1998-2000. ISBN 3-540-41608-0

Volume 74: D. Henry, A. Bergeon (eds.): Continuation Methods in Fluid Dynamics. Contributions to the ERCOFTAC/EUROMECH Colloquium 383, Aussois, France, 6.-9. September 1998. ISBN 3-540-41556-4

Volume 73: M. Griebel, S. Margenov, P. Yalamov (eds.): Large-Scale Scientific Computations of Engineering and Environmental Problems II. Proceedings of the Second Workshop on „Large-Scale Scientific Computations", Sozopol, Bulgaria, June 2-6, 1999. ISBN 3-540-41551-3

Volume 72: W. Nitsche, H.-J. Heinemann, R. Hilbig (eds.): New Results in Numerical and Experimental Fluid Mechanics II. Contributions to the 11th AG STAB/DGLR Symposium Berlin, Germany 1998. ISBN 3-540-41552-1

Volume 70: W. Hackbusch, G. Wittum (eds.): Numerical Treatment of Multi-Scale Problems. Proceedings of the 13th GAMM-Seminar, Kiel, January 24-26, 1997. ISBN 3-540-41562-9

Volume 69: D. P. Hills, M. R. Morris, M. J. Marchant, P. Guillen (eds.): Computational Mesh Adaptation. ECARP - European Computational Aerodynamics Research Project. ISBN 3-540-41548-3

Volume 67: C. H. Sieverding, G. Cicatelli, J. M. Desse, M. Meinke, P. Zunino: Experimental and Numerical Investigation of Time Varying Wakes behind Turbine Blades. Results of the Industrial and Material Technology Aeronautics Research Project AER2-CT92-0048, 1992-1996. ISBN 3-540-41546-7

Volume 66: E. H. Hirschel (ed.): Numerical Flow Simulation I. CNRS-DFG Collaborative Research Programme. Results 1996-1998. ISBN 3-540-41540-8

Volume 64: R. Friedrich, P. Bontoux (eds.): Computation and Visualization of Three-Dimensional Vortical and Turbulent Flows. Proceedings of the Fifth CNRS-DFG Workshop on Numerical Flow Simulation, München, Germany, December 6 and 7, 1996. ISBN 3-540-41557-2

Volume 63: M. Fiebig, N. K. Mitra (eds.): Vortices and Heat Transfer. Results of a DFG-Supported Research Group. ISBN 3-540-41558-0

Volume 62: M. Griebel, O. Iliev, S. D. Margenov, P. S. Vassilevski (eds.): Large-Scale Scientific Computations of Engineering and Environmental Problems. Proceedings of the First Workshop on „Large-Scale Scientific Computations" Varna, Bulgaria, June 7-11, 1997. ISBN 3-540-41560-2

Volume 57: H. Deconinck, B. Koren (eds.): Euler and Navier-Stokes Solvers Using Multi-Dimensional Upwind Schemes and Multigrid Acceleration. Results of the BRITE/EURAM Projects AERO-CT89-0003 and AER2-CT92-00040, 1989-1995. ISBN 3-540-41535-1

GPSR Compliance

The European Union's (EU) General Product Safety Regulation (GPSR) is a set of rules that requires consumer products to be safe and our obligations to ensure this.

If you have any concerns about our products, you can contact us on ProductSafety@springernature.com

In case Publisher is established outside the EU, the EU authorized representative is:

Springer Nature Customer Service Center GmbH
Europaplatz 3
69115 Heidelberg, Germany

Batch number: 10755440

Printed by Printforce, the Netherlands